AF780911

Neutron Imaging

From applied materials science to industry

Online at: https://doi.org/10.1088/978-0-7503-3495-2

Neutron Imaging

From applied materials science to industry

Edited by
Markus Strobl
Laboratory of Neutron Scattering and Imaging, Paul Scherrer Institut, Villigen, Switzerland

Niels Bohr Institute, University of Copenhagen, Copenhagen, Denmark

Eberhard Lehmann
Laboratory of Neutron Scattering and Imaging, Paul Scherrer Institut, Villigen, Switzerland

IOP Publishing, Bristol, UK

© IOP Publishing Ltd 2024

All rights reserved. No part of this publication may be reproduced, stored in a retrieval system or transmitted in any form or by any means, electronic, mechanical, photocopying, recording or otherwise, without the prior permission of the publisher, or as expressly permitted by law or under terms agreed with the appropriate rights organization. Multiple copying is permitted in accordance with the terms of licences issued by the Copyright Licensing Agency, the Copyright Clearance Centre and other reproduction rights organizations.

Certain images in this publication have been obtained by the author(s) from the Wikipedia/ Wikimedia website, where they were made available under a Creative Commons licence or stated to be in the public domain. Please see individual figure captions in this publication for details. To the extent that the law allows, IOP Publishing disclaim any liability that any person may suffer as a result of accessing, using or forwarding the image(s). Any reuse rights should be checked and permission should be sought if necessary from Wikipedia/Wikimedia and/or the copyright owner (as appropriate) before using or forwarding the image(s).

Permission to make use of IOP Publishing content other than as set out above may be sought at permissions@ioppublishing.org.

Markus Strobl and Eberhard Lehmann have asserted their right to be identified as the editors of this work in accordance with sections 77 and 78 of the Copyright, Designs and Patents Act 1988.

ISBN 978-0-7503-3495-2 (ebook)
ISBN 978-0-7503-3493-8 (print)
ISBN 978-0-7503-3496-9 (myPrint)
ISBN 978-0-7503-3494-5 (mobi)

DOI 10.1088/978-0-7503-3495-2

Version: 20240501

IOP ebooks

British Library Cataloguing-in-Publication Data: A catalogue record for this book is available from the British Library.

Published by IOP Publishing, wholly owned by The Institute of Physics, London

IOP Publishing, No.2 The Distillery, Glassfields, Avon Street, Bristol, BS2 0GR, UK

US Office: IOP Publishing, Inc., 190 North Independence Mall West, Suite 601, Philadelphia, PA 19106, USA

Contents

Part VIII Advanced methodology

Preface

Neutron imaging was explored early in the history of neutrons, perhaps inspired by the immediate production of images when x-rays were found by W C Röntgen. While neutron imaging found early use in non-destructive testing, the use of neutrons for material science was pioneered in the 1950s by B N Brockhouse and C G Shull, who received the Nobel Prize in Physics in 1994 for observing 'where the atoms are and what the atoms do' through the detection of neutrons scattered from crystal lattices.

The significant difference of length scales and phenomena targeted by neutron scattering techniques evolving when opposed to neutron imaging has led to separate developments with scattering instruments dominating large-scale neutron facilities such as research reactors and later also spallation neutron sources.

However, the development of neutron imaging and the evolution of science cases driven by societal needs have slowly integrated the neutron scattering and imaging communities, real space and reciprocal space techniques probing a large range of length scales, required in order to understand many materials, phenomena and processes up to the regime of devices such as engines, batteries and fuel cells.

Generally, it appears that imaging has gained an outstanding role in science and engineering today, providing us with a view onto the smallest to longest length scales ever seen in history, from atoms to black holes and beyond. While work and research in the medical field can hardly be imagined without imaging techniques over many decades, the same appears to hold true for material science today. Neutron imaging has a place in the latter field, a niche of applications where other, easier available, imaging techniques fail to provide the desired insights.

This book focusses on providing an overview of the diverse applications and research fields in which neutron imaging can make an impact on understanding, e.g. structure, failure, history, optimization, decay, distribution and functionality of investigated materials, artefacts, systems and devices. As such, the topics include engineering and manufacturing from construction materials, nuclear engineering and safety to advanced and additive manufacturing studies contributing to the ongoing industrial revolution. Natural materials such as wood, rocks, soil, and plants are opening contexts like understanding, predicting and avoiding natural disasters, food safety and environmental concerns. The latter also inspire energy research and contributions to novel concepts of energy conversion, hydrogen economy etc. Hard and soft condensed matter investigations are typically in the center of neutron scattering applications but can also profit increasingly from neutron imaging investigations. Additionally, neutron imaging makes an impact in targeting directly industrial applications but also cultural and natural heritage science and conservation.

This volume strives to provide insights and inspiration to researchers and engineers as well as other professionals not familiar with neutron imaging and the potential it contributes to their field and profession in order to provide knowledge and spur interest in the method and how it might be the right tool of choice to answer their research or technology questions and gives answers to problems which are hard to gain otherwise.

To this end the book provides a wide range of examples of a large variety of different applications and complements these with some introduction to the basic principles, contrast modalities and techniques, including some instrumentation and image data processing details at the end, which also cater for those readers interested in these practical aspects and realizations. Therefore, this volume should also be suited to offer information and stimulation to those closer to the field, in neutron science and decision makers considering to engage in neutron imaging and neutron imaging instrumentation at their facilities.

The editors are grateful to a large number of fellow scientists either working at neutron sources and operating neutron imaging instrumentation or being scientific users of these facilities. We want to thank all of them for their dedicated effort in walking this extra mile which made this book a reality. By sharing their experience and results from their particular research fields this book can excite new ideas to utilize neutron imaging to answer scientific questions and tackle problems that readers are concerned with. Thus, we hope the book inspires a new generation of researchers to unlock fresh insights according to our intention to create curiosity and interest in the power of neutron imaging.

Editor biographies

Markus Strobl

Markus Strobl is the Head of the Applied Materials Group operating the neutron imaging and engineering instruments at the Swiss Neutron Source SINQ of the Laboratory for Neutron Scattering and Imaging at the Paul Scherrer Institute (PSI) in Switzerland. He studied at the Technical University of Vienna and received his master diploma in engineering physics for a work on ultra-small angle neutron scattering of artificial structures, and his PhD in physics from his Viennese University for the exploration of alternative contrast signals in neutron tomography at the Helmholtz Zentrum Berlin. After operating, designing and building instruments for neutron imaging, ultra-small angle neutron scattering, neutron reflectivity and neutron time-of-flight techniques for various applications in applied materials, engineering, hard and soft matter science in Berlin he moved to Scandinavia working for the European Spallation Source (ESS) developing several instrument projects, in particular the neutron imaging instrument ODIN, acted as coordinator of the science focus areas of imaging, geology, engineering and cultural heritage sciences and as deputy head of the Neutron Instrument Division. He became an affiliated Professor for Neutron and x-ray Imaging Techniques at the Niels Bohr Institute of the University of Copenhagen in 2014 and held an Otto Mønsted visiting professorship at the Danish Technical University in fall 2022. Between 2017 and 2021 he also held a senior scientist position at the Nuclear Physics Institute of Czech Academy of Science in parallel to his main affiliation with ESS and later PSI. In the meantime, in 2017, he had moved to Switzerland and took over the lead of the Neutron Imaging and Activation Group at PSI from his predecessor Dr Eberhard H Lehmann. Markus is furthermore acting as elected Vice President of the International Society of Neutron Radiography (ISNR) in his second term, as well as second term board member of the Swiss Society of Neutron Science, and serves/d on numerous review and advisory boards around the globe.

Eberhard H Lehmann

Eberhard H Lehmann is a retired nuclear scientist, formerly founder and head of the Neutron Imaging and Activation Group (NIAG) at the Paul Scherrer Institute (PSI), Switzerland. He studied physics at the University of Leipzig, Germany, and got his diploma in physics about a work on molecular dynamics simulations of protein molecules. At the Central Institute for Nuclear Research Rossendorf, near Dresden, Germany, he dealt with the physics of the fast breeder nuclear reactors and obtained his PhD degree for studies about cross-section data of structural materials, based on special reactor configurations he designed and built for this purpose. In the framework of an international collaboration, these data became important for the design of uranium- and plutonium-driven fast breeder reactor configurations. His move to the PSI in Switzerland was accompanied with the responsibility for operation and utilization of the 10 MW research reactor there, dealing with the reactor core layout and new applications, including neutron radiography. Based on this knowledge, he initiated the technique of neutron imaging at the spallation neutron source SINQ at PSI and was responsible for irradiation test facilities. Leading a team of neutron imaging experts, he was also elected as the president of the 'International Society for Neutron Radiography', holding the '10th World Conference on Neutron Radiography' in Grindelwald, Switzerland in 2014. As author and contributor for more than 300 scientific papers, he assisted in bringing up neutron imaging towards a well-accepted and used technology. Recently, he contributed to the design and construction of a neutron imaging detector with up to now highest spatial resolution (neutron microscope project).

List of contributors

Pascal Benard
ETH Zurich, Universitätstrasse 16, 8092 Zürich, Switzerland

Pierre Boillat
Paul Scherrer Institut, Forschungsstrasse 111, 5232 Villigen, Switzerland

Andreas Borgschulte
EMPA Dübendorf, Ueberlandstrasse, Dübendorf, Switzerland

Matteo Busi
Paul Scherrer Institut, Forschungsstrasse 111, 5232 Villigen, Switzerland

Francesco Cantini
Università degli Studi di Firenze, Florence, Italy

Andrea Carminati
ETH Zurich, Universitätstrasse 16, 8092 Zürich, Switzerland

Veerle Cnudde
Ghent University, Krijgslaan 281, S8, 9000 Ghent, Belgium
and
Utrecht University, Utrecht, Belgium

Liliana I Duarte
Paul Scherrer Institut, Forschungsstrasse 111, 5232 Villigen, Switzerland

Francesco Grazzi
Consiglio Nazionale delle Ricerche, 50019 Florence, Italy

Anders Kaestner
Paul Scherrer Institut, Forschungsstrasse 111, 5232 Villigen, Switzerland

Eberhard Lehmann
Laboratory of Neutron Scattering and Imaging, Paul Scherrer Institut, Villigen, Switzerland

Florencia Malamud
Paul Scherrer Institut, Forschungsstrasse 111, 5232 Villigen, Switzerland

David Mannes
Paul Scherrer Institut, Forschungsstrasse 111, 5232 Villigen, Switzerland

Peter Niemz
ETH Zurich, Stefano-Franscini-Platz 3, 8093 Zurich, Switzerland

Efthymios Polatidis
Paul Scherrer Institut, Forschungsstrasse 111, 5232 Villigen, Switzerland

Michael Schulz
Technical University Munich, Lichtenbergstrasse 1, Garching, Germany

Walter Sonderegger
Swiss Wood Solutions AG, Giessenstrasse 10, 6460 Altdorf, Switzerland

Markus Strobl
Laboratory of Neutron Scattering and Imaging, Paul Scherrer Institut, Villigen, Switzerland
and
Niels Bohr Institute, University of Copenhagen, Copenhagen, Denmark

Alessandro Tengattini
Institut Laue-Langevin, 71 avenue des Martyrs, 38000 Grenoble, France
Université Grenoble Alpes, Grenoble INP, 3SR, 38000 Grenoble, France

Sigita Trabesinger
Paul Scherrer Institut, Forschungsstrasse 111, 5232 Villigen, Switzerland

Pavel Trtik
Paul Scherrer Institut, Forschungsstrasse 111, 5232 Villigen, Switzerland

Okan Yetik
Paul Scherrer Institut, Forschungsstrasse 111, 5232 Villigen, Switzerland

Mohsen Zare
Technical University Munich, Hans-Carl-von-Carlowitz-Platz 2, 85354 Freising, Germany

Robert Zboray
EMPA Dübendorf, Ueberlandstrasse, Dübendorf, Switzerland

Part I

Introduction to neutron imaging

Chapter 1

History and basics of neutron imaging

Markus Strobl

1.1 First neutron images

The first neutron images (figure 1.1) are reported to have been recorded in Berlin between 1935 and 1938 by Kallmann and Kuhn using Ra–Be sources and a small D–D neutron generator. They developed a first converter-film system and vacuum cassette for radiographic neutron imaging. It resembles closely systems used later, and in some rare cases still today. This work was published only in 1948, but a first US patent entitled 'Photographic Detection of Slowly Moving Neutrons' had already been awarded to them in 1940 [1]. In 1946, however, Peter reported neutron imaging experiments utilizing the penetration power of neutrons transmitting massive metal objects and thus enabling observation of the interior of devices and materials opaque to other forms of radiation. Peter, also working in Berlin, profited from an advanced neutron source as compared to Kallmann and Kuhn. He judged that the neutron flux from Ra–Be sources was too weak and contaminated by Ra-gammas and cyclotron-based sources yield too high x-ray background. Thus, he used a high-voltage generator where x-rays are produced only in the discharge tube, which is remote from the neutron source and therefore corresponding background can be shielded. The optimization of flux versus background is a key task carrying on till today, although neutron sources have taken a remarkable development since then. Peter obtained the neutron images communicated as the first of their kind till today. An example of these very early images is reported in reference [2] and shown in figure 1.1(a), comparing Ra-gamma and neutron images of objects such as a massive faucet but also two ampules filled with light and heavy water, respectively. The choice is remarkable in demonstrating two key features making neutrons unique probes for many material investigations. The high penetration power, in particular for dense metallic materials, is demonstrated in the image of the faucet, while the very different contrast for H_2O and D_2O underlines the unique isotope sensitivity of neutrons, yielding much better penetration for heavy water than for light water.

doi:10.1088/978-0-7503-3495-2ch1

© IOP Publishing Ltd 2024

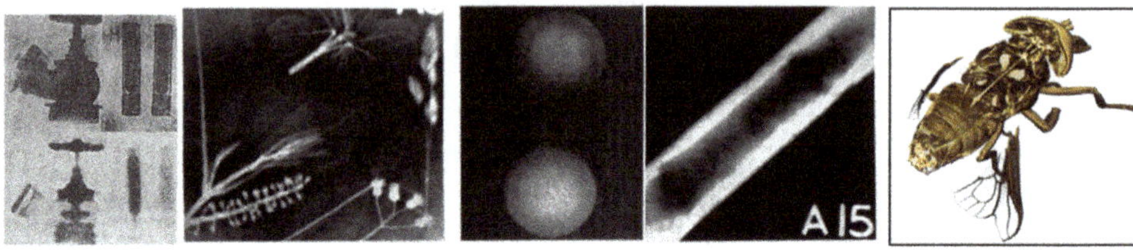

Figure 1.1. (Left) Very first neutron (bottom) and gamma (top) images fire hydrant valve and H_2O and D_2O from (before) 1946. Reproduced with permission from [2], copyright 1946–2014: Verlag der Zeitschrift für Naturforschung. (mid) Neutron images of plants, uranium and steel discs (1 inch thickness) and defective uranium cylinder from 1956. Reproduced from [4], copyright IOP Publishing Ltd. All rights reserved. (right) Neutron tomography of fly with the PSI micro-setup detector (2014). Reprinted from [13], copyright (2011), with permission from Elsevier.

An excellent review of these first years of neutron imaging between 1935 and 1944 in Berlin, Germany, can be found in reference [3].

The availability of nuclear fission reactors for imaging purposes from the mid fifties marks a new era. The orders of magnitude improved flux conditions enabled the first practical applications and subsequently a broad portfolio of utilization of neutron imaging. The BEPO reactor at Harwell was the first large-scale neutron source exploited for neutron imaging [4]. Neutron imaging developed first into a recognized and quantitative inspection and non-destructive testing technique with applications for industry and in particular also the nuclear sector (figure 1.1). Starting with programs in Europe and the United States the technique spread also to Asian countries and a small community formed around a series of newsletters from 1964 to 1977 [5] and a first dedicated conference in 1973 [6]. In 1975 the first ASTM standards for neutron imaging (E545) were published followed by the 1st World Conference on Neutron Radiography (WCNR) in 1981 in San Diego that resulted in a book of proceedings counting 1000 pages with 140 articles from authors from 20 countries [7]. Meanwhile, time-resolved neutron imaging had also become feasible using TV cameras [8], and in the late seventies, early eighties the first 3D tomographies were reported with film and TV cameras, respectively [9, 10].

In 1990 a second international conference series was added to the meanwhile regular WCNR series, namely the International Topical Meeting on Neutron Radiography ITMNR, both still held every four years offset by two years. In the meantime, a wide range of applications has been explored from non-destructive testing to plant science, studies of two-phase flow, shock-waves, hydrogen transport in electrochemical applications, nuclear fuel studies, aerospace applications, corrosion, building materials, ceramic composite materials, liquid metals, porous media and cultural heritage etc. Many of these topics remain of high relevance today.

Finally, in 1992 the formation of the International Society for Neutron Radiography (initially Radiology) (ISNR) started, and the society was formally established during the 1996 WCNR-5 in Berlin, Germany, at the very place of the initial roots of the technique [11]. This early history is well summarized in a proceedings paper of the ITMNR-7 in 2012 by J S Brenizer [12].

At this point, in the 1990s, a second leap in evolution, after the availability of high flux from nuclear fission reactors in the 1950s, was the introduction of digital imaging detectors that revolutionized neutron imaging through straightforward access to 3D resolution and quantitative image analyses. The impact on methodical development was enormous and thus the introduction of digital detectors marks the beginning of **modern neutron imaging** (figure 1.1). This modern and further advanced neutron imaging, and in particular its application in science and technology, is the core topic of this book. Subsequently, wavelength-resolved neutron imaging gained importance through a number of advanced contrast methods which will be described in chapter 2. This inspired an additional workshop series of the neutron imaging community called NEUWAVE created by B Schillinger and E Lehmann, which was held ten times since the first workshop 2008 in Munich. This format has significantly supported corresponding developments and in particular the advent of neutron imaging instruments at state-of-the-art pulsed spallation neutron sources.

1.2 Basics: the neutron

Remarkably, the history of neutron imaging started only three years after the discovery of the neutron by Chadwick in 1932. In fact, Bothe and Becker in Germany had found already in 1930 that alpha particles impinging on light elements including Be, B, or Li create a radiation that is highly penetrating, even more than earlier known gamma radiation, which it was first hypothesized to be. However, Joliot-Curie and Joliot demonstrated in 1932 in Paris that when this radiation is incident on hydrogenous material, protons of high energy are released. Finally, Chadwick after some additional experiments proposed that it was uncharged particles with about the mass of protons that were emitted [14]. These particles, which were predicted since 1920 based on the existence of isotopes, were called 'neutrons', and Chadwick was credited with the discovery and, eventually, in 1935 the Nobel prize in physics.

Chadwick had used, like Kallmann and Kuhn at some point two years later, a Ra–Be source. Neutrons emitted from such a source have a spectrum peaking at around a few MeV [15]. Also, neutrons from a D–D neutron generator, first used by Kallmann and Kuhn for imaging, and neutrons produced by bombarding beryllium by a deuteron beam from a high-voltage generator, as Peter did, provide initially fast neutrons. These are ranging in the MeV regime in the same way that neutrons produced by fission have energies peaking at around 1 MeV.

Neutrons can be categorized by their energy and are referred to as *fast neutrons* and *medium fast neutrons* above 0.5 MeV and between 0.1 and 0.5 MeV, respectively. At such energies neutrons have extremely high penetration and are thus hard to stop, to collimate and to detect, limiting imaging applications to low spatial resolution of, however, massive objects.

Such neutrons are moderated to lower energies by scattering in thermal media such as, in particular, in hydrogenous material with large inelastic scattering but low absorption cross-section. Kallman and Kuhn's neutron source was surrounded by water. Also, in fission reactors neutrons are thermalized by moderators such as water to enable further fission reactions. Between the *thermal neutron* energy region at around 0.025 eV and the medium fast neutron regime are the *epithermal neutron* regime up to 1 eV and the *resonance neutron* regime between 1 and 100 eV. The resonance regime is interesting for elemental imaging based on the characteristic resonance absorption of specific elements. The vast majority of neutron imaging applications is, however, found in the thermal and *cold neutron* regime (obtained with moderators at only a few Kelvin), where the neutron energy matches excitation energies in condensed matter and their wavelengths match the inter-atomic distances. In this regime neutrons still provide excellent penetration, and are relatively easy to collimate and to detect, enabling high spatial resolution for non-destructive testing and material science applications. Below the cold neutron energies are the *very cold and ultra-cold neutron* regimes around 0.1 meV and 0.2 μeV, respectively. No imaging applications are to date found in this regime of e.g. neutron lifetime experiments. Neutrons below the medium fast regime can be summarized as *slow neutrons*.

Today neutrons are most efficiently produced by fission reactors and spallation neutron sources (figure 1.2), which host the world-leading neutron imaging instruments. These are found at medium and high flux reactors as well as at several 100 kW to a few MW accelerator driven spallation sources. Note that in figure 1.2 the peak flux is given for pulsed sources, because for wavelength-resolved methods, this is the value that counts. And only for such methods do the existing advanced pulsed spallation sources outperform high flux reactors. For conventional neutron imaging this is not the case, and thus the time integral flux has to be considered. However, pulsed sources have accelerated the development of advanced wavelength dispersive neutron imaging methods, as will be outlined in subsequent chapters.

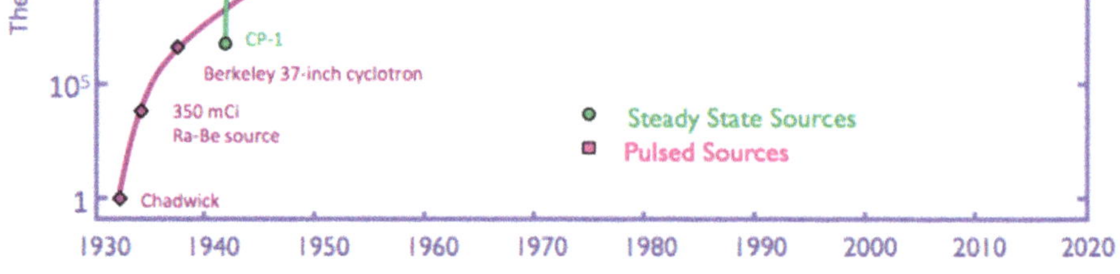

Figure 1.2. History of neutron sources. The early neutron sources starting at Chadwick, the fission reactor sources starting in the 1940s and saturating in brightness with the high flux reactors HFIR and ILL, and finally the pulsed spallation sources breaking even with the high flux reactors with the ISIS neutron source and reaching a currently projected maximum with the European Spallation Source ESS under construction in Sweden. Reproduced from [16]. Copyright IOP Publishing Ltd. CC BY 3.0.

It should be noted that small- and medium-scale neutron sources like D–D neutron generators or compact accelerator driven neutron sources (CANS) are also utilized for specific imaging methods and applications.

Neutrons are subject to gravity, weak, strong and electromagnetic interaction. Free neutrons have a mass of $m = 1.675\ 10^{-27}$ kg and a lifetime of 881.5 s, after which a beta minus decay based on the weak force converts them into a proton resulting in the emission of an electron and an electron anti-neutrino. Through the strong and electromagnetic interaction neutrons interact with matter in various ways. While neutrons hardly scatter from electrons they possess a magnetic moment ($\mu = 0.966\ 10^{-26}$ J T^{-1}) through which they interact with electric and mainly magnetic fields and moments. The magnetic moment of these fermions, which as Baryons consisting of three quarks (1 up, 2 down), is aligned anti-parallel to their spin ½. Based on the strong force neutrons are captured, i.e. absorbed, by a nucleus or scattered. Both processes contribute to the attenuation of a transmitted neutron beam, display significant energy dependence and are isotope specific. Their probabilities are described by corresponding cross-sections, where the total microscopic cross-section $\sigma_t = \sigma_a + \sigma_s$ is the sum of the absorption cross-section σ_a and the scattering cross-section σ_s. Together with the atomic density N the macroscopic cross-section, or the linear attenuation coefficient $\Sigma = N\sigma_t$ is derived. Extensive tabulated cross-section data can be found in the literature [17] and online [18] and some examples highlighting particularly effects in specific spectral regions are presented in figure 1.3. In the cold neutron regime the total cross-section is for quite some relevant structural materials dominated by the coherent elastic cross-section through diffraction at the crystal lattice. This has triggered the development of wavelength-resolved cold neutron imaging.

The need for wavelength resolution in condensed matter research with neutrons made pulsed neutron sources an efficient alternative to high flux reactors, which they were already gradually starting to outperform before the Millennium. Consequently, the time-of-flight (ToF) approach, where the wavelength λ is encoded according to de Broglie [21] $\lambda = h/p = h/vm = ht/Lm$, with h being the Planck constant, $p = vm$ the

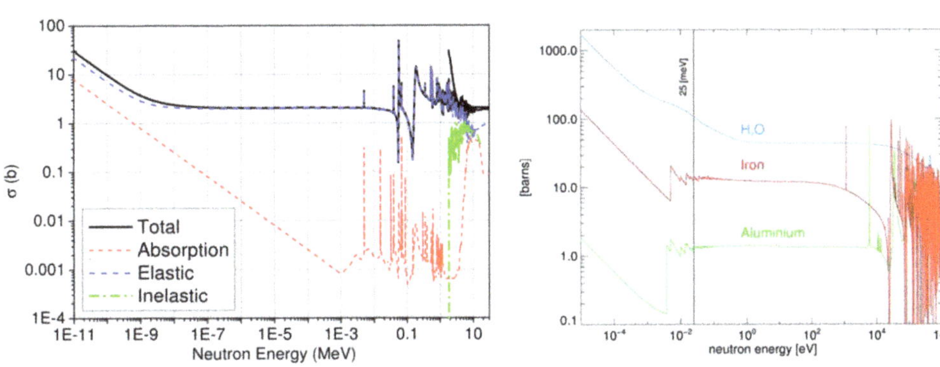

Figure 1.3. Energy dependent microscopic cross-sections of Si-28, reprinted from [19], copyright (2017), with permission from Elsevier, and total microscopic cross-sections of water, iron and aluminium, reproduced with permission from [20].

momentum with m the mass and $v = L/t$ the velocity measured as ToF t over a length L, has also been introduced to neutron imaging and corresponding instruments are available at modern state-of-the-art pulsed neutron spallation sources.

1.3 Context of neutron imaging

1.3.1 Neutron scattering

Clifford Shull developed neutron diffraction, i.e. elastic neutron scattering, together with Ernest Walton at Oak Ridge National Laboratory [22] after the Second World War. Bertram Brockhouse at the Chalk River Laboratory in Canada, operating since 1946 the first nuclear reactor outside the USA, developed inelastic neutron scattering and the first neutron triple axis spectrometer. These developments established thermal and cold neutrons as invaluable probes for condensed matter science (figure 1.4) and earned both scientists the Nobel prize. Neutron scattering [23] becomes a driving force for the establishment of various neutron sources and user laboratories around the world. The scattering techniques diversify quickly and besides diffraction, exploring atomic and magnetic structures of crystalline materials, small-angle scattering techniques extend the range of structural investigations to the nano- and finally micrometer range, exploring macromolecules, nanoparticles, micelles, polymers, porosities and multilayer structures in soft- and hard-matter research. Similarly, in inelastic scattering besides phonon dispersions spinwaves and magnons, ToF spectrometers explore quasi-elastic scattering and, thus, relaxation and diffusion processes and backscattering and spin-echo spectrometers polymer reptation, glassy dynamics and libration.

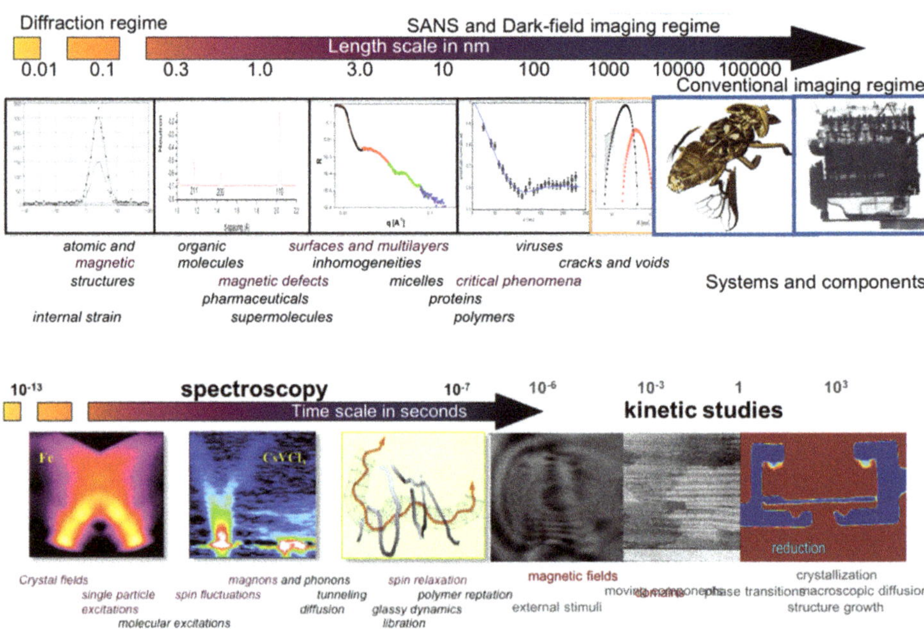

Figure 1.4. Neutron scattering and imaging in the context of probed length and time scales.

In contrast to neutron imaging, scattering techniques probe structures and dynamics in reciprocal space and on reciprocal time scales. Correspondingly, these techniques also operate on complementary scales and despite neutron imaging probing real times and space, recent developments in resolution of both imaging and scattering techniques today enable some overlaps of scales. In particular, some small-angle scattering methods explore length scales up to a few 10 μm [24, 25] and neutron imaging on the other hand reaches down to a few micrometer real space resolution nowadays [26]. Thus, in this regime the microscopic scattering information and macroscopic imaging information meet [27]. However, resolution is fundamentally different between the reciprocal and real space techniques. Only in real space are individual structures resolved, while reciprocal space scattering techniques provide average information of all illuminated structures.

1.3.2 Complementarity to other imaging methods

Neutron imaging is indeed a real-space technique amongst many. Imaging is performed with all kinds of radiation, in particular electromagnetic ones in the ranges of visible light, x-rays and gamma rays. But other particle rays like electrons are also used of course. Some of these methods are limited to 2D imaging of a sample surface or are destructive in order to assess the third dimension (light optical and scanning electron microscopy (SEM), atom probe tomography (APT)). A number of imaging methods provide resolutions down to the microscopic regime of scattering techniques, like transmission electron microscopy (TEM) and APT reaching down to the Angstrom domain (figure 1.5). However, such resolution comes with a trade-off concerning sample size and often the question remains to be answered how representative a local micrometer or even sub-micrometer-sized volume can be for bulk material behavior and structure. Other methods such as

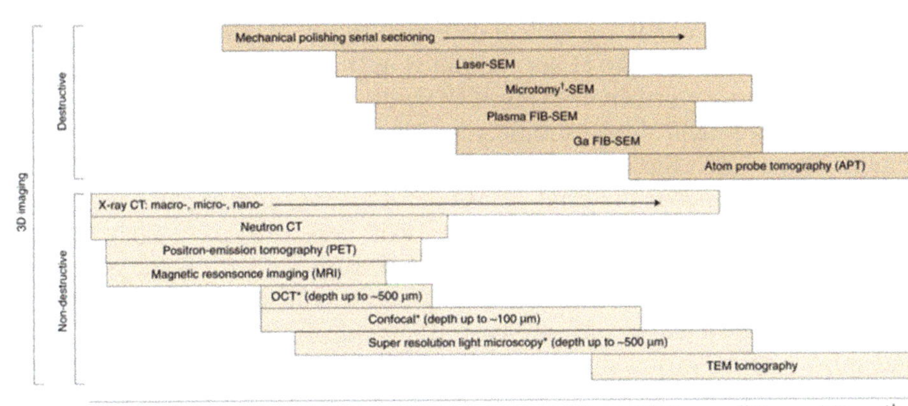

Figure 1.5. Two figures illustrating the regime and resolution versus sample size, respectively, for different imaging methods. Reprinted from [28], copyright (2019), with permission from Springer Nature. Notably, in the first case neutron tomography (CT) is assumed with a resolution of only 100 mm, while in reality resolutions of a few micrometer are reached.

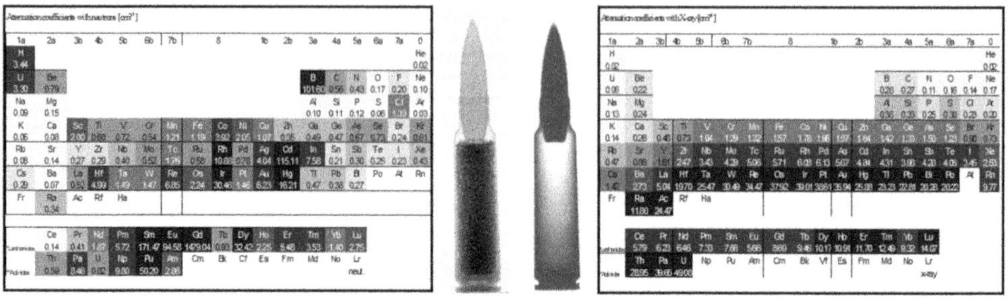

Figure 1.6. Attenuation coefficients for thermal neutrons and x-rays (100 keV) represented as grey scale with respect to the table of elements and corresponding radiographies of a cartrige underlining complementary contrast of neutron and x-rays. Reprinted from [29], copyright (2022), with permission from Springer Nature.

positron emission tomography (PET) or magnetic resonance imaging (MRI) operate on similar length scales to neutrons, but with limitations in size, resolution and contrast. Similar as for neutron scattering techniques, the overlap is particularly big with x-ray techniques probing similar size ranges, often with superior maximum resolutions, as obvious in diagrams.

The need for neutron imaging, despite operating in a similar regime to x-ray imaging arises mainly from the complementarity in contrast and transmission properties between neutrons and x-rays. This complementarity is illustrated especially well when regarding the representation of attenuation coefficients as grey scale values throughout the table of elements like those provided in figure 1.6 and the corresponding images taken of a cartridge.

Several differences and characteristics become obvious immediately. Some light elements such as hydrogen, lithium and boron are strong attenuators for neutrons, while they are rather invisible for x-rays, where attenuation is increasing with atomic number, due to the increasing number of electrons. No such regular behavior is found for neutrons. Li and B are typical detection materials (in particular via their isotopes Li-6 and B-10) and like H are contained in shielding materials for neutrons. It is clear that while x-rays easily transmit soft tissue, neutrons are not suited for such application through their limited ability to transmit thick layers of hydrogen containing material. On the other hand, indeed, they are well suited to detect tiny ammounts of hydrogeneous material, invisible for x-rays, and in particular in bulk materials, that x-rays cannot, but neutrons can penetrate well, such as some important structural materials like metals. This already hints for some of the most important application fields of neutron imaging in the hydrogen economy of modern energy concepts, hydrogen uptake and embrittlement of metal components, corrosion and most importantly water management in fuel cells, moisture transport in materials such as concrete, wood and soils, water uptake of roots and plants, coolant and lubricant transport and distribution in machines and devices to name but a few. Li, on the other hand, is another key element of today's energy research, namely in Li-ion batteries (compare chapter 11) and boron is often well suited as tracer material. Indeed large ojects and thick material layers can often also be penetrated by x-rays, in particular when tuning to higher energies. However, increasing energy hampers contrast and in addition some resolution advantages that x-rays are generally assumed to provide, however, on different, i.e. smaller scales. The images of the cartridge are a striking example, where x-rays, when penetrating the metal containment are not sensitive enough to observe the explosive powder consisting of light elements, and denser metal layers and a lead bullet they do not penetrate in contrast to neutrons which provide all relevant details in this case.

Another obvious feature of neutron attenuation is the irregularity of attenuation values throughout the table of elements, in particular in contrast to the respective x-ray attenuation. Neighbouring elements, thus, hardly provide contrast in x-ray imaging, while they often do for neutrons. It depends strongly on the materials involved and sample geometries, in particular sizes and resolution requirements, which radiation is preferred and where neutrons are the best, often the only means of choice.

Additionally, a remarkable characteristic of neutron interaction with matter is the isotope sensitivity, as a consequence of the interaction with the nuclei, which not only enables visualization of isotope distributions per se, but also enables contrast variation. In particular, the exchange of hydrogen with deuterium is an invaluable opportunity in neutron scattering and imaging (compare, e.g. chapter 7) to trace certain deuterated components contrasting with hydrogenated ones. In neutron imaging especially, the variable employment of water and heavy water is a powerful tool for observations of water transport and exchange in complex structures such as fuel cells or even living organisms like plants. But also other isotopes, e.g. Li (compare chapter 11) and other elements, can be key for unique observations.

The low attenuation of some structural materials, especially Al, Zr, C or Mg, also allows realization of complex and stable sample environments and cells for a large number of investigations of processes under realistic conditions under which materials need to be studied to understand their function in service and operando.

Thermal neutrons are not suited in general to alter or damage sample materials in a way, that would render observations questionable as is the case in some high intensity x-ray studies. However, potential activation of samples has to be considered and the occurrence of certain elements in a sample might rule out further use of the sample material for significant times, depending on the half-life of created isotopes.

On the other hand, the activation can be utilized for alternative imaging techniques, which are beyond the scope of this book. Prompt gamma activation imaging (PGAI), based on the bulk PGA method, enables 2D and 3D elemental imaging in combination with thermal neutron imaging and is employed, e.g. in cultural heritage applications. A detailed description and examples can be found in reference [30]. Another spatially resolved technique based on neutron activation is neutron activated autoradiography (NAAR), which allows for the analyses of pigments in paintings based on element specific radiation emmitance which is captured on film [31].

Spatial resolution and images of certain physical properties are also reported from neutron scattering techniques, in particular engineering diffractometers which are referred to as strain scanners for example provide strain maps in 2D and 3D through scanning a gauge volume [32]. Such scanning is also referred to as gauging, which is however not dealt with in this book, which focusses on full-field imaging methods, which are typically subsumed as neutron imaging. Spatial scanning is also reported, e.g. in small-angle scattering, but rarely used with neutrons.

1.3.3 *In situ* combinations with other (imaging) methods

An elegant way to utilize the complementary contrast behaviour of neutrons and x-rays in imaging lies in combining both techniques in order to achieve the full wealth of information on the inner structures, which only a synergy of both modalities provides. Due to the fact that lab x-ray sources can provide similar image quality and resolution to neutron imaging, the choice to install movable x-ray sources at large-scale neutron imaging beamlines became obvious and was pioneered at the thermal neutron imaging instrument NEUTRA at the Paul Scherrer Institut in the early 2000s [33]. The by then established digital detectors enabled registration and various ways to directly exploit the modalities' complementarity. Various and numerous applications in material research including, in particular, studies of concrete, reinforced concrete and corrosion but also in geology, soil and plant science and, last but not least, in natural and cultural heritage research have been reported since these beginnings. Many of these will be reported in the corresponding chapters of this book. While at NEUTRA neutron and x-ray images were still recorded subsequently, but in the very same geometry, later realizations have enabled simultaneous *in situ* x-ray imaging during the neutron experiment, which in contrast to the earlier approach enabled additionally time-resolved studies, which also flourished with the digital detectors and data handling. Today several state-of-the-art instruments worldwide feature the option of *in situ* x-ray imaging [34–36].

Also, other combinations have been reported, e.g. with MRI, or in particular with neutron diffraction, which has moved into the focus with diffraction contrast and diffraction-based imaging. A notable example is a combination of conventional attenuation contrast with diffraction contrast (compare chapter 2) and *in situ* diffraction applied to the research of batteries [37] (chapter 11). This is also an attractive complementary method combination for, e.g. engineering materials research (chapter 5) and is already foreseen for implementation in the next generation pulsed neutron source instruments. This combination allows combination of spatially resolved structural information from a macroscopic scale to be directly combined with detailed crystallographic data in operando studies.

References

[1] Kallmann H and Kuhn E 1940 Photographic detection of slowly moving neutrons *United States Patent 2,186,757*

[2] Peter O 1946 Neutronen-Durchleuchtung *Z. Naturforsch.* A **1** 557–9

[3] Fischer C 1994 The history of the first neutron radiographs in Berlin 1935–1944 *Neutron Radiography (4), Proceedings of the Fourth World Conference*; J P Barton (San Francisco, CA: Gordon and Breach Science Pub) pp 3–9

[4] Thewlis J 1956 Neutron radiography *Br. J. Appl. Phys.* **7** 345–50
[5] Neutron Radiography Newsletters 1964–1977 ed J P Barton, soft cover book 214 pages, archives only
[6] Hawesworth M R 1973 *Radiography with Neutrons* (London: British Nuclear Energy Society)
[7] Barton J P (ed) 1981 *Proceedings of the 1st World Conference on Neutron Radiography* (Dordrecht: D. Reidel Publishing Company)
[8] Bossi R H, Robinson A H and Barton J P 1983 High frame-rate neutron radiography of dynamic events *Proceedings of the 1st World Conference on Neutron Radiography* ed J P Barton and P Von Der Hardt (San Diego, CA: D. Reidel Pub. Co.) pp 643–51
[9] Schlapper G A, Brugger R M, Seydel J E and Larsen G N 1977 Neutron tomography investigations at the Missouri University Research Reactor *Trans. Am. Nucl. Soc.* **26** 39
[10] Matsumoto G and Krata S 1983 The neutron computed tomography *Neutron Radiography, Proc. of the 1st World Conf.* ed J P Barton and P Von Der Hardt (San Diego, CA: D. Reidel Pub. Co.) pp 899–906
[11] Barton J P 1996 International Society for Neutron Radiology Foundation *5th World Conference on Neutron Radiography* ed C O Fischer *et al* (Berlin: DGZfP) pp 765–9
[12] Brenizer J S 2013 A review of significant advances in neutron imaging from conception to the present *Phys. Proc.* **43** 10–20
[13] Kaestner A P, Hartmann S, Kuehne G, Frei G, Gruenzweig C, Josic L, Schmid F and Lehmann E H 2011 The ICON beamline—a facility for cold neutron imaging at SINQ *Nucl. Instrum. Methods Phys. Res. Sect.* A **659** 387–93
[14] Chadwick J 1932 Existence of a neutron *Proc. R. Soc.* A **136** 692–708
[15] Geiger K W, Hum R and Jarvis C J D 1964 Neutron spectrum of A Ra–Be(α, n) source *Can. J. Phys.* **42** 1097–100
[16] Andersen K H and Carlile C J 2016 A proposal for a next generation European neutron source *J. Phys. Conf. Ser.* **746** 012030
[17] McLanne V, Dunford C L and Rose P F 1988 *Neutron Cross Sections, Volume 2: Neutron Cross Section Curves* (New York: Academic)
[18] https://ncnr.nist.gov/resources/n-lengths/ https://wwwndc.jaea.go.jp/NuC/index.html https://iaea.org/resources/databases/neutron-cross-section-standards
[19] Hodgson M, Lohstroh A, Sellin P and Thomas D 2017 Neutron detection performance of silicon carbide and diamond detectors with incomplete charge collection properties *Nucl. Instrum. Methods Phys. Res.* A **847** 1–9
[20] https://psi.ch/en/niag/neutron-interaction-with-matter
[21] de Broglie 1925 Recherches sur la théorie des quanta (Researches on the quantum theory), Thesis, Paris, 1924 *Ann. Phys.* **10** 22
[22] Shull C G, Wollan E O and Marney M C 1948 Neutron diffraction studies *Oak Ridge National Laboratory (ORNL), United States Department of Energy (through predecessor agency the Atomic Energy Commission)*
[23] Willis B T M and Carlile C J 2009 *Experimental Neutron Scattering* (Oxford: Oxford University Press)
[24] Treimer W, Strobl M and Hilger A 2001 Development of a tuneable channel cut crystal *Phys. Lett.* A **289** 151–4
[25] Bouwman W G 2021 Spin-echo small-angle neutron scattering for multiscale structure analysis of food materials *Food Struct.* **30** 100235

[26] Ghasemi-Tabasi H *et al* 2021 Neutron microtomography reveals spatial distribution of porosity in an additively manufactured gold alloy *Appl. Sci.* **11** 1512

[27] ZhouZhou *et al* 2016 From nanopores to macropores: fractal morphology of graphite *Carbon* **96** 541–7

[28] Burnett T L and Withers P J 2019 Completing the picture through correlative characterization *Nat. Mater.* **18** 1041–9

[38] Lehmann E H, Mannes D, Henss M and Speidel M 2022 Ancient Buddhist metal statues using neutron tomography *Handbook of Cultural Heritage Analysis* (Springer) pp 273–303

[30] Kis Z, Szentmiklósi L, Schulze R and Abraham E 2017 Prompt gamma activation imaging (PGAI) *Neutron Methods for Archaeology and Cultural Heritage. Neutron Scattering Applications and Techniques* ed N Kardjilov and G Festa (Cham: Springer)

[31] Denker A, Kardjilov N and Schröder-Smeibidl B 2017 Neutron activation autoradiography *Neutron Methods for Archaeology and Cultural Heritage. Neutron Scattering Applications and Techniques* ed N Kardjilov and G Festa (Cham: Springer)

[32] Withers P J and Webster P J 2001 Neutron and synchrotron x-ray strain scanning *Strain* **37** 19–33

[33] Lehmann E H, Mannes D, Kaestner A P, Hovind J, Trtik P and Strobl M 2021 The XTRA option at the NEUTRA facility—more than 10 years of bi-modal neutron and x-ray imaging at PSI *Appl. Sci.* **11** 3825

[34] Kaestner A P, Hovind J, Boillat P, Muehlebach C, Carminati C, Zarebanadkouki M and Lehmann E H 2017 Bimodal imaging at ICON using neutrons and x-rays *Phys. Proc.* **88** 314–21

[35] Tengattini A, Lenoir N, Andò E, Giroud B, Atkins D, Beaucour J and Viggiani G 2020 NeXT-grenoble, the neutron and x-ray tomograph in grenoble *Nucl. Instr. Meth.* A **968** 163939

[36] LaManna J M, Hussey D S, Baltic E and Jacobson D L 2017 Neutron and x-ray tomography (NeXT) system for simultaneous, dual modality tomography *Rev. Sci. Instrum.* **88** 113702

[37] Lăcătuşu M-E *et al* A multimodal operando neutron study of the phase evolution in a graphite electrode arXiv:2104.03564 [cond-mat.mtrl-sci]

IOP Publishing

Neutron Imaging
From applied materials science to industry
Markus Strobl and Eberhard Lehmann

Chapter 2

State-of-the-art

Markus Strobl

This chapter gives a brief insight into the status of neutron imaging today, outlining the capabilities that today's technology provides as well as the meanwhile numerous interactions of neutrons with matter that provide contrast and information far beyond the conventional attenuation contrast neutron imaging. This will provide the reader with a brief familiarization with some basic concepts that are utilized in applications of neutron imaging, the description and discussion of which constitute the heart of this book. Technical and methodical explanations with respect to contemporary neutron imaging techniques, including computational approaches and basics for data reduction and analyses are provided in the methods and instrumentation sections (chapters 19 and 20) at the end of the book.

2.1 State-of-the-art resolution

The state-of-the-art of an imaging method is intrinsically characterized foremost by its **spatial resolution** capability. The highest resolution limit of neutron imaging has seen a remarkable development since the introduction of digital detectors around the onset of the new Millennium. While analogue neutron imaging on film already enabled spatial resolutions of the order of a few 10 μm, digital detectors achieved such values only since the second half of the first decade of the Millennium. Today, with the leading instruments, resolutions of a few micrometer (<5 μm) are available for practical applications [1–4], while single demonstration experiments claim to reach down to a limit of around a single micron and aim beyond [5].

Spatial resolution capabilities in neutron imaging are governed by the geometric resolution provided by the instrumental set-up and the detector technology. The available neutron flux is indeed another key factor and, thus, cutting-edge resolutions are only available at instruments at large-scale neutron sources such as nuclear research reactors and spallation neutron sources. Another limitation to date is the limited performance of neutron optics that prohibits the benefits of lenses in neutron imaging. Indeed, a number of attempts, tests and considerations have been

doi:10.1088/978-0-7503-3495-2ch2

© IOP Publishing Ltd 2024

made with regards to efficient lens optics for neutron imaging [6], the latest development being some hope placed in nested Wolter optics [7, 8], but no seminal solution was hitherto provided. Therefore, neutron imaging still relies on the basic principle of pinhole geometry to achieve high spatial resolution. Also, more sophisticated approaches to improve the efficiency of such a pinhole set-up like, e.g. coded source imaging [9, 10] in the past or the computationally intense alternative of what is dubbed ghost imaging [11] utilizing a bucket detector imaging principle today, did not so far change this situation.

Modern instruments in general feature pinhole (diameter D) exchangers and multiple measurement positions, at distances L from the pinhole, providing access to different collimation ratios L/D and, hence, different trade-offs between geometrical resolution (blur) d at the detector position at a minimized distance l downstream of the sample

$$d = l*D/L \tag{2.1}$$

and incident neutron flux $I_0 \propto (D/L)^2$. Improving the geometrical resolution implies a flux penalty scaling with $(L/D)^2$ which forbids high collimation at low flux sources and for keeping reasonable exposure times.

For well optimized imaging instruments at powerful neutron sources (compare chapter 19) the limiting factor up to now was detector technology. The dominating digital neutron imaging detector technology is based on the combination of neutron sensitive scintillators with digital cameras for visible light, coupled by optical components, in general a mirror and objective lens system [12]. All these components have seen a significant technological development, which enabled resolution improvements with time on a logarithmic scale throughout the last two decades [1, 12–15] (compare section 19.3). Scintillators diversified for different applications and gained improved long-term performance stability, efficiency, resolution and light output. Different camera technologies based on CCD and CMOS chips, with decreased noise, smaller pixel sizes, higher pixel numbers, higher frame rate, faster shutters, amplifiers etc matured and improved in performance. Similarly, optics were further tuned for higher light efficiency and definition [1, 14, 15]. The small micrometer-sized pixel dimensions for highest resolution naturally further aggravate the flux deficiency, and so do the thin scintillator screens required [16].

On the other end of the spatial resolution scale is the **field of view** (FoV) enabled by neutron imaging and in particular for certain resolutions. Especially, for the mainly discussed scintillator/camera detectors FoV and spatial resolution are coupled by the pixel number of the camera chip as the maximum pixel size coincides in conventional sensing with half the resolution. Hence, e.g. 2 μm resolution requires at least 1 μm pixel size and thus with a 2k chip the FoV is limited to only 2 mm in each direction, while 200 μm resolution can be achieved with such a system on a FoV of 20 cm. Only certain scanning strategies enable investigations of samples with larger dimensions with the same detector system at a given resolution. The flexibility of trading resolution and FoV is one strength of the flexible scintillator/camera detector technology (compare chapter 19.3, figure 19.3). Most other technologies to date find applications mainly in niches. This includes the enduring use of film due to standards (e.g. American Society for Testing and Materials, ASTM), but also, due to, in contrast to most digital detectors, an independence of FoV and resolution.

Further, e.g. imaging plates and amorphous Si detectors are in use [14] and neutron imaging detectors based on timepix chips, like above all the seminal micro-channel plate (MCP) detector [17] and meanwhile also the TPX-Cam technology [18], gained significant importance not only in the field of time-of-flight (ToF) neutron imaging, but also e.g. for stroboscopic imaging [19], where high time resolution, quasi-noiseless imaging performance is required and large FoV is often not an absolute need. More details on instrumentation are provided in chapter 19 discussing instrumentation and methods.

Time resolution is indeed another performance category of high interest. Again, resolution is defined by flux and technology and an obvious trade-off between spatial and temporal resolution is implied by the limited flux availability (compare chapter 1, figure 1.2). This limitation can largely be bypassed by what is referred to as stroboscopic neutron imaging in the case of repetitive processes [19], thus providing the highest temporal resolution, with the limitation, however, that each time slice is an average of numerous repetitions of the process and, thus, not a real-time image at any individual moment. Besides the neutron flux, the efficiency and frame rate capability of the detector system is the bottleneck for pushing the limits of time resolution. However, at the current state of development flux is the limiting factor. While average exposure times range in the order of sub-seconds to minutes, and some specific measurements might require hours, the best time resolutions in neutron imaging as of today are reported in the literature within the range of milliseconds [20, 21], in some cases even microseconds [22]. It is microseconds and beyond for stroboscopic measurements, but also for novel capabilities of ToF neutron imaging (compare chapter 19). In the latter case the time resolution does not refer to the sample system, but to the arrival time of neutrons at the detector from a pulsed source in order to determine their wavelengths. Time resolution with regards to the sample system is in such a case rather limited and depends on the accumulated statistics in the ToF channels over extended exposure times with large numbers of individual pulses.

The best reported process time resolution for ToF diffraction contrast is rather in the order of minutes so far [23], while in general exposure, times often range in the order of hours. Not only for ToF imaging but also for stroboscopic applications, novel detector technologies with intrinsic microsecond time resolution (and beyond) have been a game changer, because in contrast to before, when one image was taken per period [24], all phases (time bins) of a process can now be probed continuously [19]. This implies more than an order of magnitude efficiency gains for stroboscopic imaging, becoming a rather quasi-stroboscopic accumulative imaging method.

While time resolution lends itself as a third dimension to 2D-spatial resolution, 3D resolution commonly refers to volumetric tomographic imaging. **Tomography** is indeed a standard technique also in neutron imaging and can be applied for the full range of spatial resolutions and related fields-of-view. Unlike temporal resolution which has been introduced with television cameras [25] already in the 1960s, neutron tomography was even more limited and tedious when introduced in the 1970s using film for detection [26]. However, today digital detectors have removed such

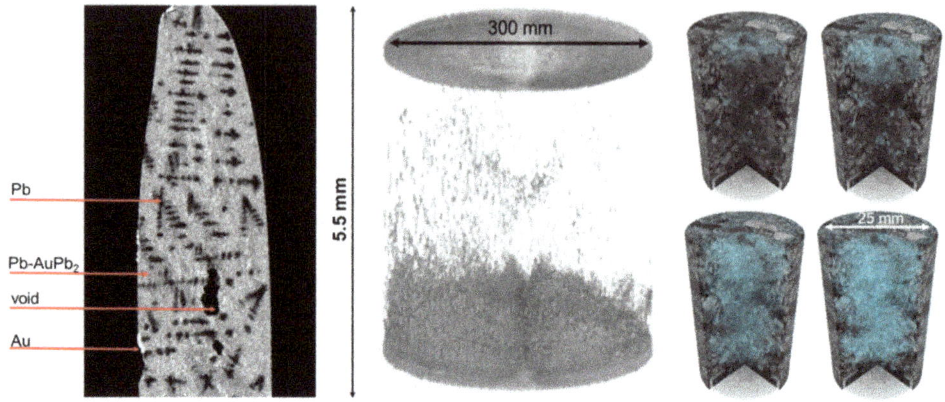

Figure 2.1. Examples of different size, resolution and dimensionality of neutron images: High resolution (4 µm) image of a lead–gold eutectic with a sample height of 5.5 mm, reproduced from [3] CC BY 3.0, (left), neutron tomography volume of the soot and ash distribution in a diesel particulate filter of a truck with 30 cm diameter, reproduced from [28], copyright SAE International (mid); and an on-the-fly time resolved bimodal n/x tomography set of water flow through a soil sample of 25 mm diameter (courtesy A Kaestner, D Mannes *et al*).

obstacles and all state-of-the-art instruments are prepared to perform computed tomography. In tomography the best reported 3D spatial resolutions range in the order of about 5 to 10 µm [2, 27] (figure 2.1(left)).

The combination of 3D spatial resolution and significant time resolution, thus **4D neutron imaging**, has been introduced only relatively recently. While already in 2005 tomographic time resolution of less than 10 s was reported at a high flux reactor [29], only the utilization of on-the-fly tomography introduced to neutron imaging in 2015 [30] enabled overcoming the 90% time overhead reported in the initial rapid tomography approach. This development, where the sample is turned continuously while taking projection images, thus, paved the way to high speed 4D imaging applications in today's state-of-the-art instruments. Tomographic time resolutions for investigations of sample evolutions in 3D are approaching 1 s and beyond today [21]. Indeed, here in particular, achievable time resolutions are subject to compromises on spatial resolution impacting the factors of collimation (flux), pixel (bin) sizes (statistics) and number of projections (Nyquist sampling theorem [31]). Depending on the process, speed motion artefacts have to be considered and depending on the rotation speed, also potential influence of the centrifugal force on the system. In order to increase the number of time bins of a 4D image, individual projections are utilized for several consecutive 3D reconstructions within shifting time frames, i.e. projection sets (similar to a running average).

Finally, the dimensionality of neutron imaging can be further increased through advanced techniques of, e.g. multi-parameter sample environments or multimodality (figure 2.1(right)) including advanced contrast modalities beyond attenuation, which will be introduced in the next section (figure 2.2.).

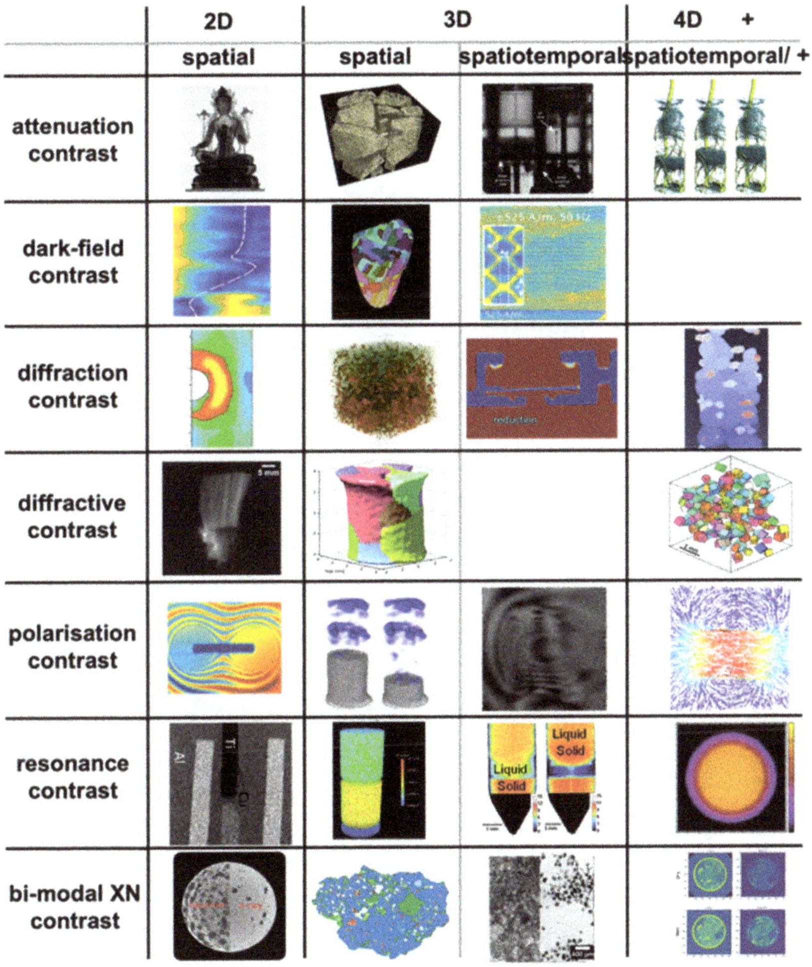

Figure 2.2. Matrix of neutron imaging contrast modalities and dimensions of neutron imaging: row 1 attenuation contrast radiography of a Buddha statue (courtesy of E Lehmann); tomography of porous rock, reproduced from [32] CC BY 3.0; time resolved imaging of a running engine, reprinted from [33], copyright (2006), with permission from Elsevier; time resolved tomography of root/soil/water system, reproduced from [34] CC BY 3.0; row 2 dark-field (DF) contrast radiography study of sedimented micro-particles, reproduced from [35] CC BY 4.0; DF tomography of magnetic domain walls [135]; time resolved DF imaging of moving domain walls in an AC field, reproduced from [19] CC BY 4.0; row 3: diffraction contrast strain map of cold expanded whole in steel plate, reproduced from [37], copyright (2002), with permission from Springer Nature; tomography of residual austenite phase distribution in quenched steel cube; time resolved radiography of Ni oxidation and reduction in SOFC anode [38]; 3D grain morphology and orientation map [39]; row 4: projection image in diffraction geometry showing variations in single crystal turbine blade, reprinted from [40], copyright (2016), with permission from Elsevier; a 3D grain reconstruction, reproduced from [41] with permission from the Royal Society of Chemistry; and a 3D map of crystal grain orientations, reprinted from [42], copyright (2022), with permission from Elsevier; row 5: polarized neutron radiopgraphy of magnetic dipole field [43]; depolarization contrast tomography of ferromagnetic phase. distribution, reproduced from [44], copyright IOP Publishing, all rights reserved; time resolved imaging of the magnetic field of an AC coil, reproduced from [45], copyright IOP Publishing, all rights reserved; 3D magnetic vector field of an electric coil

[46]; row 6: resonance absorption contrast radiography, reproduced from [47] CC BY 2.0; and tomography of ^{235}U in nuclear fuel pellets, reproduced from [48] CC BY 4.0; time resolved study of elemental distribution in solidification process of scintillator material, reproduced from [49] CC BY 4.0; 4D resonance imaging revealing elemental and temperature distribution, reproduced from [50] CC BY 4.0; row 7: bimodal neutron and x-ray imaging of concrete displaying complementary contrast, reprinted from [51], copyright (2018), with permission from Elsevier; bimodal tomography revealing constituting material phases of an impactite, reproduced from [52] CC BY 4.0; N/X time resolved study of lithiation in a Li-ion battery with Di electrode, reprinted from [53], copyright (2017), with permission from Elsevier;concurrent on-the-fly time resolved study of water uptake in soil (courtesy of A Kaestner *et al*).

2.2 Applied contrast modalities

Neutron imaging is usually associated with attenuation contrast, sometimes wrongly referred to as absorption contrast, and often considering this image contrast as the only available modality, in particular, when generalizing. However, neutron imaging has, apart from taking more and more dimensions at ever better resolution, developed a great variety of contrast modalities, which allow completely different and new insights into many materials and components (figure 2.2). 'New insights' here refers not only to better accuracy but also to different length scales, to completely different characteristics probed and different conceptualizations of investigated structures. This includes probing crystalline strains and microstructures, magnetic fields and structures, temperature, aggregate state, or structural correlation functions rather than individual structures in the micro- and nanometer regime.

2.2.1 Conventional attenuation contrast

Conventional attenuation contrast neutron imaging is typically performed utilizing a broad wavelength spectrum in order to maximize the available flux and to minimize exposure times. The measured attenuation contrast image $I(x,y)$ is generally well described by simply approximating the effect of the sample on the beam transmission by Beer–Lambert's law reading

$$I(x, y) = I_0(x, y)e^{-\int \mu(x, y, z)dz} \tag{2.2}$$

Here $I_0(x,y)$ is the incident beam and $I(x,y)$ is the transmitted beam (attenuation contrast image), z points in the beam direction, while $\mu(x,y,z)$ is the local attenuation coefficient. Here the attenuation coefficient denotes the spectrum weighted average of the local macroscopic cross-section $\Sigma(x,y,z,\lambda)$, and λ denotes the neutron wavelength/energy. The wavelength dependent macroscopic cross-section is a product of the wavelength independent material characteristic of the atomic number density N, and the material and wavelength/energy dependent total microscopic cross-section $\sigma = \sigma_a + \sigma_s$. For materials consisting of numerous elements the contributions of the macroscopic cross-sections are simply summed. The total microscopic cross-section is the sum of the absorption cross-section σ_a and the scattering cross-section σ_s. The latter is itself the sum of various contributions of elastic and inelastic coherent and incoherent scattering cross-sections. Notably, for many relevant materials, the scattering cross-section is competitive or even larger than the absorption cross-section and thus pure absorption contrast

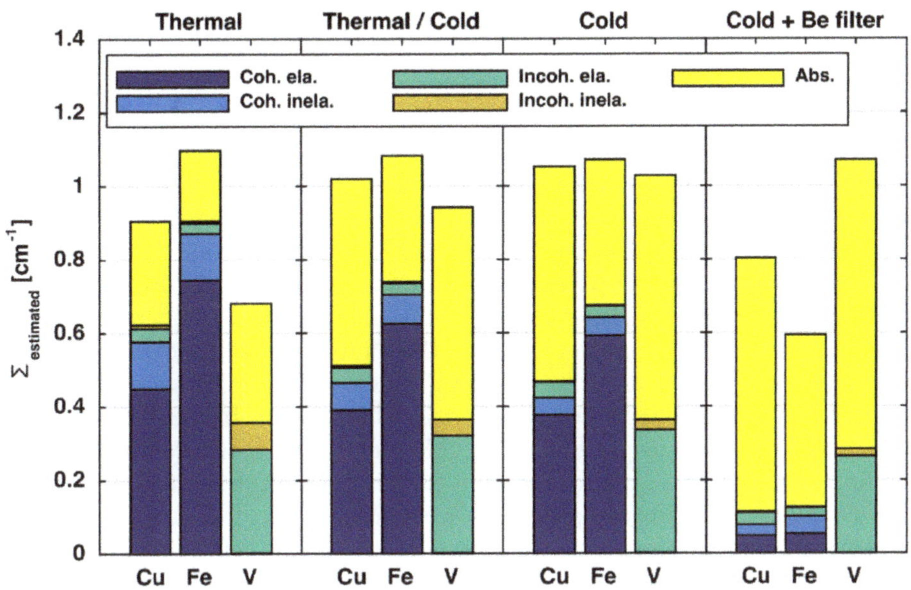

Figure 2.3. A comparison of the different contributions to the attenuation of some relevant isotropic powder-like polycrystalline materials for different utilized neutron spectra. Reproduced from [54] CC BY 2.0.

is rarely measured (figure 2.3). The scattering components on the other hand do not only contribute to attenuation, but often scattering too close to the forward direction contributes to background and thus bias in quantification of the actual attenuation as described in Beer–Lambert's law [54, 55]. Only recently, an efficient experimental and software approach has been introduced to overcome such bias largely [56, 57]. The specific value of neutron attenuation characteristics has been introduced earlier in chapter 1 (figure 1.6). The scattering characteristics of neutrons can, however, not only lead to bias through background contributions but their footprint in the attenuation can carry significant information, which becomes accessible, e.g. as diffraction contrast in wavelength-resolved neutron imaging, as the next section will detail.

Conventional attenuation contrast imaging is still the most applied technique and finds applications in a wide variety of fields, which will become obvious in the core chapters 3–18 of this book dealing with applications in applied materials science.

2.2.2 Diffraction contrast

When considering the wavelength dependence of attenuation, in particular the peculiar features of the coherent elastic scattering, cross-section of crystalline materials become obvious, especially for some important metals. These are related to diffraction at the crystal lattice. However, examining the scattering cross-section σ_s in detail, four different additive components can be identified, namely two coherent, the elastic and inelastic coherent cross-sections, $\sigma_{\text{el,coh}}$ and $\sigma_{\text{inel,coh}}$, and two incoherent, again an elastic and an inelastic incoherent cross-section, $\sigma_{\text{el,incoh}}$ and $\sigma_{\text{inel,incoh}}$. While all contributions but the elastic coherent manifest as smooth functions of wavelength, the latter creates a distinct pattern with discontinuities referred to as Bragg edges, especially if the crystalline material has a relatively high bound coherent scattering length b, a parameter denoting the scattering power of nuclei and being proportional to the amplitude of the spherical scattered waves. Calculated cross-sections of the different components as well as the total cross-section for α-iron powder are illustrated in figure 2.4 alongside with two measured and one calculated reference total cross-sections for austenitic steel. Note, that in the cold neutron regime the coherent elastic contribution does not only display key features but is also absolutely dominant with an increasing contribution of absorption towards longer wavelengths.

Nuclear scattering lengths and cross-sections are tabulated and can be found, e.g. for thermal neutrons on a webpage of the National Institute of Science and Technology (NIST) [60] or from National Nuclear Data Center at Brookhaven National Lab of the Unites States of America. These are the basis for calculating the total cross-section, and in particular the coherent elastic cross-section of a specific crystalline material taking into account crystal lattice parameters and involved nuclei. Today several software programs are available to calculate such wavelength dependent cross-section spectra, as shown in figure 2.4 left-hand side, which is calculated with one of the first of these, namely BETMAn, developed in the framework of a PhD thesis by S Vogel [58]. Today, most used programs are e.g. nx-plotter [61, 62] and ncrystal [63].

The coherent elastic cross-section in the ideal case of an isotropic polycrystalline material can be expressed by

$$\sigma_{\text{el,coh}}(\lambda) = \frac{\lambda^2}{2V_0}\sum\nolimits_{hkl}|F_{hkl}|^2 d_{hkl} \tag{2.3}$$

where V_0 is the volume of the unit cell of the crystal lattice, d_{hkl} are the lattice constants for the crystal lattice families denoted by their Miller indices h,k,l, while F_{hkl} is the corresponding structure factor, which can be written as

$$F_{hkl} = w_{hkl}\sum\nolimits_n o_n b_n \exp(2\pi i(hx_n + ky_n + lz_n)) \exp\left(\frac{-B_{\text{iso},\,n}}{4d_{hkl}^2}\right) \tag{2.4}$$

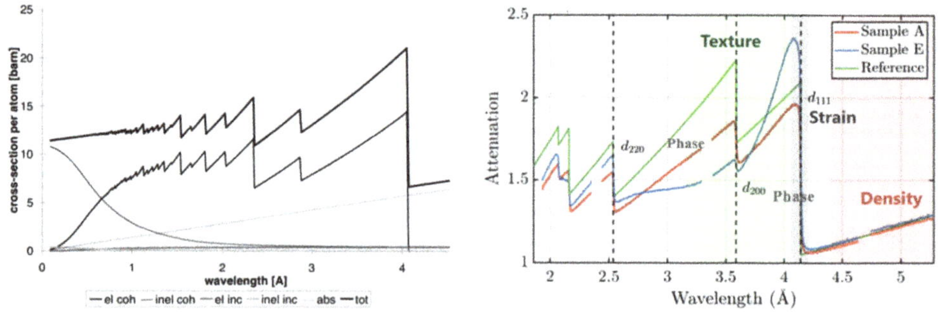

Figure 2.4. Contributions of different cross-sections to total cross-section for α-iron (BCC), reproduced with permission from [58] (left); theoretical powder attenuation spectrum and two measured attenuation spectra for textured FCC steel samples, reprinted from [59], copyright (2021), with permission from Elsevier (right).

In this equation ω denotes the multiplicity associated with the specific *hkl* lattice planes, o is the site occupation factor and (x_n, y_n, z_n) is the position of the n-th atom within the unit cell in fractional coordinates. The second exponential can be referred to as the Debye–Waller factor accounting for thermally induced displacements of atoms in the lattice, and B_{iso} is the isotropic atomic displacement factor containing in particular temperature, Debye temperature and the mass of the nucleus.

The Bragg edges in transmission spectra appear due to the fact that when λ reaches a value where the angle of diffraction θ is 90° for a certain lattice plane family *hkl* the Bragg equation

$$\lambda = 2d_{hkl} \sin\theta \tag{2.5}$$

reduces to

$$\lambda = 2d_{hkl} \tag{2.6}$$

and no wavelength longer than $2d_{hkl}$ can be diffracted by those lattice planes anymore. Thus, a discontinuity appears in the spectrum, because the attenuation for longer wavelengths drops significantly. However, this also implies that Bragg edges are a measure for the lattice constants d_{hkl} for an isotropic polycrystalline material. Consequently, **crystalline phases** can be identified and quantified according to Bragg edge positions and heights [58, 64] and **lattice strain** variations can be detected [37], if the Bragg edge position can be measured with sufficient accuracy [65].

However, several things have to be considered for valid Bragg edge imaging and analyses, as one specific but the most utilized form of diffraction contrast neutron imaging:

- The transmission signal in terms of Bragg edge parameters corresponds to an average along the beam trajectory.
- Lattice distortions, i.e. strains, can only be detected in beam direction ($\theta = 90°$).
- Grain sizes have to be considered and for Bragg edge analyses have to be significantly smaller than spatial resolution.
- The situation is only as simple as described for isotropic polycrystalline materials and texture can falsify or prevent strain and even phase analyses.

These points as well as the tensorial nature of stresses limit straightforward application in particular for 3D tomographic imaging of such parameters significantly. Nevertheless, *a priori* knowledge and/or substantial modelling have been shown to enable overcoming such limitations for 3D strain mapping in some highly relevant cases [66, 67] and potentially also generally [68]. For 3D phase mapping often isotropy is assumed even when a mild texture has to be considered or is found [65].

Texture is a relevant factor for many potential applications, because material manufacturing and service conditions often contribute to corresponding anisotropies, which in turn imply anisotropic material behavior. While texture variations within a specimen can easily be identified by the local response of the approach of Bragg edge imaging, quantitative conclusions are a sophisticated challenge. So far mainly forward modelling from *a priori* diffraction measurements was used to

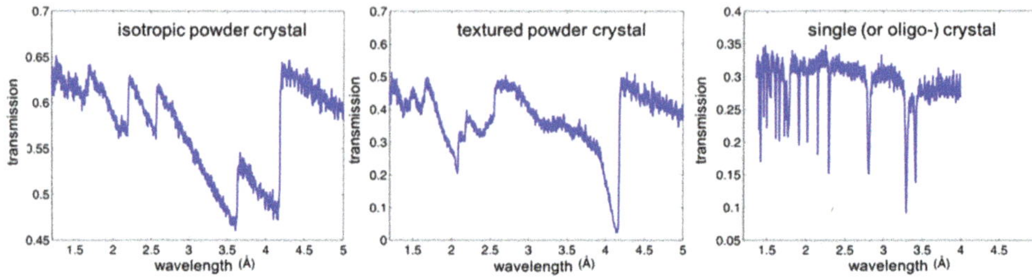

Figure 2.5. Transmission spectra of perfectly random powder-like polycrystalline Cu (left), of a textured polycrystalline Cu sample (mid) and an single/oligo-crystalline Cu sample displaying the signature of individual Bragg peaks. Reprinted from [65], copyright (2018), with permission from Elsevier.

demonstrate the quantitative nature of the transmission spectra in this regards [69–71], but significant progress is made in analyzing texture aspects that can in fact be extracted [72]. The influence of texture on the transmission spectrum (figure 2.5 mid) can be considered by an additional factor P in describing the wavelength dependent elastic coherent cross-section.

$$\sigma_{\text{el,coh}}(\lambda) = \frac{\lambda^2}{2V_0} \sum_{hkl} |F_{hkl}|^2 \, d_{hkl} P_{hkl} \tag{2.7}$$

The March–Dollase formalism utilized in diffraction [73] is often adapted for transmission applications [74–76] to parametrize the additional factor. A more quantitative approach has been introduced lately [77] using the expansion of the orientation distribution function (ODF) into a generalized Fourier series, which enables iterative fitting solutions, especially for some integrated ODF values, when lattice symmetry and a respective number of measured projections are taken into account. This has been demonstrated, e.g. for a hexagonal lattice where Kearns factors [78] have been extracted for αZr and excellent full pattern fitting was achieved [72]. As these kind of analyses are most relevant in engineering applications, a detailed discussion in particular with regards to texture investigations is given in chapter 5.

In general, the approach to analyzing full transmission spectra, in particular with physical parametrization, is still rare and the analyses of specific individual parameters of single edges prevails for strain and phase mapping applications. Full physical modelling and fitting of transmission spectra in a general case has to consider even more parameters and grain size-dependent extinction effects have to be accounted for, bearing on the other hand potential to extract grain size information. Such an approach is implemented, e.g. in the RITS code [75]. Extinction effects have been attempted to be modelled by yet another factor in the cross-section formulation according to

$$\sigma_{\text{el,coh}}(\lambda) = \frac{\lambda^2}{2V_0} \sum_{hkl} |F_{hkl}|^2 d_{hkl} P_{hkl} E_{hkl} \tag{2.8}$$

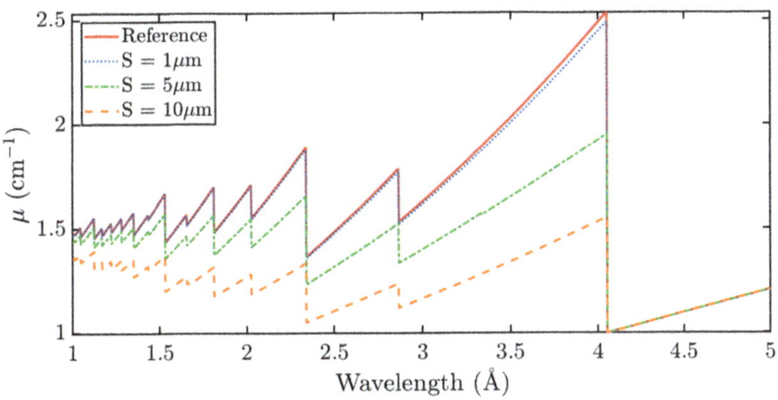

Figure 2.6. Calculated cross-section spectra displaying the impact of extinction with regards to different crystallite sizes of a powder-like polycrystalline α-iron sample (figure courtesy M Busi).

where the extinction factor E has been expressed primarily again by adapting an approach of powder diffractometry [79], the effect of which is illustrated in figure 2.6. Given the wealth of parameters involved, the problem of quantitatively fitting a transmission spectrum suffers from significant under-sampling, which in general cannot be overcome even by tomographic exposure, i.e. recording spatially resolved spectra for numerous projections. This again implies the need for extensive *a priori* information being available in order to focus analyses and modeling on specific isolated aspects, where in turn, again, often single-edge analyses or analyses of other isolated aspects suffices. Note also that the cross-section is often provided in literature with an additional factor (often denoted 'R') describing the instrumental resolution. While it is indeed important to consider instrumental influences on the measured cross-section, these are not considered part of the cross-section itself, and thus are not included here (compare chapter 19).

The situation again changes very much, when grain sizes approach the achievable real space image resolution, when instead of partially smooth Bragg edge patterns individual Bragg reflections are resolved as peaks in the cross-section, thus inverse peaks, so called Bragg dips, in transmission (figure 2.5 right). In this case the highly anisotropic cross-section has to be rewritten as a linear combination for n individual single crystal contributions as [65]

$$\sigma_{\text{el,coh}}(\lambda) = \frac{\lambda^2}{2V_0}\sum\nolimits_{\theta_{n,hkl}} |F_{hkl}|^2 d_{hkl}\delta(\lambda - 2d_{hkl}\sin\theta_{n,hkl}). \tag{2.9}$$

This not only enables elucidating the quality and orientation of single crystals [80], but also orientation distributions in mosaic crystals [81, 82] and other large grained oligo-crystals [83]. It could be shown that recording tomographic scans enables indexing and morphological reconstruction of the grain network of coarse-grained materials in 3D [39]. Figure 2.7 displays several examples of diffraction contrast imaging including a grain map, a 3D crystalline phase map as well as 2D texture factor, orientation and grain size maps.

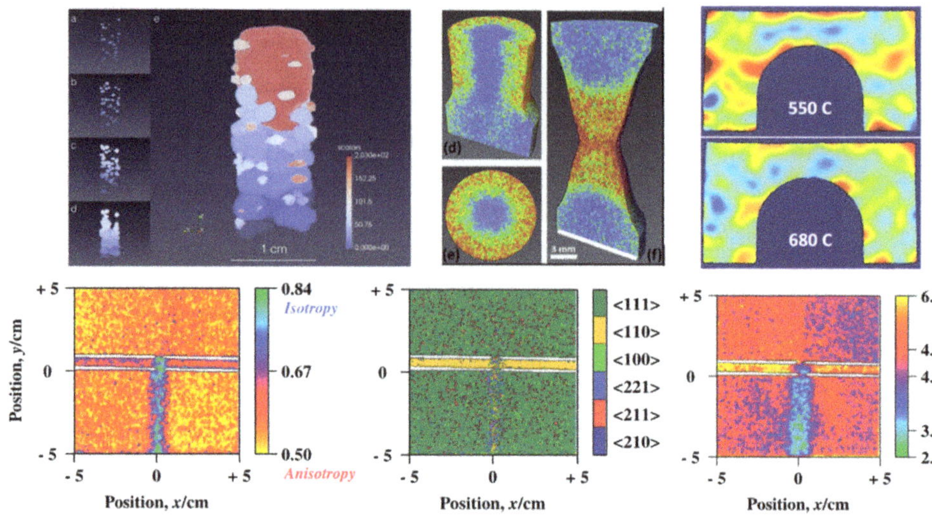

Figure 2.7. Examples of diffraction contrast neutron imaging: grain map reconstructed from ToF transmission tomography, reproduced from [39] CC BY 4.0 (left top); martensite/austenite phase map from wavelength selective pair of monochromatic tomographies [84] John Wiley & Sons, copyright 2014 WILEY-VCH Verlag GmbH & Co. KGaA, Weinheim (top mid); strain maps from *in situ* stress relaxation studies with ToF diffraction contrast imaging (Bragg edge imaging), reprinted from [85], copyright (2021), with permission from Elsevier; bottom: crystallographic anisotropy map (left), preferred orientations with respect to the beam direction (mid) and crystallite size distribution of (welded) steel plates placed in different orientations, reproduced with permission from [75], copyright 2011 The Japan Institute of Metals and Materials.

2.2.3 Inelastic scattering contrast

Lately it has been shown that inelastic scattering contributions can also be utilized in wavelength-resolved neutron imaging studies to provide valuable information on local states of condensed matter. Indeed, on the one hand, inelastic scattering also has an influence in diffraction contrast imaging, and whereas the elastic contributions decrease with temperature, driven by the Debye–Waller factor in the case of coherent elastic scattering, also the incoherent elastic cross-section and the inelastic contributions increase at elevated temperatures in certain wavelength ranges [86]. Thus, contributions have to be considered carefully for observations depending on heating such as annealing and phase transformations. On the other hand, in particular the latter contributions can be utilized to sense the temperature in bulk materials efficiently [87].

However, more importantly even in particular aggregate and binding states of water and hydrogen, respectively, can be targeted by analyzing the subtle differences in attenuation spectra based on inelastic scattering. Typically, the ratio of attenuation at shorter and longer wavelengths enables characterizing a particular state when spectral effects can be modeled sufficiently. The excitation energy for the water molecule with respect to translational oscillations associated with diffusion lies within the same order of magnitude as the kinetic energy of cold neutrons. Therefore, cold neutrons are highly sensitive to such excitations and the incoherent inelastic cross-section decreases with temperature and corresponding to slower diffusion in ice than in liquid water, e.g. ice and super-cooled water can be distinguished based on their cold neutron cross-section [88]. These effects can not only be modelled [89] but have been used in imaging to remotely distinguish distinct local aggregate states of water within operating devices, such as in particular fuel cells [90] (figure 2.8), where research is driven by the need to overcome issues of, e.g. fuel cell powered vehicles in cold climates.

Also, beyond water hydrogen bonds play an important role in the chemistry of processes. Hydrogen is in focus not only because it is involved in numerous materials and processes of outstanding interest and importance, but because its total cross-section is dominated by an extraordinarily high incoherent cross-section of around 80 barn, while other important elements, e.g. in organic solvents, such as carbon or oxygen have negligible incoherent cross-sections in the range of 10^{-3} barn. Attenuation spectra contain information on phonons and diffusional motion, on inter-molecular interactions and hydrogen bonds. At the lower wavelength end of the spectra, in particular the vibrational modes of a molecule contribute, and thus molecules with different hydrogen bond characteristics can be distinguished, like e.g. different hydrocarbon functional groups in organic solvents. Changes in concentration and diffusional motions on the other hand are rather observed through the development of the slope of longer wavelength cross-sections. This implies that analyses and extraction of information is not straightforward and requires extensive

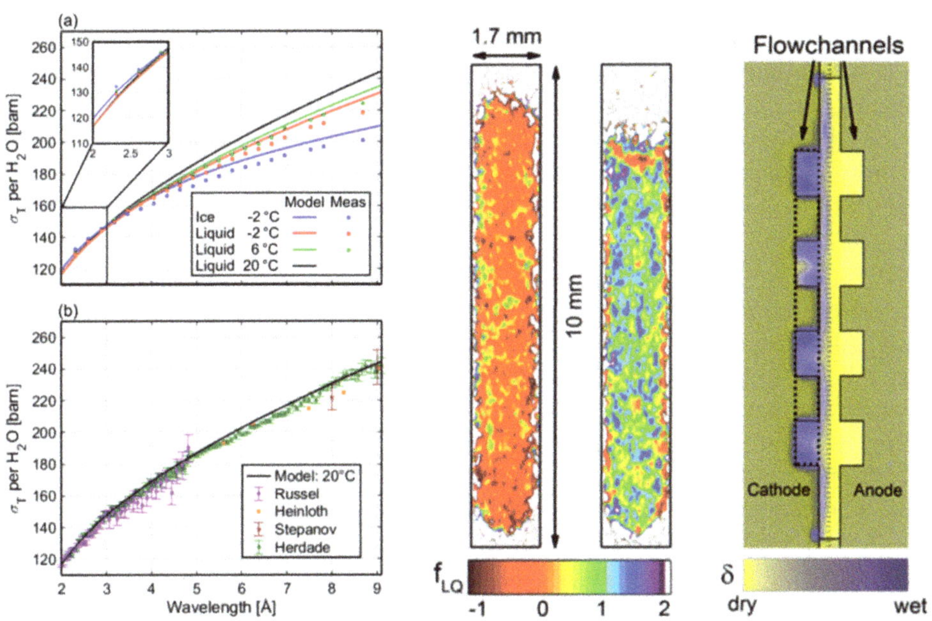

Figure 2.8. Cross-section of water for different temperatures modelled and measured, reprinted from [88], with the permission of AIP Publishing (left); mapping water phases (liquid water fraction) at different temperatures (−1/+1 °C) in flow channels of a fuel cell, reprinted with permission from [90], copyright (2014) by the American Physical Society.

modelling. However, models are being developed such as the average functional group approximation (AFGA) [91] and are available from quasi-elastic/inelastic neutron scattering such as the Egelstaff–Schofield model for diffusion in liquids [92] and are embedded in programs like in particular NJOY2016 [93] calculating neutron cross-sections.

This is a promising new field for neutron imaging probing local states of hydrogen dynamics and thus physicochemical changes and processes with spatial resolution. A prominent example in energy research will be introduced in the Energy Research part IV in chapter 11.

2.2.4 Phase contrast

Phase contrast has also been a prominent topic in neutron imaging from the beginning of the Millennium [94]. Phase contrast imaging takes advantage of the real part of the refractive index of materials, in contrast to the attenuation coefficient, in conventional radiographic imaging, but also in diffraction and inelastic contrast neutron imaging. The real part of the refractive index can be written as

$$n = 1 - \delta \approx 1 - Nb_c\lambda^2/2\pi \tag{2.10}$$

where Nb_c is the neutron scattering length density (SLD) with N being the particle density and b_c the bound coherent scattering length, while λ is the wavelength. In full analogy to light optics the refractive index leads to refraction and total reflection phenomena also for neutrons. However, the neutron refractive index is generally very close to unity, and, thus, involved angles are rather small. For example, the total reflection angle for natural Ni, used for reflecting neutron guides is 0.1 deg Å^{-1}. Depending on the algebraic sign of b_c the refractive index can be smaller or bigger than unity for neutrons. The phase shift φ of a plane neutron wave with wavelength λ transmitting a material with SLD = Nb_c and a thickness of t can be calculated by

$$\varphi = -k \times \delta \times t = -2\pi/\lambda \times Nb_c\lambda^2/2\pi \times t = -Nb_c\lambda t \tag{2.11}$$

based on the integral of the real part of the refractive index δ over the path length through the material. However, the phase of a neutron wave cannot be assessed directly but measured for example in a Mach–Zehnder interferometer utilizing a reference beam. It has been shown that in principle phase imaging with a Mach–Zehnder neutron interferometer can be performed [95], but is tedious and limited in application due to the high requirements of such an interferometer with regards to spatial and temporal coherence as well as alignment and stability. However, the distortion of the wave front due to variations in φ across the beam induced by penetration of an irregularly shaped object implies corresponding angular deviations in local beam paths leading to detectable intensity variations beyond the conventional attenuation footprint. In general, distinct patterns at inclined interfaces are superimposed to the attenuation image and can be described by refraction and total reflection effects, referred to as differential phase effects [96]. First demonstrations of this effect amplifying the usual artefacts through high collimation and extended

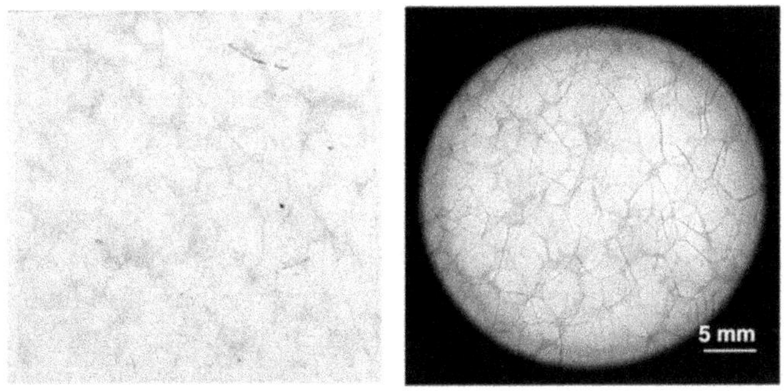

Figure 2.9. Attempted application of neutron phase contrast imaging to metal foams. Reproduced with permission from [104], copyright Trans Tech Publications Ltd.

sample to detector distances, both supporting the resolution of the refraction contributions, which is highly regarded as a breakthrough for a new contrast with higher sensitivity based on the neutron's phase [94]. However, practical application remained limited (figure 2.9) and against the background of the development of increasing spatial resolution refraction effects were rather constituting a drawback in the form of artefacts hindering quantification and correct interpretation, in particular close to interfaces [97].

Refraction is also referred to as a differential phase effect because the deflection ϕ of a beam at a refractive interface can be described by the transversal phase gradient $\Delta_{\perp}\varphi$ determining the local orientation of the wave front and thus the angle of refraction as

$$\phi = \Delta_{\perp}\varphi/k = \Delta_{\perp}(\delta^{*}t). \tag{2.12}$$

Therefore, differential phase contrast methods are those, measuring the local refraction angle of a beam after it has transmitted a sample, which can be referred to as a phase object. Initial attempts have been made with double crystal diffractometer (DCD) instrumentation otherwise utilized for ultra-small-angle scattering (USANS) [98, 99]. While the method was well suited to study the details and quantification of differential phase contrast neutron imaging, again, unlike its x-ray counterpart, the method proved too slow, often requiring tedious scanning with a highly monochromatic beam, to qualify for real applications. However, experimenting with phase contrast in DCDs was extended to the use of polarized neutrons, demonstrating that the contribution of a magnetic field to the real part of the refractive index corresponding to

$$\delta_{\mu} \approx \pm\mu_{\mathrm{n}}B/(2E_0) = \pm2\mu_{\mathrm{n}}Bm\lambda^2/h^2 \tag{2.13}$$

can be utilized for differential phase contrast imaging of magnetic fields and structures [100]. Here $\mu_n = -9.6623 \times 10^{-27}$ J T^{-1} is the magnetic moment of the neutron, B represents a magnetic field and E_0, m and h are the unperturbed, i.e. incident kinetic energy of the neutron, the mass of the neutron and the Planck constant, respectively. The ± sign indicates the difference with regards to the spin-up and spin-down states of neutrons.

Finally, neutron grating interferometry (NGI) was introduced [101] in analogy to its introduction to x-ray imaging shortly before. Neutron grating interferometry (compare chapter 19) based on the Talbot–Lau effect enabled an unprecedented efficiency in differential phase contrast imaging due to very relaxed requirements to spatial and temporal coherence of the beam. Grating interferometers have been demonstrated to be compatible with conventional imaging instruments with pinhole geometry straightforwardly, although alignment can be sophisticated and time consuming. In contrast to a DCD where the local angle of refraction is resolved directly by an angle dispersive instrumentation, a grating interferometer utilizes a spatially modulated intensity, which encodes local refraction into spatial phase shifts of the modulation pattern. Differential phase imaging has subsequently also been reported with polarized neutrons applied to magnetic fields [102]. However, in contrast to x-rays, where phase contrast provided access to, e.g. soft tissue investigations and lower doses in medical applications, the method found little real scientific application with neutrons so far.

On the other hand, magnetic phase contrast was heavily exploited, without the use of polarized neutrons, where the opposite phase shift for spin-up and spin-down neutrons rather induced a local loss of visibility of the modulation of the grating interferometer due to the superposition of the phase shift of both spin states [103], rather than a clear differential phase contrast signal. The loss of visibility is separate from attenuation and differential phase the third contrast modality provided by DCDs and grating interferometry, or more general modulated beam techniques (compare chapter 19), and shall be discussed in the subsequent section on dark-field contrast.

2.2.5 Dark-field contrast (small-angle scattering contrast)

The term dark-field contrast was coined in the context of grating interferometry, first with x-rays and shortly after with neutrons [105]. It refers, in analogy to dark-field microscopy, to the detection of intensity in areas that initially, without scattering from the sample, are dark as they are outside of the peaking intensity of the unperturbed beam. In grating interferometry, these are indeed the periodic minima of the spatially modulated beam. Scattering from a sample, more specifically small-angle scattering, will redistribute intensity from the maxima into the dark-field region of the minima and, thus, induce a loss of modulation amplitude, i.e. visibility

$$V = (I_{\max} - I_{\min})/(I_{\max} + I_{\min}). \tag{2.14}$$

The relative local change of visibility measured with sample and normalized by the unperturbed modulation of the open beam is referred to as dark-field contrast, which is, in general, a measure of small-angle scattering within the regime of sensitivity of the specific set-up [106]. However, this kind of contrast has also initially been explored with DCDs and was termed ultra-small-angle scattering (USANS) contrast [107], although also here contrast was based on detecting scattering contributions in the dark-field. The dark field in this case is the intensity depleted angular area around the peaking distribution of the unperturbed beam. Again, for this contrast modality grating interferometry proved superior in efficiency and is paving the way for practical applications.

While for DCDs the interpretation of small-angle scattering was straightforward, this was not the case for spatial modulation visibility data of dark-field contrast in grating interferometers. It can, however, be shown, that the dark-field contrast of modulated beam experiments has analogy to another established small-angle scattering technique, namely spin-echo SANS (SESANS) [108] and even more to SE-modulated SANS (SEMSANS) [109]. The latter has been demonstrated to have potential for dark-field contrast imaging as well [110]. It can be shown, that the probed correlation length ξ related to a specific set-up of a spatially modulated beam technique such as neutron grating interferometry (NGI) or SEMSANS can be defined as

$$\xi = \lambda L_s / p \tag{2.15}$$

where L_s is the distance of the sample to the position where the modulation with period p is analyzed [110, 111]. This definition is closely related to the definition of the spin-echo length in SESANS, and just like in SESANS, where the beam polarization is measured, here the visibility can be shown to be related to the small-angle scattering function $S(q)$, with q being the modulus of the scattering vector. The convolution of the cosine modulation with the small-angle scattering function yields [111]

$$V(\xi) = V_0(\xi) \int S(q) \cos(\xi q) dq \tag{2.16}$$

where

$$S(q) = \int G(\xi) \cos(\xi q) d\xi \tag{2.17}$$

is the cosine Fourier transform of $G(\xi)$, the scattering length density correlation function of the scattering structure [108]. Thus, considering the inverse transformation

$$G(\xi) = \int S(q) \cos(\xi q) dq \tag{2.18}$$

it can be concluded that the relative visibility

$$V(\xi) / V_0(\xi) = G(\xi). \tag{2.19}$$

Taking into account the total small-angle scattering probability Σ, the thickness of the sample t and multiple scattering, this takes the form

$$\frac{V(\xi)}{V_0(\xi)} = e^{\Sigma t (G(\xi) - 1)} \tag{2.20}$$

and describes the dark-field signal measured depending on the probed correlation length ξ [111].

Thus, probing the dark-field contrast at a specific instrumental setting with respect to λ, L_s and p provides an image based on the local small-angle scattering related to the corresponding probed correlation length ξ. While in the first place such images can reveal the presence and distribution of structures, which are not directly resolved in the attenuation contrast image as they range on a shorter length scale, systematic scans of ξ even allow obtaining the related structural parameters through modelling $G(\xi)$ [112–114]. Thus, quantitative dark-field imaging can provide unique insights into structural features in the nanometer-to-micrometer range paired with the direct spatial image resolution in the micrometer-to-centimeter range. Quantitative structural studies can be found in various fields including engineering materials [113] and soft matter research [114], but are still a young field exploring various applications.

Previously dark-field contrast has foremost been applied in the field of magnetism (chapter 13), in particular for the visualization of domain walls (figure 2.10) and investigations of domain structures [19]. However, the dark-field contrast from magnetic domain walls of large magnetic domains is rather not related to conventional small-angle scattering, but to spin-dependent differential phase contrast (compare section 2.2.4). The superposition of the phase shifted intensities of spin-up and spin-down neutrons induces a loss of visibility [103] at the locations of domain walls aligned close to parallel to the incoming beam. This signal provides a direct visualization of domain walls of large domains, namely such that can be resolved spatially by the imaging set-up, and a qualitative measure of the density of domain walls of smaller domains [115]. A coherent path towards quantification and a size-dependent relation to small-angle scattering dark-field contrast has yet to be explored [116].

Also, other highly relevant applications, e.g. in energy research and engineering materials science provided unique insights without quantitative small-angle scattering analyses. Dark-field contrast imaging has been utilized, e.g. to observe water accumulation in the gas diffusion layer of fuel cells despite the shading effect of water transport channels [118] and for detecting precipitates [119] and micro-cracks [119, 120] in metals, to name but a few.

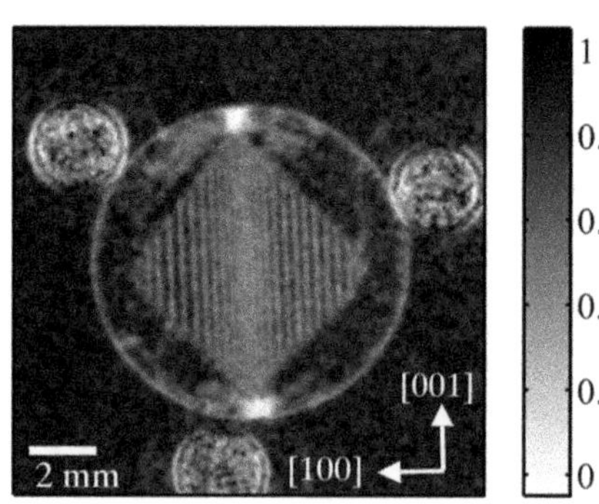

Figure 2.10. The first dark-field contrast image of magnetic domain walls (FeSi crystal disk). Reprinted from [117], with the permission of AIP Publishing.

Dark-field contrast in grating interferometers is generally probed together with attenuation and differential phase contrast. However, it has been shown that it can also be combined with diffraction contrast in wavelength dispersive measurements as has been nicely demonstrated in a load-dependent phase transition study in transformation induced plasticity (TRIP) steel [113].

2.2.6 De/polarization contrast

The utilization of polarized neutrons is key in exploiting neutrons as a probe in condensed matter science. While the neutron is charge-neutral it carries a magnetic moment related to its spin. Thus, neutrons can transmit deep into bulk materials and are sensitive to magnetic interactions, which make them a key tool for condensed matter magnetism studies and exploration. While applications exploiting the magnetic moment of the neutron have a long history in neutron scattering, polarized neutrons were introduced to neutron imaging only about a decade into the new Millennium [43]. In polarized neutron imaging [121] the impact of a sample on the polarization of a polarized incident beam is detected with spatial resolution. Polarization contrast can be subdivided in depolarization contrast imaging and polarization contrast imaging, where the latter is mostly associated with the term polarized neutron imaging in contrast to neutron depolarization imaging for the former. Depolarization contrast is achieved, when locally, related to the spatial resolution, the beam has to be considered depolarized to a certain degree, and the degree of depolarization provides the image contrast [122]. In contrast, polarization contrast is achieved, when polarization is in principle sustained, but the local polarization direction is altered. In this case the final state can be assumed reversible, which is less the case for depolarization contrast. Polarization contrast imaging requires, except when a spin-echo principle is applied [123], a monochromatic beam, because it fundamentally builds on the phenomenon of Larmor precession of the spin around a magnetic field vector, where the final precession angle φ_{p} is wavelength dependent due to the relation

$$\varphi_{\mathrm{p}} = \omega_L t = \frac{\gamma}{v}\int_{\mathrm{path}} B ds = \frac{\gamma m \lambda}{h}\int_{\mathrm{path}} B ds \tag{2.21}$$

where

$$\omega_L = \gamma B \tag{2.22}$$

is the Larmor frequency, t is the time spent in the field B, v is the neutron velocity and $\gamma = -4\pi\mu_n/h = 1.832 \times 10^8$ rad s^{-1} T^{-1} is the gyromagnetic factor of the neutron.

More generally, when a beam is described by its polarization vector $\boldsymbol{P} = (P_x, P_y, P_z)$ then

$$\frac{d\boldsymbol{P}}{dt} = \gamma \boldsymbol{P} \times \boldsymbol{B} \tag{2.23}$$

and the polarization is characterized by $|\boldsymbol{P}| \leqslant 1$, while the vector components Pi take values of $-1 \leqslant \mathrm{Pi} \leqslant 1$ and are measured as $\mathrm{Pi} = (I^+ - I^-)/(I^+ + I^-)$ with I^+ and I^- being the measured intensities for spin-up and spin-down intensities with respect to the spin analyzation direction i. Thus, one can define, that polarization contrast is measured when $|\mathbf{P_0}| \approx |\mathbf{P}| \approx 1$ although P_x, P_y and P_z depend on x and y, while for depolarization contrast $|\boldsymbol{P}(x,y)| \leqslant 1$ is a function of the coordinates x and y of the image. As the resolution of x and y is defined by the imaging system, this implies that polarization contrast is probing magnetic fields and structures which can be resolved spatially and, thus, change rather slow and depolarization contrast is characterized by smaller structures and faster changes of fields on short length scales, beyond the resolution limit of the imaging system. The latter applies in general for ferromagnetic materials with small domains, where e.g. phase transformations between para- and ferromagnetic state, or in general the magnetic phase fraction, can be studied with spatial resolution [122, 124]. The former, polarization contrast, is rather useful to study magnetic fields, and in contrast to other methods allows remote sensing even within bulk materials and components [43, 121] (compare chapter 13). It has furthermore been demonstrated that in extreme cases polarization contrast can also be achieved for electric fields [125] (figure 2.11).

A significant challenge for polarization contrast imaging is constituted by the vector field nature of magnetic fields and the periodicity of the response signal, which hinder tomography and straightforward quantification without *a priori* knowledge and modelling. However, recent literature has introduced solutions for 3D magnetic vector field reconstructions [127] (figure 2.11) and approaches to overcome the phase wrapping limit [123, 126], i.e. limitations to small fields preventing precessions beyond π. 3D vector field measurements are based on a polarimetric neutron imaging technique that has been proposed by Strobl *et al* [128] to probe the polarization matrix

$$P(x, y) = \begin{pmatrix} P_{x,x}(x, y) & P_{x,y}(x, y) & P_{x,z}(x, y) \\ P_{y,x}(x, y) & P_{y,y}(x, y) & P_{y,z}(x, y) \\ P_{z,x}(x, y) & P_{z,y}(x, y) & P_{z,z}(x, y) \end{pmatrix} \tag{2.24}$$

through experimental variation of the incoming (first index) and analyzed (second subscript) polarization direction by a symmetric set of spin turners before and after the sample (compare section 19.6).

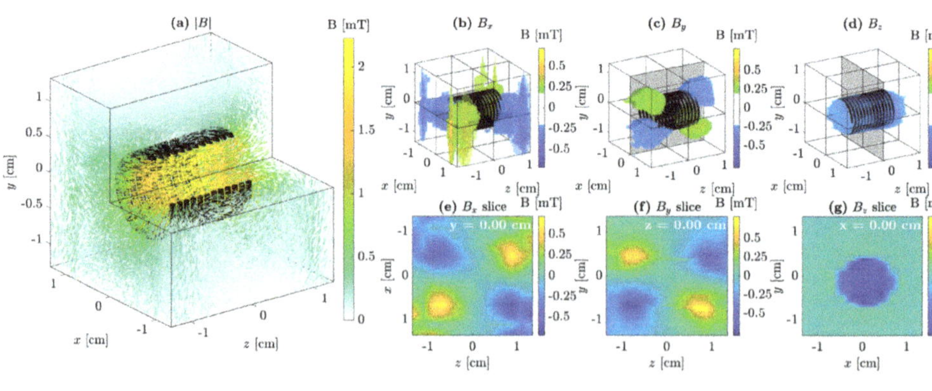

Figure 2.11. Magnetic vector field reconstruction from polarimetric tomography of an electric coil. Reproduced from [126], copyright IOP Publishing Ltd. All rights reserved.

2.2.7 Resonance absorption contrast

When extending the spectral range of neutron imaging towards higher energies, apart from high transmission low resolution fast neutron attenuation contrast studies of large objects that cannot be penetrated by thermal neutrons, the specific nuclear resonances of elements in the epithermal energy region provide unique contrast for elemental imaging. Resonances are found in the neutron energy range of about a few eV to MeV (compare figure 1.3, chapter 1). Thus, the method requires access to the corresponding neutron energy range as well as means for energy-resolved measurements. Resonance absorption contrast imaging was initially mainly explored at small-scale sources such as D–D neutron generators, where the energy can be selected by the choice of angle of view on the deuterium gas target. The ratio of images acquired at discrete energies in the MeV range enables one to differentiate and quantify the content of elements such as H, C, N, O and commercial as well as security applications were envisaged [129, 130]. However, the limited flux and detection capabilities severely limited the spatial resolution and image quality of this approach.

The advent of powerful short pulse neutron sources and fast ToF imaging detectors created new opportunities for high resolution neutron resonance absorption contrast imaging [131]. At these sources ToF imaging with sub-microsecond ToF resolution can provide access to a range of about 1 to some 100 eV, which provides access to resonances of some heavier elements such as Au, Ag, Gd, Pb etc and bears potential for studies not only in cultural heritage (figure 2.12) but also, e.g. in energy materials research and other fields of material science. Furthermore, the energy ranges of resonances and diffraction contrast, i.e. Bragg edges, can be combined to retrieve simultaneous complementary information about contained nuclides, their distribution and crystallographic microstructure of examined materials [132].

In addition to element mapping, it can be shown that the subtle broadening of resonance peaks with temperature can be utilized for remote temperature sensing with regards to specific constituents in bulk materials and devices.

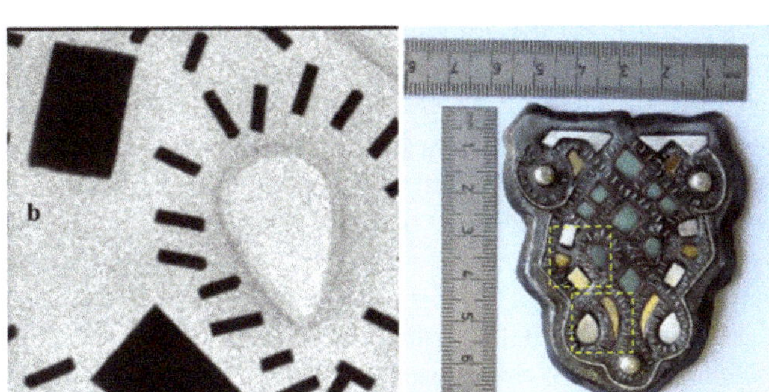

Figure 2.12. Silver resonance (5.3 eV) absorption contrast image of detail and photograph of historic belt mount replica. Copyright 2012 IEEE. Reprinted with permission from [131].

2.3 Scattering imaging: scanning, gauging, 3DND—principles

Other techniques providing real space neutron images, deviating from the full field transmission principle of the contrast mechanisms outlined above convey scanning techniques such as used rarely in small-angle scattering [133], gauging in strain diffractometers [134] or advanced approaches of 3D diffraction imaging (compare section 2.2). While neutron–matter interactions exploited might coincide with such as discussed in the context of the contrast modalities described above, here they do not manifest as an immediate image contrast, but numerous reciprocal space measurements have to be combined, analyzed, cross-correlated and computed together in order to obtain spatially resolved maps or images of specific physical properties, such as strain, grain and orientation maps. While scanning small-angle scattering is rather done with high intensity synchrotron x-ray beams than with neutrons, strain mapping in engineering diffractometers is performed regularly at least in 2D in order to visualize and analyze strain fields. Both latter methods are not typically considered neutron imaging techniques, but rather specific neutron scattering approaches.

A more recent approach is Laue 3-dimensional neutron diffraction tomography (Laue 3DNDT) [41, 42], where the entirety of diffraction spots measured in a tomographic scan with a white beam, i.e. utilizing the full available thermal source spectrum, on detectors covering a large fraction of solid angle around the sample, is used to identify the individual crystal grains and their orientations, which generate the measured diffraction patterns. Analyses of such data sets is done in an extensive forward modelling approach, which is computationally expensive and time consuming, but provides unique insights into the bulk grain network of, e.g. engineering and smart materials.

The subsequent chapters will deal with applications of neutron imaging in various fields and will show the highly versatile nature of neutron imaging which is expanding steadily with the introduction of novel contrast mechanisms and techniques to probe such to extract ever more information content from a large variety of materials, processes and devices best probed by neutron imaging.

References

[1] Trtik P and Lehmann E H 2016 Progress in high-resolution neutron imaging at the Paul Scherrer Institut—The Neutron Microscope project *J. Phys.: Conf. Ser.* **746** 012004

[2] Trtik P, Scheuerlein C, Alknes P, Meyer M, Schmid F and Lehmann E H 2017 Neutron microtomography of MgB_2 superconducting multifilament wire *Phys. Proc.* **88** 95–9

[3] Trtik P, Meyer M, Wehmann T, Tengattini A, Atkins D, Lehmann E H and Strobl M 2020 PSI 'Neutron Microscope' at ILL-D50 beamline—first results *Mater. Res. Forum* **15** 23

[4] Tengattini A *et al* 2022 Compact and versatile neutron imaging detector with sub-4 μm spatial resolution based on a single-crystal thin-film scintillator *Opt. Express* **30** 14461–77

[5] Hussey D S, LaManna J M, Baltic E and Jacobson D L 2017 Neutron imaging detector with 2 μm spatial resolution based on event reconstruction of neutron capture in gadolinium oxysulfide scintillators *Nucl. Instrum. Methods Phys. Res.* A **866** 9–12

[6] Leemreize H, Knudsen E B, Birk J O, Strobl M, Detlefs C and Poulsen H F 2019 Full-field neutron microscopy based on refractive optics *J. Appl. Crystallogr.* **52** 1299–311

[7] Mildner D F R and Gubarev M V 2011 Wolter optics for neutron focusing *Nucl. Instrum. Methods* A **634** 7–11

[8] Abir M, Hussey D S and Khaykovich B 2020 Design of neutron microscopes equipped with wolter mirror condenser and objective optics for high-fidelity imaging and beam transport *J. Imaging* **6** 100

[9] Bingham P, Santos-Villalobos H and Tobin K 2011 Coded source neutron imaging *Proc. SPIE 7877, Image Process.: Mach. Vision Appl. IV* **78770M**

[10] Coakley K J and Hussey D S 2007 Feasibility of single-view coded source neutron transmission tomography *Meas. Sci. Technol.* **18** 3391

[11] Kingston A M, Myers G R, Pelliccia D, Salvemini F, Bevitt J J, Garbe U and Paganin D M 2020 Neutron ghost imaging *Phys. Rev.* A **101** 053844

[12] Kaestner A P, Hartmann S, Kühne G, Frei G, Grünzweig C, Josic L, Schmid F and Lehmann E H 2011 The ICON beamline—a facility for cold neutron imaging at SINQ *Nucl. Instrum. Methods Phys. Res.* A **659** 387–93

[13] Hilger A, Kardjilov N, Strobl M, Treimer W and Banhart J 2006 The new cold neutron radiography and tomography instrument CONRAD at HMI Berlin *Physica* B **385–386** 1213–5

[14] Lehmann E H, Frei G, Kuehne G and Boillat P 2007 The micro-setup for neutron imaging: a major step forward to improve the spatial resolution *Nucl. Instrum. Methods Phys. Res.* A **576** 389–96

[15] Williams S H, Hilger A, Kardjilov N, Manke I, Strobl M, Douissard P A, Martin T, Riesemeier H and Banhart J 2012 Detection system for microimaging with neutrons *J. Instrum.* **7** P02014

[16] Crha J, Vila-Comamala J, Lehmann E, David C and Trtik P 2019 Light yield enhancement of 157-gadolinium oxysulfide scintillator screens for the high-resolution neutron imaging *MethodsX* **6** 107–14

[17] Tremsin A S and Vallerga J V 2020 Unique capabilities and applications of Microchannel Plate (MCP) detectors with Medipix/Timepix readout *Radiat. Meas.* **130** 106228

[18] Losko A S, Han Y, Schillinger B A, Tartaglione M, Morgano M, Strobl J, Long A S, Tremsin and Schulz M 2021 New perspectives for neutron imaging through advanced event-mode data acquisition *Sci. Rep.* **11** 21360

[19] Harti R P, Strobl M, Schäfer R, Kardjilov N, Tremsin A S and Grünzweig C 2018 Dynamic volume magnetic domain wall imaging in grain oriented electrical steel at power frequencies with accumulative high-frame rate neutron dark-field imaging *Sci. Rep.* **25** 15754

[20] Zboray R and Trtik P 2019 In-depth analysis of high-speed, cold neutron imaging of air-water two-phase flows *Flow Meas. Instrum.* **66** 182–9

[21] Tötzke C, Kardjilov N, Lenoir N, Manke I, Oswald S E and Tengattini A 2019 What comes NeXT?—High-speed neutron tomography at ILL *Opt. Express* **27** 28640–8

[22] Tremsin A, Kardjilov N, Strobl M, Manke I, Dawson M, McPhate J B, Vallerga J V, Siegmund O H W and Feller W B 2015 Imaging of dynamic magnetic fields with spin-polarized neutron beams *New J. Phys.* **17** 043047

[23] Makowska M, Strobl M, Lauridsen E M, Kabra S, Kockelman W, Tremsin A, Frandsen H L and Theil Kuhn L 2016 In-situ time-of-flight neutron imaging of NiO-YSZ reduction under influence of stress *J. Appl. Crystallogr.* **49** 1674–81

[24] Schillinger B, Abele H, Brunner J, Frei G, Gähler R, Gildemeister A, Hillenbach A, Lehmann E and Vontobel P 2005 Detection systems for short-time stroboscopic neutron imaging and measurements on a rotating engine *Nucl. Instrum. Methods* A **542** 142–7

[25] Berger H 1966 Characteristics of a thermal neutron television imaging system *Mater. Eval.* **24** 475–81

[26] Schlapper G A, Brugger R M, Seydel J E and Larsen G N 1977 Neutron tomography investigations at the Missouri university research reactor *Trans. Am. Nucl. Soc.* **26** 39

[27] Ghasemi-Tabasi H, Trtik P, Jhabvala J, Meyer M, Carminati C, Strobl M and Logé R E 2021 Neutron microtomography reveals spatial distribution of porosity in an additively manufactured gold alloy *Appl. Sci.* **11** 1512

[28] Gruenzweig C, Mannes D, Schmid F and Rule R 2016 Neutron imaging: a non-destructive testing method to investigate canned exhaust after-treatment system components for the three dimensional soot, ash, urea and coating distributions *SAE Technical Paper 2016-01-0985*

[29] Dierick M, Vlassenbroecka J, Masschaeleb B, Cnuddeb V, Van Hoorebekea L and Hillenbach A 2005 High-speed neutron tomography of dynamic processes *Nucl. Instrum. Methods* A **542** 296–301

[30] Zarebanadkouki M, Carminati A, Kaestner A, Mannes D, Morgano M, Peetermans S, Lehmann E and Trtik P 2015 On-the-fly neutron tomography of water transport into lupine roots *Phys. Proc.* **69** 292–8

[31] Kak A C and Slaney M 2001 *Principles of Computerized Tomographic Imaging* (Society for Industrial and Applied Mathematics)

[32] Kaestner A P, Trtik P, Zarebanadkouki M, Kazantsev D, Snehota M, Dobson K J and Lehmann E H 2016 Recent developments in neutron imaging with applications for porous media research *Solid Earth* **7** 1281–92

[33] Schillinger B, Brunner J and Calzada E 2006 A study of oil lubrication in a rotating engine using stroboscopic neutron imaging *Physica* B **385–6** 921–3

[34] Toetzke C, Kardjilov N, Manke I and Oswald S E 2017 Capturing 3D water flow in rooted soil by ultra-fast neutron tomograph *Sci. Rep.* **7** 6192

[35] Harti R, Strobl M, Betz B, Jefimovs K, Kagias M and Grünzweig C 2017 Sub-pixel correlation length neutron imaging: spatially resolved scattering information of micro-structures on a macroscopic scale *Sci. Rep.* **7** 44588

[36] Manke I *et al* 2010 Three-dimensional imaging of magnetic domains *Nat. Commun.* **1** 125

[37] Santisteban J R, Edwards L, Fizpatrick M E, Steuwer A and Withers P J 2002 Engineering applications of Bragg-edge neutron transmission *Appl. Phys.* A **74** 1433–6

[38] Makowska M, Kuhn L T, Frandsen H L, Lauridsen E M, De Angelis S, Cleemann L N, Morgano M, Trtik P and Strobl M 2017 Coupling between creep and redox behavior in nickel—yttria stabilized zirconia observed *in situ* by monochromatic neutron imaging *J. Power Sources* **340** 167–75

[39] Cereser A *et al* 2017 Time-of-flight three dimensional neutron diffraction in transmission mode for mapping crystal grain structures *Sci. Rep.* **7** 9561

[40] Peetermans S and Lehmann E H 2016 Simultaneous neutron transmission and diffraction imaging investigations of single crystal nickel-based superalloy turbine blades *NDT E Int.* **79** 109–13

[41] Peetermans S, King A, Ludwig W, Reischig P and Lehmann E H 2014 Cold neutron diffraction contrast tomography of polycrystalline materia *Analyst* **139** 5765–71

[42] Samothrakitis S, Larsen C B, Čapek J, Polatidis E, Raventós M, Tovar M, Schmidt S and Strobl M 2022 Microstructural characterization through grain orientation mapping with Laue three-dimensional neutron diffraction tomography *Mater. Today Adv.* **15** 100258

[43] Kardjilov N, Manke I, Strobl M, Hilger A, Treimer W, Meissner M, Krist T and Banhart J 2008 Three-dimensional imaging of magnetic fields with polarized neutrons *Nat. Phys.* **4** 399–403

[44] Schulz M, Neubauer A, Masalovich S, Mühlbauer M, Calzada E, Schillinger B, Pfleiderer C and Böni P 2010 Towards a tomographic reconstruction of neutron depolarization data *J. Phys. Conf. Ser.* **211** 012025

[45] Tremsin A S, Kardjilov N, Strobl M, Manke I, Dawson M, McPhate J B, Vallerga J V, Siegmund O H W and Feller W B 2015 *New J. Phys.* **17** 043047

[46] Sales M, Strobl M, Shinohara T, Tremsin A, Kuhn L T, Lionheart W R B, Desai N M, Dahl A B and Schmidt S 2018 Three dimensional polarimetric neutron tomography of magnetic fields *Sci. Rep.* **8** 2214

[47] Tremsin A S, Ganguly S, Meco S M, Pardal G R, Shinohara T and Feller W B 2016 Investigation of dissimilar metal welds by energy-resolved neutron imaging *J. Appl. Crystallogr.* **49** 1130–40

[48] Nelson R O *et al* 2018 Neutron Imaging at LANSCE—from cold to ultrafast *J. Imaging* **4** 45

[49] Tremsin A, Perrodin D, Losko A, Vogel S C, Bourke M A M, Bizarri G A and Bourret E D 2017 Real-time crystal growth visualization and quantification by energy-resolved neutron imaging *Sci. Rep.* **7** 46275

[50] Kiyanagi Y 2018 Neutron imaging at compact accelerator-driven neutron sources in Japan *J. Imaging* **4** 55

[51] Dauti D, Tengattini A, Dal Pont S, Toropovs N, Briffaut M and Weber B 2018 Analysis of moisture migration in concrete at high temperature through *in situ* neutron tomography *Cem. Concr. Res.* **111** 41–55

[52] Fedrigo A, Marstal K, Bender Koch C, Andersen Dahl V, Bjorholm Dahl A, Lyksborg M, Gundlach C, Ott F and Strobl M 2018 Investigation of a monturaqui impactite by means of bi-modal x-ray and neutron tomography *J. Imaging* **4** 72

[53] Sun F, Markötter H, Manke I, Hilger A, Alrwashdeh S S, Kardjilov N and Banhart J 2017 Complementary x-ray and neutron radiography study of the initial lithiation process in lithium-ion batteries containing silicon electrodes *Appl. Surf. Sci.* **399** 359–66

[54] Raventos M *et al* 2018 A Monte Carlo approach for scattering correction towards quantitative neutron imaging of polycrystals *J. Appl. Crystallogr.* **51** 386–94

[55] Hassanein R, de Beer F, Kardjilov N and Lehmann E 2006 Scattering correction algorithm for neutron radiography and tomography tested at facilities with different beam characteristics *Physica* B **385–6** 1194–6

[56] Boillat P, Carminati C, Schmid F, Grünzweig C, Hovind J and Kaestner A *et al* 2018 Chasing quantitative biases in neutron imaging with scintillator-camera detectors: a practical method with black body grids *Opt. Express* **26** 15769

[57] Carminati C, Boillat P, Schmid F, Vontobel P, Hovind J and Morgano M *et al* 2019 Implementation and assessment of the black body bias correction in quantitative neutron imaging *PLoS One* **14** e0210300

[58] Vogel S 2000 *PhD Thesis* Christian-Albrechts-Universität zu Kiel

[59] Busi M, Kalentics N, Morgano M, Griffiths S, Tremsin A S, Shinohara T, Loge R, Leinenbach C and Strobl M 2021 Non-destructive characterization of laser powder bed fusion parts with neutron Bragg edge imaging *Addit. Manuf.* **39** 101848

[60] https://ncnr.nist.gov/resources/n-lengths/

[61] https://xsplot.com/

[62] Boin M 2012 *nxs*: a program library for neutron cross section calculations *J. Appl. Crystallogr.* **45** 603–7

[63] Cai X-X and Kittelmann T 2020 NCrystal: a library for thermal neutron transport *Comp. Phys. Commun.* **246** 106851

[64] Steuwer A, Santisteban J R, Withers P J and Edwards L 2004 Pattern decomposition and quantitative-phase analysis in pulsed neutron transmission *Physica* **B 350** 159–61

[65] Woracek R, Santisteban J, Fedrigo A and Strobl M 2018 Diffraction in neutron imaging—a review *Nucl. Inst. Meth.* A **878** 141–58

[66] Kirkwood H J, Zhang S Y, Tremsin A S, Korsunsky A M, Baimpase N and Abbey B 2015 Neutron strain tomography using the radon transform *Mater. Today: Proc.* **2S** 414–23

[67] Busi M *et al* 2022 Bragg edge tomography characterization of additively manufactured 316L steel *Phys. Rev. Mater.* **6** 053602

[68] Hendricks J N, Gregg W T, Jackson R R, Wensrich C M, Wills A, Tremsin A S, Shinohara T, Luzin V and Kirstein O 2019 Tomographic reconstruction of triaxial strain fields from Bragg-edge neutron imaging *Phys. Rev. Mater.* **3** 113803

[69] Xie Q, Song G, Gorti S, Stoica A D, Radhakrishnan B, Bilheux J C, Kirka M, Dehoff R, Bilheux H Z and An K 2018 Applying neutron transmission physics and 3D statistical full-field model to understand 2D Bragg-edge imaging *J. Appl. Phys.* **123** 074901

[70] Santisteban J R, Vicente-Alvarez M A, Vizcaino P, Banchik A D, Vogel S C, Tremsin A S, Vallerga J V, McPhate J B, Lehmann E and Kockelmann W 2012 Texture imaging of zirconium based components by total neutron cross-section experiments *J. Nucl. Mater.* **425** 218–27

[71] Malamud F, Santisteban J R, Vicente Alvarez M A, Busi M, Polatidis E and Strobl M 2023 An optimized single-crystal to polycrystal model of the neutron transmission of textured polycrystalline materials *J. Appl. Crystallogr.* **56** 143–54

[72] Vicente Alvarez M A, Laliena V, Malamud F, Campo J and Santisteban J 2021 A novel method to obtain integral parameters of the orientation distribution function of textured polycrystals from wavelength-resolved neutron transmission spectra *J. Appl. Crystallogr.* **54** 903–13

[73] Dollase W A 1986 Correction of intensities for preferred orientation in powder diffractometry: application of the March model *J. Appl. Crystallogr.* **19** 267–72

[74] Santisteban J R, Edwards L and Stelmukh V 2006b Characterization of textured materials by TOF transmission *Physica B: Cond. Matter.* **385–386 Part 1** 636–8

[75] Sato H, Kamiyama T and Kiyanagi Y 2011 A rietveld-type analysis code for pulsed neutron Bragg-edge transmission imaging and quantitative evaluation of texture and microstructure of a welded alpha-iron plate *Mater. Trans.* **52** 1294–302

[76] Boin M, Wimpory R C, Hilger A, Kardjilov N, Zhang S Y and Strobl M 2012 Monte Carlo simulations for the analysis of texture and strain measured with Bragg edge neutron transmission *J. Phys.: Conf. Ser.* **340** 012022

[77] Laliena V, Vicente Alvarez M A and Campo J 2020 Monte Carlo simulation of neutron scattering by a textured polycrystal *J. Appl. Crystallogr.* **53** 512–29

[78] Kearns J J 2001 On the relationship among '*f*' texture factors for the principal planes of zirconium, hafnium and titanium alloys *J. Nucl. Mater.* **299** 171–4

[79] Sabine T M, Von Dreele R B and Jørgensen J-E 1988 Extinction in time-of-flight neutron powder diffractometry *Acta Crystallogr.* A **44** 374–9

[80] Strickland J *et al* 2020 2D single crystal Bragg-dip mapping by time-of-flight energy-resolved neutron imaging on IMAT@ISIS *Sci. Rep.* **10** 20751

[81] Malamud F and Santisteban J R 2016 Full-pattern analysis of time-of-flight neutron transmission of mosaic crystals *J. Appl. Crystallogr.* **49** 348–65

[82] Malamud F, Santisteban J R, Gao Y, Shinohara T, Oikawa K and Tremsin A 2022 Non-destructive characterization of the spatial variation of γ/γ' lattice misfit in a single-crystal Ni-based superalloy by energy-resolved neutron imaging *J. Appl. Crystallogr.* **55** 228–39

[83] Sato H *et al* 2017 Inverse pole figure mapping of bulk crystalline grains in a polycrystalline steel plate by pulsed neutron Bragg-dip transmission imaging *J. Appl. Crystallogr.* **50** 1601–10

[84] Woracek R, Penumadu D, Kardjilov N, Hilger A, Boin M, Banhart J and Manke I 2014 3D mapping of crystallographic phase distribution using energy-selective neutron tomography *Adv. Mater.* **26** 4069–73

[85] Tremsin A S, Gao Y, Makinde A, Bilheux H Z, Bilheux J C, An K, Shinohara T and Oikawa K 2021 Monitoring residual strain relaxation and preferred grain orientation of additively manufactured Inconel 625 by *in situ* neutron imaging *Addit. Manuf.* **46** 102130

[86] Al-Falahat A M *et al* 2022 Temperature dependence on Bragg edge neutron transmission measurements *J. Appl. Crystallogr.* **55** 919–28

[87] Sato H, Miyoshi M, Ramadhan R S, Kockelmann W and Kamiyama T 2023 A new thermography using inelastic scattering analysis of wavelength-resolved neutron transmission imaging *Sci. Rep.* **13** 688

[88] Siegwart M, Woracek R, Márquez Damián J I, Tremsin A, Manzi V, Strobl M, Schmidt T J and Boillat P 2019 Distinction between super-cooled water and ice with high duty cycle time-of-flight neutron imaging *Rev. Sci. Instrum.* **90** 103705

[89] Márquez Damián J I, Granada J R and Malaspina D C 2014 CAB models for water: a new evaluation of the thermal neutron scattering laws for light and heavy water in ENDF-6 format *Ann. Nucl. Energy* **65** 280–9

[90] Biesdorf J, Oberholzer P, Bernauer F, Kaestner A, Vontobel P, Lehmann E H, Schmidt T J and Boillat P 2014 Dual spectrum neutron radiography *Phys. Rev. Lett.* **112** 248301

[91] Romanelli G *et al* 2021 Thermal neutron cross sections of amino acids from average contributions of functional groups *J. Phys.: Condens. Matter* **33** 285901

[92] Springer T 2006 Quasielastic neutron scattering for the investigation of diffusive motions in solids and liquids *Springer Tracts in Modern Physics, Volume 64* (Berlin: Springer) 1–100

[93] Macfarlane R, Muir D W, Boicourt R, Kahler A C and Conlin J L 2017 The NJOY nuclear data processing system, version 2016 *Tech. Rep.* (Los Alamos National Lab.(LANL), Los Alamos, NM (United States))

[94] Allman B E, McMahon P J, Nugent K A, Paganin D, Jacobson D L, Arif M and Werner S A 2000 Phase radiography with neutrons *Nature* **408** 158–9

[95] Řeháček J, Hradil Z, Zawisky M, Bonse U and Dubus F 2005 Phase-contrast tomography with low-intensity beams *Phys. Rev.* A **71** 023608

[96] Strobl M, Treimer W, Kardjilov N, Hilger A and Zabler S 2008 On neutron phase contrast imaging *Nucl. Instrum. Meth.* B **266** 181–6

[97] Strobl M, Kardjilov N, Hilger A, Kühne G, Frei G and Manke I 2009 High-resolution investigations of edge effects in neutron imaging *Nucl. Instr. and Meth.* A **604** 640–5

[98] Treimer W, Strobl M, Hilger A, Seifert C and Feye-Treimer U 2003 Refraction as imaging signal for computerized (neutron) tomography *Appl. Phys. Lett.* **83** 398–400

[99] Strobl M, Staack K, Treimer W and Hilger A 2008 Quantitative neutron phase contrast tomography *Meas. Sci. Technol.* **19** 034020

[100] Strobl M, Treimer W, Keil S, Walter P and Manke I 2007 Magnetic field induced differential neutron phase contrast imaging *Appl. Phys. Lett.* **91** 254104

[101] Pfeiffer F, Grünzweig C, Bunk O, Frei G, Lehmann E and David C 2006 Neutron phase imaging and tomography *Phys. Rev. Lett.* **96** 215505

[102] Valsecchi J *et al* 2019 Visualization and quantification of inhomogeneous and anisotropic magnetic fields by polarized neutron grating interferometry *Nat. Commun.* **10** 3788

[103] Valsecchi J, Makowska M G, Gruenzweig C, Piegsa F M, Kim Y, Lee S W, Thijs M A, Plomp J and Strobl M 2021 Decomposing magnetic dark-field contrast in spin analyzed Talbot-Lau interferometry—a Stern–Gerlach experiment without spatial beam splitting *Phys. Rev. Lett.* **126** 070401

[104] Kardjilov N, Lee S W, Lehmann E, Lim I C, Steichele E, Sim C M and Vontobel P 2004 Applied phase-contrast with polychromatic thermal neutrons *Key Eng. Mater.* **270–3** 1330–6

[105] Strobl M, Grünzweig C, Hilger A, Manke I, Kardjilov N, David C and Pfeiffer F 2008 Neutron dark-field tomography *Phys. Rev. Lett.* **101** 123902

[106] Strobl M, Harti R P, Grünzweig C, Woracek R and Plomp J 2018 Small angle scattering in neutron imaging—a review *J. Imaging* **3** 64

[107] Strobl M, Treimer W and Hilger A 2004 Small angle scattering signals for (neutron) computerized tomography *Appl. Phys. Lett.* **85** 488–90

[108] Andersson R, van Heijkamp L F, de Schepper I M and Bouwman W G 2008 Analysis of spin-echo small-angle neutron scattering measurements *J. Appl. Crystallogr.* **41** 868–85

[109] Strobl M, Bouwman W G, Wieder F, C h, Duiff , Hilger A, Kardjilov N and Manke I 2012 Using a grating analyser for SEMSANS investigations in the very small angle range *Physica* B **407** 4132–5

[110] Strobl M, Sales M, Plomp J, Bouwman W G, Tremsin A S, Kaestner A, Pappas C and Habicht K 2015 Quantitative neutron dark-field imaging through spin-echo interferometry *Sci. Rep.* **5** 16576

[111] Strobl M 2014 General solution for quantitative dark-field contrast imaging with grating interferometers *Sci. Rep.* **4** 7243

[112] Strobl M, Betz B, Harti R P, Hilger A, Kardjilov N, Manke I and Gruenzweig C 2016 Wavelength dispersive dark-field contrast: micrometer structure resolution in neutron imaging with gratings *J. Appl. Crystallogr.* **49**

[113] Bacak M *et al* 2020 Neutron dark-field imaging applied to porosity and deformation-induced phase transitions in additively manufactured steels *Mater. Design* **195** 109009

[114] Kim Y, Valsecchi J, Oh O, Kim J, Lee S W, Boue F, Lutton E, Busi M, Garvey C and Strobl M 2022 Quantitative neutron dark-field imaging of milk: a feasibility study *Appl. Sci.* **12** 833

[115] Grünzweig C *et al* 2010 Visualizing the propagation of volume magnetization in bulk ferromagnetic materials by neutron grating interferometry *J. Appl. Phys.* **107** 09D308

[116] Valsecchi J, Kim Y, Lee S W, Saito K, Gruenzweig C and Strobl M 2021 Towards spatially resolved magnetic small-angle scattering studies by polarized and polarization-analyzed neutron dark-field contrast imaging *Sci. Rep.* **11** 8023

[117] Grünzweig C, David C, Bunk O, Dierolf M, Frei G, Kühne G, Schäfer R, Pofahl S, Rønnow H M R and Pfeiffer F 2008 Bulk magnetic domain structures visualized by neutron dark-field imaging *Appl. Phys. Lett.* **93** 112504

[118] Siegwart M, Harti R P, Manzi-Orezzoli V, Valsecchi J, Strobl M, Gruenzweig C, Schmidt T J and Boillat P 2019 Selective visualization of water in fuel cell gas diffusion layers with neutron dark-field imaging *J. Electrochem. Soc.* **166** F149–57

[119] Hilger A, Kardjilov N, Kandemir T, Manke I, Banhart J, Penumadu D, Manescu A and Strobl M 2010 Revealing micro-structural inhomogeneities with dark-field neutron imaging *J. Appl. Phys.* **107** 036101

[120] Brooks A J *et al* 2018 Neutron interferometry detection of early crack cormation caused by bending fatigue in additively manufactured SS316 dogbones *Mater. Des.* **140** 420–30

[121] Strobl M, Heimonen H, Schmidt S, Sales M, Kardjilov N, Hilger A, Manke I, Shinohara T and Valsecchi J 2019 Topical review: Polarisation measurements in neutron imaging *J. Phys.* D **52** 123001

[122] Schulz M, Neubauer A, Masalovich S, Mühlbauer M, Calzada E, Schillinger B, Pfleiderer C and Böni P 2010 Towards a tomographic reconstruction of neutron depolarization data *J. Phys.: Conf. Ser.* **211** 012025

[123] Strobl M, Pappas C, Hilger A, Wellert S, Kardjilov N, Seidel S-O and Manke I 2011 Polarized neutron imaging—a spin-echo approach *Physica* B **406** 2415–8

[124] Busi M, Polatidis E, Sofras C, Boillat P, Ruffo A, Leinenbach C and Strobl M 2022 Polarization contrast neutron imaging of magnetic crystallographic phases *Mater. Today Adv.* **16** 100302

[125] Jau Y-Y, Hussey D S, Gentile T R and Chen W 2020 Electric field imaging using polarized neutrons *Phys. Rev. Lett.* **125** 110801

[126] Sales M, Shinohara T, Sørensen M K, Knudsen E B, Tremsin A, Strobl M and Schmidt S 2019 *J. Phys. D: Appl. Phys.* **52** 205001

[127] Sales M, Strobl M, Shinohara T, Tremsin A, Kuhn L T, Lionheart W, Desai N, Dahl A B and Schmidt S 2018 Three-dimensional polarimetric neutron tomography of magnetic fields *Sci. Rep.* **8** 2214

[128] Strobl M, Kardjilov N, Hilger A, Jericha E, Badurek G and Manke I 2009 Imaging with polarized neutrons *Physica* B **404** 2611–4

[129] Schrack R A, Behrens J W, Johnson R and Bowman C D 1981 Resonance neutron radiography using an electron linac *IEEE Trans. Nucl. Sci.* **28** 1640–3

[130] Dangendorf V, Bar D, Bromberger B, Feldman G, Goldberg M B, Lauck R, Mor I, Tittelmeier K, Vartsky D and Weierganz M 2009 Multi-frame energy-selective imaging system for fast-neutron radiography *IEEE Trans. Nucl. Sci.* **56** 1135–40

[131] Tremsin A S, McPhate J B, Vallerga J V, Siegmund O H W, Kockelmann W, Schooneveld E M and Rhodes N J 2012 W. Bruce Feller *IEEE Trans. Nucl. Sci.* **59** 3272–7

[132] Tremsin A S, Ganguly S, Meco S M, Pardal G R, Shinohara T and Feller W B 2016 *J. Appl. Crystallogr.* **49** 1130–40

[133] Dewhurst C D and Grillo I 2016 Neutron imaging using a conventional small-angle neutron scattering instrument *J. Appl. Crystallogr.* **49** 736–42

[134] Cottam R, Wang J and Luzin V 2014 Characterization of microstructure and residual stress in a 3D H13 tool steel component produced by additive manufacturing *J. Mater. Res.* **29** 1978–86

[135] Kardjilov N, Manke I, Hilger A, Arlt T, Bradbury R, Markötter H, Woracek R, Strobl M, Treimer W and Banhart J 2021 The Neutron Imaging Instrument CONRAD—Post-Operational Review *J. Imag.* **7** 11

Part II

Structural materials and manufacturing

IOP Publishing

Neutron Imaging
From applied materials science to industry
Markus Strobl and Eberhard Lehmann

Chapter 3

Construction materials

Pavel Trtik

Building construction materials have been used by humans since (at least) the times we stopped being hunters–gatherers and settled down to stay in dwellings. The choice of the materials for building the 'infrastructure' has developed extensively since these times. The range of current construction materials is very wide and includes cement-based materials (such as concrete and mortar), bricks and masonry, metallic materials (such as steel, aluminium, copper and others), wood and processed wood, ceramics, structural glass, structural plastics, fibres and fibre-reinforced composites, fabrics (such as textiles and even paper), foams, bituminous materials, soil and others. Consequently, the construction materials industry is a specialized domain with a large number of industrially available types of products. The usual number of material items that one can these days buy, even in a standard sized building materials supermarket, in developed countries is in the order of 10 000s and the total global construction market is worth about 10 trillions USD and expected to grow to about 15 trillion USD by 2030 [1].

Clearly, to overview the uses of neutron imaging for all the above-mentioned building material types goes beyond the possible scope of one book chapter. Therefore, this chapter will focus on the intrinsically porous building construction materials for which neutron imaging proves useful for the assessment of transport properties and durability. This chapter is thus also recommended to be read in conjunction with chapter 7 (Processes in soil and plants) and chapter 9 (Geology) that present uses of neutron imaging for related porous media. Due to the other uses of wood that go beyond the building construction materials, neutron imaging applied to wood and processed wood will be covered in its own chapter (chapter 8, Wood). The applications of neutron imaging for metallic construction/structural materials and metallurgy are dealt with in chapter 6 (Manufacturing).

doi:10.1088/978-0-7503-3495-2ch3 © IOP Publishing Ltd 2024

3.1 Cement-based materials

Materials based on hydraulic cements (collectively known these days as concretes) are not new. Already the ancient Romans used such materials and called them *opus caementicium* [2]. Just like nowadays, the Roman concretes were purposefully engineered and tailored for the application—as becomes evident by the examples of the dome of the Pantheon whose upper parts are in contrast to its lower parts made of light-weight material, and/or in the underwater Roman concrete structures in Caeserana. The modern concretes are nowadays the most man-made engineering materials that form the backbone of the infrastructure of our society. Concretes and hydraulic cement-based materials, however, not only constitute a majority of materials used in roads and pavements, bridges, tunnels and tunnel linings, oil rigs, retaining walls, (load-bearing) structures of (not only high-rise) buildings [3], they can be found also in, for laypersons, slightly unexpected applications, such as hulls of ships [4] and even submarines [5], temporary portable harbours [6], gas-tight pressure seals of underground natural gas storage [7] and even statues [8] and jewellery [9].

Consequently, the production of concrete structures bears relatively large CO_2 footprint [10], however, it should be borne in mind that the CO_2 footprint of such infrastructure would be in all likelihood much higher should it be produced from any other available construction materials. Deterioration of concrete may occur due to a large number of reasons—corrosion, structural damage, water infiltration, etc. Being a porous material, concrete (may) degrade over time due to exposure to an aqueous environment. Concrete durability is the keyword here and it has a profound influence on the service life of concrete structures.

A neutron beam is much more attenuated by hydrogen than by virtually all other elements present in cement-based materials and therefore neutron imaging provides a very efficient non-destructive tool allowing for quantitative investigation of water content and water movement in concretes. Two relevant reviews for this chapter were published a few years ago. In 2017, Peng *et al* [11] summarized the applications of neutron imaging for durability of cement-based materials, while even more recently Tengattini *et al* provided a review of neutron imaging applications in geomechanics [12] that naturally includes a number of relevant investigations of cement-based materials. Hereinafter, the author of this chapter attempts to partially update on these reviews and focus a bit more on recent applications towards concrete and other porous construction materials.

3.1.1 Cement-based materials in hardened state

The uptake/transport of water in hardened concrete is one of the most often investigated phenomena using neutron imaging. As early as in the 1970s, Zeilinger and Hübner observed the moisture transport in samples of concrete from the containment of SNR-300 (Kalkar, Germany) nuclear reactor. Many other investigations followed. Prazak *et al* [13] utilized neutron radiography in the Rez reactor near Prague, Czech Republic, for a number of construction materials (including aerated concrete). Justnes *et al* [14] visualized water uptake in cylindrical

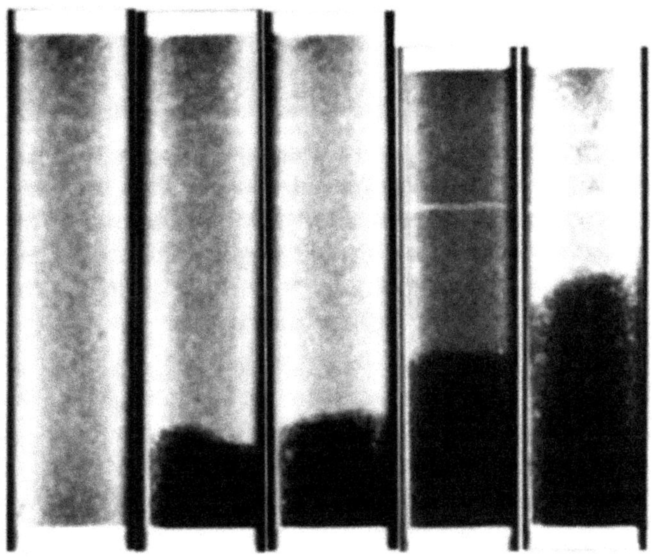

Figure 3.1. Water uptake in cement mortar after 0, 2.5, 4, 24, and 48 h. Reproduced with permission from [14], copyright ICE Publishing.

samples of cement mortar and predicted the use of isotopic contrast of D_2O for future experiments (figure 3.1).

Later, Hanzic and Ilic [15] applied neutron radiography to investigate capillarity (the height of the liquid front) and sorptivity (the volume of absorbed liquid) for three different types of concretes (mortars) and two liquids (water and fuel oil). They observed that while for fuel oil both the sorptivity and the capillarity follow rather closely the linear relationship against the square root of time, for water this was not the case and explained this anomaly by additional hydration which takes place in the presence of additional water and thus hinders the water movement through the porous structure of concrete.

Durability of cement-based materials is strongly influenced by the ability to withstand the penetration of liquids into them. de Beer *et al* [16] and later Peng *et al* [17] compared the water uptake in concretes that differed strongly in their respective water–cement ratios (wcr) (0.5 to 0.7 and 0.4 versus 0.6, respectively) proving the faster water uptake of the higher wcr materials.

As hardly any materials made of hydraulic cements remain crack free, the natural interest is in the investigations of the influence of cracks on the water transport as the cracks function as preferential routes for the water ingress. Kanematsu *et al* applied neutron radiography for visualization and quantification of water penetration into concrete through cracks [18]. Snoeck *et al* [19] visualized the water uptake in cracked cement paste samples with and without superabsorbent polymers (SAP) demonstrating the positive influence of the addition of SAP on the transport properties of the material. They concluded that cracks in samples without SAP filled very fast and complete filling of the crack with water occurred within 15 min, while for samples

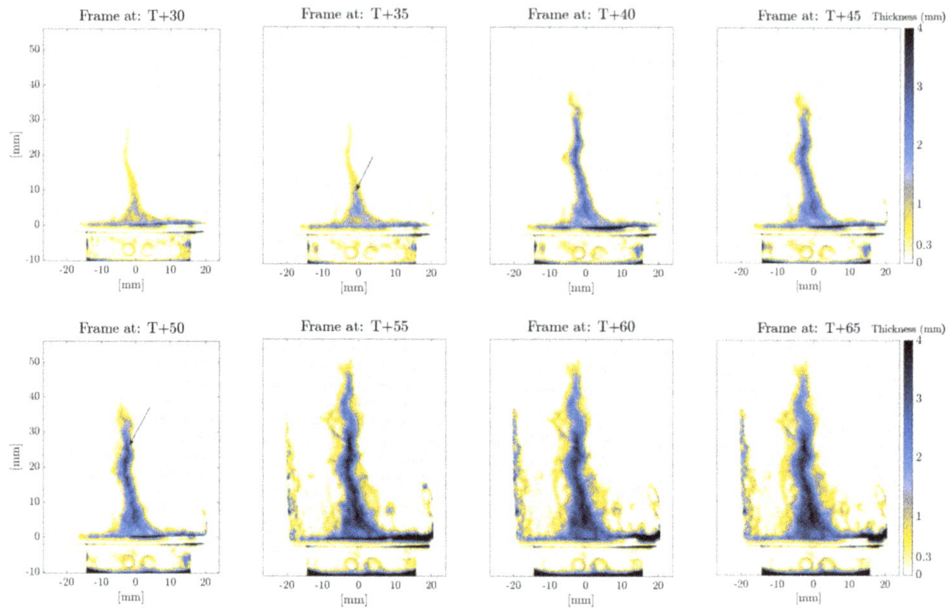

Figure 3.2. A sequence of the water thickness maps in dynamic measurement of vapour injection and condensation within fractured concrete. The maps highlight the sudden changes in moisture movement preceded by a moisture accumulation. Reprinted from [20], copyright (2021), with permission from Elsevier.

with SAP the water penetration height was less than the total crack height even after prolonged water absorption testing.

Recently, Lukic *et al* [20] utilized neutron imaging for dynamic measurements of vapour injection and condensation within a fractured concrete (see figure 3.2). A crack (of approximately 150 micrometers opening) was induced in a cylindrical specimen of 40 mm in diameter and 50 mm in height. The process of vapour injection and condensation has been visualized using high temporal resolution imaging (acquisition rate of 30 Hz) and revealed that the movement of condensed moisture along and transversely to the crack plane is non-monotonic in both space and time showing occasional sub-second jumps.

Neutron imaging is also useful for investigations of crack self-healing and crack-sealing in cement-based materials. Crack healing was first studied by Van Tittelboom *et al* [21]. From the moisture distribution profiles derived from the neutron radiographies the encapsulation of polyurethane-based healing agents was shown to be very suitable for autonomous healing of cracks. Van den Heede *et al* [22] studied the effectiveness of incorporating encapsulated polyurethane-based healing agents by investigating water ingress in healed materials using neutron radiography. Neutron imaging has also been utilized extensively for visualization of the influence of repellent agents on water ingress in concretes [23].

Similar to other materials utilizing an additive manufacturing process, the 3D printing process allows for production of extremely geometrically complex concrete/

mortar structures that would be difficult or even impossible to be produced using standard casting procedures. However, the intrinsically layered structure exhibits the distinct layer interfaces (sometimes called ‘cold joints’). The formation of these joints in 3D printing with concrete still remains a not fully resolved processing issue. It is known that these joints form and can lead to a degradation in strength as well as to the higher permeabilities [24]. The two phenomena do not necessarily correspond, but observations generally have shown that high permeabilities are a more sensitive criterion than decreased strength in determining the existence of a distinct interface. Additionally, the actual conditions of their formation remains controversial, with two competing (not necessarily exclusive) hypotheses being (a) superficial drying between layers and (b) hardening of the substrate layer preventing mechanical intermixing of the layer being placed on top [25]. Neutron imaging has proved to be a very useful tool for the assessment of permeability in these materials alongside the layer-to-layer interfaces.

The preferential water uptake by the interlayer interfaces is among other parameters clearly influenced by the layer-to-layer deposition interval (also called interlayer time gap). While Schroefl *et al* showed clear preferential water uptake for the concretes printed with rather long interlayer time gaps (up to 24 h) and only very moderate preferential uptake for relatively short time gaps of several minutes [26], van der Putten *et al* report no preferential water ingress through the interlayer for the time gaps of 15 s [27] (see figure 3.3).

Despite these first attempts on quantification of permeability of the interlayer interfaces in 3D printed concrete, the literature on how the ‘cold joints’ influence the durability of the 3D printed materials remains scattered and it seems important to perform a true systematic investigation into the nature of the formation of the ‘cold

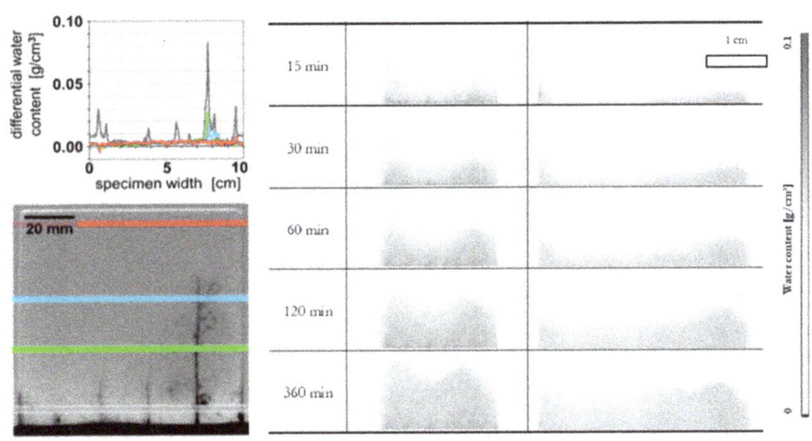

Figure 3.3. Water content distribution in 3D printed concrete. (Left) clear preferential water uptake after 3 h of immersion in water for up to 24 h layer-to-layer deposition interval, reproduced with permission from [26], copyright RILEM. (Right) hardly any preferential water uptake by the interlayer regions over 6 h of immersion in water with the layer-to-layer deposition interval of 15 s, reprinted from [27], copyright (2020), with permission from Elsevier.

joints'. Recently, Bran-Anleu *et al* [28] combined micro-x-ray fluorescence (μXRF) to image chloride ingress into layer interfaces of 3D printed fine-grained concrete specimens produced with varying layer deposition time intervals with neutron imaging of moisture uptake. Also, Ghantous *et al* [29] observed drying behaviour of 3D printed pastes containing cellulose nanocrystals. Generally, neutron radiography provides a very good tool for investigations of drying of cementitious materials [30].

One of the largest challenges for the digital fabrication (3D printing) of concrete is the inclusion of reinforcement in the manufactured elements. There is fast progress in the field of 3D printed reinforced concrete [31] and neutron imaging will undoubtedly provide a very useful tool for assessment of the 3D printed reinforced concrete materials not only from the perspective of the distribution of the water uptake, but also regarding the investigations of reinforcement corrosion. Another application of neutron imaging focussed on the interface between two layers of concretes was performed by Xue *et al* [32]. In their study, the dynamic water transport (imbibition) with focus on the interface between old and repaired mortar was investigated by time-resolved radiography.

Another example of corrosion resistant properties is represented by concrete with inclusions of anaerobic granular sludge for applications in sewers. Neutron tomography clearly revealed the distribution and volumes of the granular sludge inclusions [33]. The corrosion resistance of these types of concretes can be mainly attributed to sulfate-reducing bacteria from the granular sludge.

Damage in (cementitious) materials does not necessarily have to be originated by mechanical loading. Concrete—being a water-containing material—can degrade due to exposure to high temperature, in particular in the case of exposure to fire. High-performance concretes are particularly susceptible to explosive spalling when exposed to highly elevated temperatures. Spalling the concrete cover may directly expose the reinforcement of a concrete structure to the fire and may lead to the structure failure. The moisture transport during heating plays an important role in the mechanism of spalling. Neutron imaging thus can well contribute to the understanding of mechanism concrete spalling. Toropovs *et al* [34] utilized neutron radiography for derivation of moisture profiles in high-performance concretes that were from one side exposed to temperature of up to 550 °C. Water loss profiles of the concretes exposed to high temperatures could be derived and were coupled with results of simultaneous measurements of temperature and pressure by embedded thermocouples and pressure sensors. The concretes reinforced with polypropylene fibres exhibited no spalling. Naturally, providing there is enough neutron flux the process of relatively fast moisture migration due to the exposure to high temperature can be investigated also in 4D (3D + time). Dauti *et al* [35] managed to perform a series of neutron tomographies with the total acquisition time of 60 s for one tomography of 500 projections. Even though heavily binned in the spatial domain, the tomographic sequences clearly revealed the development of the drying front in time and showed that the coarser grained material was prone to faster moisture migration than the finer grained material.

Opposite to high temperatures, concrete can suffer damage at the other side of the thermometer as well. In particular, exposure to the cyclic freeze–thaw environment may lead (and in the long term leads) to the occurrence of cracks and thus to the deterioration of mechanical properties and durability. Peng *et al* [36] utilized neutron radiography to assess the influence of freeze–thaw cycles on capillary water absorption and chloride penetration into concretes of different wcr and concretes with and without air-entrainment. Neutron radiography revealed that the exposure to freeze–thaw cycles led to a damage gradient from the surface to the centre of the samples. Peng *et al* expect that more realistic predictions of service life of reinforced concrete structures when exposed to combined freeze–thaw cycles and chloride penetration—a combination often encountered in real concrete structures (e.g. roads, bridges)—can be provided. Future *in situ* investigations of freeze–thaw cycles applied to concretes and cementitious materials are foreseen to profit from utilization of climatic chambers (e.g as developed at Paul Scherrer Institute (PSI) [37]) for *in situ* experiments. Such chambers allow for application of a wide range of temperature and relative humidity environments and can thus add to the understanding of freezing mechanisms in concrete. For such investigations, the utilization of high-duty cycle time-of-flight (ToF) neutron imaging that can readily provide information on the distribution of supercooled water versus ice [38] based on the difference in their inelastic neutron contrasts is also foreseen.

Another aspect of concrete degradation that can be investigated by neutron radiography is concrete carbonation. As early as 1972, Reijonen and Pilhajavaara published the results of investigation that visualized the carbonated layers of concrete exposed to different carbonation regimes [39]. The images published in this paper are truly pioneering in the sense of neutron radiography of concrete; they have—at the same time—indicated the progress of neutron radiography since the early days. The recent publication by Moradllo *et al* [40] reported on the drying and carbonation processes in calcium silicate-based carbonated mortars and developed a methodology to quantify the degree of carbonation in carbonated calcium silicate-based specimens. The degree of carbonation derived from neutron radiography has been compared with the values of degree of carbonation based on thermogravimetric analyses (TGA) with both methods providing rather similar results. Naturally, the neutron radiography can provide (unlike the TGA) the spatial distribution of degree of carbonation over the entire sample size.

As neutrons are sensitive to hydrogen, neutron imaging may also be utilized for identification of water rich material phases in concretes. An example of such an approach is provided by the work of Livingston *et al* [41] who used neutron tomography to identify the 3D distribution of ettringite in concrete (see figure 3.4).

Ettringite [$Ca_6Al_2(SO_4)_3(OH)_{12}\cdot 26H_2O$] is a calcium aluminate sulfate hydrate mineral that may develop in concrete significantly later after casting and after the original hydration process (in the order of months to years). Its delayed formation causes internal stresses when the material is already fully hardened and can thus cause cracking and reduction in strength. It is foreseen that other water rich minerals/gels and connected concrete deterioration processes, such as silica gel in

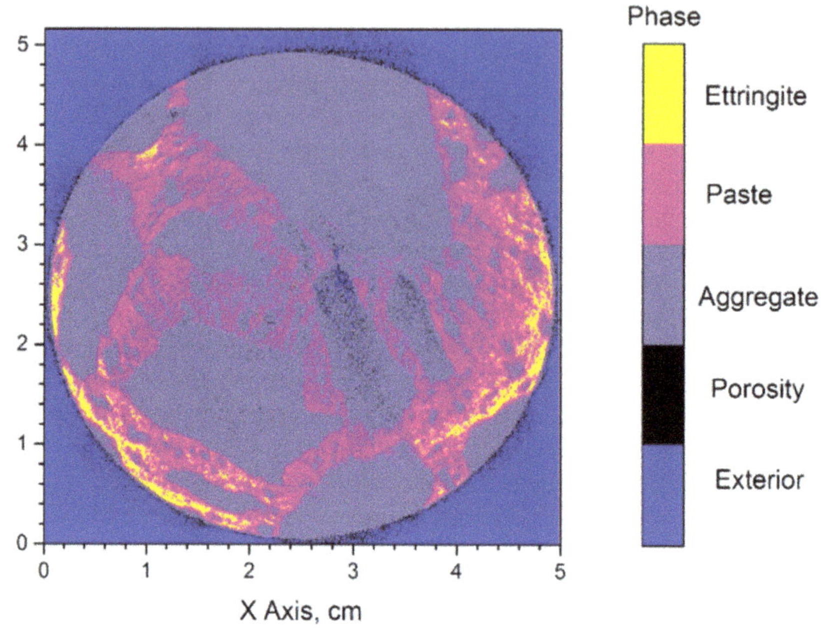

Figure 3.4. Horizontal slice from neutron tomography of sample affected by delayed ettringite formation; ettringite rich areas (in yellow). Image adapted from [41], copyright (2015), with permission from Elsevier.

alkali-silica reaction or thaumasite in thaumasite sulfate attack, might be well investigated by neutrons and/or combined N/X approach in the future. Following up on this outlook, the future application of ToF neutron imaging techniques will likely include investigations focussing on the utilization of inelastic neutron scattering cross-section of hydrogen from which the information about the composition of the hydrogen containing species in concrete might well be derived.

Last but not the least, an application to be mentioned in this section deals with the migration of ions dissolved in water through microstructure of hardened cementitious materials. Recent development of ToF neutron imaging technique allows for selection of different neutron energies, thus offering elemental sensitivity. Losko *et al* [42] investigated the liquid uptake in concrete for water saturated with uranyl acetate (depleted uranium) and for 10 weight percent iodine solution. Thanks to the ToF approach and its focus on energy selection in the epithermal regime (i.e. on the resonance frequencies of iodine and uranium), the authors were able to separate the uptake of ions from that of the uptake of water in the porous medium of cementitious mortars. In other words, the iodine concentration could have been successfully mapped independent of the distribution of the absorbed water. Their first results indicated that the uptake of the observed ionic species into the microstructure of cement-based mortar is significantly slower than the uptake of water itself. Studies of the uptake of ions in the material microstructure are generally interesting for understanding the behaviour of various porous media [43]. The clearly important application domain lies, however, in the field of nuclear materials/

engineering where the separation of moisture and ion transport through various barriers [44] would provide very useful input for the design of systems for the storage of radioactive waste.

3.1.2 Early age cement-based materials

Long-term performance of concrete and concrete structures is largely affected by the properties and treatment of concrete at an early age. In general, the early age is considered to be the first few hours or days after casting concrete. During this time, the major part of the cement hydration usually takes place. By the formation of hydration products, the setting of cementitious material occurs in which isolated particles are connected together. In analogy to the macroscale, the cement hydration bridges the originally depercolated (cement) particles into a load-bearing network, while at the same time it 'builds dams' in the originally water (pore solution) filled network of pores [45]. The process of hydration is intrinsically connected with the change in material volume (i.e. shrinkage). In particular, for high-strength concretes (concretes of low wcr) there is an increased shrinkage caused by self-desiccation and hence increased risk of early-age cracking. To mitigate this problem, the internal water curing is often utilized. The mostly used internal curing reservoirs are pre-wetted light-weight aggregates (LWA) and SAPs. The internal water curing reservoirs provide additional water and its availability within the concrete micro-structure is directly dependent on the distance which water is able to travel from them. Naturally, the questions about the distance of the water availability and the kinetics of the water transport from the internal curing reservoirs do arise and can to a certain extent be successfully answered by neutron imaging.

In this regards, Maruyama *et al* [46] used neutron radiography to image water transport from a single saturated 2D slice of LWA embedded in a hardening cement paste. The authors observed that part of the water in the LWA diffused to the cement paste during the first few hours after casting. Further, the water redistribution from the internal curing reservoirs in hydrating cement pastes were investigated by Trtik *et al* [47] in 4D (3D + time) using neutron tomography and later combined with x-ray tomography. The combined information from neutron and x-ray tomographies proved crucial for the correct establishment of boundaries of LWA reservoirs and cement pastes. This study confirmed that internal curing water is able to travel relatively far from the internal curing reservoirs and hardly any gradient in the additional water content within the cement paste with the distance from the LWA could be observed.

In the follow-up study, the internal curing with LWAs produced from biomass-derived waste was investigated by Lura *et al* [48]. Figure 3.5 shows examples of the slices of two samples of two different cement-based mortars showing domains of a loss (red regions) or gain (blue regions) of water content between the times of approximately 1 and 15 h after mixing

Internal curing based on superabsorbent polymers was recently investigated by Snoeck *et al* [49] who documented the kinetics of release of water from two types of

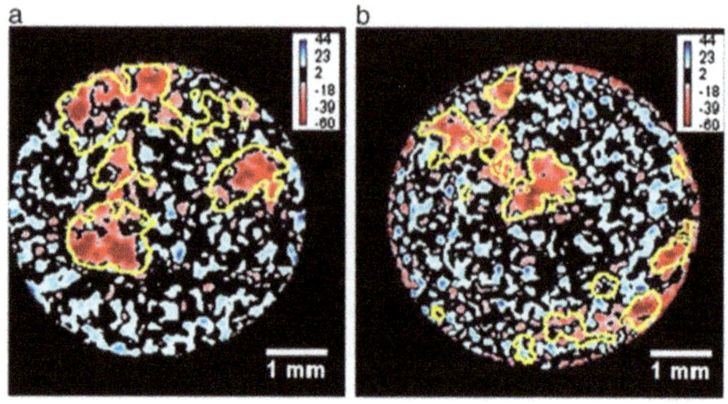

Figure 3.5. Example of slices of two samples of two different cement-based mortars based on neutron tomographic datasets showing domains of a loss (red regions) or gain (blue regions) of water content between the times of approximately 1 and 15 h after mixing. The yellow contours refer to LWA boundaries. Reprinted from [48], copyright (2014), with permission from Elsevier.

SAP—a copolymer of acrylamide and sodium acrylate, a cross-linked potassium salt polyacrylate. The investigation proved that the cross-linked potassium salt poyacrylate SAPs released the water prematurely while the copolymer of acrylamide SAPs was effective for the mitigation of autogeneous shrinkage.

Early-age drying (immediately after casting) of mortars and the corresponding plastic shrinkage were investigated by combined N/X tomography computed tomography by Wyrzykowski *et al* [50]. The study concentrated on the effect of a paraffin-based curing compound. Their results confirmed that when the curing compound was applied directly onto the drying surface in a sufficient amount, both the evaporation rate and the rate of vertical displacement (settlement) were substantially reduced.

The curing of concrete is an important material treatment immediately following the casting that ensures that sufficient moisture content is attained for a sufficient period of time during which the desired material properties develop. The influence of curing time on material properties of concrete was investigated by de Beer [16] who derived sorptivity coefficient based on neutron radiographies for samples of concrete cured between 1 and 21 days.

Moradllo *et al* [51] utilized neutron radiography to quantify the degree of hydration at various distances from the finished surface. They demonstrated how different curing approaches affect cement hydration in terms of both time and distance from the surface. They showed that in the case of early drying of the samples of plain mortars the degree of hydration in close-to-surface areas of the sample can differ markedly from the areas far from surface. A similar trend was also demonstrated for the mortar samples produced with superabsorbent polymers.

The (micro)-structure of cement-based materials is highly complex and despite continuous research not fully understood yet. Concrete is a multiscale material exhibiting inhomogeneities on several length scales with corresponding sizes of the representative volume elements for each inhomogeneity level. Another level of inhomogeneity has been discovered in cement-based materials by Diamond who reported on the occurrence of so called 'patchy microstructure' in cement [52]—the sharply delineated areas of dense and porous cement paste appearing in electron microscopy images of samples prepared for standard back-scattered electron microscopy investigations [53]. Diamond has shown that the patchiness has been present in the microstructure of cement-based mortar irrespective of the length of the mixing time [52] and irrespective of the superplasticizer dosage [54]. The size of the patchy areas has been reported in the order of several hundreds of micrometres, they seem to occur quasi-randomly in the material and are therefore not to be identified as an interfacial transition zone (ITZ) occurring around the aggregates in about one order of magnitude smaller scale. The porous patch features were observed using neutrons by the author of this chapter for the first time during the above-mentioned combined neutron/x-ray imaging investigation that was dedicated to the early-age water migration from curing water reservoirs (LWA) into adjacent cement paste [47]. Likewise, the porous patches were clearly revealed in one sample of pure cement pastes of the bulk average wcr of 0.28 investigated by combined neutron/x-ray imaging investigation. The porous patches were distributed rather randomly throughout the sample. The porous patches were clearly sharply delineated from the remaining bulk material. The combination of the neutron tomography from early-age samples, neutron tomography of the same sample after setting and the x-ray tomography allowed for quantification of the local wcr with the spatial resolution given by the neutron tomography technique. With the original bulk wcr of 0.28, it was found (based on the quantitative values of neutron tomographic datasets) that the original wcr within the areas of porous patches was about 0.31 (see figure 3.6).

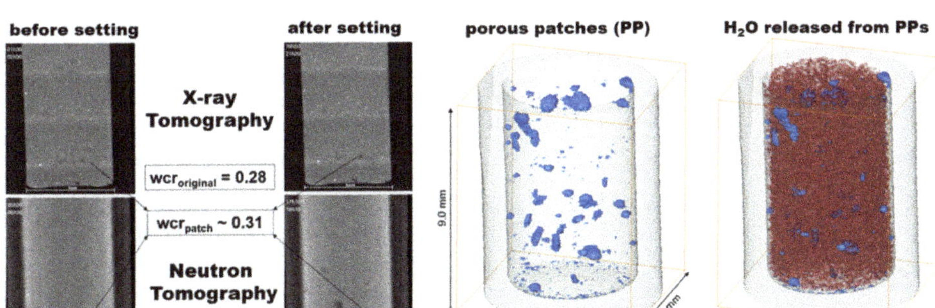

Figure 3.6. (left) Corresponding vertical slices of a sample of hardening cement paste from (upper row) synchrotron x-ray and (lower row) neutron tomographies. The comparison of the images (left column) before and (right column) after setting allows for the quantification of the original local wcr (e.g. that inside the porous patch). The white scalebars correspond to 5 mm. (right) 3D rendering of porous patches within a sample of cement paste, (right) visualization of the pore solution transported from the patches (in brown) into the bulk cement paste.

3.1.3 Reinforced concrete

All the above-mentioned examples of the use of neutron imaging for cementitious materials were performed using unreinforced concrete (i.e. without steel reinforcement). Concretes are quasi brittle materials exhibiting a large discrepancy between their compressive and tensile strengths. The reinforcement embedded in concrete forms a composite material, in which the concrete provides strength against compressive stresses while the reinforcement provides strength against tensile stresses. Thanks to the nearly identical coefficient of thermal expansion of steel and concrete, the steel rebars are by far the most common types of structural reinforcement used in concretes. The water uptake in cracked steel-reinforced concrete has been seminally investigated using thermal neutrons by Wittmann *et al* [55] showing fast penetration of water alongside the steel/concrete interface damaged by the specimen fracture (figure 3.7).

The water uptake into and kinetics of drying of composite specimens of steel-reinforced concrete (RC) strengthened with strain hardening cement-based composite (SHCC) were investigated by Schröfl *et al* [56] by time series of neutron radiographies. The image analysis revealed that capillary suction was very fast and within approximately 1 min the cracks in both SHCC and RC were filled fully deep into the sample interior.

Apart from the water ingress to the interfacial zones of the reinforcing bars, the corrosion of reinforcing steel bars represents one of the gravest threats to the durability of structures made of reinforced concrete. Thanks to its highly alkaline environment (pH usually between 12.5 and 13) [57], sound concrete provides very decent protection against corrosion. The corrosion initiation is triggered by rebar depassivation (drop in the pH) usually due to the ingress of foreign species into the porous structure of concrete, such as chlorides [58]. As the corrosion products are

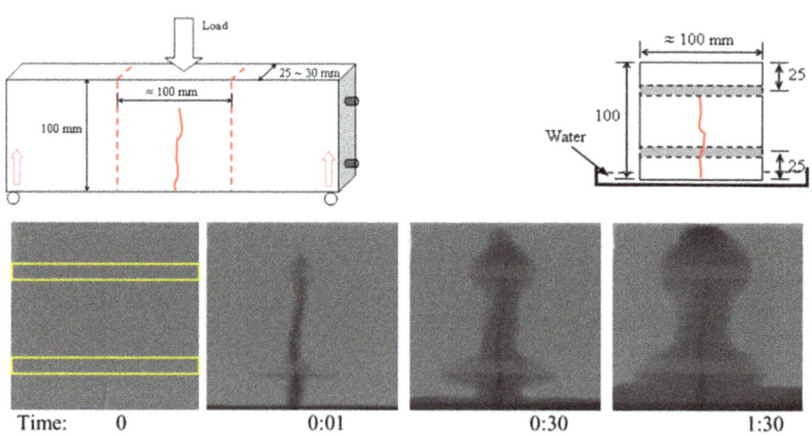

Figure 3.7. Neutron radiography of water penetration into the cracked steel-reinforced concrete, (top) the schematic of the sample preparation and the neutron radiography sample, (bottom) neutron images at the times from 0 to 1 h and 30 min from the start of the test, the yellow boxes represent the approximate positions of the reinforcing bars. Reproduced with permission from [55], copyright RILEM.

more voluminous than the steel, the consequence of the rebar corrosion includes the volume expansion leading to exertion of internal pressure and eventual cracking and even spalling of concrete. Likewise, corrosion pits in rebars decrease their effective cross-section which is a serious issue for the safety of concrete structures. Numerous models for both the rebar corrosion and for the corrosion-induced concrete cracking/spalling have been developed in the last decades (e.g. [59, 60]). Despite the availability of these models, rebar corrosion remains one of the main causes of deterioration of concrete structures.

There is a twofold reason for the use neutron imaging for the visualization of the internal structure of corroded reinforced concrete [61]. First, the more voluminous corrosion products often contain a significant amount of hydrogen (e.g. $Fe(OH)_2$, $Fe(OH)_3$, $Fe(OH)_3{\cdot}3H_2O$) thus providing advantageous contrast for neutrons with respect to the rest of the material microstructure of concrete. Second, the interfacial regions between concrete and steel rebar are intrinsically difficult to be observed using x-rays due to the presence of metal (streak) artifacts. Such artifacts are particularly present in the natural concavities of ribbed reinforcing bars as well as in the prospective corrosion pits. At the same time, the combined approach of N and X provides a very useful mechanism for revelation of various material phases in corroded concrete.

First attempts at investigations of rebar corrosion were performed at the PSI [62] using neutron tomography at the NEUTRA beamline (see figure 3.8).

The more advanced bimodal approach towards characterisation of corrosion in concrete has been pioneered at PSI by Boschmann and colleagues [63]. The scientific question to be answered by the bimodal N/X approach was if the air voids in concrete at the rebar/concrete interface represent the weak spots for the corrosion initiation. For this purpose, the samples were cored out of existing concrete structures (two bridges in rural Switzerland) and the corrosion has been initiated in them in the laboratory. For details of the methodology, please see figure 3.9.

Figure 3.8. 3D renderings of corrosion products (in red) in cylindrical samples of reinforced concrete subjected to corrosion initiation (image courtesy Florian Schmid, PSI).

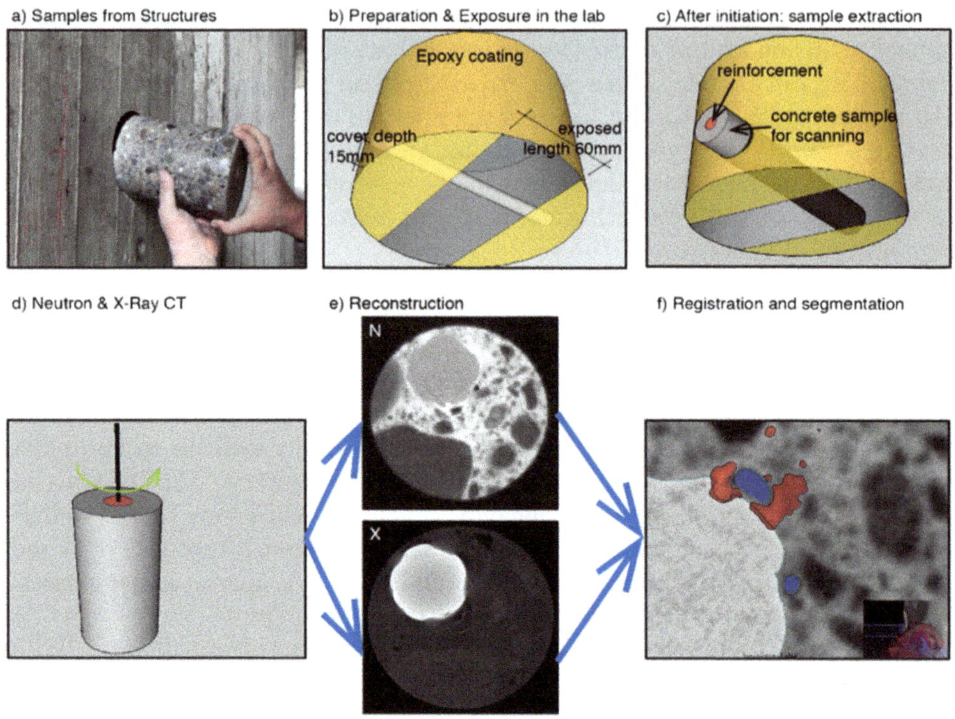

Figure 3.9. (Left) Methodological summary for investigation of chloride induced corrosion mechanisms of rebars in concrete (reproduced with permission from [63]). The mutual positions of the corrosion products (in red) and air voids (in blue) was successfully derived using bimodal N/X imaging.

The mutual positions of the corrosion products and air voids were derived from the registered N/X datasets revealing that vast majority of air voids in the direct vicinity of concrete/steel interface actually do not show any sign of corrosion initiation.

More recently, Robuchi *et al* [64] gave a closer look at two concrete samples with embedded reinforcement using bimodal N/X imaging. While one sample has been subjected to natural corrosion, the other one was corroded artificially in lab-based accelerated conditions. The authors concluded that the bimodal N/X imaging is a promising non-destructive technique for qualitative and quantitative investigation of corrosion products in reinforced concrete samples. The corrosion distribution was observed to be substantially different between the naturally corroded and artificially corroded samples. In the naturally corroded sample, corrosion products were mostly observed in the macroscopic voids close to interface, while in the artificially corroded sample, corrosion products were almost uniformly distributed along the length of the bar.

Future experiments using bimodal N/X imaging of corrosion will certainly focus on the time-resolved studies. As concrete rebar corrosion is a relatively slow process, the investigations will not necessarily have to be performed *in situ*. Instead, experiments in which samples are to be imaged in a pristine state followed by *ex situ* corrosion initiation and concluded with the imaging of the same samples in corroded state are foreseen [58].

Another recent publication worth mentioning here is focussed on the identification of iron oxides using a wide-energy neutron imaging [65]. In this combined fast neutron/resonance neutron imaging and simulation study, the close-to-0.1 mm rust layers of iron oxide inside 20 cm thick corroded reinforced concrete sample could apparently be identified. This result provides an interesting avenue for identification of corrosion in entire concrete segments (e.g. revealing corrosion/defects in post-tensioned pre-stressed elements in precast concrete bridges).

The author wishes to conclude this section with the discussion of the issue of production concrete/cement-based material specimens for neutron imaging using heavy water. The contrast of heavy (D_2O) and light water (H_2O) in neutron images had been shown already in the very first scientific paper about neutron imaging [66], however, heavy water for production of samples for neutron radiography has been hitherto used relatively sparsely. The advantage of such D_2O-based samples is naturally in their lower opacity for neutrons in comparison with their light-water-based counterparts. Consequently and not surprisingly, larger (i.e. more representative) samples of cement-based materials can be successfully investigated [67]. Caution should, however, be applied in particular for the investigations of hydration processes as the hydration kinetics differs markedly for the pastes produced with H_2O and D_2O [68].

3.2 Other construction materials

Apart from cement-based materials, neutron imaging has also been used for other construction materials. This last section will therefore be dedicated to investigation of masonry and other porous construction materials including asphalts. It should be borne in mind that some of the applications presented here are on the fringe with applications presented at the chapter 9. The reader is therefore kindly also requested to consult chapter 9.

3.2.1 Masonry and other porous construction materials

Similar to the investigation of the cementitious materials, the water capillary uptake in masonry materials can be conveniently investigated using neutron radiography. Guizzardi *et al* [69] studied water uptake in clay bricks and other masonry materials equipped with a system for monitoring water penetration by means of electrical conductivity measurements. Behaviour of water along the interfaces in building materials is of intrinsic interest to building physicists. The water transport in masonry brick/cement mortar interface was investigated by Zhou *et al* [70] who applied neutron radiography for acquisition of time series of water uptake. The moisture transport in masonry could be strongly affected by the interface resistance, when the interface is in proximity to moisture source, however, it was shown that the orientation of the interface between brick and mortar (be it vertical or horizontal) does not have an influence on the interface resistance.

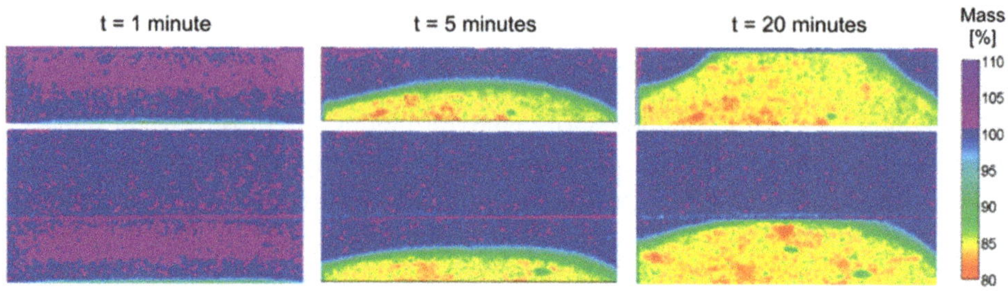

Figure 3.10. Change in the total mass in gypsum plaster boards (top) single layer, (bottom) double layer. Reprinted from [71], copyright (2013), with permission from Elsevier.

Also focussed on the interfacial areas, the development of porosity at a cement paste/clay interface relevant for the nuclear waste storage applications was investigated over the period of two years by Shafizadeh *et al* [44]. The water content derived from neutron radiography revealed a porosity decrease in the clay and its increase in the cement paste.

Analogously to the earlier-mentioned experiments of concrete spalling [34, 35], the release and migration of moisture in gypsum plaster boards due to one-sided exposure to high temperatures had been earlier investigated by Sedighi-Gilani *et al* [71]. The *in situ* exposure of the plaster boards to temperature as high as 600 °C triggered gypsum endothermic dehydration and release of moisture that diffuses through the sample in the direction away from the heating source.

The *in situ* neutron radiography of the dehydration process revealed not only the mass change due to dehydration, but also the moisture accumulation in the not yet dehydrated areas of the gypsum plaster board ahead of the dehydration front (see the over 100% areas in the figure 3.10).

Also deserving a mention here is the study of the absorption of water droplets impacting surfaces of porous stones by Lee *et al* [72]. Neutron imaging with acquisition times of 3 s revealed absorption, droplet spreading and evaporation as well as the droplet mass depletion and the moisture content distributions in the porous medium. Further advances in temporal resolution [73] of neutron imaging may offer truly real-time observation of the droplet impingement on surfaces of porous construction materials in the future. Likewise, the progress in spatial resolution of neutron imaging [74] allows for observation of moisture distribution within relatively thin (~100 μm) films [75].

Last but not least, another example that neutron imaging can be utilized not only for the investigation of moisture transport but also for identification of other deleterious processes in building construction materials, are the zones of salt precipitation in limestone samples subjected to repeated wetting and drying cycles [76]. The 2D maps of halite crystal quantitative distributions were derived from neutron radiographies performed at the NEUTRA beamline. Likewise, the uptake of water and salt solution in the bespoke replicate samples has been compared and it was shown that the liquid penetration and diffusivities were markedly lower for the salt solutions than for water.

3.2.2 Bitumens and asphalts

At PSI, primarily Poulikakos *et al* [77] and shortly later Lal *et al* [78] successfully investigated water uptake in porous asphalt—a highly porous hydrophobic composite material. A nifty test set-up that allowed for immersion of cylindrical samples of porous asphalt into a basin of water utilized for acquisition of *in situ* neutron radiographies was developed for this study. Both aged and unaged samples were investigated, moisture content distribution was derived and in conjunction with complementary x-ray computed tomography datasets, saturation degree distributions could have been plotted from which the regions of active saturated flow could have been identified (figure 3.11).

Another interesting aspect of material science that has been investigated by neutron imaging is the self-healing of fatigue cracks in the structure of bitumen. Time series of neutron tomographies was recently performed on samples with different diameters, hydrated lime (HL) filler concentration, notch size, and contact area of the simulated micro-cracks and their respective healing ability being analyzed [79].

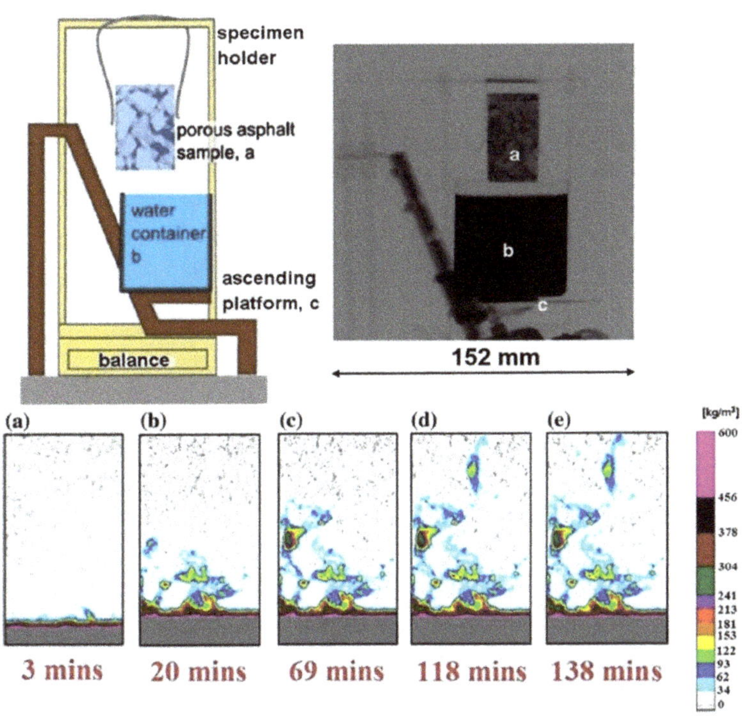

Figure 3.11. Neutron imaging of water uptake by porous asphalt: (top left) schematics of the experimental set-up, (top right) neutron radiography showing the porous asphalt specimen before immersion in the water container, (below) moisture content distribution in five time steps (adapted from [78], copyright (2014), with permission from Springer Nature.

References

[1] Robinson G (https://ice.org.uk/ICEDevelopmentWebPortal/media/Documents/News/ICE%20News/Global-Construction-press-release.pdf)

[2] Xu K *et al* 2021 Microstructure and water absorption of ancient concrete from Pompeii: an integrated synchrotron microtomography and neutron radiography characterization *Cem. Concr. Res.* **139** 106282

[3] Aldred J 2010 Burj Khalifa—a new high for high-performance concrete *Proc. Inst. Civ. Eng. Civ. Eng.* **163** 66–73

[4] Concrete ships (https://en.wikipedia.org/wiki/Concrete_ship)

[5] Concrete submarine (http://imulead.com/tolimared/concretesubmarine/picturegallery/itle)

[6] Trtik K 1994 Stavba umelych pristavu Mulberry—50 let (in Czech) *Staveb. Obz.* **7** 202–5

[7] Bartak J 1999 Underground gas storage of Pribram-Haje *Tunnel* **8** 2–10

[8] Concrete statues (https://italcement.com/cement/st-056/)

[9] Concrete earrings (https://nausnice.heureka.sk/gravelli-nausnice-z-betonu-block-gjew-naa004un/#prehlad/)

[10] Gregory J, AzariJafari H, Vahidi E, Guo F, Ulm F J and Kirchain R 2021 The role of concrete in life cycle greenhouse gas emissions of US buildings and pavements *Proc. Natl Acad. Sci. USA* **118** 1–9

[11] Zhang P, Wittmann F H, Lura P, Müller H S, Han S and Zhao T 2018 Application of neutron imaging to investigate fundamental aspects of durability of cement-based materials: a review *Cem. Concr. Res.* **108** 152–66

[12] Tengattini A, Lenoir N, Andò E and Viggiani G 2021 Neutron imaging for geomechanics: a review *Geomech. Energy Environ.* **27** 100206

[13] Pražák J, Tywoniak J, Peterka F and Šlonc T 1990 Description of transport of liquid in porous media-a study based on neutron radiography data *Int. J. Heat Mass Transf.* **33** 1105–20

[14] Justnes H, Bryhn-Ingebrigtsen K and Rosvold G O 1994 Neutron radiography: an excellent method of measuring water penetration and moisture distribution in cementitious materials *Adv. Cem. Res.* **6** 67–72

[15] Hanžič L and Ilić R 2003 Relationship between liquid sorptivity and capillarity in concrete *Cem. Concr. Res.* **33** 1385–8

[16] De Beer F C, Le Roux J J and Kearsley E P 2005 Testing the durability of concrete with neutron radiography *Nucl. Instrum. Methods Phys. Res.* A **542** 226–31

[17] Zhang P, Wittmann F H, Zhao T-J, Lehmann E H and Vontobel P 2011 Neutron radiography, a powerful method to determine time-dependent moisture distributions in concrete *Nucl. Eng. Des.* **241** 4758–66

[18] Kanematsu M, Maruyama I, Noguchi T, Iikura H and Tsuchiya N 2009 Quantification of water penetration into concrete through cracks by neutron radiography *Nucl. Instrum. Methods Phys. Res.* A **605** 154–8

[19] Snoeck D, Steuperaert S, Van Tittelboom K, Dubruel P and De Belie N 2012 Visualization of water penetration in cementitious materials with superabsorbent polymers by means of neutron radiography *Cem. Concr. Res.* **42** 1113–21

[20] Lukić B, Tengattini A, Dufour F and Briffaut M 2021 Visualising water vapour condensation in cracked concrete with dynamic neutron radiography *Mater. Lett.* **283** 1–4

[21] Van Tittelboom K, Snoeck D, Vontobel P, Wittmann F H and De Belie N 2013 Use of neutron radiography and tomography to visualize the autonomous crack sealing efficiency in cementitious materials *Mater. Struct. Constr.* **46** 105–21

[22] Van den Heede P, Van Belleghem B, Alderete N, Van Tittelboom K and De Belie N 2016 Neutron radiography based visualization and profiling of water uptake in (un)cracked and autonomously healed cementitious materials *Materials* **9** 311

[23] Zhang P, Wittmann F H, Zhao T-J and Lehmann E H 2014 Investigations into the water repellent surface near layer in concrete by neutron radiographytle *Restor. Build. Monum.* **20** 79–84

[24] Kruger J and van Zijl G 2021 A compendious review on lack-of-fusion in digital concrete fabrication *Addit. Manuf.* **37** 101654

[25] Roussel N 2018 Rheological requirements for printable concretes *Cem. Concr. Res.* **112** 76–85

[26] Schröfl C, Nerella V N and Mechtcherine V 2019 Capillary water intake by 3D-printed concrete visualised and quantified by neutron radiography *RILEM Bookseries* **19** 217–24

[27] Van Der Putten J, Azima M, Van den Heede P, Van Mullem T, Snoeck D, Carminati C, Hovind J, Trtik P, De Schutter G and Van Tittelboom K 2020 Neutron radiography to study the water ingress via the interlayer of 3D printed cementitious materials for continuous layering *Constr. Build. Mater.* **258** 119587

[28] Bran-Anleu P, Wangler T, Nerella V N, Mechtcherine V, Trtik P and Flatt R J 2023 Using micro-XRF to characterize chloride ingress through cold joints in 3D printed concrete *Mater. Struct. Constr.* **56** 51

[29] Ghantous R M, Valadez-Carranza Y, Reese S R and Weiss W J 2022 Drying behavior of 3D printed cementitious pastes containing cellulose nanocrystals *Cement.* **9** 100035

[30] Hu Z, Cajuhi T, Toropovs N, Griffa M, Wyrzykowski M, Kaestner A, De Lorenzis L and Lura P 2023 A neutron radiography study on the drying of cement mortars : effect of mixture composition and crack length *Cem. Concr. Res.* **172** 107245

[31] Perrot A 2019 *3D Printing of Concrete: State of the Art and Challenges of the Digital Construction Revolution* (Wiley)

[32] Xue S, Meng F, Zhang P, Wang J, Bao J and He L 2021 Influence of substrate moisture conditions on microstructure of repair mortar and water imbibition in repair-old mortar composites *Meas. J. Int. Meas. Confed.* **183** 109769

[33] Song Y, Chetty K, Garbe U, Wei J, Bu H, O'moore L, Li X, Yuan Z, McCarthy T and Jiang G 2021 A novel granular sludge-based and highly corrosion-resistant bio-concrete in sewers *Sci. Total Environ.* **791** 148270

[34] Toropovs N, Lo Monte F, Wyrzykowski M, Weber B, Sahmenko G, Vontobel P, Felicetti R and Lura P 2015 Real-time measurements of temperature, pressure and moisture profiles in high-performance concrete exposed to high temperatures during neutron radiography imaging *Cem. Concr. Res.* **68** 166–73

[35] Dauti D, Tengattini A, Dal Pont S, Toropovs N, Briffaut M and Weber B 2018 Analysis of moisture migration in concrete at high temperature through *in situ* neutron tomography *Cem. Concr. Res.* **111** 41–55

[36] Zhang P, Wittmann F H, Vogel M, Müller H S and Zhao T 2017 Influence of freeze-thaw cycles on capillary absorption and chloride penetration into concrete *Cem. Concr. Res.* **100** 60–7

[37] Mannes D, Schmid F, Wehmann T and Lehmann E 2017 Design and applications of a climatic chamber for *in situ* neutron imaging experiments *Phys. Procedia* **88** 200–7

[38] Siegwart M, Woracek R, Márquez Damián J I, Tremsin A S, Manzi-Orezzoli V, Strobl M, Schmidt T J and Boillat P 2019 Distinction between super-cooled water and ice with high duty cycle time-of-flight neutron imaging *Rev. Sci. Instrum.* **90** 103705
[39] Reijonen H and Pihlajavaara S E 1972 On the determination by neutron radiography of the thickness of the carbonated layer of concrete based upon changes in water content *Cem. Concr. Res.* **2** 607–15
[40] Khanzadeh Moradllo M, Ghantous R M, Quinn S, Aktan V, Reese S and Weiss W J 2022 Quantifying drying and carbonation in calcium silicate-cement systems using neutron radiography *ACI Mater. J.* **119** 231–42
[41] Livingston R, Feuze S, Amde A, Hussey D S and Jacobson D 2015 Neutron tomography measurement of delayed ettringite formation in concrete *14th Int. Symp. Nondestruct. Charact. Mater. (NDCM 2015) (Marina Del Rey, CA)* ww.ndt.net/app.NDCM2015
[42] Losko A S, Daemen L, Hosemann P, Nakotte H, Tremsin A, Vogel S C, Wang P and Wittmann F H 2020 Separation of uptake of water and ions in porous materials using energy resolved neutron imaging *JOM* **72** 3288–95
[43] Butcher T A, Prendeville L, Rafferty A, Trtik P, Boillat P and Coey J M D 2021 Neutron imaging of paramagnetic ions: electrosorption by carbon aerogels and macroscopic magnetic forces *J. Phys. Chem.* C **125** 21831–9
[44] Shafizadeh A, Gimmi T, Van Loon L R, Kaestner A P, Mäder U K and Churakov S V 2020 Time-resolved porosity changes at cement-clay interfaces derived from neutron imaging *Cem. Concr. Res.* **127** 105924
[45] Bentz D P 2007 Cement hydration: building bridges and dams at the microstructure level *Mater. Struct. Constr.* **40** 397–404
[46] Maruyama I, Kanematsu M, Noguchi T, Iikura H, Teramoto A and Hayano H 2009 Evaluation of water transfer from saturated lightweight aggregate to cement paste matrix by neutron radiography *Nucl. Instrum. Methods Phys. Res.* A **605** 159–62
[47] Trtik P, Münch B, Weiss W J, Kaestner A, Jerjen I, Josic L, Lehmann E and Lura P 2011 Release of internal curing water from lightweight aggregates in cement paste investigated by neutron and x-ray tomography *Nucl. Instrum. Methods Phys. Res. A* **651** 244–9
[48] Lura P, Wyrzykowski M, Tang C and Lehmann E 2014 Internal curing with lightweight aggregate produced from biomass-derived waste *Cem. Concr. Res.* **59** 24–33
[49] Snoeck D, Goethals W, Hovind J, Trtik P, Van Mullem T, Van den Heede P and De Belie N 2021 Internal curing of cement pastes by means of superabsorbent polymers visualized by neutron tomography *Cem. Concr. Res.* **147** 106528
[50] Wyrzykowski M, Ghourchian S, Münch B, Griffa M, Kaestner A and Lura P 2021 Plastic shrinkage of mortars cured with a paraffin-based compound—bimodal neutron/x-ray tomography study *Cem. Concr. Res.* **140** 106289
[51] Moradllo M K, Montanari L, Suraneni P, Reese S R and Weiss J 2018 Examining curing efficiency using neutron radiography *Transp. Res. Rec.* **2672** 13–23
[52] Diamond S 2005 The patch microstructure in concrete: effect of mixing time *Cem. Concr. Res.* **35** 1014–6
[53] Diamond S 2004 The microstructure of cement paste and concrete—a visual primer *Cem. Concr. Compos.* **26** 919–33
[54] Diamond S 2006 The patch microstructure in concrete: the effect of superplasticizer *Cem. Concr. Res.* **36** 776–9

[55] Wittmann F, Zhang P, Zhao T, Lehmann E and Vontobel P 2008 Neutron radiography, a powerful method for investigating water penetration into concrete *Adv. Civ. Eng. Mater. 50-Year Teach. Res. Anniv. Prof. Sun Wei* pp 61–70

[56] Schröfl C, Mechtcherine V, Kaestner A, Vontobel P, Hovind J and Lehmann E 2015 Transport of water through strain-hardening cement-based composite (SHCC) applied on top of cracked reinforced concrete slabs with and without hydrophobization of cracks—investigation by neutron radiography *Constr. Build. Mater.* **76** 70–86

[57] Behnood A, Van Tittelboom K and De Belie N 2016 Methods for measuring pH in concrete: a review *Constr. Build. Mater.* **105** 176–88

[58] Angst U M *et al* 2019 The effect of the steel–concrete interface on chloride-induced corrosion initiation in concrete: a critical review by RILEM TC 262-SCI *Mater. Struct. Constr.* **52** 88

[59] Zhang J, Ling X and Guan Z 2017 Finite element modeling of concrete cover crack propagation due to non-uniform corrosion of reinforcement *Constr. Build. Mater.* **132** 487–99

[60] Molina F J, Alonso C and Andrade C 1993 Cover cracking as a function of rebar corrosion: part 2—numerical model *Mater. Struct.* **26** 532–48

[61] Garbe U, Ahuja Y, Ibrahim R, Li H, Aldridge L, Salvemini F and Paradowska A Z 2017 Industrial application experiments on the neutron imaging instrument DINGO *Phys. Procedia* **88** 13–8

[62] Zhang P, Liu Z, Wang Y, Yang J, Han S and Zhao T 2018 3D neutron tomography of steel reinforcement corrosion in cement-based composites *Constr. Build. Mater.* **162** 561–5

[63] Boschmann C 2019 Chloride-induced reinforcement corrosion in concrete : the role of the steel-concrete interface and implications for engineering *PhD Thesis* ETH Zurich

[64] Robuschi S, Tengattini A, Dijkstra J, Fernandez I and Lundgren K 2021 A closer look at corrosion of steel reinforcement bars in concrete using 3D neutron and x-ray computed tomography *Cem. Concr. Res.* **144** 106439

[65] Tian B, Wang S, Jing H, Yan M, Gao X and Yang X 2021 Iron-oxide-identifying imaging method based on a wide-energy neutron beam for corrosion inspection in reinforced concrete structures *Rev. Sci. Instrum.* **92** 123703

[66] Peter O 1946 Neutronen-Durchleuchtung *Z. Naturforschg.* **726278** 557–9

[67] Goodwin M N, Ghantous R M, Weiss W J and Reese S R 2022 Neutron radiography of cement paste made with light and heavy water *J. Radioanal. Nucl. Chem.* **331** 5113–21

[68] Thomas J J and Jennings H M 1999 Effects of D_2O and mixing on the early hydration kinetics of tricalcium silicate *Chem. Mater.* **11** 1907–14

[69] Guizzardi M, Derome D, Mannes D, Vonbank R and Carmeliet J 2016 Electrical conductivity sensors for water penetration monitoring in building masonry materials *Mater. Struct. Constr.* **49** 2535–47

[70] Zhou X, Desmarais G, Vontobel P, Carmeliet J and Derome D 2020 Masonry brick–cement mortar interface resistance to water transport determined with neutron radiography and numerical modeling *J. Build. Phys.* **44** 251–71

[71] Sedighi-Gilani M, Ghazi Wakili K, Koebel M, Hugi E, Carl S and Lehmann E 2013 Visualizing moisture release and migration in gypsum plaster board during and beyond dehydration by neutron radiography *Int. J. Heat Mass Transf.* **60** 284–90

[72] Lee J B, Radu A I, Vontobel P, Derome D and Carmeliet J 2016 Absorption of impinging water droplet in porous stones *J. Colloid Interface Sci.* **471** 59–70

[73] Zboray R and Trtik P 2018 800 fps neutron radiography of air-water two-phase flow *MethodsX* **5** P96–102

[74] Trtik P and Lehmann E H 2016 Progress in high-resolution neutron imaging at the Paul Scherrer Institut—the neutron microscope project *J. Phys. Conf. Ser.* **746** 012004

[75] Ott N, Cancellieri C, Trtik P and Schmutz P 2020 High-resolution neutron imaging: a new approach to characterize water in anodic aluminum oxides *Mater. Today Adv.* **8** 100121

[76] Derluyn H *et al* 2013 Characterizing saline uptake and salt distributions in porous limestone with neutron radiography and x-ray micro-tomography *J. Build. Phys.* **36** 353–74

[77] Poulikakos L D, Sedighi Gilani M, Derome D, Jerjen I and Vontobel P 2013 Time resolved analysis of water drainage in porous asphalt concrete using neutron radiography *Appl. Radiat. Isot.* **77** 5–13

[78] Lal S, Poulikakos L D, Sedighi Gilani M, Jerjen I, Vontobel P, Partl M N, Carmeliet J C and Derome D 2014 Investigation of water uptake in porous asphalt concrete using neutron radiography *Transp. Porous Media* **105** 431–50

[79] Grange M, Kaestner A, Kringos N and Poulikakos L 2021 Self-healing study of bitumen mastic mixtures using time series neutron tomography (University of Rennes)

Chapter 4

Nuclear materials

Pavel Trtik, Robert Zboray, Liliana I Duarte, Okan Yetik and Florencia Malamud

Modern society is highly dependent on electricity and an exponential growth in energy consumption has been observed in recent decades. To develop sustainable technologies for energy production contributing to a reduction in the carbon footprint is crucial to our society. Nuclear energy is a zero-emission clean energy source and produces more electricity on less land than any other clean-air source. However, despite several countries recently reducing their reliance on nuclear power, about 2500 TWh yearly (10% of the world's electricity) is still generated by more than 400 nuclear power reactors worldwide [1]. It is therefore clear that nuclear power will remain in the palette of energy sources at least for many decades to come. At the same time, the problem of safety of materials used in the entire process of energy production from nuclear power plants has to be addressed and the understanding of the material properties and processes occurring in nuclear materials is thus the paramount information for safety regulators, nuclear fuel manufacturers, utilities and final storage organizations.

Nuclear material, according to IAEA, is used to designate uranium, plutonium, or thorium metals-based materials. All are used as fuel material in nuclear reactors at power plants, with uranium being used most widely. However, nuclear materials do not only include the fuel (uranium) but also the fuel cladding, control rods, and structural materials. Structural nuclear materials include the piping, core barrel, fuel structural, internals, pumps, turbines, and heat exchanger tubing. That means all the materials which underline a conversion into radioactive ones. They have to be stored carefully after the operation cycle. Further, nuclear fuel rods are used mainly to protect and keep the fuel in a well-defined geometry while the fuel rod provides the first barrier separating the fission products from the environment. Preserving the nuclear fuel rod integrity is therefore the primary goal of fuel design and safety rules during nuclear reactor operation [2]. Nuclear fuel rods are periodically replaced, typically their lifetime in the reactor is about 3–6 years [3]. About once a year,

doi:10.1088/978-0-7503-3495-2ch4

© IOP Publishing Ltd 2024

25%–30% of the fuel is unloaded and replaced with fresh fuel. Other fuel elements with limited burnup are placed at different locations in the reactor core, according to a sophisticated equal burnup procedure. After their useful life of 3–6 years, fuel assemblies are removed from the reactor. However, they are expected to last for the life of the plant and longer as desired lifetimes are extended (about six years in a reactor, until the fission process uses up uranium fuel. Fuel is replaced after being in the core for six years, so every two years a third of the fuel is replaced and the other two thirds are moved around to make for even burning.

As a non-destructive technique, neutron imaging provides a very good characterization tool suitable for a large number of characterization tasks that can be hardly performed by other probes in the field of nuclear materials. As is the case in other non-nuclear materials, neutron imaging here often utilizes the favourable ratio between the attenuation contrasts of light and heavy elements (e.g. H versus Zr). However, in the case of nuclear materials (and in particular in the case of highly activated materials) there is yet another advantage that neutron imaging holds against other imaging modalities. As neutron beams used for neutron imaging (or generally for any neutron science) activate the investigated materials, the neutron imaging beamlines are placed inside controlled zones and are designed with appropriate shielding against radiation. Thus, these facilities are intrinsically well equipped to accommodate (highly) radioactive materials.

The chapter is divided into four sections. In the first part we are reviewing the use of neutron imaging for characterization of nuclear fuels and highly activated objects. In the second part, we focus on the use of neutron imaging for characterization of other structural materials in nuclear applications with a particular emphasis on the issue of safety of nuclear fuel claddings. The penultimate section is dedicated to the use of high-temporal resolution neutron imaging for the characterization of two-phase flows along the mock-ups of nuclear fuel rods and the chapter ends with a short section neutron imaging of nuclear waste.

4.1 Nuclear fuels

The capability of neutrons to transmit uranium, in particular both low enriched and depleted, while at the same time being sensitive to hydrogen allows characterization of nuclear fuel and cladding. Likewise, due to the activation of the samples via neutron irradiation, the neutron imaging facilities are intrinsically best suited for investigation of nuclear materials. Consequently, the investigation of nuclear materials by neutron imaging has become one of the rather early applications of the technique.

Since the investigated irradiated objects are strong sources of gamma radiation, dedicated techniques are required for samples handling and for performing neutron imaging with high gamma-radiation background.

The neutron imaging of radioactive specimens has therefore been limited to indirect transfer methodology using films and foils (e.g. cadmium-wrapped indium foils). As early as the 1970s, neutron imaging was used in the United States by John Barton [4] who evaluated neutron radiography as a method for inspection of bundles of fast breeder reactor fuel contained in steel test vehicles for safety experiments. The indirect technique of the choice of resonance energy neutron radiography using cadmium-wrapped indium foil allowed for penetration of fuel masses up to the equivalent of 217 pins of fuel and even allowed for the computed tomographies based on the foil-based images to be performed.

Alternatively to films, dysprosium-based imaging plates [5] utilizing relatively long self-exposure by radiation from the intrinsically neutron-activated dysprosium (namely ^{165}Dy isotope decay) were introduced later. The operating principle of this type of neutron imaging of the active samples is based on a sample transport container that is placed on a dedicated shielding block into which the samples are lowered from the bespoke container. Recently, investigations into the possibility of performing direct digital neutron imaging have been performed either using micro-channel plate (MCP) detectors [6] or dysprosium-based scintillator screens [7].

4.1.1 Fresh fuel elements

Fresh fuel elements and mock-ups can be investigated using standard attenuation neutron imaging because their inherent activity level is still low. Neutron radiography and tomography are utilized for investigations of structural integrity of the pellets and serve as quality control of the fabricated fuels. Likewise, homogeneity of the burnable poisons (e.g. Gd) [8] that are loaded directly into the fuel pellets and ingress of water into the fuel pin can be well revealed by neutron radiography. The capability of neutron imaging to discriminate some adjacent high-Z elements proves very useful for discerning of plutonium oxide agglomerates as inclusions in PuO_2–UO_2 mixed oxide fuel (MOX) pellets [9] (see figure 4.1).

Thanks to relatively low neutron cross-section for natural and low enriched uranium, the internal structure of even relatively large objects can be investigated non-destructively. An example of such structure is shown in figure 4.2 that depicts a 6 cm in diameter fuel pebble consisting of many thousands of small 0.5 mm diameter low enriched uranium oxide fuel particles with a TRISO coating, embedded in a graphitic matrix.

4.1.2 Post-irradiation examinations

Neutron imaging can well be considered a very useful tool for the post-irradiation examination technique of nuclear fuel and examination of other highly radioactive objects. Material alterations from the burnup process and the fuel-cladding

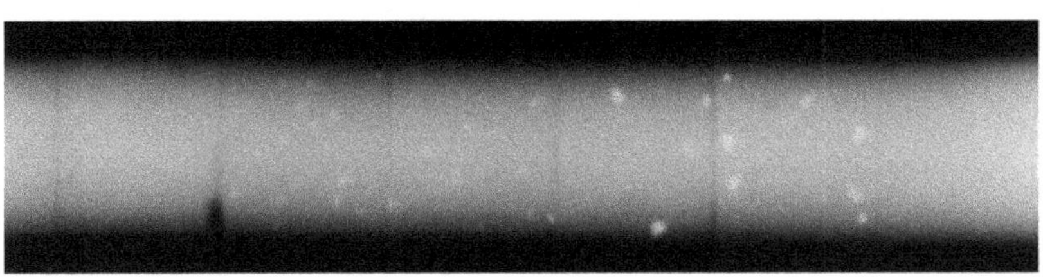

Figure 4.1. Neutron radiography showing the plutonium oxide agglomerates (white spots) inside mixed oxide UO_2–PuO_2 pellets. Reprinted with permission from [9], copyright IAEA.

Figure 4.2. (left) 6-cm diameter fuel element from the High Temperature Reactor (HTR) containing approximately 8500 individual TRISO fuel pebbles of 0.5 mm in diameter. Reprinted from [8], copyright (2003), with permission from Elsevier.

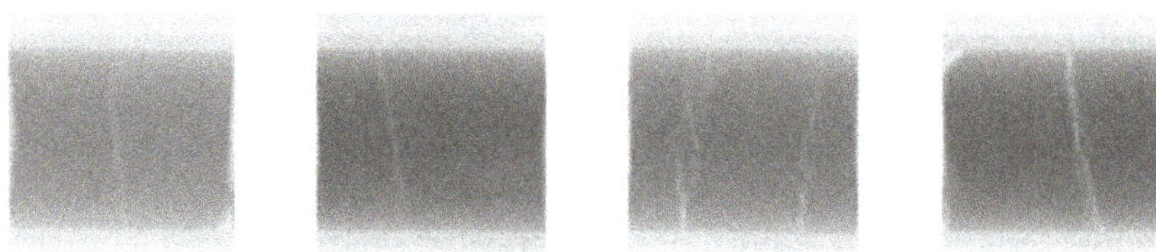

Figure 4.3. Cracks in fuel pellets. The pellets are enclosed in zirconium-based tubes of approximately 10 mm outer diameter (bottom). Reprinted from [11] CC BY 4.0.

interaction behaviour can be thus well investigated non-destructively. However, the number of facilities in which such investigations can be performed is rather limited.

At the Paul Scherrer Institute (PSI), the highly radioactive samples can be investigated by the NEURAP insert at the NEUTRA beamline [10]. Figure 4.3 shows several fuel elements after their operation in a Swiss nuclear power plant and exhibiting cracks.

Apart from the spent nuclear fuel, NEURAP insert is frequently used for investigation of the PSI's lead-filled Zircaloy rods that form the PSI's neutron spallation target. The investigations of these rods (both in radiographic [12] and in tomographic modes [13]) are proving very useful for the assessment of the safety and efficiency of the neutron target.

In the Halden reactor project in Norway, neutron radiography was established to be a very useful tool for the examination of nuclear fuels irradiated in the loss-of-coolant accident (LOCA) conditions and fuel degradation experiments. Jenssen *et al* [14] performed radiographies and SART tomographies based on a low number of projection of the irradiated fuel rods.

In particular, hydrogen uptake of the fuel cladding (including blistering), fuel fragmentation (see figure 4.4), and other features such as particle size distributions obtained from neutron tomography data were quite relevant for reactor core safety study.

A set-up similar to the one in PSI NEURAP has been utilized for investigation of failed fuel rods [15] at the CIRUS reactor at Bhabha Atomic Research Centre in

Figure 4.4. Neutron radiographs of nuclear fuel rod (in six angular orientations) irradiated in the loss-of-coolant series showing clear fuel fragmentation. Reprinted from [14], copyright (2014), with permission from Elsevier.

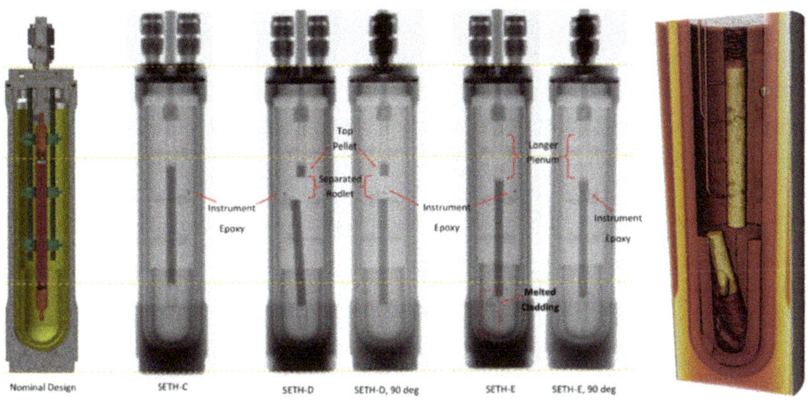

Figure 4.5. (Left) Post Transient Neutron Radiographs of Separate Effects Test Holder (SETH) capsule for the neutron irradiation test at TREAT research reactor at INL, reprinted from [18], copyright (2020), with permission from Elsevier; (right) 3D rendering of the neutron tomography of SETH capsule clearly showing the broken fuel rod and blistering and relocation of the cladding, reproduced from [19] CC BY 4.0.

India. The very advanced PIE studies can be performed in the United States, namely at Idaho National Laboratory (INL) and Los Alamos National Laboratory (LANL). The early pioneering work (e.g. [16]) using NRAD (250kW TRIGA) reactor in 1980 has been recently rejuvenated [17]. The NRAD reactor specifically installed for neutron radiography of irradiated nuclear fuels and materials benefits from its close proximity to TREAT (19 GW transient research reactor), and from its location underneath the main hot cell of INL's Hot Fuel Examination Facility (HFEF), with a direct transfer system from the HFEF to NRAD.

Both TREAT and NRAD reactors offer neutron radiography investigations for fuel fabrication; fuel characterization; fuel recycling and nuclear material management; reactor materials characterization, transient irradiation testing fuel. Separate Effects Test Holder (SETH) capsules for neutron irradiation test were subjected to Gaussian shaped reactivity-initiated accident type transient and were afterwards imaged first at the TREAT reactor in the post-transient state [18] (see figure 4.5 left).

Later, the capsule was transferred to NRAD reactor for neutron tomographic investigation that revealed only the fuel rod breakage and blistering/relocation of the cladding [19].

At LANL, LANSCE spallation neutron source provides a broad range of advanced PIE modalities for the nuclear fuel [20]. Energy-resolved neutron imaging (applications are discussed in detail in section 4.1.3) provides a particularly interesting option for the assessment of nuclear fuel samples. Naturally, the transport of highly radioactive samples poses a rather difficult problem for any nuclear fuel investigations. Therefore, very interesting near future synergies reside in the current development of the transport cask SHERMAN (Vogel *et al* [21]) that will allow for the transport of the nuclear fuel between INL and LANL facilities.

4.1.3 Fuel enrichment and isotopic distribution

As the attenuation coefficient of the uranium isotopes, ^{235}U and ^{238}U, differ significantly (700 and 12 barn, respectively), the fuel enrichment, i.e. the ^{235}U content can be derived non-invasively. Figure 4.6 presents the enrichment level of fresh uranium oxide fuel pellets that were investigated at PSI by Lehmann *et al* [8].

Noteworthy advances regarding the fuel enrichment and isotopic/elemental distributions were recently achieved using energy-resolved neutron imaging. Tremsin *et al* [22] investigated a set of mock-up rodlets containing pellets of depleted uranium dioxide (dUO_2) with defects similar to that seen in irradiated fuel rodlets using an MCP detector. By summing up the images around the tungsten resonance energies (17.57–19.74 eV), the tungsten inclusions in the investigated pellet assemblies were clearly resolved.

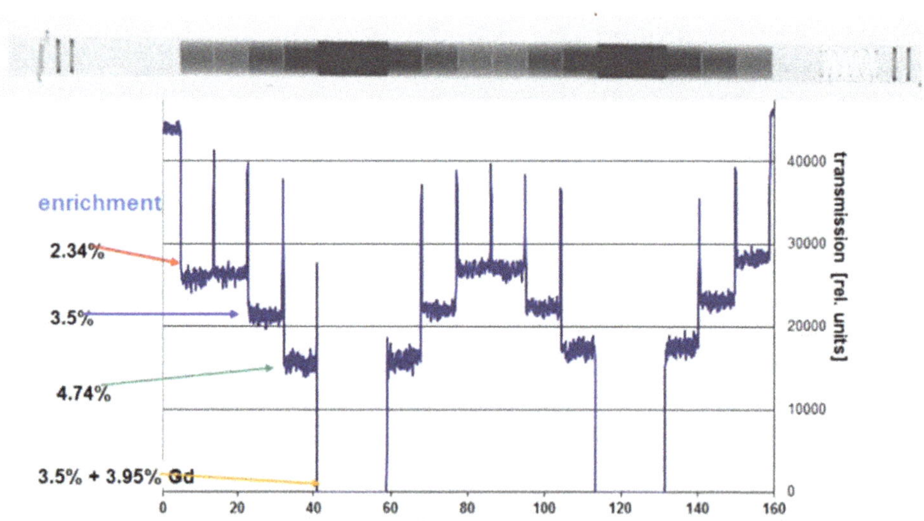

Figure 4.6. Transmission contrast and thus derived enrichment along a fuel pin also containing two sections with gadolinium as a burnable poison. Reprinted from [8], copyright (2003), with permission from Elsevier.

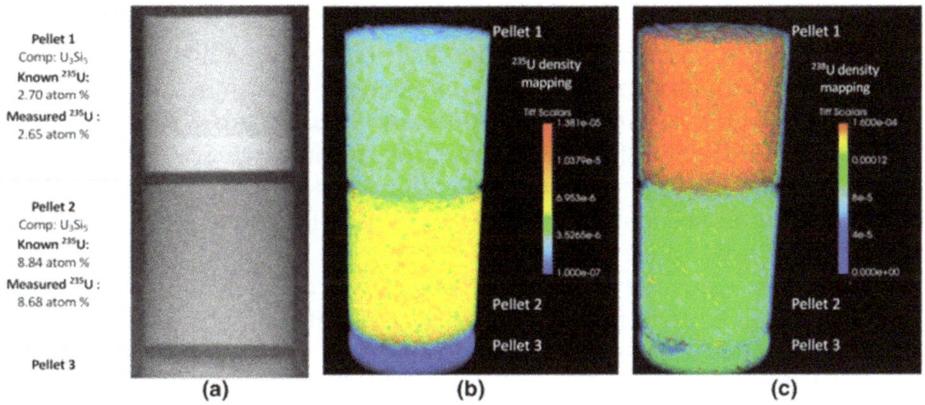

Figure 4.7. (a) Thermal neutron radiography and reconstructions of ^{235}U and ^{238}U isotope densities of in UN/U_3Si_5 and U_3Si_5 pellets. Reprinted from [20] CC BY 4.0.

The enrichment level can be very precisely investigated at pulsed spallation sources using neutron resonance transmission analysis. Figure 4.7 presents 3D isotope density maps of UN/U-Si composite fuels that were investigated by Nelson *et al* [20], who showed that the average enrichment levels measured by neutron resonance transmission analysis in this way were within 0.1% of the nominal enrichment levels.

Even more recently, Losko and Vogel [23] utilized energy-resolved neutron imaging for characterization of 3D isotopic distribution in a U-20/Pu-10/Zr-3/Np-2/Am sample of an alloy researched for transmutation nuclear fuels and compared their results with mass spectrometry.

4.2 Structural materials for nuclear applications

4.2.1 Zirconium alloys

Zirconium alloys are used worldwide in light water reactors as fuel cladding (in this chapter they are called Zr-claddings) due to their very low neutron capture cross-section, excellent mechanical stability, and superior corrosion resistance in the nuclear reactor environment [24, 25]. In particular, Zircaloy-2 (1.5%–1.7%Sn, 0.07%–0.20%Fe, 0.05%–0.15%Cr, 0.03%–0.08%Ni) and Zircaloy-4 (1.5%–1.7%Sn, 0.18%–0.24%Fe, 0.07%–0.13%Cr) are the most common zirconium alloys used for the fabrication of fuel cladding rods, with tin and iron as the main alloying elements. Zr–Nb (0%–0.99%Sn, 0.11%Fe, 0.98%Ni) alloys are used in fuel cladding rods, as well as in the pressure tubes used in pressurized heavy water reactors.

In those materials, oxidation and hydrogen embrittlement-related degradation are well-known and inevitable problems during the in-reactor time in nuclear power plants and during the storage period after in-reactor service [26]. During in-reactor operation, oxidation happens at the surface of cladding with coolant water, and hydrogen, as a side product, is released according to the reaction: $Zr + 2H_2O \rightarrow ZrO_2 + 4H$. Zirconium metal absorbs a part of the hydrogen [24, 27]. The absorbed hydrogen degrades the mechanical properties (e.g. the reduction of ductility) and may become safety-significant for Zr-claddings or pressure tubes. Besides the total amount of hydrogen absorbed by the component, the zirconium hydride orientation and its distribution also influence the mechanical performance (e.g. toughness) of Zr-cladding [28]. Specifically, radially oriented hydrides are more prone to reduce the mechanical integrity of cladding material as they act like an ideal crack growth path along the cross-section. That increases the risk of failure during the handling of Zr-claddings in dry storage conditions [27, 29].

In conclusion, the oxidation and hydrogen uptake of Zr-claddings are considered some of the main limiting factors of the fuel rod's lifetime [30, 31]. Therefore, the understanding of hydrogen accumulation, hydride reorientation and distribution along the cladding cross-section as a function of hydrogen composition, temperature gradient, stress load, microstructure, and radiation damage is an important research question for regulators, manufacturers, and academia.

Using the advantage of large neutron attenuation difference between hydrogen and zirconium, makes neutron imaging a powerful tool for the investigation of the locally resolved hydrogen concentrations in Zr alloys (nuclear fuel claddings and pressure tubes). In this respect, many studies seeking an answer to hydrogen/hydride-related degradation questions in Zr have been conducted using neutron imaging techniques to quantify hydrogen concentrations and their spatial distributions in Zr-claddings and pressure tubes [32–34].

The early attempts of using neutron imaging as a tool for the investigation of hydrogen in zirconium claddings go back to 2002. Yasuda and collaborators [32, 35] used neutron-sensitive imaging plates (NIP). These studies showed that neutron imaging can be a tool to investigate quantitatively the hydrogen distribution in the zirconium claddings. They claimed a linear relation of attenuation contrast with hydrogen concentration in the Zr metal in reasonable agreement with a theoretical formula although the NIP method was not adequate yet examining realistic hydrogen concentration in the Zr matrix, where the spatial resolution and quantitative precision were not sufficient, 0.1×0.1 mm^2 and 1000 wppm, respectively. In similar years, Lehmann *et al* [8, 34] reported the uses of neutron imaging in the case of H investigation in Zr. Unlike Yasuda *et al* [32, 35], they used a CCD-camera set-up and claimed it is possible to measure hydrogen concentration of 100 wppm for 2 mm sample thickness by the achieved signal-to-noise behaviour of the detector. Moreover, they also claimed both thermal and cold neutrons give high contrast for small amounts of hydrogen detection. Another early investigation on hydrogen in zirconium fuel cladding (Zr–1%Nb) was conducted by Svab *et al* [33] by using neutron diffraction for crystallographic characterization of hydrides and hydrogen distribution. They also claimed a linear relation between nominal hydrogen concentration and the grey levels of the neutron images with a minimum estimation limit of 273 wppm as well as the capability of H-quantification and spatial resolution is quite limited. Grosse *et al* [36] focussed on optimization of neutron radiography for the investigation of the nuclear fuel and control rod cladding behaviour during steam oxidation under severe nuclear accident conditions. They quantified the amount of the absorbed hydrogen and determined oxide layer thickness in their early studies contrary to Yasuda *et al* [32]. A relation for a quantitative determination of the hydrogen content of the different phases in steam-oxidized zircaloy is presented which is related to the total cross-section measured depending linearly on the oxide layer thickness [36].

4.2.2 Oxidation—hydrogen uptake mechanism

Neutron imaging makes the hydrogen diffusion during the oxidation of Zr-claddings. Hydrogen is released during the cladding oxidation and a fraction of hydrogen diffuses into the cladding. Oxidation and hydrogen uptake events take place together and thus understanding the hydrogen uptake mechanism during oxidation is a high consideration for nuclear safety such as the loss-of-coolant accidents (LOCA) and severe fuel damage (SFD). In the case of a loss of coolant scenario, the reactor core temperature dramatically increases which leads to overheating of the fuel roads. The primary accident management measure in such cases is reflooding the reactor with water. However, this enhances steam generation and, as a consequence, the fuel claddings are subjected to high-temperature oxidation in a steam-containing atmosphere [37].

Grosse *et al* [37–40] investigated hydrogen diffusion behaviour with various commercial zirconium-based fuel claddings (Zircaloy-4, DUPLEX (DX-D4) are Sn rich and E110, M5 are Nb rich) under different oxidation atmosphere, temperature and quenching regimes simulating LOCA conditions. They focussed on the hydrogen diffusion kinetic and quantified diffused hydrogen by using neutron imaging. These studies provided experimental knowledge for the already known hydrogen absorption theory thanks to the advantages of neutron imaging [37]. Their investigations indicate that the hydrogen diffusion kinetic not only depends on time and temperature, but also the oxide and metal morphology and local hydrogen concentration play a significant role [37–40].

Hydrogen uptake occurs only in the initial phase controlled by hydrogen diffusion through the growing oxide layer in the temperature range of 1000 °C–1400 °C where state hydrogen release is determined by decreasing hydrogen partial pressure in the environmental gas phase and thus the hydrogen concentration in the metal depends on time [37]. The transition time from the initial phase to the equilibrium phase depends on the chemical composition of Zr-alloy (whether it is Zr–Sn or Zr–Nb) and also changes with temperature. Zr–Sn alloys reach the equilibrium phase earlier than Zr–Nb alloys at the temperature range of 1000 °C–1400 °C [37]. On the other hand, an instant and significant hydrogen peak appear as a consequence of a large amount of hydrogen absorbed by the metal that is released at once produced by the strong oxidation. The local gas conditions at the metal or oxide surface are the driving factor for the hydrogen uptake of the zirconium alloy, which is strongly related to the macroscopic oxide appearance. However, this important finding had never been taken into account in computer codes simulating severe accidents previously. Strong differences from an Arrhenius-like temperature dependence were found for Zr–Nb alloy, E110, used in a large-scale bundle simulation test [38].

Radial cracks in the oxide layer act like a 'hydrogen pump' as they are good diffusion paths for hydrogen and, as a consequence, hydrogen enrichment close to the metal surface occurs. The investigations provide good evidence that the temperature and time data are insufficient alone to estimate the hydrogen uptake of zirconium alloys during steam oxidation. Additionally, the monoclinic—tetragonal phase transition in the oxide should be taken into account for the prediction of hydrogen absorption and release during LOCA and severe accident scenarios [39].

4.2.3 Kinetics of hydrogen uptake and diffusion, *in situ* investigations

As previously explained, the hydrogen uptake into Zr metal happens only in the initial phase of the oxidation. Thanks to the non-destructive and sufficiently short acquisition time requirement of neutron imaging, an *in situ* investigation of hydrogen uptake during oxidation with a spatial resolution of approx. 30 μm and time resolution up to 7.8 s was possible [41]. Grosse *et al* [41] developed a high temperature vertical reaction tube furnace (INRRO) which has transparent windows for entry and exit of neutron beam through the furnace. In their attempt [42], the kinetics of the hydrogen uptake during steam oxidation and the kinetics of the hydrogen diffusion was obtained for the first time.

The activation energy for the diffusion of the hydrogen in Zircaloy-4 was determined as 55.4 kJ mol^{-1} K^{-1} in the steam atmosphere. Rapid hydrogen absorption takes place as long as the oxide layer does not cover the metallic surface of the cladding tube. However, the relation of $C_H^{(m)}$ and E_{total} was determined by isothermal experiments at various temperatures and hydrogen concentrations in the furnaces gas. A calibration curve from the relation of hydrogen content and total macroscopic neutron cross-section cannot be determined using hydrogen preloaded specimens due to the fact that hydrogen is very volatile and would be released at elevated temperatures (1123 and 1623 K in [42]). Moreover, a standard calibration curve may not be accurate for high temperatures as hydrogen is precipitated into hydrides at room temperature, whereas it is dissolved in β-Zr phase at temperatures of about 800 K and above [41, 42].

In conclusion, however, no significant temperature dependence of the total microscopic cross-sections of hydrogen and oxygen was found as long as the zirconium and the oxide layer do not change their structures. Grosse *et al* [43] determined activation energy of 48.3 kJ mol^{-1} K^{-1} at the temperature in which the β-Zr is stable for Zircaloy-4 in the steam atmosphere. Moreover, the hydrogen diffusion at the α- to β-Zr transition temperature (823 and 1123 K) is dependent on the distribution of α- to β-Zr percentage, since the α-phase can absorb less hydrogen than the β-Zr (figure 4.8). In addition, it can be a case of more hydrogen absorption in the β-Zr region than the former α-phase as not only the hydrogen in the gas phase diffuses into β-phase but also hydrogen from the neighbouring α-Zr locations diffuses into the transformed regions. However, it should be mentioned that the absorption of hydrogen extends the temperature range at which the β-Zr is stable to lower temperatures.

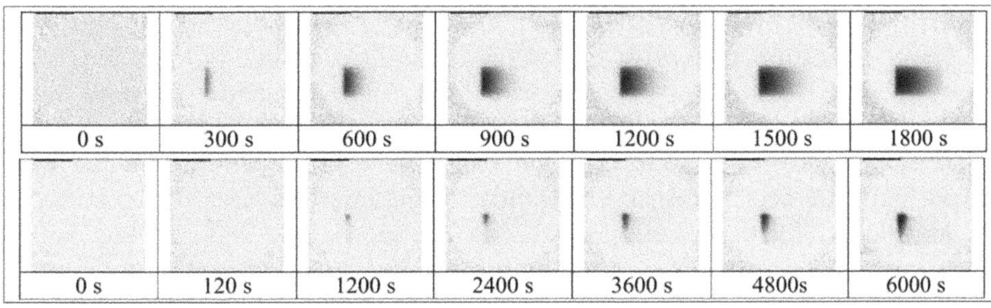

Figure 4.8. Neutron radiography image sequence of the hydrogen diffusion into zircaloy-4 at 1273 K(top) and 823 K (bottom) measured at the ICON facility (SINQ, PSI), LOCA. Reproduced from [43], copyright IOP Publishing Ltd.

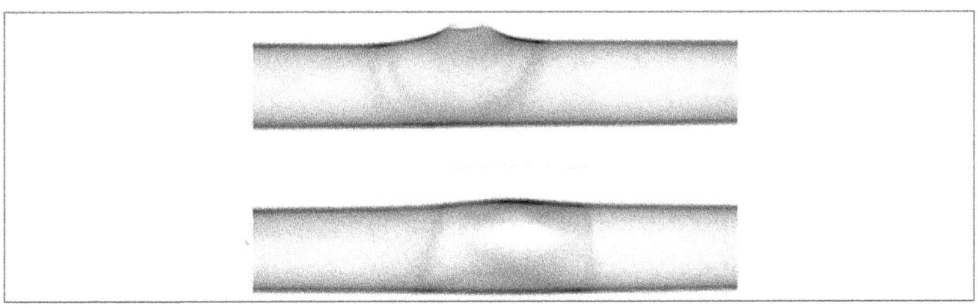

Figure 4.9. Neutron radiographs of the burst zone of the fuel rod simulator cladding measured at ICON making the hydrogen distribution visible [60] (2015), reprinted by permission of the publisher (Taylor & Francis Ltd, http://www.tandfonline.com).

Another subject of investigation using neutron imaging on hydrogen behaviour in Zr-cladding is the LOCA scenarios. In the case of LOCA, the pressure equality is broken down between the interior and exterior of Zr-cladding resulting in higher pressure inside the cladding. As a result, the gaseous fission products cause the ballooning phenomenon which ends up in the failure of the cladding rod. During LOCA conditions, the cladding tube is oxidized more than a standard operating condition due to the increased temperature and, as a consequence, intensive hydrogen diffusion occurs which reduces the toughness and thermo-shock resistance of the cladding. Therefore, understanding hydride distribution and concentration during LOCA is important to improve the safety of fuel rods.

For the first time, the hydrogen bands surrounding the area of the burst opening were imaged by Grosse *et al* [55] using the neutron imaging method. The formation of the bands is related to the time between burst and quenching ΔT_{b-q} and the temperatures during this time (figure 4.9). They also used neutron imaging to validate the QUENCH-LOCA simulation model [56], which is developed to understand the band-shaped hydrogen enrichment at both sides of the ballooning area (figure 4.9). The neutron image results prove that the model has to be improved in a way where also the material parameters are considered such as the oxidation kinetics, the temperature of the monoclinic-to-tetragonal phase transition in the zirconium oxide, and the yield strength and plasticity at high temperatures.

The studies [52] and [56] showed that neutron imaging is a special method for not only the investigation of hydrogen behaviour in Zr-cladding systems but also an efficient method to validate and improve the computer codes used in nuclear application due to the possibility of spatially resolved quantification of hydrogen in Zr-claddings. Moreover, Grosse *et al* concluded that depending on the LOCA scenario, band-shaped hydrogen enrichment in the interface area of the ballooning zone and the oxidation of the inner tube surface can be possible above a threshold temperature. The temperature of the monoclinic to the tetragonal transition of the zirconium oxide may determine the threshold temperature as the hydrogen diffusion kinetics changes by the change of the crystalline structure of the oxide [57, 58]. The alloying composition of cladding and the initial hydrogen concentration in the matrix have no significant influence on the secondary hydrogenation during LOCA [57].

4.2.4 High-resolution neutron imaging, HR-NI

High mobility of hydrogen atoms and thus their non-uniform distribution in the zirconium matrix mostly causes brittle hydride regions which can be detrimental to the fuel rod integrity. Therefore, the assessment of hydrogen migration behaviour is of great importance when considering its potential detrimental effects during spent fuel storage conditions [44].

As discussed in the previous sections, hydrogen diffusion and quantification have been investigated with several neutron imaging techniques, achieving a spatial resolution on the order of about 0.1 mm. However, the achieved spatial resolution had not been sufficient to provide a quantitative spatially resolved assessment of the hydrogen/hydride accumulation in Zr-claddings. The achievement in the spatial resolution with the PSI neutron microscope detector [45] enables much higher resolution in neutron imaging—so-called high-resolution neutron imaging (HR-NI), investigations of hydride distribution in cladding cross-sections. The detector has a 5.4 μm spatial resolution which could achieve an effective spatial resolution of 9.6 μm for the Zr-cladding samples—for the length of 4.5 mm samples—used in the study of Gong *et al* [44]. The effective resolution is lower than the inherent spatial resolution of the detector itself by the fact that the increasing sample-scintillator distance with the sample thickness gives limitations due to the beam divergence. However, the achieved effective spatial resolution is sufficient to image hydride packets in the cladding cross-section. Several studies thus have attempted to investigate the quantitative spatially resolved assessment of the hydrogen/hydrides redistribution induced by diffusion-precipitation in commercially used PWR claddings (DXD4) and BWR claddings (LTP, LTP2, LK3/L) by Gong *et al* [44] and Duarte *et al* [46], respectively.

The study of Gong *et al* [44] showed that the liner of the duplex claddings acts like a hydride sink and thus the hydrogen depletion over the entire substrate and consequent accumulation of hydrogen in the liner becomes more significant for the slowest cooling rates (figure 4.10). This result was supported by the quantitative hydrogen measurement with neutron imaging that showed that the hydrogen accumulation in the liner near the interface can reach up to 1500 wppm in samples

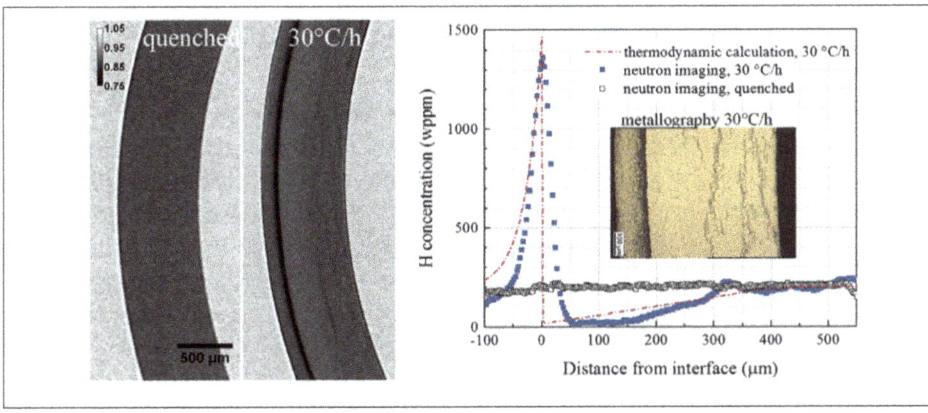

Figure 4.10. Neutron radiographs, radial neutron transmission profile and metallography image of the duplex cladding tubes (200 wppm hydrogen) with various cooling rates. Reprinted from [44], copyright (2019), with permission from Elsevier.

with bulk concentrations of 200 wppm depending on the cooling rate. Furthermore, the hydrogen concentration in the substrate is 20–30 wppm higher than in the liner area at equilibrium conditions. An implication of this is the possibility that the reference chemical potential is higher in the liner, which is supposed to be responsible for hydrogen redistribution in the duplex cladding.

Duarte *et al* [46] presented similar hydrogen diffusion behaviour for zircaloy-2 based alloys with liner. The inner liner influences on the radial hydrogen diffusion and precipitation were imaged by high-resolution neutron radiography in non-irradiated zircaloy-2 claddings. The findings contributed to the previous results showing that more than 99% of the hydrogen diffuses to the liner–cladding interface with the cooling rate of 10 °C h^{-1} whereas there was no considerable additional influence on the diffusion of hydrogen to the interface with slower cooling rates (for instance 0.3 °C h^{-1}). However, an increase in the amount of hydrogen was observed in the liner interior with such lower cooling rates.

4.2.5 Improvement of hydrogen quantification sensitivity

Another important achievement with high-resolution neutron imaging, in parallel to the improvement of the effective spatial resolution of 9.6 μm reported by Gong *et al* [44], is the improvement of the quantification sensitivity which was also investigated by Gong *et al* [44]. They claimed that the relationship between hydrogen composition and attenuation should be exponential according to Beer–Lambert Law. Moreover, they proposed using 1D concentration resolution instead of concentration resolution by pixel. Quantitative sensitivity for a pixel volume ($0.0027 \times 0.0027 \times 4.5$ mm^3) was calculated at ~100 wppm from typical standard deviation of 0.0089 due to the high uncertainty in the high-resolution neutron imaging. Quantitative sensitivity, however, was improved to 9 wppm for 1D concentration resolution. It refers to the radial profile and defines the hydrogen sensitivity that can be detected by an average of pixels over the radial direction and that corresponds to the radial profile. Buitrago *et al* [25] also supported the 1D concentration resolution of Gong *et al* [44]. They claimed the directional sensitivity of hydrogen detection can be remarkably improved to 5 wppm.

The sensitivity in the comparison of the two approaches is significant. However, it should be highlighted that high sensitivity can be only achievable in the case where the hydrogen concentration field of interest is often a unidimensional distribution, either due to sample geometry or symmetry of driving force, since the average operation for 1D concentration resolution suppresses the statistical uncertainty in pixels. Moreover, this is the relative resolution of hydrogen that can be detected in the same neutron image for the given volumes of material.

4.2.6 The characterization of irradiated Zr-claddings

Although a significant milestone has been achieved by neutron imaging on the understanding of hydrogen diffusion and accumulation in Zr-claddings, none of the published studies has been subjecting irradiated samples. Consequently, any of the investigations subjected to the effect of irradiation-induced damage, happen during the in-reactor service time, on hydrogen diffusion and hydride precipitation behaviour in Zr-claddings.

Imaging of an irradiated cladding was not possible due to the concern of radiation safety and contamination risk during the handling, transportation and imaging of a sample. However, the mentioned concerns were overcome by introducing a specially sealed container so-called 'Active Box' (figure 4.11—left) which allows neutron imaging investigation of highly radioactive Zr-claddings (e.g. 10 mSv h^{-1} in contact [48]). The neutron imaging studies of irradiated Zr-claddings were published by Trtik *et al* [47] and Duarte *et al* [48, 49].

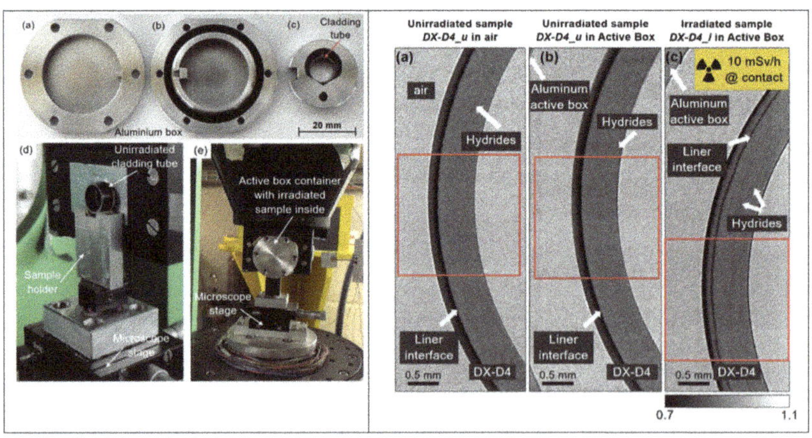

Figure 4.11. (Left) aluminum active box with a diameter of 45 mm: (a) backside, (b) front side, and (c) inside sample holder. Photograph of the NM stage with the sample holder for the unirradiated sample (d) measured in air and with the (e) active box container for irradiated samples, (right) high-resolution neutron radiographs of segments of Zr-cladding. Inactive sample imaged, irradiated sample exhibiting about 10 mSv h^{-1} dose rate at contact distance enclosed in the active box. Reprinted from [47] with the permission of AIP Publishing.

The published results are pioneers in the case of the investigation of hydrogen in-reactor irradiated Zr-claddings with neutron imaging. The key aspect of these studies, therefore, was to introduce a tool for the characterization of the irradiated samples with 'Active Box' by showing hydride distribution in a segment of various in-reactor irradiated Zr-claddings. Thus, a detailed investigation of irradiation effect on hydride distribution was beyond the scope of these studies.

The true spatial resolution of 12.8 μm for the DXD4 sample in the air was reported while it was calculated at 13.4 μm in the 'Active Box' [48, 49] (figure 4.11—right). The thin central part of the 'Active Box', where the neutron beam interacts with the sample as well, attenuates the neutron beam by only ~1%. That shows the Active Box has a negligible influence on the contrast-to-noise ratio of the resulting images of the sample and has a limited impact on true spatial resolution. The reduction in true spatial resolution is the consequence of the increase in the distance between the sample and the scintillator distance.

The samples used in these studies are extracted from zircaloy-4 [48, 49] and DXD4 [47–49] claddings (irradiated in a pressurized water reactor, PWR) to LK3/L [49] claddings (irradiated in a boiling water reactor, BWR), which are undergone in approx. five reactor cycles and their average burnup range is about 60 and 70 MWd/kgU. The packed circumferential hydrides are well visible in the neutron images of the irradiated zircaloy-4 and DX-D4 samples [49]. The hydrides' distribution in the circumferential orientation is relatively uniform along the cladding cross-section for the irradiated zircaloy-4, whereas the hydrides in the irradiated DXD4 sample can be characterized in three identical regions, which are similar to the non-irradiated cladding sample described in figure 4.10 [49, 50]. The strong preferential accumulation of circumferentially oriented hydrides is well seen and thus has strong contrast in the neutron images, whereas the possible accumulation of hydrides in the radial direction is not visible. This is probably related to the typically shorter in the axial direction morphology, with a smaller accumulation size, of the radial hydrides. The reason for the shorter in-depth radial hydrides and the indicated behaviour is probably based on the influence of the texture [49].

4.2.7 Characterization of Zr microstructure and texture with neutron imaging

Zr alloy components are employed as tubes or plates produced by conventional metallurgical procedures. Those processes, such as wire drawing, sheet rolling or tube reduction, produce several characteristic textures and residual stress distributions due to the thermo-mechanical processes involved during manufacturing. Typically, in the final piece texture, the basal planes of the hexagonal unit cells and [1010] direction align with the direction of the main deformation, while the basal poles are parallel to the direction of the effective compression. Moreover, texture gradients usually appear across the specimens due to the differences in the forces operating on the layers during these deformation processes and inevitably appear after the welding of such components, due to the temperature gradients. The resulting crystallographic texture variations and the distribution of defects of the final component influence not only the mechanical properties of the material but also the degradation mechanisms that occur in-service.

Gong *et al* [44] reported that it is possible to characterize the different microstructure of cladding material on neutron transmission. Interestingly, they observed ~0.004 lower neutron transmission for a stress-relieved structure compared to a recrystallized Zircaloy-4 sample. Moreover, the neutron transmission in the as-received substrate of the DXD4 sample was slightly lower than the hydride-depleted regions of the same cladding material where the presence of the residual hydrides should lead to a decrease in the transmission. Two factors that are expected to play a role in this transmission drop are the diffuse scattering caused by the distorted microstructure containing crystallographic defects and the Bragg-edge inflection induced by the slight change of α-Zr texture during recrystallization [44].

Santisteban *et al* [58] have investigated the crystallographic texture variation that results from typical manufacturing processes of tubes, plates and welds of Zr alloys employing energy-dispersive neutron imaging. They showed that the wavelength dependence total cross-section of Zr-2.5%Nb pressure tubes and rolled zircaloy-4 plates departs considerably from the total cross-section of an isotropic specimen and that it is possible to predict such changes through the knowledge of the orientation distribution function (ODF) of the crystallites composing the material. In particular, they proposed a parametric model of the ODF using a limited number of unimodal texture components and showed the sensibility of the wavelength-dependent total cross-section to different ODF models.

Since the texture characterization is a determinant for defining the operational performance of pressure tubes, the crystallographic texture of those components is specified by design and controlled after manufacturing, through the Kearns factors. These parameters evaluate the proportion of unit cells having their hexagonal *c*-axes projected along the principal direction of the specimens, and Zr tubes are calculated for the hoop, radial and axial directions (fH, fR and fA, respectively). Recently, Vicente and collaborators [59] showed that it is possible to obtain integral parameters of the ODF, as Kearns factors, from energy-resolved neutron transmission experiments, measuring the transmission along a reduced number of orientations. In particular, they obtained the Kearns factors for the three principal directions of a Zr-2.5%Nb pressure tube, from neutron transmission measurements performed along two different directions.

4.2.8 Effect of microstructure on the diffusion of hydrogen

The effect of zirconium microstructure, alloying composition and lattice orientation on hydrogen diffusion in zirconium lattice have been investigated by neutron imaging method which is indeed limited to investigating with other characterization methods. Different neutron attenuation values were calculated for the liner and substrate of an as-received duplex cladding (DXD4) by Gong *et al* [44]. More neutron attenuation of 0.0026 was measured for the substrate than the liner where the standard deviation was 0.0006. A possible impact of the crystallographic texture of the metal lattice on the attenuation coefficient should be also considered [25].

These investigations also support that the crystallographic orientation of the zirconium lattice has an insignificant influence of the hydrogen diffusion behaviour in the zirconium matrix as α-Zr has a weak anisotropy and β-Zr is isotropic, whereas the crystallographic orientation plays a critical role on hydrates precipitation, formation and growth. However, the elongation of the grains influences the hydrogen diffusion rate since fast diffusion mainly occurs along the β filaments or α/β grain boundaries [50]. Neutron imaging techniques made possible the experimental investigation on the influence of microstructural factors on hydrogen diffusion in Zr alloys. Santisteban *et al* [50] made several noteworthy contributions to the determination of the diffusion coefficient of hydrogen in Zr alloys. The hydrogen content profile outside-in direction was investigated for different alloys and thermal and mechanical treatments by combining microstructural characterization (figure 4.12). They claimed that a similar diffusion behaviour in zircaloy-2 plates was observed along the rolling and normal direction under different metallurgical conditions (cold rolled, recrystallized) and diffusion temperatures (250 °C, 300 °C and 350 °C) conditions since the geometry and orientation distribution of the grains are most likely similar. However, a clear divergence of diffusion rate exists along the rolling and normal direction of the hot rolled Zr-2.5%Nb plates that has α +β phase microstructure. The diffusion in the rolling direction is faster than the normal direction since hydrogen diffuses along the continuous network of β filaments.

It should be noted here that while neutron imaging is utilized predominantly for the observation of hydrogen diffusion in zirconium alloys, the diffusion of other elements of high neutron cross-section (such as erbium that can be incorporated as a

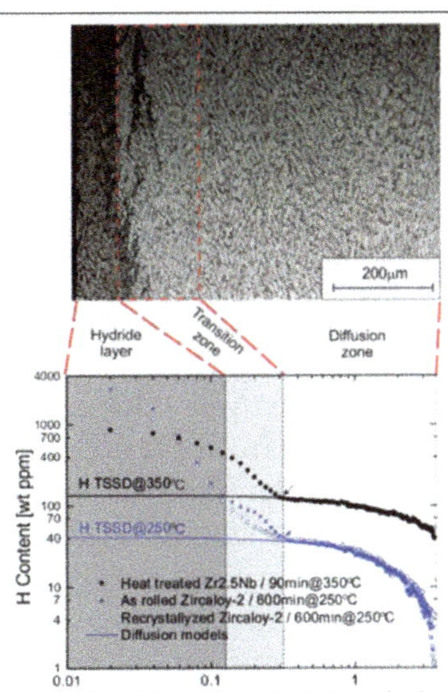

Figure 4.12. Typical H content profiles measured by neutron imaging on the zircaloy-2 and Zr-2.5 Nb, and microscopy of Zr-2.5%Nb NT specimen showing the partial decomposition of hydrides in the transition zone. Reprinted from [50], copyright (2022), with permission from Elsevier.

burnable poison in a three-layered Zr-cladding) can successfully be quantified using neutron imaging as well [51].

4.2.9 Influence of stress on hydrogen diffusion and hydrides precipitation

Zr-claddings have preferentially oriented texture, thanks to the special manufacturing process, the so-called pilgering process—it is a cold rolling process which reduces the diameter and wall thickness of the zirconium tube and leads to preferential orientation of zirconium lattice, as explained broadly in section 4.2.7. As a consequence, hydrides precipitate in the circumferential direction as hydrides preferentially precipitate in the basal plane of zirconium HCP crystal. However, during the cool-down period after in-reactor service time, hoop stress generated by the internal rod pressures may cause reorientation of the hydrides in the radial direction of the cladding. In addition to that, radiation damage may influence the diffusion behaviour of hydrogen in zirconium by increased dislocation and interstitial density. When the hydrides are reoriented in the radial direction, they act as a proper crack propagation path or they diffuse to the tip of a crack due to the change in the local stress gradient (e.g. delayed hydride cracking, DHC). The undesired reorientation of hydrides, therefore, can enhance hydride-induced failure mechanisms. The fact that the given summary of stress–hydride relation, makes understanding of hydride diffusion under the stress important to understand the mechanical integrity of Zr-claddings. Spatially resolved quantitative neutron imaging, as a sufficient tool, has been taken into account in the investigation of hydride orientation under stress conditions.

Gong *et al* [52] used HR-NI to assess the stress effect on hydrogen diffusion. The results were then correlated with finite element analysics (FEM) in order to provide a better prediction of DHC failure. They used a Zircaloy-4 sample with 600 wppm pre-charged hydrogen concentration and a notch used in order to simulate a non-uniform stress effect on hydrogen distribution. Then, a three-point bending test was applied (with the explained experimental plan in detail [52]). The aim was to investigate stress-driven hydrogen redistribution under external loading using quantitatively spatially resolved characterization, in other words neutron radiography. The transmission fluctuation of the neutron image was calculated ~0.02/100 μm at the notch vicinity, whereas the value was recorded as ~0.006/100 μm after the three-point bending test. The attenuation value returned to the same value as the pre-test sample at about 700 mm away from the notch ground. A maximum drop of 0.014 in neutron transmission was identified in the notch vicinity, equalling an elevation of ~130 wppm in hydrogen content at the distance of 100–200 um from the notch (figure 4.13). The FEM analysis was mainly based on Puls's DHC model where the effect of external stress on hydrogen solubility was neglected and hydride dissolution was not taken into account during the computation. The FEM analysis is consistent with the neutron imaging result although the peak position of hydrogen concentration curves is slightly shifted (figure 4.13). The good correlation between the neutron image and the FEM model verified the validity of the developed model.

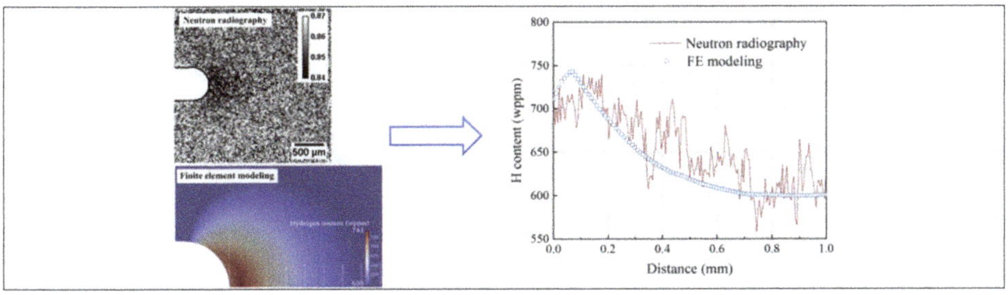

Figure 4.13. Hydrogen concentration field under stress calculated by FEM (scale bar spacing 100 mm) with the precipitation rate $a^2 = 1.2 \times 10^{-2}\text{s}^{-1}$ and the calculated hydrogen concentration profiles along the sample symmetry axis in comparison with the measurement of neutron radiography. Reprinted from [52], copyright (2018), with permission from Elsevier.

Another investigation, carried out by Gong *et al* [53], focussed on the effect of hoop stress on hydride distribution in a duplex cladding (DXD4). The study showed the efficiency of neutron imaging in terms of the investigation of hydride distribution in claddings when it is compared to the metallography since the information from metallography is limited to the imaged surface and the hydrides can be artificially affected during the etching procedure. The hydride redistribution and orientation along the circumferential direction are sensitive to the local inhomogeneous tensile and compression stress factors.

4.2.10 Delayed hydride cracking

Colldeweih *et al* [54] used HR-NI to better understand hydrogen diffusion around the crack tip under delayed hydride cracking (DHC) conditions. The results showed that the liner has a significant impact where the hydrogen concentration on the crack tip is less than without liner sample although it is dependent on the test temperature. For a cladding without a liner (LK3), the hydrogen accumulates within a crack-tip-centred small radius where the concentration was related to test temperature (figure 4.14). The hydrogen distribution, however, converges about in 120 µm radius regardless of test temperature. The test temperature has a clear influence on the concentration where the concentration is measured in the range of 1000–3300 wppm H in the first 10–20 µm radius from the crack tip and it varies about 100–500 wppm H in the converged regions depending on the test temperature. Interestingly, extreme hydrogen accumulations around the crack tip were not observed for the cladding with a liner (LK3/L) although hydrogen concentration showed a similar reducing trend away from the crack tip. The low hydrogen concentration could only be observed with HR-NI (figure 4.14). The HR-NI investigation and quantification results further support the diffusion first model (DFM) model to better explain DHC kinetics.

4.2.11 Other structural materials

There are a large number of other structural materials (e.g. stainless steel, nickel and titanium based alloys, etc) used in nuclear applications. As their application field

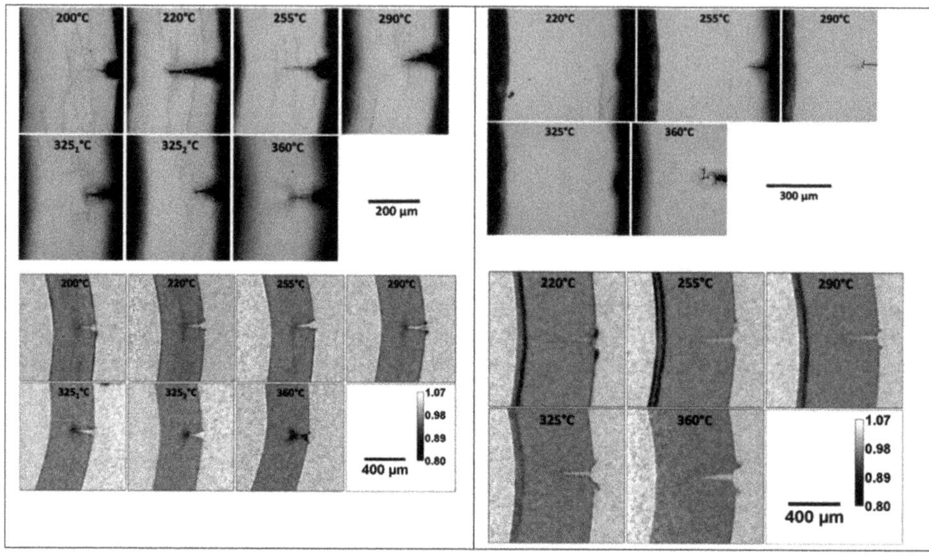

Figure 4.14. Light optical micrographs top series and high-resolution neutron image bottom series (HR-NI) of different DHC samples, (left column) the material used is zircaloy-2 (LK3) with an integral concentration of 200 wppm H, (right column) the material used is zircaloy-2 with an inner liner (LK3/L) with an integral concentration of 247 wppm H. Reprinted from [54] CC BY 4.0.

goes well beyond the nuclear applications only, the utilization of neutron imaging is covered in chapters 5 and 6.

4.3 Reactor cooling and safety

Neutron imaging offers interesting options for studying different types of two-phase flows relevant for nuclear safety and thermal hydraulics. Given the high sensitivity to even minute amounts of water, aqueous flows distributed into thin films can be conveniently investigated by cold and thermal neutrons. In particular, in the context of examining thin film or annular flows in simulated nuclear fuel bundle mock-ups, involving thick metallic casings to withstand even pressurized flow conditions, neutrons offer clear benefits. The attenuation of neutrons in metals versus water offers superior contrast for the water film compared to x-ray imaging, and is certainly the technique of choice for such problems. Furthermore, 3D neutron tomography can resolve the liquid film even on spatially very complicated, curved and entangled geometries involving multiple length scales that might not be accessible by conventional instrumentation.

One major application area of neutron imaging in this respect is examining two-phase flow in boiling water reactor (BWR) nuclear fuel bundles. Their detailed understanding is of utmost importance for the safe operation of nuclear reactors. Nuclear bundle safety focusses on optimizing the mixing and heat transfer in the bundle to safely increase margins to thermal (boiling) crisis, e.g. dry-out (see figure 4.4). Thus, it not only contributes to safe operation but also to the economy of nuclear fuel utilization. To get detailed experimental data of flow evolution and important two-phase parameters like air/water volume fraction distribution, liquid film thickness distribution etc, research is typically done on electrically heated full-sized or scaled, complete or partial bundle mock-ups. This data serves, besides better understanding of the basic two-phase flow phenomena under different operating and accident conditions, the validation of nuclear safety computational tools. Boiling two-phase flow in fuel bundles is a very complicated dynamic process, involving a multitude of spatial and temporal scales and multiple flow regimes including single-phase liquid at the bundle bottom all the way to annular and droplet flows at its top (see figure 4.15). The first neutron imaging studies done in the late 1990s and early 2000s aimed at the visualization of the flow at different parts of the bundle with different flow regimes [62–64]. Later more quantitative measurement of the gas volume fraction and including also novel fuel bundle lattice geometries have been performed [65–67]. A series of cold neutron tomographic studies from around 2010 focussed specifically on the problem of high-quality boiling crises, i.e. film dry-out of annular flow in the head of boiling water reactor fuels and the influence of different types of generic functional spacer design on these phenomena [68–73]. In a broader industrial context, annular flow is very important being one of the most frequently observed flow patterns in evaporators, heat exchangers or in many other gas–liquid two-phase systems. Annular flow is characterized by the presence of a continuous liquid film flowing on the channel wall surrounding a central gas core laden with liquid droplets entrained from the liquid film (see figure 4.15). The gas–liquid

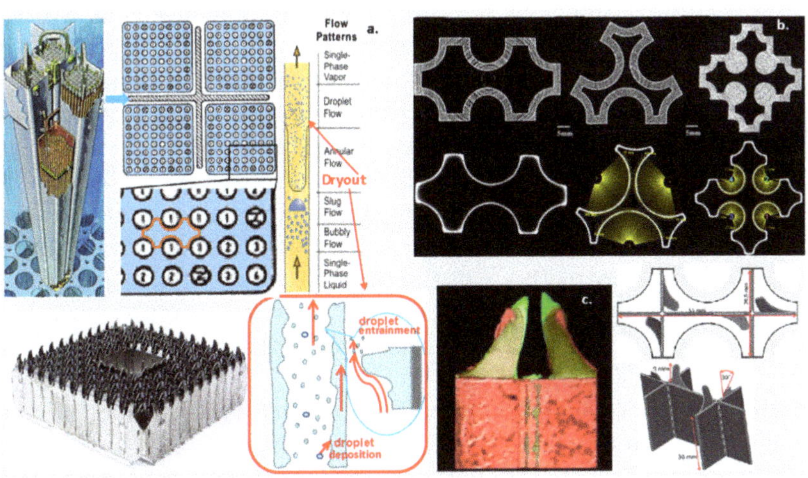

Figure 4.15. (a) Representation of the different flow regimes, dry-out, the main physical phenomena involved determining the film thickness and a photo of a typical functional spacer grid (source AREVA). (b) Cold neutron tomographic slices across different partial fuel bundle mock-ups representing a few neighbouring subchannels (see in orange shape in panel (a)) showing the empty channels (upper row) and the liquid film inside (lower row). (c) 3D rendering of the water film (green) on the two vanes of a generic model spacer and spacer body (red) obtained by cold neutron tomography. A persistent small liquid jet directed towards the liquid film is observable at the vane tips. The model spacer geometry is also shown. Reprinted from [71], copyright (2013), with permission from Elsevier.

interface is very dynamic with erratic waves travelling on it. As is shown in figure 4.15, film dry-out can occur under certain conditions at the upper part of the fuel bundle, where a huge portion of the coolant prevails in the form of droplets carried by vapour in the channel centre not contributing to the cooling of the fuel pins. Promoting droplet deposition onto the liquid film to compensate droplet entrainment from the film is therefore the preferred way in BWRs to enhance liquid film thickness (LFT) and dry-out margins. This is done by using functional spacers and spacer vanes in the fuel bundle (see figure 4.15). Spacer grids enhance the stability of the fuel pins while spacer vanes are used to influence the flow. These studies quantitatively investigated with high-resolution in 3D time-average liquid film thickness distribution in different partial subchannel geometries (see figure 4.15) [69–71]. The method has been optimized, and its performance regarding bias, statistical accuracy, upper and lower detection limits (down to ca. 25 μm) has been thoroughly analyzed using computational tools and benchmarking with other techniques [71]. The data has been extensively used to bechmark computational fluid dynamic tools [72]. Many of the studies have been carried out on adiabatic air–water two-phase flows but were later extended to boiling two-phase flows using model fluids, e.g. chloroform, in a specially designed subchannel mock-up enabling steady dry-out to occur [73].

Whereas the time-averaged two-phase flow characteristic can be thoroughly evaluated in 3D by cold neutron tomography, due to the long exposure times needed, the dynamic characteristic of the two-phase flow is not captured. The latter was, however, tested by high-speed neutron radiography typically looking at basic flow settings in simple channels [75, 76]. The developments in sCMOS detector technology allowed neutron imaging of fast processes with high temporal resolution. The negligible readout time and low readout noise of sCMOS cameras allow for the 'continuous' observation of non-cyclic processes with high-temporal resolution. Combined with the use of efficient scintillators and the high steady flux at a continuous spallation source beamline, imaging speeds up to 800 fps have been obtained [74] at an image quality that enables determining basic two-phase flow parameters like gas and liquid volume fraction, bubble sizes [75] and by applying sophisticated optical flow methods of the bubble and liquid velocity fields [76, 77]. Using the pulsing properties of a TRIGA nuclear test reactor, ultra-high speed bright-flash imaging of two-phase flows has been recently demonstrated up to 4000 fps for the first time [78, 79].

The utility of imaging with fast neutrons has been demonstrated for two-phase flow settings and regimes, where significant amounts of water have to be radiated through in the range of a few centimeters [64, 80]. Such arrangements would be opaque for cold or thermal neutrons. Dynamic two-phase flow imaging using fast neutrons has been demonstrated up to 300 fps for thick flow channels [79]. Fast neutrons represent also a promising modality to test full-sized complete fuel bundle mock-ups with flow regimes of high liquid content [80].

Another exciting area that has been studied by neutron imaging with relevance to safety of innovative types of nuclear reactors is liquid metal two-phase flows. To this end, the high penetration power of neutrons through heavy metals makes them much more applicable and beneficial compared to x-ray techniques. Such an innovative reactor type involving lead–bismuth liquid metal cooling and a nitrogen–molten lead–bismuth (Pb–Bi) mixture in a rectangular pool as a mock-up has been investigated [82, 83] and the void fraction distribution and time-averaged liquid velocity field using neutron radiography has been measured. A related research topic in nuclear severe accident safety is stem-water explosion playing a very important role in cooling and constraining molten corium. High frame rate neutron radiography was applied to study molten stainless steel particles in a heavy water bath [83].

Last but not the least, the recent work of Long *et al* [84] should be mentioned here. Molten salt reactors (MSRs) differ considerably from the conventional reactors and are from that respect closer to the above-mentioned liquid metal cooled reactors. In their work, Long *et al* proposed a new methodology and succeeded in the measurement of densities of molten salts as a function of temperature in the molten salt systems.

4.4 Nuclear waste studies

The last part of the nuclear fuel life cycle includes nuclear waste disposal. The storage of the radioactive waste is one of the fundamental aspects of the life cycle. Radioactive waste can be (and also is) produced outside the nuclear fuel industry (e.g. in medicine). High energy neutrons can be well utilized for the inspection of the large containers for radioactive waste. Fission neutron radiography [85] and tomography [86] at the NECTAR beamline of FRM-II reactor were utilized for the investigations of large mock-up waste drums.

The other aspects of the safety of nuclear waste (e.g. investigations of water/ions transport through cementitious materials and clays that for barriers for migration of radionuclides) will be treated in chapters 9 and 3.

References

[1] Nuclear Power in the World Today, https://world-nuclear.org/information-library/current-and-future-generation/nuclear-power-in-the-world-today.aspx, accessed 15th March 2024

[2] Wiesenack W 2012 *Nuclear Fuel Assembly Design and Fabrication* (Woodhead Publishing Limited)

[3] IAEA Bulletin, June 2019, https://www.iaea.org/sites/default/files/publications/magazines/bulletin/bull60-2/6020405.pdf; accessed 15th March 2024

[4] Barton J 1978 Feasibility of neutron radiography for large bundles of fast reactor fuel *Final Report for US Department of Energy*

[5] Tamaki M, Iida K, Mori N, Lehmann E H, Vontobel P and Estermann M 2005 Dy-IP characterization and its application for experimental neutron radiography tests under realistic conditions *Nucl. Instrum. Methods Phys. Res.* A **542** 320–3

[6] Tremsin A S *et al* 2019 On the possibility to investigate irradiated fuel pins non-destructively by digital neutron radiography with a neutron-sensitive microchannel plate detector with Timepix readout *Nucl. Instrum. Methods Phys. Res.* A **927** 109–18

[7] Chuirazzi W *et al* 2022 Performance testing of dysprosium-based scintillation screens and demonstration of digital transfer method neutron radiography of highly radioactive samples *Nucl. Technol.* **208** 455–67

[8] Lehmann E H, Vontobel P and Hermann A 2003 Non-destructive analysis of nuclear fuel by means of thermal and cold neutrons *Nucl. Instrum. Methods Phys. Res.* A **515** 745–59
[9] Chandrasekharan K N and Kamath H S 2007 Neutron radiography of advanced nuclear fuels *2007 Characterization and Testing of Materials for Nuclear Reactors, Proc. of a Technical Meeting Held in Vienna, 29 May–2 June 2006, IAEA TECDOC (CD-ROM) No. 1545* (Vienna: IAEA) pp 95–108
[10] Lehmann E H, Vontobel P and Wiezel L 2001 Properties of the radiography facility NEUTRA at SINQ and its potential for use as European reference facility *Nondestruct. Test. Eval.* **16** 191–202
[11] Lehmann E, Thomsen K, Strobl M, Trtik P, Bertsch J and Dai Y 2021 NEURAP—a dedicated neutron-imaging facility for highly radioactive samples *J. Imaging* **7** 57
[12] Vontobel P *et al* 2006 Post-irradiation analysis of SINQ target rods by thermal neutron radiography *J. Nucl. Mater.* **356** 162–7
[13] Trtik P *et al* 2022 Neutron tomography of a highly irradiated spallation target rod *J. Radioanal. Nucl. Chem.* **331** 5129–35
[14] Jenssen H K, Oberländer B C, Beenhouwer J D, Sijbers J and Verwerft M 2014 Neutron radiography and tomography applied to fuel degradation during ramp tests and loss of coolant accident tests in a research reactor *Prog. Nucl. Energy* **72** 55–62
[15] Singh J L *et al* 2011 Non destructive evaluation of irradiated nuclear fuel pins at cirus research reactor by neutron radiography *NDE 2011, Proc. of the National Seminar and Exhibition on Non/Destructive Evaluation* pp 8–12
[16] Betten P and Tow D M 1984 Fuel bundle distortion characterization using neutron tomography and potting *Mater. Eval.* **42** 816–21
[17] Craft A E *et al* 2015 Neutron radiography of irradiated nuclear fuel at Idaho National Laboratory *Phys. Proc.* **69** 483–90
[18] Schulthess J *et al* 2020 Non-destructive post-irradiation examination results of the first modern fueled experiments in TREAT *J. Nucl. Mater.* **541** 152442
[19] Chuirazzi W, Kane J, Craft A and Schulthess J 2022 Image fusion for neutron tomography of nuclear fuel *J. Radioanal. Nucl. Chem.* **331** 5223–9
[20] Nelson R O *et al* 2018 Neutron Imaging at LANSCE—from cold to ultrafast *J. Imaging* **4** 45
[21] Vogel S C *et al* 2022 Progress report on mockup irradiation capsule fuel measurements at LANSCE *Technical Report* LA-UR-22-30854 Los Alamos National Laboratory (LANL
[22] Tremsin A S *et al* 2013 Non-destructive studies of fuel pellets by neutron resonance absorption radiography and thermal neutron radiography *J. Nucl. Mater.* **440** 1–3
[23] Losko A S and Vogel S C 2022 3D isotope density measurements by energy-resolved neutron imaging *Sci. Rep.* **12** 1–7
[24] Simon P C A *et al* 2021 Investigation of δ zirconium hydride morphology in a single crystal using quantitative phase field simulations supported by experiments *J. Nucl. Mater.* **557** 153303
[25] Buitrago N L *et al* 2018 Determination of very low concentrations of hydrogen in zirconium alloys by neutron imaging *J. Nucl. Mater.* **503** 98–109
[26] Motta A T *et al* 2019 Hydrogen in zirconium alloys: a review *J. Nucl. Mater.* **518** 440–60
[27] Bair J, Asle Zaeem M and Tonks M 2015 A review on hydride precipitation in zirconium alloys *J. Nucl. Mater.* **466** 12–20
[28] Ells C E 1968 Hydride precipitates in zirconium alloys (a review *J. Nucl. Mater.* **28** 129–51
[29] Gojan A 2016 *Advanced Modeling of Pellet-Cladding Interaction* (Kth Royal Institute of Technology)

[30] Diniasi D, Golgovici F, Marin A H, Negrea A D, Fulger M and Demetrescu I 2021 Long-term corrosion testing of Zy-4 in a LiOH solution under high pressure and temperature conditions *Materials* **14** 4586

[31] Min S J, Kim M S and Kim K T 2013 Cooling rate- and hydrogen content-dependent hydride reorientation and mechanical property degradation of Zr–Nb alloy claddings *J. Nucl. Mater.* **441** 306–14

[32] Yasuda R, Nakata M and Matsubayashi M 2003 Application of hydrogen analysis by neutron imaging plate method to Zircaloy cladding tubes *J. Nucl. Mater.* **320** 223–30

[33] Sváb E, Mészáros G, Somogyvári Z, Balaskó M and Körösi F 2004 Neutron imaging of Zr–1%Nb fuel cladding material containing hydrogen *Appl. Radiat. Isot.* **61** 471–7

[34] Lehmann E H, Vontobel P and Kardjilov N 2004 Hydrogen distribution measurements by neutrons *Appl. Radiat. Isot.* **61** 503–9

[35] Yasuda R, Matsubayashi M, Nakata M and Harada K 2002 Application of neutron radiography for estimating concentration and distribution of hydrogen in zircaloy cladding tubes *J. Nucl. Mater.* **302** 156–64

[36] Grosse M, Lehmann E, Vontobel P and Steinbrueck M 2006 Quantitative determination of absorbed hydrogen in oxidised zircaloy by means of neutron radiography *Nucl. Instrum. Methods Phys. Res.* A **566** 739–45

[37] Grosse M, Steinbrueck M, Lehmann E and Vontobel P 2008 Kinetics of hydrogen absorption and release in zirconium alloys during steam oxidation *Oxid. Met.* **70** 149–62

[38] Grosse M, Kuehne G, Steinbrueck M, Lehmann E, Stuckert J and Vontobel P 2008 Quantification of hydrogen uptake of steam-oxidized zirconium alloys by means of neutron radiography *J. Phys.: Condens. Matter* **20** 104263

[39] Große M, Lehmann E, Steinbrück M, Kühne G and Stuckert J 2009 Influence of oxide layer morphology on hydrogen concentration in tin and niobium containing zirconium alloys after high temperature steam oxidation *J. Nucl. Mater.* **385** 339–45

[40] Steinbru M 2011 Oxidation of advanced zirconium cladding alloys in steam at temperatures in the range of 600–1200 °C *Oxid. Met.* **76** 215–32

[41] Grosse M *et al* 2010 *Ex situ* and *in situ* neutron radiography investigations of the hydrogen uptake of nuclear fuel cladding materials during steam oxidation at 1000 °C and above *MRS Proc.* **1262** 03–9

[42] Grosse M, Van Den Berg M, Goulet C, Lehmann E and Schillinger B 2011 In-situ neutron radiography investigations of hydrogen diffusion and absorption in zirconium alloys *Nucl. Instrum. Methods Phys. Res.* A **651** 253–7

[43] Grosse M, van den Berg M, Goulet C and Kaestner A 2012 In-situ investigation of hydrogen diffusion in Zircaloy-4 by means of neutron radiography *J. Phys.: Conf. Ser.* **340** 012106

[44] Gong W *et al* 2019 Hydrogen diffusion and precipitation in duplex zirconium nuclear fuel cladding quantified by high-resolution neutron imaging *J. Nucl. Mater.* **526** 151757

[45] Trtik P and Lehmann E H 2016 Progress in high-resolution neutron imaging at the Paul Scherrer Institut—the neutron microscope project *J. Phys. Conf. Ser.* **746** 012004

[46] Duarte L I, Fagnoni F, Zubler R, Gong W, Trtik P and Bertsch J 2021 Effect of the inner liner on the hydrogen distribution of zircaloy-2 nuclear fuel claddings *J. Nucl. Mater.* **557** 153284

[47] Trtik P, Zubler R, Gong W, Grabherr R, Bertsch J and Duarte L I 2020 Sample container for high-resolution neutron imaging of spent nuclear fuel cladding sections *Rev. Sci. Instrum.* **91** 056103

[48] Duarte L I, Fagnoni F, Zubler R, Gong W, Trtik P and Bertsch J 2019 Hydrides in irradiated liner cladding—local and in-depth concentration determination by neutron radiography *Top Fuel 2019* 928–32

[49] Duarte L I, Yetik O, Fagnoni F, Colldeweih A, Zubler R and Bertsch J 2021 Neutron radiography imaging: a tool for determination of hydrogen distribution in unirradiated and irradiated fuel claddings (unpublished)

[50] Santisteban J R *et al* 2022 Diffusion of H in Zircaloy-2 and Zr–2.5%Nb rolled plates between 250 °C and 350 °C by off-situ neutron imaging experiments *J. Nucl. Mater.* **561** 153547

[51] Carricondo J, Soria S R, Santisteban J R, Kardjilov N, Iribarren M and Corvalán-Moya C 2021 Analysis of erbium diffusion in zirconium-niobium alloys using neutron imaging and laser-induced breakdown spectroscopy *J. Nucl. Mater.* **549** 152869

[52] Bertsch J 2018 Hydrogen diffusion under stress in Zircaloy: high-resolution neutron radiography and finite element modeling **508** 459–64

[53] Gong W, Trtik P, Zubler R and Bertsch J 2019 Hydrides reorientation and redistribution under non-uniform stress *Top Fuel 2019* 354–9

[54] Colldeweih A W, Fagnoni F, Trtik P, Zubler R, Pouchon M A and Bertsch J 2022 Delayed hydride cracking in Zircaloy-2 with and without liner at various temperatures investigated by high-resolution neutron radiography *J. Nucl. Mater.* **561** 153549

[55] Grosse M K, Stuckert J, Steinbrück M, Kaestner A P and Hartmann S 2013 Neutron radiography and tomography investigations of the secondary hydriding of zircaloy-4 during simulated loss of coolant nuclear accidents *Phys. Proc.* **43** 294–306

[56] Grosse M *et al* 2014 Neutron imaging investigations of the secondary hydriding of nuclear fuel cladding alloys during loss of coolant accidents *Phys. Proc.* **69** 436–44 2015

[57] Grosse M, Schillinger B, Trtik P, Kardjilov N and Steinbrück M 2020 Investigation of the 3D hydrogen distribution in zirconium alloys by means of neutron tomography *Int. J. Mater. Res.* **111** 40–6

[58] Santisteban J R *et al* 2012 Texture imaging of zirconium based components by total neutron cross-section experiments *J. Nucl. Mater.* **425** 218–27

[59] Alvarez M A V, Laliena V, Malamud F, Campo J and Santisteban J 2021 A novel method to obtain integral parameters of the orientation distribution function of textured polycrystals from wavelength-resolved neutron transmission spectra *J. Appl. Crystallogr.* **54** 903–13

[60] Grosse M 2015 Quantitative analysis of hydrogen uptake, diffusion and distribution in nuclear fuel rod claddings made of zirconium alloys *Neutron News* **26** 31–3

[61] Mishima K, Hibiki T, Saito Y, Nakamura H and Matsubayashi M 1999 Review of the application of neutron radiography to thermal hydraulic research *Nucl. Instrum. Methods Phys. Res.* A **424** 66–72

[62] Matsubayashi M 1998 Three-dimensional visualization of void fraction distribution in steady two-phase flow by thermal neutron radiography *Nucl. Eng. Design* **184** 203–12

[63] Takenaka N, Asano H, Fujii T, Mizubata M and Yoshii K 1999 Application of fast neutron radiography to three-dimensional visualization of steady two-phase flow in a rod bundle *Nucl. Instrum. Methods Phys. Res.* A **424** 73–6

[64] Takenaka N and Asano H 2005 Quantitative CT-reconstruction of void fraction distributions in two-phase flow by neutron radiography *Nucl. Instrum. Methods Phys. Res.* A **542** 387–91

[65] Kureta M 2007 Development of a neutron radiography three-dimensional computed tomography system for void fraction measurement of boiling flow in tight lattice rod bundles *J. Power Energy Syst.* **1** 211–24
[66] Kureta M 2007 Experimental study of three-dimensional void fraction distribution in heated tight-lattice rod bundles using three-dimensional neutron tomography *J. Power Energy Syst.* **1** 225–38
[67] Kickhofel J L, Zboray R, Damsohn M, Kaestner A, Lehmann E H and Prasser H M 2011 Cold neutron tomography of annular coolant flow in a double subchannel model of a boiling water reactor *Nucl. Instrum. Methods Phys. Res.* A **651** 297–304
[68] Zboray R, Kickhofel J, Damsohn M and Prasser H M 2011 Cold-neutron tomography of annular flow and functional spacer performance in a model of a boiling water reactor fuel rod bundle *Nucl. Eng. Des.* **241** 3201–15
[69] Zboray R and Prasser H M 2013 Measuring liquid film thickness in annular two-phase flows by cold neutron imaging *Exp. Fluids* **54** 1596
[70] Zboray R and Prasser H M 2013 Neutron imaging of annular flows in a tight lattice fuel bundle model *Nucl. Eng. Des.* **262** 589–99
[71] Zboray R and Prasser H M 2013 Optimizing the performance of cold-neutron tomography for investigating annular flows and functional spacers in fuel rod bundles *Nucl. Eng. Des.* **260** 188–203
[72] Prasser H M *et al* 2016 Bubbly, slug, and annular two-phase flow in tight-lattice subchannels *Nucl. Eng. Technol.* **48** 847–58
[73] Zboray R, Bolesch C and Prasser H M 2018 Development of neutron and x-ray imaging techniques for nuclear fuel bundle optimization *Nucl. Eng. Des.* **336** 24–33
[74] Zboray R and Trtik P 2018 800 fps neutron radiography of air-water two-phase flow *MethodsX* **5** P96–102
[75] Zboray R and Trtik P 2019 In-depth analysis of high-speed, cold neutron imaging of air-water two-phase flows *Flow Meas. Instrum.* **66** 182–9
[76] Liu T, Zboray R, Trtik P and Wang L P 2021 Optical flow method for neutron radiography flow diagnostics *Phys. Fluids* **33** 101702
[77] Lani C and Zboray R 2020 Development of a high frame rate neutron imaging method for two-phase flows *Nucl. Instrum. Methods Phys. Res.* A **954** 161707
[78] Zboray R, Lani C and Portanova A 2020 Development of kfps bright flash neutron imaging for rapid, transient processes *Mater. Res. Proc.* **15** 154–9
[79] Zboray R, Dangendorf V, Mor I, Bromberger B and Tittelmeier K 2015 Time-resolved fast-neutron radiography of air-water two-phase flows in a rectangular channel by an improved detection system *Rev. Sci. Instrum.* **86** 075103
[80] Zboray R, Adams R, Cortesi M and Prasser H M 2014 Development of a fast neutron imaging system for investigating two-phase flows in nuclear thermal-hydraulic phenomena: a status report *Nucl. Eng. Des.* **273** 10–23
[81] Saito Y *et al* 2005 Application of high frame-rate neutron radiography to liquid-metal two-phase flow research *Nucl. Instrum. Methods Phys. Res.* A **542** 168–74
[82] Saito Y, Mishima K, Tobita Y, Suzuki T and Matsubayashi M 2005 Measurements of liquid-metal two-phase flow by using neutron radiography and electrical conductivity probe *Exp. Therm. Fluid Sci.* **29** 323–30

[83] Saito Y, Mishima K, Hibiki T, Yamamoto A, Sugimoto J and Moriyama K 1999 Application of high-frame-rate neutron radiography to steam explosion research *Nucl. Instrum. Methods Phys. Res.* A **424** 142–7

[84] Long A M *et al* 2021 Remote density measurements of molten salts via neutron radiography *J. Imaging* **7** 88

[85] Bücherl T, Kalthoff O and Von Gostomski C L 2017 A feasibility study on reactor based fission neutron radiography of 200-l waste packages *Phys. Proc.* **88** 64–72

[86] Bücherl T, von Gostomski C L and Baldauf T 2020 Fission neutron tomography of a 280-l waste package *Mater. Res. Proc.* **15** 299–304

Chapter 5

Engineering

Efthymios Polatidis, Florencia Malamud and Markus Strobl

Engineering centres around the utilization of a wide range of scientific principles and multidisciplinary knowledge with the objective to create, enhance and implement practical solutions to various technological challenges. An engineer encompasses aspects such as design, manufacturing, quality control and performance analysis to reach the project objectives. These steps in the realization of a solution, are essential to guarantee the safety, reliability and cost-efficiency of the products that are designed, produced and put into operation. Usually such investigations are not conducted necessarily experimentally, but a synergy with computational methods is used. In the latter case, however, experiments provide valuable information to validate the constitutive laws that describe the computational tools. To this end, neutron imaging for engineering applications has been one of the key fields that utilized the methodological developments in the field of neutron imaging.

Due to its electric neutrality, the neutron has a weak interaction with matter, allowing it to penetrate deeply into materials. As such, neutrons with their high penetrating capabilities are used to study the internal structure of engineering components and corresponding investigations, including materials like metals/alloys, ceramics and composites for a wide range of applications. Neutron imaging can provide information about the internal structures, composition, microstructure and defects of materials. Besides static and *ex situ* studies, neutron imaging is particularly important for engineering sciences because it can be used to study the behaviour of materials under various conditions such as magnetic, stress, strain and temperature. This information can be used for fundamental research like materials science and engineering, for optimizing materials for applications, such as aerospace, automotive, and energy industries or for quality control investigations. This information can help detect defects that could compromise the performance or reliability of an engineering component. The following paragraphs foster examples of investigations in the engineering community where neutron imaging played a key role to reach the goals.

doi:10.1088/978-0-7503-3495-2ch5

© IOP Publishing Ltd 2024

5.1 Hydrogen uptake in steels

The discovery of the detrimental effect of hydrogen in iron dates from the year 1875, when W H Johnson observed a remarkable change in the toughness and breaking-strain of iron after immersing it in hydrochloric and sulfuric acid [1]. More remarkably, it was found that this change is temporary, as the metal was able to regain its toughness and strength. This was the first evidence of hydrogen embrittlement in iron which also evidenced that hydrogen is diffusible and the degradation of the physical properties of iron is reversible. The role of nascent hydrogen, i.e. not molecular, in the embrittlement has been, since then, investigated extensively not only in iron and steel, but in many metallic alloys such as nickel, aluminum, titanium [2] or zirconium [3].

Understanding in depth the detrimental effect of hydrogen uptake by iron and steel alloys has an important technological impact and nowadays there two main theories of hydrogen embrittlement: (1) the decohesion model [4] and (2) the hydrogen enhanced local plasticity [5]. Sources of hydrogen causing embrittlement have been encountered during manufacturing by processes such as phosphating, pickling, electroplating, welding, casting, carbonizing, surface cleaning, electrochemical machining, hot rolling or heat treatments. Hydrogen can also be dissolved into the metal during service as a by-product of general corrosion or during storing liquid or gaseous H_2. Blisters and internal cracks are two of the most common types of damage caused by exposure of a material to a hydrogen-rich environment. The hydrostatic stress field at the vicinity of a propagating crack is an especially vulnerable location for hydrogen uptake. The diffusible hydrogen will accumulate at sites of increased hydrostatic stress due to dilatation of the lattice. Hydrogen also accumulates at trapping sites caused by local plastic deformation. In a straining experiment of stainless steel charged with gaseous hydrogen, an increasing amount of hydrogen in the austenitic phase in front of the crack tip was observed compared to the austenite islands in other parts of the matrix [6].

Understanding the underlying mechanisms of hydrogen embrittlement is a complex multi-faceted challenge. As such, knowing how hydrogen enters the metal, how it diffuses through the lattice or along the grain boundaries and how it interacts with or is trapped by microstructural features are important pieces of the puzzle. To unravel these processes observations the characterization tools need to cover a wide range length scales; from macro- to the nano-scopic length scales.

Neutrons have a strong interaction with the proton in the core of hydrogen, which is significantly higher than the interaction of neutrons with the isotopes of iron. This allows obtaining stronger contrast in hydrogen-enriched regions than in hydrogen-depleted regions and hence neutron imaging is ideal for mapping hydrogen-enriched regions not only qualitatively, but also quantitatively, where early efforts were undertaken for quantifying the hydrogen concentration, reaching as low as 20 ppm_H sensitivity [7]. Neutron imaging allows the radiographic and tomographic reconstruction of the hydrogen distribution in blisters or internal cracks non-destructively and it has been employed in several investigations.

Griesche *et al* [8] utilized a 10 μm thick Gd_2O_2S scintillator as light producer which was then transferred via a macro lens optic to a highly sensitive CCD camera

(ANDOR DW436N-BV with 16-bit grey value resolution). The pixel size of this set-up was ~6.4 µm reaching a spatial resolution of approximately 20–30 µm. They used electrochemically charged iron coupons and tomographically reconstructed the hydrogen distribution in the samples, as shown in figures 5.1(a)–(c). The blisters at the outer surface where clearly imaged, while the internal cracks were found to be enriched with hydrogen. The hydrogen amount in the sample was quantified by registering and substracting the tomographic volumes of hydrogen-charged and hydrogen-discharged to correct for the attenuation of iron. The grey level was then converted into linear attenuation coefficients of remaining hydrogen. The 3D distribution of the hydrogen density ρ was obtained by:

$$\rho = \sum u_{\mathrm{H}}/N_A\ \sigma_{\mathrm{total}} \tag{5.1}$$

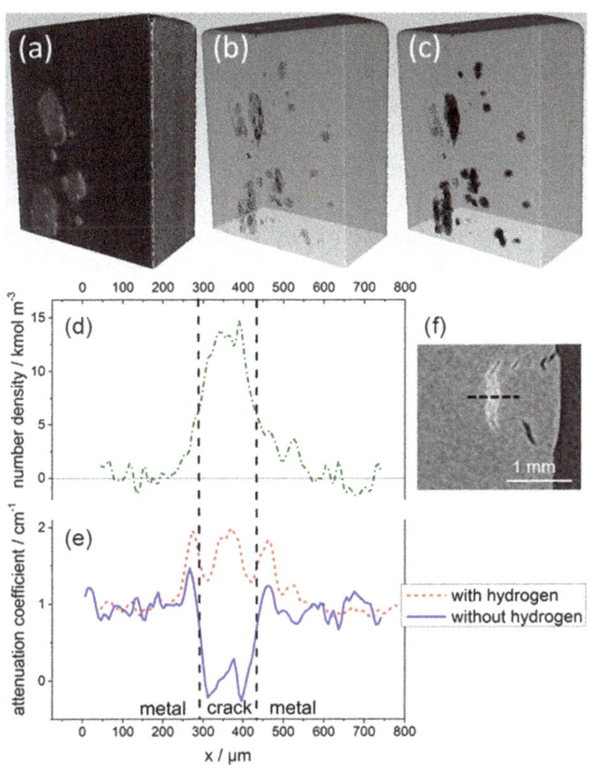

Figure 5.1. Reconstructed 3D model of a hydrogen-charged iron sample. (a) Blisters on the sample surface. (b) The crack distribution in the interior and (c) the hydrogen distribution. Most of the cracks are filled with hydrogen except some of the cracks underneath blisters [8]. (d) Profile of the hydrogen number density and (e) profile of the attenuation coefficient across a crack indicated in (f) by the dashed line in the tomographic image. The solid blue line in (e) denotes the attenuation coefficient profile across the same crack after the sample was heat treated and hydrogen diffused out. Reprinted from [8], copyright (2014), with permission from Elsevier.

where u_H is the atomic mass, N_A the Avogadro constant and σ_{total} the total cross-section of hydrogen for cold neutrons, which is experimentally determined for the imaging beamline CONRAD with 110 barns. By this, it was possible to quantify the hydrogen concentration across the cracks, as shown in figures 5.1(d)–(f).

By employing neutron tomography measurements, Griesche and collaborators also studied the 3D hydrogen distributions in supermartensitic stainless steel under tensile load [9]. The neutron tomography results of the electrochemical hydrogen-charged tensile specimen after a tensile test showed an enriched hydrogen distribution at the edges of the notch and close to the fractured surface.

5.2 Adhesive connections

Hybrid structures comprised of metallic and polymer parts have been developed by the need for lightweight, functionally optimized and aesthetically improved components in the many industrial applications, like, e.g. automotive applications. Adhesives and glues contain hydrocarbon compounds, like epoxy or polyurethane resins, and produce high neutron imaging contrast due to their high hydrogen content. Hence radiographic inspections of adhesive bonds are highly sensitive to thin layers of hydrogenous material. As such, neutron imaging is an ideal non-destructive method for imaging their distribution at the bond and for assessing the adhesion quality [10]. An example is shown in figure 5.2, where a cube is composed of 14 glued metal layers with different thicknesses [11]. The results of neutron imaging in the form of the volume rendered data are shown in figure 5.2, where the distribution of glue between the aluminum layers is imaged. The individual slices of the neutron tomography data of the adhesive joints are used to investigate the adhesive layer thickness and to assess the gluing quality. This particular tomography was reconstructed from 625 projections over an angular range of 360° with an exposure time of 25 s per projection. The resolution was approximately 150 μm/px and the field of view 150 × 150 mm^2. In applications of larger field of view, several radiographies can be stitched in order to investigate bulky objects, such an example

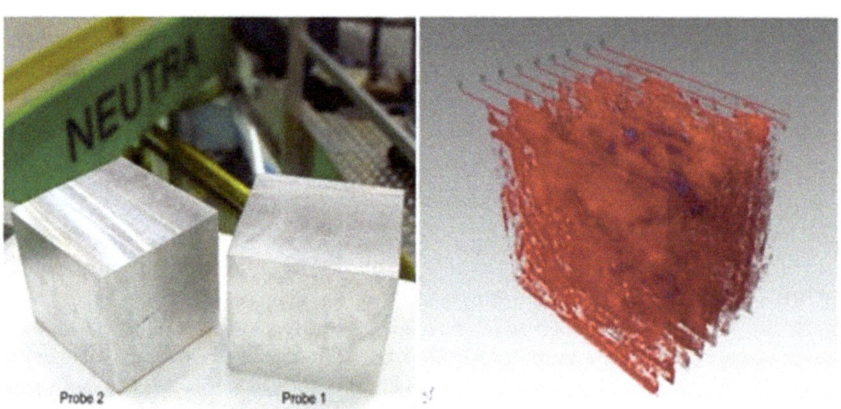

Figure 5.2. Top left: glued multilayer aluminum cube. Top right: segmented and rendered neutron tomography data showing nine interlayers of glue that bond together the aluminum plates (unpublished).

is described in [12], where the entire car door is imaged by stitching 4×4 images of 350×350 mm^2 field of view. More specific examples imaging adhesives and gluing for industrial applications are given in chapter 18.

Besides static, imaging of adhesives can be also done dynamically. In the application shown in [12], carbon fibre textile layers were put into moulds and resin was injected at different locations under high pressure. The dynamic neutron radiography measurements of resin flow through a plane model mould provided quantitative data about the resin permeabilities in different types and layer combinations of such fabrics are used as verification of models on the flow of the adhesive resins through carbon fibre fabrics [13].

5.3 Cracks in engineering components

Being able to detect cracks non-destructively as early as possible, is crucial for the structural integrity of engineering components and to prevent premature failure upon the application of load or during fatigue. The earlier a crack is detected, the smaller the crack has propagated and the less risk for catastrophic failure can occur. However, the spatial resolution of neutron attenuation contrast imaging is the bottleneck in detecting small cracks. Neutron computed tomography was used to visualize the crack formed on pre-fatigued samples and it was compared to x-ray CT showing the superiority of x-ray CT in the achieve spatial resolution [14]. The use, however, of energy-selective neutron imaging, in contrast to x-rays which exhibited uniformly attenuated x-ray beam, was successful in revealing contrast originating from the crystallographic phases of the studied duplex steel [14].

Where neutron attenuation contrast imaging fails to detect small cracks, dark-field imaging can provide superior spatial resolution and reveal cracks or defects 'unseen' by conventional attenuation contrast methods. This was demonstrated by, e.g. revealing a crack in an Al fatigue test sample as well as porosity and precipitates in different engineering material samples [15]. In a later dedicated study of fatigued 316 stainless steel samples until their half fatigue life, using Talbot-Lau interferometry crack formation was revealed [16].

5.4 Internal channels and complex structures

The operation and internal structures of several metallic components that deliver or distribute fluids, such as fuel, coolants or components that contain or distribute combustion engine products can be imaged with neutron diffraction. Insight into the 'hidden' internal channels is very important for understanding the function or degradation by, e.g. corrosion of the internal structure. An example of imaging the corrosion damage in the internal cooling channels in turbine blades made of Ni-superalloys due to corrosion, utilized Gd tagging and neutron imaging [17]. It was shown that the detection of corrosion hidden in the cooling channels of turbine blades is a useful tool for maintaining the integrity turbine blades for aerospace applications non-destructively. The non-destructive character of the method lies in the fact that neutrons penetrate deep through the structural materials, e.g. automotive components of steel or aluminum alloys, silicon carbides in catalysts or diesel particulate filters (DPFs) etc. In the latter the strong neutron contrast derives from the hydrogen that exists in the fluids or combustion products or the different elemental composition. Neutron radiography and tomography are useful for visualizing small amounts of hydrogenous fluids enclosed in dense material. For instance, a diesel injection nozzle can be imaged while in operation, here the fuel becomes visible with high contrast due to the existence of hydrogen and phenomena like cavitation can be unravelled [18, 19]. An extensive overview of imaging internal channels using neutrons for industrial applications is given in chapter 18.

5.5 Deformation tests

Conventional neutron and x-ray diffraction techniques are established and powerful tools for the non-destructive determination of internal strain and microstructure evolution during deformation tests. In both techniques, the elastic change in interplanar spacing with respect to the unloaded state, or the lattice strain, of the crystalline material is followed upon application of external load or presence of locked, in the microstructure, residual stresses. Conventional diffraction is well-suited for following the micromechanical response during loading in either tensile sample geometries, where the strain is relatively uniform in the gauge section, or in the centre of cruciform-shaped multiaxial loading samples [20, 21]. In this case the gauge volume, i.e. the volume of material that is interacting with the neutron or x-ray beam, is embedded within a volume with relatively uniform strain distribution. Neutron Bragg edge imaging has demonstrated that it can be utilized for following the lattice stain evolution during uniaxial tensile or compression deformation [22–24]. Apart from the simple case of tensile testing, when spatially-resolved lattice strain measurements are required in complex geometries, neutron Bragg edge imaging can provide unique insights into the distribution of strain and strain concentrations at weak points of the specimens. Deformation tests on notched tensile specimens, represent a case where strong strain gradients appear and the assessment of the lattice strains require high spatial resolution. Therefore, the position dependence of the elastic strain during tensile deformation can only be achieved by Bragg edge imaging [24]. In combination with the attenuation information where the crack propagation can be followed, Bragg edge imaging is a useful tool for investigating the strain along the crack path with high spatial resolution during *in situ* fatigue tests [14, 25].

Apart from strains, Bragg edge is particularly sensitive in detecting spatially-resolved changes in the Bragg edge height due to crystallographic texture variations. In polycrystalline materials also, when the grain size is comparable to the spatial resolution of the imaging resolution, it is possible to detect grains that alter their orientation upon plastic deformation. As such, a dedicated compression deformation rig has been developed and tested at the IMAT beamline at ISIS, UK on Gilsocarbon graphite under diametral compression [26]. The feasibility case study showcased the possibility to capture contrast in the transmission images when the sample is under load, which is attributed to grain reorientation, i.e. deformation texture, due to plasticity. A prime example of steep microstructural changes is bending tests, where the sign of the stress state gradually changes from tensile to compression. In such loading state, Bragg edge imaging allows characterizing the texture, crystalline size, phase volume fraction and residual elastic strain, along bent ferritic steel and duplex stainless steels [27, 28].

With the development of multiaxial loading rigs, the spatially varied strain and deformation-induced phase transformations in, e.g. transformation-induced-plasticity steels, Bragg edge imaging has been utilized to map strain concentration in 304 stainless steels [29, 30]. Besides *ex situ* studies on pre-deformed samples, a dedicated multiaxial deformation rig is available for *in situ* Bragg edge imaging investigations at the J-PARC neutron source in Japan [31]. The machine is used for biaxially testing ISO-standard specimens of thin sheets.

5.6 Unveiling phases by advanced neutron imaging

For materials science and engineering material investigations, knowing the microstructural properties of a material/component is key for the understanding of its mechanical or functional properties. Often the microstructure is not homogeneous and local variations in, e.g. grain size, crystallographic texture and phases are encountered. In particular, the presence of classical methods for characterizing the distribution of phases within a sample, usually involves viewing an image from a sectioned surface, where the area of interest is polished and viewed in an optical or scanning electron microscope. However, the spatial phase distribution and interface boundaries cannot be inferred from 2D sections accurately. Such knowledge is not only essential for understanding the material's properties, but it is very useful for validating macroscopic analytical tools which predict phase distributions within the simulated volume. Examples of spatially varying phase distributions in single components are: (i) materials which undergo complex thermal cycles, such as additive manufacturing [32]; (ii) complex stress state resulting in different extent of TRIP effects [30, 33]; and (iii) heterogeneous deformation such as the Lüders bands in superelastic NiTi alloys [34].

Similar to conventional x-ray or neutron diffraction, Bragg edges associated with the presence of certain polycrystalline phases occur at several wavelengths, and can justify the existence of a specific phase. For more details about the Bragg edge, also known as diffraction contrast imaging, readers are referred to chapter 2. Therefore, often studies utilizing Bragg edge imaging focus on qualitative phase characterization [35], rather than quantitative. With the development of quantitative investigations, it was possible to follow phase transformations *in situ*, such as austenite-to-bainite formation during isothermal annealing in grey iron [36], martensite reversion during annealing of cold-rolled austenitic steel [37] and the transformation-induced plasticity effect in duplex steels [38]. With the development of dedicated time-of-flight (ToF) instruments, phase analysis was enabled efficiently in, e.g. *in situ* investigations of the reduction of solid oxides cells [39, 40] or by utilizing Rietveld-type approaches to quantify phase fractions in mixtures of powders with different crystal structure [41]. Besides ToF instruments, monochromatic beam measurements were able to quantitatively estimate phase fractions [33, 42]. In this case the attenuation ratio between two wavelengths corresponding in the Bragg positions of austenite and martensite in a TRIP steel was utilized for quantifying the martensite fraction in a deformed sample [43]. Such an efficient measurement approach can be exploited for measuring a series of projections and for reconstructing topographically the phase distribution in three dimensions. More recent approaches were utilized by least square fitting the measured attenuation spectra to the theoretical combination of pure FCC and BCC to estimate the phase fraction in metastable austenitic steels [29, 30] or Rietveld-type analysis to quantify the fraction between fcc/bcc in steels [44–46].

Besides direct observations exploiting the crystallographic differences of successive phases with a single sample, there have been indirect observation methods which can exploit the interaction of neutrons with magnetic fields. By exploiting the ferromagnetic properties of the BCC-iron phase it was demonstrated that dark-field neutron imaging can reveal the presence of martensitic, α′phase (ferromagnetic) within an austenitic, γ phase matrix (non-magnetic) due to the difference of the magnetic scattering length density of the two phases [47]. The detection limit of ferromagnetic phase fractions is usually limited to a few percent in the latter, while a recent development by Busi *et al* [48] using polarized neutrons, has been demonstrated to be able to detect sub-percent fractions of ferromagnetic phases. The method leverages on the utilization of a polarized neutron beam in radiographic or even tomographic modes while the depolarization signal can be combined simultaneously with attenuation of the neutron beam being able to reveal defects or cavities. Each measurement only requires exposure times of only a few tens of seconds to a few minutes, making it very promising for *in situ* and operando time-resolved studies. By doing so, Busi *et al* [49] were able to map, in three dimensions, the deformation-induced martensite in a high-Mn steel lattice structure that was compressed to different extends (figure 5.3). For more details about the polarization contrast neutron imaging, readers are referred to chapters 2 and 19.

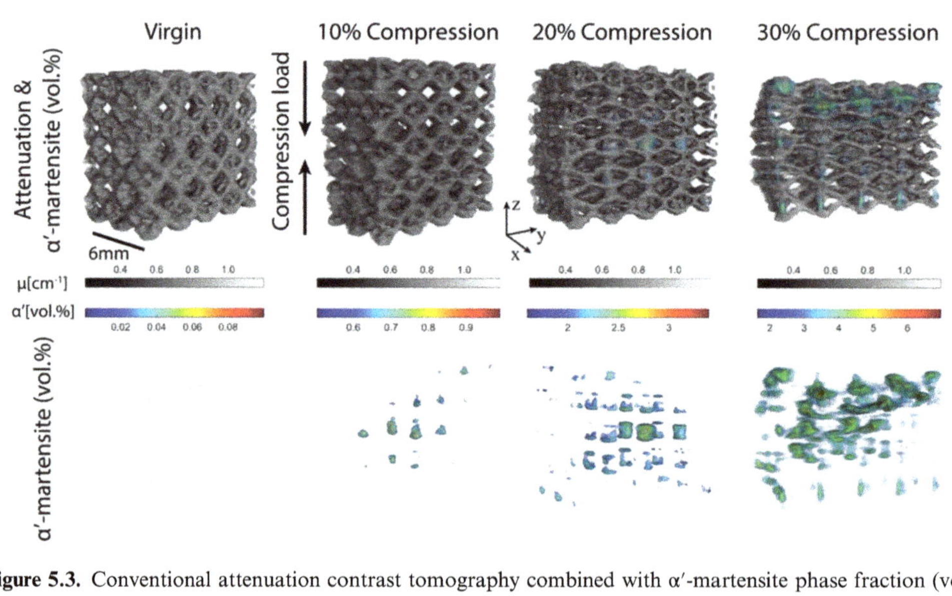

Figure 5.3. Conventional attenuation contrast tomography combined with α′-martensite phase fraction (vol. %) reconstruction of the samples in virgin state and after undergoing relative compression of 10%, 20%, and 30%. The bottom row shows the 3D phase fraction maps alone. Reproduced from [49] CC BY 4.0.

5.7 Residual strains in engineering components

Residual stresses are typically hard to predict and if undetected can add up to the external loads and lead to plastic deformation, or premature failure due to fatigue during operation. To avoid therefore such unexpected behavior of engineering materials or components, characterizing the magnitude and distribution of residual stress in engineering materials is very important. For example, residual stresses play an important role in achieving the desired properties of induction hardened gear products. Due to the selective heating induced by temperature gradients in the samples during induction hardening, the stress generation and the final stresses distributions are very complex. Su and collaborators [50] employed Bragg edge imaging to study the stress distribution and the crystallographic phases in a hardened steel gear developed by a two-step induction-heating method. The precursor sample (Sample 1) has a deep hardened layer in which martensite forms on the entire teeth, while the gear subjected to double induction hardening (Sample 2), displays fine-grained martensite at the tooth surface and a ferrite-pearlite microstructure at the core. The residual strain maps, shown in figures 5.4(c) and (d), revealed that compressive strains exist in the martensite hardened zone of both gears, while in the core near-zero residual strains are present. The rapid induction-heating and quenching process of Sample 2 produces a steep compressive strain in the fine-grained martensite at the tooth surface.

Among others, Bragg edge imaging was employed to study strain distributions in plastically bent ferritic steels and duplex stainless steel plates [28], to follow micro- and macro-strains in a plastically deformed α-iron plate during tensile tests [52], to map residual strain on an AlSiC metal matrix composite [51] and the residual strain in punched electrical steels [53]. The composite material represents an

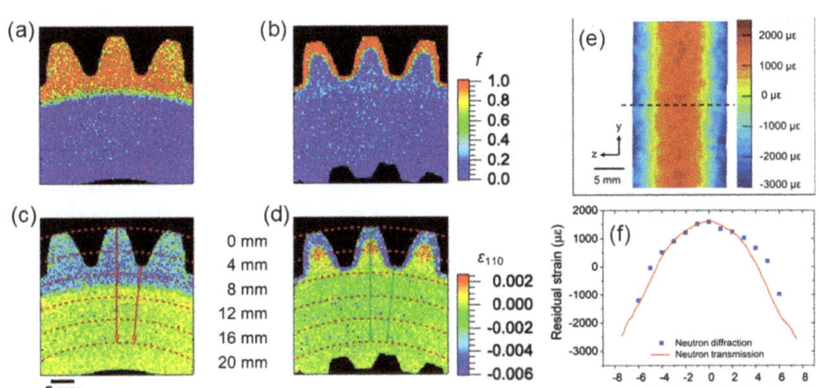

Figure 5.4. (a and b) Martensite volume fraction, *f*, and (c and d) residual strain of 110 lattice plane, ε110, where (a and c) are for Sample 1 and (b and d) for Sample 2. The pixel size is ~0.165 mm^2. Reproduced from [50] CC BY 4.0. (e) Residual strain map produced by Bragg Edge imaging, showing compressive strains at the outer surfaces of the a AlSiCp metal matrix composite and tensile strains in the middle of the sample. (f) Comparison between Bragg edge imaging (red line) and neutron diffraction (blue symbols) results. Reproduced from [51] with permission from the International Union of Crystallography.

attractive option for engineering and structural applications, and its production includes a heat-treatment process followed by cold water quenching. As shown in figure 5.3(e) and (f), the steep parabolic residual strain variation through the thickness of the plate, produced during the material manufacturing route is well captured by Bragg edge imaging with high spatial resolution. The 'ground truth' measurements from conventional neutron diffraction, with inferior spatial resolution, appear to corroborate well with the Bragg edge imaging. Bragg edge imaging has not only been used for characterizing residual strains which are locked in the body of the material, but also elastic strains that build up during engineering applications. Exploiting the high penetration of neutrons, it was possible to visualize the elastic strains that build up, *in situ*, in assemblies with torqued bolts featuring two different types of thread characteristics deep in an engineering component [54].

Besides elastic strains, i.e. type-I and type-II according to the definition by Withers and Bhadeshia [55], ToF transmission signal carries information about type-III strains, i.e. strains which vary over the atomic scale and are associated with peak broadening in conventional diffraction. In a qualitative assessment of the edge broadening, a good agreement between enhanced hardness and Bragg edge broadening on the outer surface of a heat treated and quenched ferrite/martensite was observed [45, 56–59]. A more elaborate line-broadening analysis was applied to the Bragg edges and a Williamson–Hall type of analysis was undertaken to assess the plastic deformation *in situ* during tensile tests [60].

Since the deterioration of the mechanical properties due to residual stresses is mainly attributed to the existence of strong residual tensile stresses, compressive residual strains are in several applications deliberately introduced in the components. Such beneficial application of compressive residual strains in aviation applications is the split sleeve cold expansion. Rivet joints that are used for assembling aircraft components are susceptible to fatigue failure due to stress concentrations. Hence, the strain distribution surrounding a cold expanded hole is of significant engineering interest and it can be very precisely characterized by Bragg edge imaging, as has been shown in [61].

The determination of residual strain by diffraction-based methods, whether these are conventional diffraction or Bragg edge imaging, is the determination of a reliable reference lattice spacing, deriving from a reliable determination of the Bragg edge position in a completely strain free material. The advantage of Bragg edge imaging is that it offers superior spatial resolution, compared to conventional diffraction, which is the possibility of measure reference d0 values in materials that exhibit sharp chemical gradients, which are often encountered in the engineering community such as dissimilar welds, or multi-materials. The precursor for the determination of the reference d0 value, is nevertheless, achieving complete stress relaxation, without altering the microstructure, compared to the 'strained material'. The latter is particularly important when annealing treatments are employed, leading to, besides strain relaxation, compositional changes in the solute content which affect the lattice parameter.

A common dispute in the engineering community employed in the determination of residual strains with diffraction-based methods is the choice of diffraction peaks or Bragg edges for reliable strain characterization. This raises from the accumulation of intergranular strain in specific lattice plane families which result in erroneous residual stress estimations [62]. In the case of conventionally manufactured fcc materials, the grain families {111}, {422} and {311} are considered to fulfil these requirements. The {111} lattice plane family is a popular edge as it is the strongest, in powder transmission signals due to high multiplicity, and it provides the least uncertainty in Bragg position determination and hence in the determination of residual strain.

5.8 Crystallographic texture

Crystallographic texture describes the preferential orientations of the crystallites within a polycrystalline material. When there is no preferred crystallographic texture, one refers to it as random or nearly random crystallographic texture. There are, however, cases, especially after thermomechanical treatments, where the material does not exhibit a random crystallographic texture, but a strong crystallographic texture. Meaning that the crystallites preferentially orient themselves with respect to the sample coordinate system. In such cases the strongly textured material can exhibit strong elastic and plastic mechanical anisotropy, abnormal phase transformations and slip activity or it can even affect their magnetic properties.

Energy-selective neutron imaging has been used to analyze crystallographic texture in different materials over the years [63–67]. The analysis is based on the wavelength-dependent sensibility of the neutron transmission signal of polycrystalline materials with the crystallographic texture of the samples and the orientation of the specimen in the neutron beam. The elastic coherent cross-section of a textured material departs largely from that expected, i.e. an isotropic sample, and from the changes on the Bragg edges heights it is possible to extract texture information. In particular, the strength of neutron imaging is the visualization of spatial variation of the Bragg edge heights across the samples, allowing the identification of texture inhomogeneities and *in situ* observation of those changes during processes. As an example, figure 5.5 shows the neutron radiographs of Zr-2.5%Nb pressure tube specimens measured along the transversal and the axial direction [68]. The left images were produced from all the transmitted neutrons between 1.4 and 2 Å

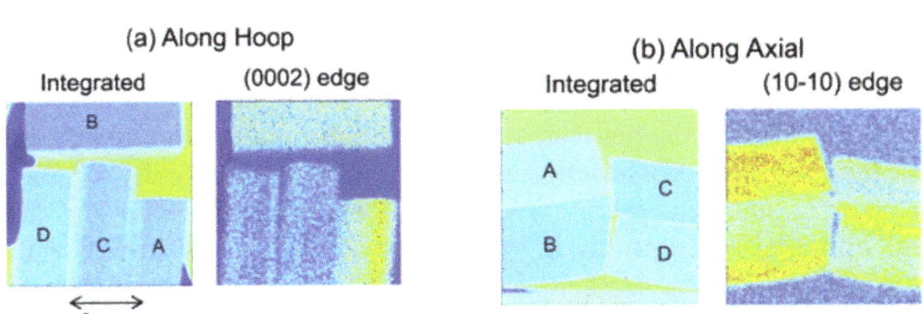

Figure 5.5. Neutron radiographs of pressure tube specimens along the radial and axial direction produced from neutrons with wavelengths between 1.4 and 2 Å (left) and using the normalized height of Bragg edges I_{hkl}. Reprinted from [68], copyright (2012), with permission from Elsevier.

(integrated) and the right images were obtained from narrow wavelength ranges near the (0002) and (10-10) Bragg edges, revealing heterogeneities of the manufacturing process of the component. Similar methodology has been recently used to study texture distributions on additively manufactured samples [69–71].

Although it is not possible to quantify a complete unknown orientation distribution function (ODF) from neutron imaging experiment, for uniformed texture, it is possible to calculate the neutron transmission spectra when the ODF of the sample is known. Santisteban and collaborators [65] showed that is possible to extend the neutron elastic coherent cross-section expression obtained by Fermi [72] for polycrystalline random materials to textured samples, by the inclusion of a correction factor for each Bragg edge. This multiplicative factor accounts for the difference of volume fraction under diffraction of a textured sample in comparison with a randomly textured material, and it is evaluated as a line integral along the pole figure of each Bragg edge. This approach has been applied in a combination of simple models for the ODF (such as individual components [68, 73] or fibre texture components [74]) and it has been used to study several materials [30, 68, 75]. In a recent paper, Laliena [76] derived a compact expression for the elastic coherent cross-section for textured materials by analytical integration of the expression proposed by Santisteban. The coherent elastic cross-section term is reduced to a summation of functions weighted by the Fourier coefficients of the ODF, and this expression has been incorporated into a method to estimate integral parameters of the ODF (such Kearns factors in hexagonal crystals) in textured polycrystals from the wavelength-resolved neutron transmission experiment [77].

On the other hand, different models of the elastic coherent cross-section have been implemented in order to obtain the crystallographic texture of a material from transmission experiments, however, the problem has not been completely solved yet. The first approaches involved the introduction the March–Dollase formulation in a Rietveld refinement analysis of transmission data [74, 78, 79]. As an anisotropy factor, these factors measure the degree of departure of the experimental data from the expected for a randomly oriented sample, and has been used to study Al welds [80, 81], as is shown in figure 5.1(c). A different approach involves using the height of the Bragg edge to quantify the fraction of crystals with their plane normal pointing to the beam direction [82]. This method, does not consider the dependence of the transmission data just after the Bragg edge, which also depends on the crystallographic texture of the material. Dessieux and collaborations [83] presented a different methodology, combining the transmission contribution of each single-crystal grain to generate the neutron total cross-section of a polycrystalline textured material. Here, the total cross-section is calculated semi-empirically as a function of crystal structure, neutron energy and crystal orientation, including the crystal mosaicity and its orientation in the neutron beam as input parameters. This approach has been applied to simulate the effect of deformation and recrystallization crystallographic textures on an additive manufactured Inconel 718 specimen [84]. Most recently, Malamud *et al* [85] developed an attenuation coefficient model that is able to describe polycrystalline materials with strong crystallographic texture. The model is based on a single-crystal-to-polycrystalline approach where it evaluates the Bragg contribution to the attenuation coefficient of polycrystalline materials. This is achieved by considering the combination of the Bragg-reflected component of a discrete number of imperfect single crystals with different orientations, weighted by the volume fraction of the corresponding component in the orientation distribution function. In addition, the model accounts for the instrument resolution and the statistical uncertainty of the experimental transmission spectra. As such, the model is able to retrieve the attenuation coefficient of polycrystalline textured materials using a reduced number of texture components and it opens the possibility of using it in least-squares fitting routines that will allow crystallographic texture characterization from wavelength-resolved transmission experiments.

Indeed only a few characteristic examples out of a wealth of neutron imaging applications in engineering materials research can be provided here in order to give a flavour of insights that neutron imaging enables in this field. Currently methodical progress in neutron imaging, in particular with view on diffraction contrast (compare chapter 2), is thriving and new instruments with ToF capabilities for strain mapping etc (compare chapter 19) are coming online in the next years. More *in situ* studies and complex sample environments for these, which can also be developed in collaboration between beamline scientists and users, will become available to enable studies being well related to materials service conditions and manufacturing processes, which will be discussed in chapter 6. Therefore, it is always worth considering and following the progress of capabilities of neutron imaging for engineering materials research.

References

[1] Johnson W H and Thomson II W 1875 On some remarkable changes produced in iron and steel by the action of hydrogen and acids *Proc. R. Soc. Lond.* **23** 168–79

[2] Birnbaum H K 1984 Mechanical properties of metal hydrides *J. Less-Common Met.* **104** 31–41

[3] Perovic V, Weatherly G C and Simpson C J 1983 Hydride precipitation in α/β zirconium alloys *Acta Metall.* **31** 1381–91

[4] Troiano A R 2016 The role of hydrogen and other interstitials in the mechanical behavior of metals *Metallogr. Microstruct. Anal.* **5** 557–69

[5] Birnbaum H K and Sofronis P 1994 Hydrogen-enhanced localized plasticity—a mechanism for hydrogen-related fracture *Mater. Sci. Eng.* A **176** 191–202

[6] Oltra R, Bouillot C and Magnin T 1996 Localized hydrogen cracking in the austenitic phase of a duplex stainless steel *Scr. Mater.* **35** 1101–5

[7] Beyer K, Kannengiesser T, Griesche A and Schillinger B 2011 Neutron radiography study of hydrogen desorption in technical iron *J. Mater. Sci.* **46** 5171–5

[8] Griesche A, Dabah E, Kannengiesser T, Kardjilov N, Hilger A and Manke I 2014 Three-dimensional imaging of hydrogen blister in iron with neutron tomography *Acta Mater.* **78** 14–22

[9] Griesche A, Dabah E and Kannengiesser T 2015 Neutron imaging of hydrogen in iron and steel *Can. Metall. Q.* **54** 38–42

[10] Michaloudaki M, Lehmann E and Kosteas D 2005 Neutron imaging as a tool for the non-destructive evaluation of adhesive joints in aluminium *Int. J. Adhes. Adhes.* **25** 257–67
[11] Grünzweig C, Mannes D, Kaestner A, Schmid F, Vontobel P, Hovind J, Hartmann S, Peetermans S and Lehmann E 2013 Progress in industrial applications using modern neutron imaging techniques *Phys. Proc.* **43** 231–42
[12] Vontobel P, Lehmann E and Frei G 2022 Neutrons for the study of adhesive connections https://www.ndt.net/article/wcndt2004/html/htmltxt/365_vontobel.htm
[13] von Manitius J 2001 Fluidströmung in Textilhalbzeugen aus Kohlenstoff *Proc. Anwender-Workshop zur Nutzung der Neutronenradiographie*
[14] Reid A, Marshall M, Kabra S, Minniti T, Kockelmann W, Connolley T, James A, Marrow T J and Mostafavi M 2019 Application of neutron imaging to detect and quantify fatigue cracking *Int. J. Mech. Sci.* **159** 182–94
[15] Hilger Kardjilov A N, Kandemir T, Manke I, Banhart J, Penumadu D, Manescu A and Strobl M 2010 Revealing microstructural inhomogeneities with dark-field neutron imaging *J. Appl. Phys.* **107** 036101
[16] Brooks A J *et al* 2018 Neutron interferometry detection of early crack formation caused by bending fatigue in additively manufactured SS316 dogbones *Mater. Des.* **140** 420–30
[17] Sim C M, Oh H S, Kim T J, Lee Y S, Kim Y K, Kwak S S and Hwang Y H 2014 Detecting internal hot corrosion of in-service turbine blades using neutron tomography with Gd tagging *J. Nondestruct. Eval.* **33** 493–503
[18] Takenaka N, Kadowaki T, Kawabata Y, Lim I C and Sim C M 2005 Visualization of cavitation phenomena in a diesel engine fuel injection nozzle by neutron radiography *Nucl. Instrum. Methods Phys. Res.* A **542** 129–33
[19] Lehmann E, Grünzweig C, Jollet S, Kaiser M, Hansen H and Dinkelacker F 2015 Visualisation of diesel injector with neutron imaging *J. Phys.: Conf. Ser.* **656** 012089
[20] Daymond M R, Bourke M a M, Von Dreele R B, Clausen B and Lorentzen T 1997 Use of Rietveld refinement for elastic macrostrain determination and for evaluation of plastic strain history from diffraction spectra *J. Appl. Phys.* **82** 1554–62
[21] Van Petegem S, Wagner J, Panzner T, Upadhyay M V, Trang T T T and Van Swygenhoven H 2016 In-situ neutron diffraction during biaxial deformation *Acta Mater.* **105** 404–16
[22] Santisteban J R, Edwards L, Steuwer A and Withers P J 2001 Time-of-flight neutron transmission diffraction *J. Appl. Crystallogr.* **34** 289–97
[23] Woracek R, Penumadu D, Kardjilov N, Hilger A, Strobl M, Wimpory R C, Manke I and Banhart J 2011 Neutron Bragg-edge-imaging for strain mapping under *in situ* tensile loading *J. Appl. Phys.* **109** 093506
[24] Iwase K, Sato H, Harjo S, Kamiyama T, Ito T, Takata S, Aizawa K and Kiyanagi Y 2012 *In situ* lattice strain mapping during tensile loading using the neutron transmission and diffraction methods *J. Appl. Crystallogr.* **45** 113–8
[25] Connolly M, Bradley P, Slifka A and Drexler E 2017 *In situ* neutron transmission Bragg edge measurements of strain fields near fatigue cracks grown in air and in hydrogen *International Hydrogen Conference (IHC 2016): Materials Performance in Hydrogen Environments* ed B P Somerday and P Sofronis (ASME Press)
[26] Zillhardt T A C, Burca G, Liu D and Marrow T J 2022 *In situ* mechanical loading and neutron Bragg-edge imaging, applied to polygranular graphite on IMAT@ISIS *Exp. Mech.* **62** 59–73
[27] Oikawa K *et al* 2017 A comparative study of the crystallite size and the dislocation density of bent steel plates using Bragg-edge transmission imaging, TOF neutron diffraction and EBSD *Phys. Proc.* **88** 34–41

[28] Su Y *et al* 2016 Time-of-flight neutron Bragg-edge transmission imaging of microstructures in bent steel plates *Mater. Sci. Eng.* A **675** 19–31
[29] Busi M, Čapek J, Polatidis E, Hovind J, Boillat P, Tremsin A S, Kockelmann W and Strobl M 2020 Frame overlap Bragg edge imaging *Sci. Rep.* **10** 14867
[30] Polatidis E, Morgano M, Malamud F, Bacak M, Panzner T, Van Swygenhoven H and Strobl M 2020 Neutron diffraction and diffraction contrast imaging for mapping the trip effect under load path change *Materials* **13** 1450
[31] Kiriyama K, Zhang S, Hayashida H, Suzuki J and Kuwabara T 2019 Development of a biaxial tensile testing machine for pulsed neutron experiments *MethodsX* **6** 2166–75
[32] Arabi-Hashemi A, Maeder X, Figi R, Schreiner C, Griffiths S and Leinenbach C 2020 3D magnetic patterning in additive manufacturing via site-specific *in situ* alloy modification *Appl. Mater. Today* **18** 100512
[33] Woracek R, Penumadu D, Kardjilov N, Hilger A, Boin M, Banhart J and Manke I 2014 3D mapping of crystallographic phase distribution using energy-selective neutron tomography *Adv. Mater.* **26** 4069–73
[34] Polatidis E, Zotov N, Bischoff E and Mittemeijer E J 2015 The effect of cyclic tensile loading on the stress-induced transformation mechanism in superelastic NiTi alloys: an *in situ* x-ray diffraction study *Scr. Mater.* **100** 59–62
[35] Dabah E, Pfretzschner B, Schaupp T, Kardjilov N, Manke I, Boin M, Woracek R and Griesche A 2017 Time-resolved Bragg-edge neutron radiography for observing martensitic phase transformation from austenitized super martensitic steel *J. Mater. Sci.* **52** 3490–6
[36] Meggers K, Priesmeyer H G, Trela W J, Bowman C D and Dahms M 1994 Real time neutron transmission investigation of the austenite-bainite transformation in grey iron *Nucl. Instrum. Methods Phys. Res.* B **88** 423–9
[37] Bourke M A M, Maldonado J G, Masters D, Meggers K and Priesmeyer H G 1996 Real time measurement by Bragg edge diffraction of the reverse ($\alpha' \rightarrow \gamma$) transformation in a deformed 304 stainless steel *Mater. Sci. Eng.* A **221** 1–10
[38] Woo W, Kim J, Kim E-Y, Choi S-H, Em V and Hussey D S 2019 Multi-scale analyses of constituent phases in a trip-assisted duplex stainless steel by electron backscatter diffraction, *in situ* neutron diffraction, and energy selective neutron imaging *Scr. Mater.* **158** 105–9
[39] Makowska M G, Strobl M, Lauridsen E M, Frandsen H L, Tremsin A S, Kardjilov N, Manke I, Kelleher J F and Theil Kuhn L 2015 Effect of stress on NiO reduction in solid oxide fuel cells: a new application of energy-resolved neutron imaging *J. Appl. Crystallogr.* **48** 401–8
[40] Makowska M G, Strobl M, Lauridsen E M, Kabra S, Kockelmann W, Tremsin A, Frandsen H L and Theil Kuhn L 2016 *In situ* time-of-flight neutron imaging of NiO–YSZ anode support reduction under influence of stress *J. Appl. Crystallogr.* **49** 1674–81
[41] Steuwer A, Withers P J, Santisteban J R and Edwards L 2005 Using pulsed neutron transmission for crystalline phase imaging and analysis *J. Appl. Phys.* **97** 074903
[42] Soria S R, Li X H, Schulz M, Boin M and Hofmann M 2020 Determination of martensite content and mapping phase distribution on Austempered Ductile Iron using energy-selective neutron imaging *Mater. Charact.* **166** 110453
[43] Woracek R, Penumadu D, Kardjilov N, Hilger A, Boin M, Banhart J and Manke I 2015 Neutron Bragg edge tomography for phase mapping *Phys. Proc.* **69** 227–36
[44] Sato H, Mochiki K, Tanaka K, Ishizuka K, Ishikawa H, Kamiyama T and Kiyanagi Y 2019 Bragg-edge neutron transmission spectrum analysis using a high-speed-camera-type time-of-flight neutron imaging detector *Nucl. Instrum. Methods Phys. Res.* A **943** 162501

[45] Sato H, Mochiki K, Tanaka K, Ishizuka K, Ishikawa H, Kamiyama T and Kiyanagi Y 2020 Crystallographic microstructure study of a Japanese sword made by noritsuna in the muromachi period by pulsed neutron Bragg-edge transmission imaging *Mater. Res. Proc.* **15** 214–20

[46] Sato H, Sato M, Su Y, Shinohara T and Kamiyama T 2021 Improvement of Bragg-edge neutron transmission imaging for evaluating the crystalline phase volume fraction in steel composed of Ferrite and Austenite *ISIJ Int.* **61** 1584–93

[47] Bacak M *et al* 2020 Neutron dark-field imaging applied to porosity and deformation-induced phase transitions in additively manufactured steels *Mater. Des.* **195** 109009

[48] Busi M, Polatidis E, Sofras C, Boillat P, Ruffo A, Leinenbach C and Strobl M 2022 Polarization contrast neutron imaging of magnetic crystallographic phases *Mater. Today Adv.* **16** 100302

[49] Busi M, Polatidis E, Samothrakitis S, Köhnen P, Malamud F, Haase C and Strobl M 2023 3D characterization of magnetic phases through neutron polarization contrast tomography *Addit. Manuf. Lett.* **6** 100155

[50] Su Y, Oikawa K, Shinohara T, Kai T, Horino T, Idohara O, Misaka Y and Tomota Y 2021 Neutron Bragg-edge transmission imaging for microstructure and residual strain in induction hardened gears *Sci. Rep.* **11** 4155

[51] Ramadhan R S, Kockelmann W, Minniti T, Chen B, Parfitt D, Fitzpatrick M E and Tremsin A S 2019 Characterization and application of Bragg-edge transmission imaging for strain measurement and crystallographic analysis on the IMAT beamline *J. Appl. Crystallogr.* **52** 351–68

[52] Kamiyama T, Iwase K, Sato H, Harjo S, Ito T, Takata S, Aizawa K and Kiyanagi Y 2017 Microstructural information mapping of a plastic-deformed α-iron plate during tensile tests using pulsed neutron transmission *Phys. Proc.* **88** 50–7

[53] Sasada S 2022 Strain distribution visualization of punched electrical steel sheets using neutron Bragg-edge transmission imaging *Jpn. J. Appl. Phys.* **9** 046004

[54] Tremsin A S, Yau T Y and Kockelmann W 2016 Non-destructive examination of loads in regular and self-locking Spiralock® threads through energy-resolved neutron imaging *Strain* **52** 548–58

[55] Withers P J and Bhadeshia H K D H 2001 Residual stress. Part 1—measurement techniques *Mater. Sci. Technol.* **17** 355–65

[56] Sato H, Sato T, Shiota Y, Kamiyama T, Tremsin A S, Ohnuma M and Kiyanagi Y 2015 Relation between Vickers hardness and Bragg-edge broadening in quenched steel rods observed by pulsed neutron transmission imaging *Mater. Trans.* **56** 1147–52

[57] Sato H, Sasaki T, Moriya T, Ishikawa H, Kamiyama T and Furusaka M 2018 High wavelength-resolution Bragg-edge/dip transmission imaging instrument with a supermirror guide-tube coupled to a decoupled thermal-neutron moderator at Hokkaido University Neutron Source *Physica* B **551** 452–9

[58] Cho S, Kim J, Kim T, Sato H, Huh I and Cho N 2021 Neutron imaging for metallurgical characteristics of iron products manufactured with ancient Korean iron making techniques *Nucl. Eng. Technol.* **53** 1619–25

[59] Sakurai Y, Sato H, Adachi N, Morooka S, Todaka Y and Kamiyama T 2021 Analysis and mapping of detailed inner information of crystalline grain by wavelength-resolved neutron transmission imaging with individual Bragg-dip profile-fitting analysis *Appl. Sci.* **11** 5219

[60] Sato H, Iwase K, Kamiyama T and Kiyanagi Y 2020 Simultaneous broadening analysis of multiple Bragg edges observed by wavelength-resolved neutron transmission imaging of deformed low-carbon ferritic steel *ISIJ Int.* **60** 1254–63

[61] Santisteban J R, Edwards L, Fitzpatrick M E, Steuwer A, Withers P J, Daymond M R, Johnson M W, Rhodes N and Schooneveld E M 2002 Strain imaging by Bragg edge neutron transmission *Nucl. Instrum. Methods Phys. Res.* A **481** 765–8

[62] Clausen B, Lorentzen T and Leffers T 1998 Self-consistent modelling of the plastic deformation of f.c.c. polycrystals and its implications for diffraction measurements of internal stresses *Acta Mater.* **46** 3087–98

[63] Lehmann E H, Peetermans S, Josic L, Leber H and van Swygenhoven H 2014 Energy-selective neutron imaging with high spatial resolution and its impact on the study of crystalline-structured materials *Nucl. Instrum. Methods Phys. Res.* A **735** 102–9

[64] Lehmann E H, Frei G, Vontobel P, Josic L, Kardjilov N, Hilger A, Kockelmann W and Steuwer A 2009 The energy-selective option in neutron imaging *Nucl. Instrum. Methods Phys. Res.* A **603** 429–38

[65] Santisteban J R, Edwards L and Stelmukh V 2006 Characterization of textured materials by TOF transmission *Phys. B: Condens. Matter.* **385–6, Part1** 636–8

[66] Sato H, Kamiyama T, Iwase K, Ishigaki T and Kiyanagi Y 2011 Pulsed neutron spectroscopic imaging for crystallographic texture and microstructure *Nucl. Instrum. Methods Phys. Res.* A **651** 216–20

[67] Shiota Y, Hasemi H and Kiyanagi Y 2017 Crystallographic analysis of a Japanese Sword by using Bragg edge transmission spectroscopy *Phys. Proc.* **88** 128–33

[68] Santisteban J R, Vicente-Alvarez M A, Vizcaino P, Banchik A D, Vogel S C, Tremsin A S, Vallerga J V, McPhate J B, Lehmann E and Kockelmann W 2012 Texture imaging of zirconium based components by total neutron cross-section experiments *J. Nucl. Mater.* **425** 218–27

[69] Busi M, Kalentics N, Morgano M, Griffiths S, Tremsin A S, Shinohara T, Logé R, Leinenbach C and Strobl M 2021 Nondestructive characterization of laser powder bed fusion parts with neutron Bragg edge imaging *Addit. Manuf* **39** 101848

[70] Song G *et al* 2017 Characterization of crystallographic structures using Bragg-edge neutron imaging at the spallation neutron source *J. Imaging* **3** 65

[71] Tremsin A S, Gao Y, Makinde A, Bilheux H Z, Bilheux J C, An K, Shinohara T and Oikawa K 2021 Monitoring residual strain relaxation and preferred grain orientation of additively manufactured Inconel 625 by *in situ* neutron imaging *Addit. Manuf.* **46** 102130

[72] Fermi E, Sturm W J and Sachs R G 1947 The transmission of slow neutrons through microcrystalline materials *Phys. Rev.* **71** 589–94

[73] Malamud F and Santisteban J R 2016 Full-pattern analysis of time-of-flight neutron transmission of mosaic crystals *J. Appl. Crystallogr.* **49** 348–65

[74] Vogel S 2000 A Rietveld-approach for the analysis of neutron time-of-flight transmission data *Proc. of the Mathematisch-Naturwissenschaftliche Fakultät, Uni Kiel, Kiel.* p 174

[75] Malamud F, Santisteban J R, Vicente Alvarez M A, Bolmaro R, Kelleher J, Kabra S and Kockelmann W 2014 Texture analysis with a time-of-flight neutron strain scanner *J. Appl. Crystallogr.* **47** 1337–54

[76] Laliena V, Vicente-Álvarez M Á and Campo J 2020 Monte Carlo simulation of neutron scattering by a textured polycrystal *J. Appl. Crystallogr.* **53** 512–29

[77] Vicente Alvarez M A, Laliena V, Malamud F, Campo J and Santisteban J 2021 A novel method to obtain integral parameters of the orientation distribution function of textured polycrystals from wavelength-resolved neutron transmission spectra *J. Appl. Crystallogr.* **54** 903–13

[78] Sato H 2017 Deriving quantitative crystallographic information from the wavelength-resolved neutron transmission analysis performed in imaging mode *J. Imaging* **4** 7

[79] Sato H, Takada O, Iwase K, Kamiyama T and Kiyanagi Y 2010 Imaging of a spatial distribution of preferred orientation of crystallites by pulsed neutron Bragg edge transmission *J. Phys.: Conf. Ser.* **251** 012070

[80] Kardjilov N, Manke I, Hilger A, Williams S, Strobl M, Woracek R, Boin M, Lehmann E, Penumadu D and Banhart J 2012 Neutron Bragg-edge mapping of weld seams *Int. J. Mater. Res.* **103** 151–4

[81] Sato H, Kamiyama T and Kiyanagi Y 2011 A Rietveld-type analysis code for pulsed neutron Bragg-edge transmission imaging and quantitative evaluation of texture and microstructure of a welded α-iron plate *Mater. Trans.* **52** 1294–302

[82] Malamud F 2016 Microstructural effects on the neutron transmission of nuclear materials (in Spanish:Efectos de la microestructura sobre la transmisión de neutrones de materiales de interés nuclear) *PhD Thesis* Instituto Balseiro

[83] Dessieux L L, Stoica A D and Bingham P R 2018 Single crystal to polycrystal neutron transmission simulation *Rev. Sci. Instrum.* **89** 025103

[84] Dessieux L L, Stoica A D, Bingham P R, An K, Frost M J and Bilheux H Z 2019 Neutron transmission simulation of texture in polycrystalline materials *Nucl. Instrum. Methods Phys. Res. Sect.* B **459** 166–78

[85] Malamud F, Santisteban J R, Vicente Alvarez M A, Busi M, Polatidis E and Strobl M 2023 An optimized single-crystal to polycrystal model of the neutron transmission of textured polycrystalline materials *J. Appl. Crystallogr.* **56** 143–54

Chapter 6

Manufacturing

Efthymios Polatidis, Florencia Malamud and Pavel Trtik

Ancient civilizations that evolved in Egypt, Mesopotamia, India, Greece, China and other regions in the Middle East and Europe, started utilizing copper to develop weapons and tools as early as 8700 B.C. By 4500 B.C., tin and copper were melted and cast to create bronze which is stronger and more durable than copper. As such, bronze art, statues, coinage, weapons and tools were manufactured. This is the historical age known as the Bronze Age. While terrestrial iron is naturally much more abundant than copper or tin, its high melting point, i.e. 1538 °C, was out of reach for the manufacturing methods developed by the Bronze Age. Therefore, it was not until the end of the second millennium BC that larger and hotter furnaces were developed, which allowed the implementation or iron in the technology. Since then, metal became a political material, influencing the rise and fall of civilizations, as it was used for crafting weapons or used as a symbol of the elite and cultural growth. As iron became available, blacksmiths became the basic metalworkers giving shape to the metal pieces, known as forging. During forging, the blacksmith heats the metal, works it briefly, returns it to the heat and continues this sequence until the piece reaches its final shape. With the emergence of metallurgy, the shaping process was often followed by final heat treatments or quenching, to provide the material with the desired mechanical properties. The traditional metal forming methods evolved through the middle ages, until a new process called rolling, wherein materials in sheet form emerged. It is not known who created the first rolling mill, but one of the earliest drawings by Leonardo da Vinci shows a rolling mill. Since the 15th century, the rolling process has evolved and nowadays immense continuous production lines combine rolling and complex heat treatment processes, while the product quality is ensured by standardization and process monitoring. In addition to metal forming, new materials and manufacturing emerged during the ages. As such, the development and implementation of advanced materials, combined with innovative manufacturing processes, have transformed society, accelerated technological advancements, and improved the lives of humanity by enabling breakthroughs in fields such as construction, transportation, electronics and medicine.

doi:10.1088/978-0-7503-3495-2ch6

© IOP Publishing Ltd 2024

Today, very few components are produced as single parts and therefore joining methods are used for building more complex parts out of constitutive smaller parts or additive manufacturing is used for rapid prototyping of near net-shape objects or composite materials that tend to replace single materials. Welding, soldering, and brazing are all techniques for joining two or more metallic parts, as well as techniques for filling gaps in metal parts. The joining method of choice depends on the material, the desired strength and the application. The perfect joint would be indistinguishable from the two parts that are joined together, however, in reality this is very rarely the case. Joints exhibit a multitude of problems during the process that often result in scrapping the material, due to distortions or defects of the end product, however, the most crucial problem is failure occurring during service of a welded component. This can be due to internal stresses that remain in the material after welding, stress corrosion cracking, weakening due to unwanted microstructure and precipitation, or ductile/brittle fractures due to loads that such components undergo during service or further processing. Therefore, non-destructive methods for accessing the quality of the joints are often utilized in the production lines. However, in order to characterize the microstructure and residual stresses deep inside the joints non-destructively and with high spatial resolution, the capabilities of neutron imaging methods are often indispensable, compared to x-rays or conventional neutron diffraction. The following paragraphs highlight some of the microstructural features that can be revealed using neutron imaging.

6.1 Soldering-brazing

Soldering and brazing are ancient joining methods that emerged for jewellery and decorative purposes. In both cases, a filler material is melted and used to join pieces of usually dissimilar materials. Brazing and soldering are similar methods, whereas brazing requires higher temperatures than soldering in order to melt the filler metal.

A particular family of brazing filler material for aerospace, high temperature and corrosive environment applications, is the eutectic alloy systems predominantly containing nickel or nickel–chromium. Such systems also contain other alloying elements, the so-called melting point depressants, such as silicon, boron and phosphorus. Boron-containing filler materials are especially useful for neutron imaging applications due to the strong neutron capture ability of boron. In the examples shown in figure 6.1, the application of neutron tomography is showcased, where information about the 3D structure of the filler material is revealed, while weak points, insufficient or excess filler material can be also captured in a complex steel component manufactured by brazing. Non-destructive investigations on solders and brazed materials are especially useful for cultural heritage objects and jewellery, which are addressed in the corresponding chapter.

6.2 Welding

Welding is a fabrication process whereby two or more parts are fused together by use of heat, pressure or a combination of both in order to form a join between the initially separate parts. During welding three regions can be distinguished: (1) the

Figure 6.1. Tomographic reconstruction of the brazing filler metal used for joining a steel cask. The boron-containing filler material provides high contrast for this purpose.

melt pool, where the material melts and resolidifies forming the joint; (2) the heat-affected zone (HAZ), where the excess heat from the melt pool affects the adjacent microstructure of the initially separate parts; and (3) the base metal, which remains practically unaffected, although the temperature rises locally, but not to such an extent that any microstructural change is introduced.

The local high heat input and the fast cooling rates during welding, introduce residual strains in the material which are detrimental for the mechanical integrity of the welded parts, or they can cause distortions. Moreover, in welding applications, where a filler metal is used and the chemical composition of which is different than the base material, or chemical segregation can occur in the melt poll and the HAZ, compositional variations are observed that affect the lattice parameter. Finally, the direction of the heat flow during cooling can result in strong crystallographic texture. All these effects of welding can be inferred through diffraction, while Bragg edge imaging can offer superior spatial resolution over conventional neutron or x-ray diffraction.

Before Bragg edge imaging became widely available in neutron imaging facilities, energy-selective radiographies were utilized for investigating the weld characteristics. Energy-selective neutron imaging in a limited range of or at fixed wavelengths is able to deliver qualitative information about variation in strain and texture within welded materials, which cannot be obtained with broad band integrating neutron imaging. Characteristic applications of energy-resolved neutron imaging of welding are described in [1–4], where a monochromatic neutron beam at specific wavelengths reveals characteristic contrast bands and streaking, in the attenuation coefficient images, which are associated with the existence of strain variation within the bulk of the weld. Utilizing a monochromatic beam at specific wavelengths can also reveal contrast occurring from gains with specific crystallographic texture which fulfil Bragg's Law. In such cases, the morphology of the grains can be revealed, when the grain size is comparable with the field of view and spatial resolution of the neutron imaging technique. Such an example is shown in figure 6.2, where the preferential scattering of the monochromatic beam, by the large grains, exhibits contrast which reveals the grain size and morphology in the melt pool. The grain size and morphology agree well with the results obtained by Electron backscatter diffraction (EBSD).

With the establishment of Bragg edge imaging (see section 2.2.2), investigations of welded components are widely undertaken to reveal residual strain [6–9], crystallographic texture [10] or crystallite size [11]. As an example, figure 6.3 shows the application of Bragg edge imaging to study the spatial distribution of thermally induced residual strains in a steel armour plate welded with a hybrid laser arc process HLAW [7]. The strain maps show tensile residual strain in the heat-affected zone and compressive strain in the weld fusion zone and the release of the stress upon tempering.

Measuring strain with diffraction methods, whether conventional diffraction or diffraction contrast imaging, strain is derived by comparing the position of an Edge or diffraction peak with respect to a reference, 'strain' free measurement [12]. As such, the determination of residual strain depends strongly on the correct

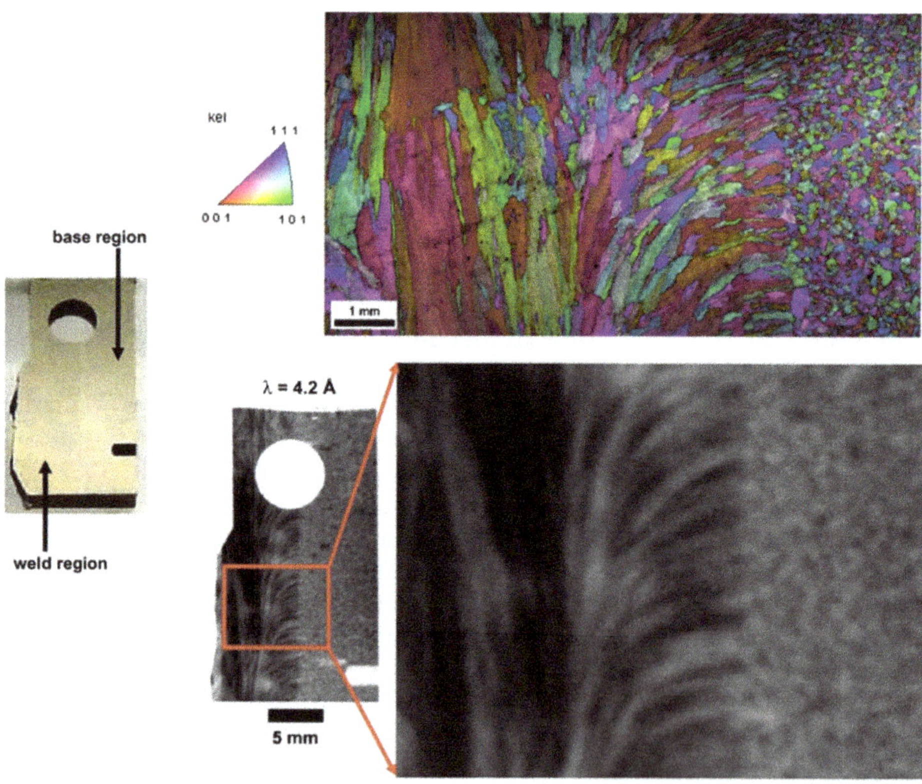

Figure 6.2. Comparison between electron backscatter diffraction results and the neutron transmission image at 4.2 Å of a welded region of an austenitic-stainless-steel sample. The large grains in the weld are verified in both investigations, while the small grains in the bulk are close to the detection limit with respect to the spatial resolution of neutron imaging. Reprinted from [5], copyright (2014), with permission from Elsevier.

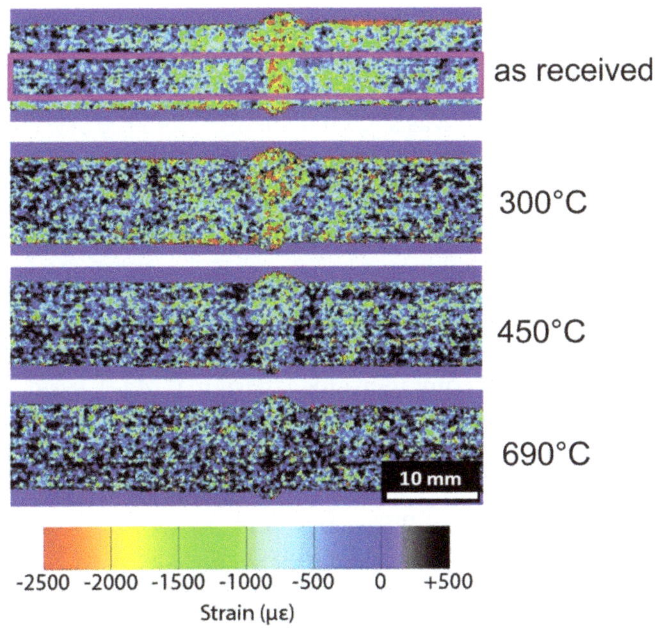

Figure 6.3. Strain maps determined from neutron Bragg edge measurement of a steel armour plate welded with a hybrid laser arc process for the as-welded sample and after stress relief heat treatments at 300 °C, 450 °C, and 690 °C. Reprinted from [7], courtesy of the National Institute of Standards and Technology, U.S. Department of Commerce.

measurement of a reference 'strain-free' lattice spacing. This becomes especially important in welds, where lattice spacing variations that are induced by chemical variations can be observed in the strain-free reference samples. In this case, using a global value of the reference lattice spacing, is not advisable and a reference sample has to be prepared and measured with the field of view, covering exactly the locations of the strain measurement itself. Bragg edge imaging has been also shown to be a very attractive method for investigating the residual strains in welds of dissimilar metals/alloys, the microstructure (e.g. texture, grain size and morphology), while neutron resonance absorption provides the possibility to map the degree of uniformity in mixing and intermetallic formation within the welds [13]. In such complex measurements, conventional diffraction becomes even more challenging due to the intermixing of the materials and compositional gradient that can exist within the gauge volume.

Bragg edge imaging, being fundamentally a diffraction method, carries information that is traditionally assessed by conventional diffraction methods, like texture and crystallite size. Therefore, several models that are able to describe the way the polycrystalline diffracts neutrons can be used, and refined in order to least-square fit the experimental data; this method is known as Rietveld refinement in the conventional diffraction community. As such, a Rietveld-type analysis was developed and incorporated in the code RITS (Rietveld Imaging of Transmission Spectra) [14], which is used to fit experimental Bragg edge profiles. The method allowed obtaining the information on the preferred orientation and crystallite size, exploiting the information on the edge shape. The analysis was demonstrated on a weld of rolled ferritic plates exhibiting a typical α-fibre along the rolling direction and γ-fibre in the RD-55° textures. The weld line was shown to change both the crystallite size and the texture with respect to the virgin material, which was revealed with the aid of Bragg edge imaging [11].

6.3 Additive manufacturing

Additive manufacturing (AM) is a revolutionary process of rapid prototyping which is often called 3D printing. The basic principle of AM is that an object, the model of which is generated using a computer-aided design (CAD) system, is fabricated directly by adding material in layers; where each layer is a thin cross-section slice of the entire object, included in the CAD file. AM has gained significant industrial and academic research interest because through AM, parts with complex geometries can be manufactured faster and more precisely in near net shape. Often, AM-produced parts do not require subsequent machining or treatments which minimizes the material loss and lead time. One of the most crucial issues in the AM community is how to control and suppress the detrimental residual stresses that build up during the layer-by-layer manufacturing process, which in turn lead to cracks, porosity and geometrical anomalies, while residual stresses can affect the final microstructure. It is apparent that the high geometrical complexity of parts processed by AM requires advanced spatially resolved characterization techniques, to analyze the micro-structure, defects and residual stresses, with the corresponding 3D spatial resolution; therefore, neutron imaging is an ideal non-destructive tool that has been employed in several studies which are summarized in the following paragraphs.

6.3.1 Imaging crystal structures

During additive manufacturing, sharp thermal gradients and complex thermal cycles occur which can strongly affect the microstructure of the processed part. A prime example is tailoring the heat input to manipulate the crystal structure. This was achieved by selectively evaporating nitrogen in a nitrogen-stabilized austenitic steel, where a different process during laser powder bed fusion resulted in heterogeneous materials comprised of different crystal structures [15]. By manipulating the process parameters it is also possible to control the phase fractions in dual-phase materials [16]. Crystal structures can be revealed with diffraction contrast imaging, as well as with imaging methods that are sensitive to the magnetic properties, when one of the phases is ferromagnetic. Neutron dark-field imaging (DFI) was used to investigate the microstructure of additive manufactured steels where, several DFI methods were combined to assess the porosity and defects over more than two orders of magnitude in size. The samples were subsequently deformed and an increased dark-field contrast was observed which was explained by accounting for the austenite (with fcc crystal structure) and martensite (with bcc crystal structure) distribution and the magnetic properties of the martensitic phase [17].

6.3.2 Imaging strain

Additive manufacturing processes are characterized by fast heating up and cooling down rates, in the order of for instance 10^5–10^6 °C s^{-1}, particularly during laser powder bed fusion (LPBF). Furthermore, very steep temperature gradients occur within small volumes of material, while during the layer-wise building, the spatially varied thermal cycles that the material undergoes result in the formation of thermal expansion mismatches and the occurrence of residual stresses. The residual stresses impose risks for the geometrical and mechanical integrity and sustainability of the AM components and significant efforts are currently made to control the appearance and relieving of the residual strain in the as-built components or by subsequent post-built heat treatments. Imaging strain in additively manufactured components is a prime use of Bragg edge imaging because it can very well reveal the steep strain gradients in complex components with good spatial resolution. Bragg edge imaging has been used for the characterization of the strain distribution in complex samples produced by AM [18] as well as the effect of annealing treatment on the residual strains [19].

Besides heat treatments that are able to relieve much of the residual stresses, laser shot peening has been shown to be an effective way to change the detrimental residual tensile strains, after additive manufacturing, into beneficial compressive strain [20]. The depth of the affect zone by a single treatment of laser shot peening (LSP) is limited to only a few hundreds of microns and high spatial resolution is required to be able to quantify the effect of the peening treatment. Bragg edge imaging was seen to very well capture the residual strain profile of compressive strain extending up to a depth of 100–300 μm, while the effect of LSP was seen to extend to as deep as 1 mm from the treated surface [21]. Further away of the bathtub-like shape of the strain profile, which is typically seen in additively manufactured components [21]. By exploiting the high performance of neutron imaging instruments, parametric investigations can be undertaken on the effect of different processes and shot peening treatments on additively manufactured components, as shown in figure 6.4 [22].

Neutron Bragg edge tomography has been utilized to produce volumetric maps of the planar strain distribution of 316L stainless steel samples processed by laser

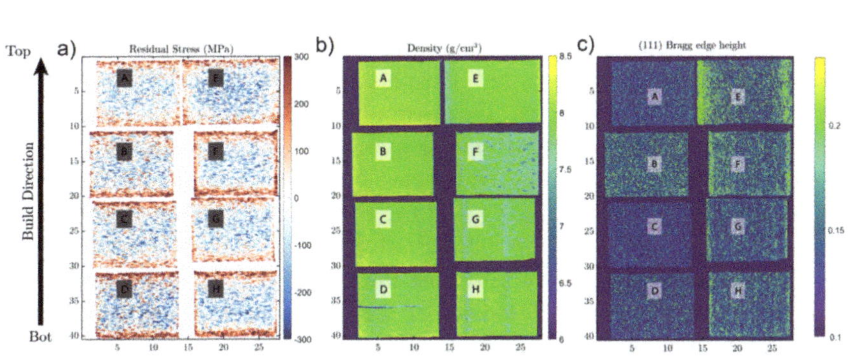

Figure 6.4. Residual stress (A), density (B) and (111) Bragg edge height maps (C), of 316L stainless steel samples built with different LPBF processing parameters. Reprinted from [22], copyright (2021), with permission from Elsevier.

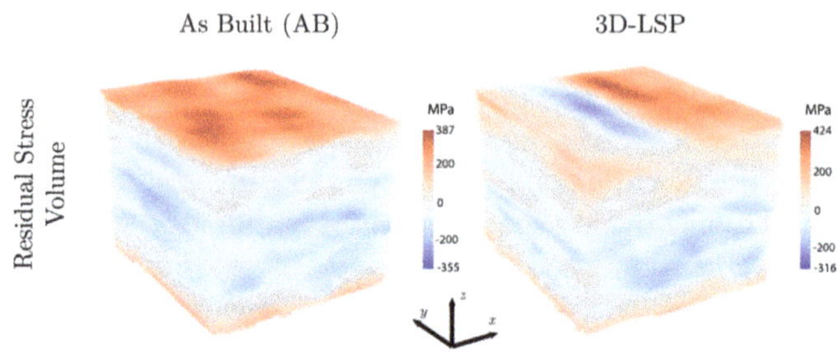

Figure 6.5. Reconstructed strain volumes corresponding to the 316L processed by laser powder bed fusion and the one partially treated with LSP, where the build direction is aligned to the *z*-axis. Reprinted with permission from [23], copyright (2022) by the American Physical Society.

powder bed fusion and laser shot peening LSP. The tomographic reconstruction was possible, due to the isotropic property of the strain in the transverse directions, which are normal to the building direction and parallel to the neutron beam direction, as shown in figure 6.5 [23].

New developments of operando machines and/or in powder bed fusion conditions requires investigations with transmission methods, where the signal is averaged through the sample and surrounding crystalline materials. A common application is the utilization of Bragg edge imaging to characterize the strain in powder bed fusion conditions, where the incoming beam passes through the building chamber material, the building piston material, a layer of powder before reaching the sample. The transmitted beam moreover, passes through symmetrically arranged materials. A study on such simulated conditions was undertaken by Sumarli *et al* [24] with the aim to disentangle the strain values in additively manufactured samples which were attached to the base plate and surrounded by their corresponding unfused powder.

6.3.3 Imaging defects

A vital criterion for the quality assurance of the AM components parts is the material density and the evaluation of the porosity size and distribution. Bragg edge imaging has proved to be a powerful technique to study the effect of different process parameters on the defects appearing in the bulk of the sample, as is shown in figure 6.5(B) for 316L LPBF-printed samples where the density maps display the presence of porosity and cracks. In particular, porosity is detrimental for the mechanical integrity, for the functional properties of materials as well as undesired for ornamental applications, in e.g. jewellery. A recent tomographic reconstruction on red-gold samples manufactured is a prime example of measurement where x-ray tomography would struggle, due to the high absorptivity of x-rays by gold, while any microscopy method is destructive and cost inefficient [25]. The 3D imaging assessment confirmed the high relative density of the AM processed red-gold sample and indicated residual pore sizes, which was verified by optical microscopy, as shown in figure 6.6. The density of two samples is dependent on the process parameters, however, they are predominantly below the threshold size provided by the quality assurance for jewellery applications.

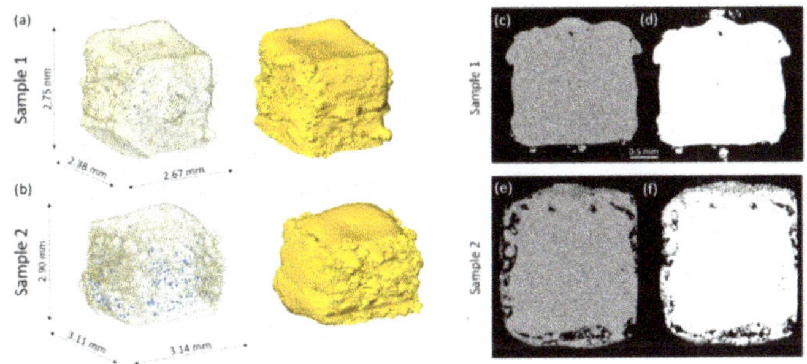

Figure 6.6. 3D renderings of two samples from red-gold process with laser powder bed fusion, revealing the final sample shape and the enclosed porosity in blue (A), (B). Comparison of the neutron tomography slices (C and E) and their equivalent optical microscopy (D and F) sections. Reproduced from [25] CC BY 4.0.

For most engineering applications and pores in the range of hundreds of micrometers in size, are crucial for the mechanical integrity. Grating interferometry or dark-field imaging provides an alternative route, mainly focusing on the lower range of size of porosity, as for instance demonstrated by dark-field imaging accessing pores in the range of μm to sub-μm in size in additively manufactured austenitic steels [17, 26]. Despite the small size of such pores, when the defects are distributed and concentrated in certain locations, they can also affect the mechanical behaviour of components. Such arrangement of porosity has been revealed in titanium parts processed with electron beam-melting additive manufacturing [27].

The sensitivity of dark-field imaging defects, compared to conventional attenuation contrast imaging can be crucial in detecting cracks, that are invisible, with conventional methods. A batch of additively manufactured samples was deformed to different extents of total elongation and dark-field imaging contrast was observed, in locations where eventually fracture occurred [28]. As such, dark-field imaging is a powerful non-destructive tool for detecting defects concentration and early crack formation in additive manufactured components.

6.3.4 Imaging crystallographic texture

The crystallographic texture is very sensitive with respect to the additive manufacturing process parameters. Studies using Bragg edge imaging on Inconel 718 processed by electron beam melting, revealed strongly heterogeneous microstructures, with strong crystallographic and morphological texture [29, 30]. Although the geometry of the investigated sample was simple, the full-field spatially resolved results were used for validating the modelling results of the heterogeneous microstructure and its theoretical transmission signals at different incident beam angles [30].

Additive manufacturing has gained a lot of attention recently not only for the possibility of producing parts with complex geometries but also for the possibility of manipulating the crystallographic texture. Texture manipulation can provide an alternative role for tailoring and controlling the deformation behaviour in, e.g. metastable transformation steels [31]. The crystallographic texture can be tailored, by the use of appropriate process parameters that achieves specific melt pool

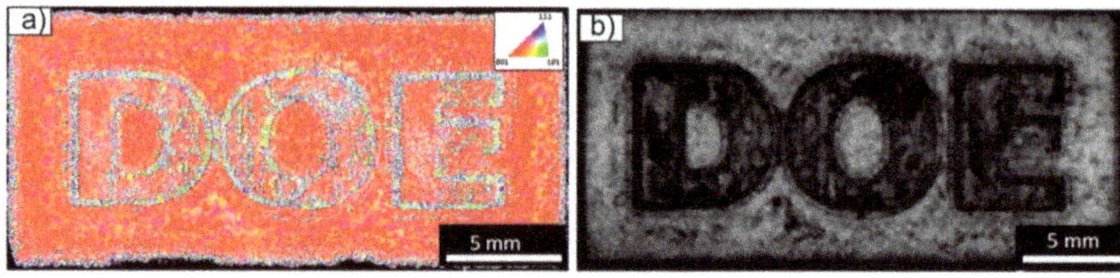

Figure 6.7. (A) Electron backscattered electron orientation map with inverse pole figure colouring obtained on cross-section of EBM build: showing preferred <001> columnar solidification grain growth; similarly, the outline of the D, O and E, letters consists of misoriented growth indicated by lack of any significant <001> components; The interior areas of letters D, O and E show a mixture of <001> growth and misoriented growth. (B) Energy-selective neutron radiograph at wavelength of 1.5 Å, with spatial resolution of 150 μm through 5 mm of material. [34] (2015), reprinted by permission of the publisher (Taylor & Francis Ltd, http://www.tandfonline.com).

dimensions and solidification dynamics [32, 33]. It is therefore apparent that spatially tailored crystallographic textures can be realized, as has been recently demonstrated [31, 34]. The possibility to reveal such variations of crystallographic textures or even encoded messages in the microstructure, hidden deep in the material, is possible by the use of neutron imaging. In the later example, a spatially varied crystallographic texture was revealed by energy-selective neutron radiography on Inconel 718 alloy in which the crystallographic texture was site-specifically manipulated by electron beam additive manufacturing. The obtained contrast in the neutron transmission signal originatied from the differently orientated grains, as shown in figure 6.7 [34].

By analyzing the Bragg edge spectra in a 316L sample processed by LPBF along different tomographic projections, it was observed that the relative intensity of the Bragg edge heights varies symmetrically. The revealed symmetry in the crystallographic texture is associated with the laser scanning strategy during the LPBF process. The qualitatively assessed symmetry in the crystallographic texture by Bragg edge imaging was verified by neutron diffraction experiments [23].

6.4 Single crystals and oligocrystals

Components consisting of single crystals are often found in electronics, optical, piezoelectric aerospace and in numerous other applications. It has been shown that Bragg edge imaging is a useful real-time diagnostic tool for visualizing the growth of single crystals in terms of *in situ* measurements of the liquid/solid interface and elemental distribution [35]. Revealing the bulk mosaicity of Ni-superalloy single crystals used for aerospace applications has also been proved to be a valuable tool, not only for quality assurance but also for misorientations induced by heat flow inconsistencies allowing optimization of the processes [36, 37]. The dendritic growth orientation in industrial Ni-superalloys turbine blades cast as single crystal was also studied using neutron imaging [38]. In such materials, the lattice misfit between the main phases is one of the most important parameters that control their mechanical properties, such as creep behaviour at high temperatures. It has been shown that

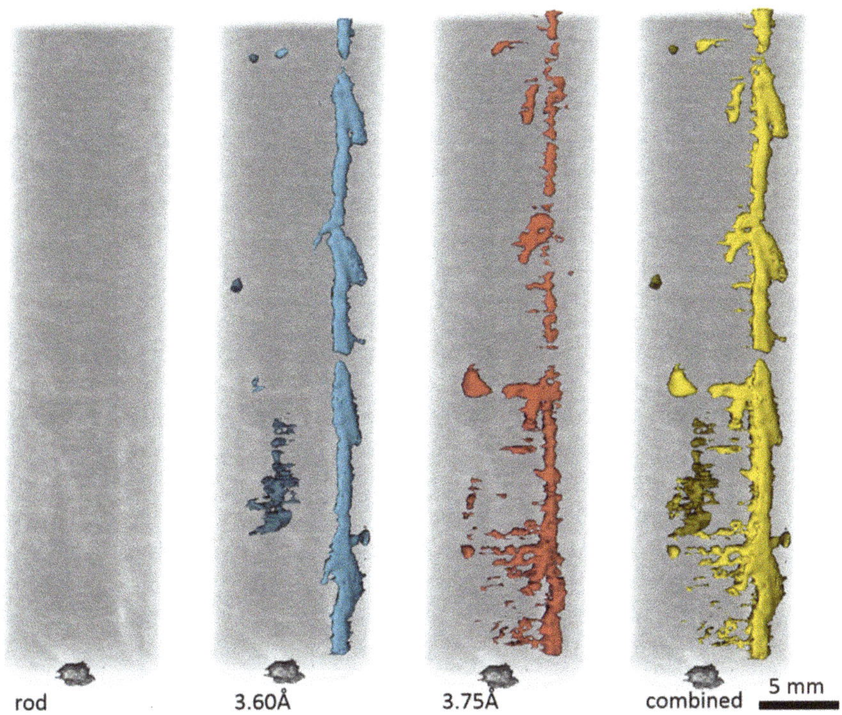

Figure 6.8. 3D reconstruction of a single-crystal copper rod. The volume of the copper rod is shown in transparency and used as a spatial reference for the two dendritic branches highlighted, respectively, in blue and orange; the combined structure consisting of the two branches is represented in yellow. Reproduced from [42] CC BY 4.0.

using wavelength resolved neutron imaging it is possible to map the lattice misfit with a spatial resolution of 500 μm × 500 μm [39]. For samples composed of a small number of grains, Bragg edge transmission imaging was proven to be a powerful tool for mapping the crystalline orientations and microstrain within the grains [40, 41]. Selective monochromatic neutron tomography has been also been shown to be a useful tool for assessing, non-destructively, misorientations that appear in the bulk of the single crystals [42]. As an example, the monochromatic tomography reconstructions of a single-crystal copper rod at two wavelengths are shown in figure 6.8, where the presence of a branched dendritic structure with different spatial localization is exhibited.

6.5 Cold forming processes and tribological manufacturing methods

Cold forming processes such as rolling, forging, deep drawing or machining result in anisotropic microstructures and stress concentrations. Such stress concentrations are especially important to be known and characterized as they can often lead to premature failure of parts because of fatigue. In a study on filleted bolts where the stress concentration bold head and the shank was revealed by Bragg edge imaging, which is prone to failures [43].

6.6 Sintered powders

Cutting inserts and drill bits are manufactured by compacting hard metallic powder blends into a predefined shape and then sintered to fully densify the part. The powder mixtures contain tungsten carbides (WC) which is a major issue for x-ray and even cold neutron studies due to absorption. It was shown that polychromatic thermal neutron imaging can be successfully utilized for characterizing the density distributions in cemented carbide compacts [44].

6.7 Liquid metals

Properties of metallic products depend crucially on their composition, microstructure and purity. In many technologies in which metals are processed in the liquid state, flow and transport processes have a decisive influence on the final product. Post-mortem analyses have proved to be inadequate to reveal the complex interrelationship between the liquid metal flow and the final product quality and/or process efficiency. Thanks to its low cross-section for many metals, neutron imaging provides a very useful tool for obtaining a better understanding of processes occurring under representative flow conditions in liquid metals. The alternative method to neutron imaging for obtaining visual information about flows in liquid metals is x-ray imaging [45], however, the applicability of it is often limited by the thickness of the melt volume under investigation.

Early attempts on the investigation of liquid metal by neutron radiography were performed by Takenaka *et al* [46], who utilized gold–cadmium tracers to visualize streak lines in slow flow processes in liquid Pb–Bi eutectics. However, the flow of liquid metals usually represents a relatively fast process, a high frame-rate neutron imaging is required and therefore further progress in this field was triggered by the provision of suitable detector systems (such as high-speed video cameras fitted with an image intensifier or CMOS detectors). One of the first attempts into the investigation of fast processes in liquid metals such as the two phase flows have been performed by Cha *et al* [47] and Saito *et al* [48] who measured void (bubble) fraction in Pb–Bi molten metal using neutron radiography with frame rate of 1000 fps.

Liquid metal stirring and the distribution of solid inclusions in molten liquid metal was investigated by researchers from the Institute of Physics in Latvia (Sarma *et al* [49], and Sčepanskis *et al* [50] using a test arrangement that utilized four counter-rotating magnets. This geometry was chosen for its representativeness for flow conditions in large-scale industrial crucibles. The distribution of particle concentrations, number of admixed particles and their velocities as functions of the magnet rotation speeds were evaluated from dynamic neutron radiography experiments with 32 Hz acquisition rate. Similarly, the particle tracking velocimetry has been successfully applied in the investigation of liquid metal flow around a cylindrical obstacle [51]. In all of the above-mentioned studies, gallium has been utilized as the model metal due to its rather low melting point (~30 °C). The experiments using a sample of molten tin (melting point 232 °C) similarly electromagnetically stirred as previously mentioned in the experiments by Sarma and Sčepanskis and left to directional solidification, revealed information on the particle trapping in a semi-solid ('mushy') zone [52]. The process of solidification of liquid metals has also been investigated by Baake *et al* [53] and Musaeva *et al* [54]. Opposite to the process of solidification, the process of melting and liquid phase separation of multicomponent high-entropy alloy (HEA) systems has been recently investigated by dynamic neutron radiography by Derimow *et al* [55, 56]. Likewise, entire devices (such as heat pipes for thermal management of hypersonic flight vehicles) that include liquid metal coolant can be imaged using neutron radiography and even tomography [57]. It is worth noting here that even though samples subjected to very high temperatures can be investigated by neutron imaging, it should be born in mind that the spatial resolution of such investigations is usually somewhat inferior due to the necessity of keeping the test piece at a larger distance from the detector (increasing image blurring due to the beam divergence) than in the case of room-temperature samples.

Many metallurgical processes, such as melt stirring, homogenization, purification, involve gas, i.e. often inert gas such as argon, flow through electrically conductive fluids. Some of the metallurgical processes (such as stirring and casting) can be potentially controlled by application of external magnetic field. The application of magnetic field on electrically conductive liquid metal has a pronounced effect on the trajectories of rising bubble in liquid metals and hence on the final efficiency of the mentioned processes. A quantum step in the detection of features from high temporal resolution neutron radiographies (hence low signal-to-noise ratio (SNR) images) has been recently achieved by Birjukovs and colleagues [58]. In their seminal work [59], the trajectories and shapes of the rising bubbles in liquid gallium vessels large enough to avoid wall effects are derived from neutron radiographic time-series of 100 fps acquisition rate, which are shown in figure 6.9.

It was demonstrated that the bubble shape parameters (such as bubble tilt) could be successfully extracted from inherently low SNR images, while the investigated

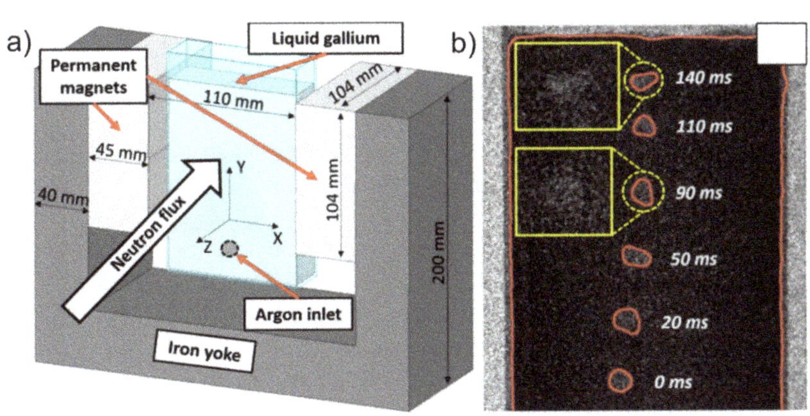

Figure 6.9. (A) Experimental setup for the neutron radiographic experiments for the investigation of rising argon bubbles in liquid gallium under applied permanent magnetic field, (B) close-ups of the bubbles in the individual radiographies (in yellow) and their derived shape (in red). Reprinted with permission from [59], copyright (2020) by the American Physical Society.

liquid metal layers were substantially thick (i.e., representative of industrial applications). The image processing pipeline has been recently published elsewhere [60]. Future experiments on the dynamics of rising objects may include investigations utilizing rising droplets of hydrogenous liquids that promise to enhance the relatively poor contrast between liquid metal and gas. Likewise, the future potential of vertical neutron beams [61, 62], for investigations of liquid metal samples deserves to be mentioned here.

The chapter presents the most important topics related to manufacturing, where neutron imaging techniques have been applied, with a focus on defect and/or microstructure characterization. As such, this section fosters a range of applications, much like engineering, using simple attenuation contrast to advanced neutron imaging techniques. In line with applications presented in chapter 5, research on manufacturing will also in the future benefit further from advances in instrumentation where improvements in flux and resolution are implied. It is much more eminent though, that research on manufacturing can benefit immensely from the development of advanced operando set-ups that allow obtaining insight under real-time manufacturing and realistic conditions. To this end, there has been significant development of, for instance, novel operando additive manufacturing set-ups that are used with x-ray imaging [63, 64], x-ray diffraction [65] or neutron diffraction [66]. It is thus anticipated that such developments will grow in neutron imaging, where the researchers can benefit from the high penetration, larger fields of view, bulk information combining structure quality and microstructure characterization that neutron imaging techniques offer.

References

[1] Schulz M, Böni P, Calzada E, Mühlbauer M and Schillinger B 2009 Energy-dependent neutron imaging with a double crystal monochromator at the ANTARES facility at FRM II *Nucl. Instrum. Methods Phys. Res.* A **605** 33–5

[2] Josic L, Steuwer A and Lehmann E 2010 Energy selective neutron radiography in material research *Appl. Phys.* A **99** 515–22

[3] Lehmann E H and Wagner W 2010 Neutron imaging at PSI: a promising tool in materials science and technology *Appl. Phys.* A **99** 627–34

[4] Josic L, Lehmann E and Kaestner A 2011 Energy selective neutron imaging in solid state materials science *Nucl. Instrum. Methods Phys. Res.* A **651** 166–70

[5] Lehmann E H, Peetermans S, Josic L, Leber H and van Swygenhoven H 2014 Energy-selective neutron imaging with high spatial resolution and its impact on the study of crystalline-structured materials *Nucl. Instrum. Methods Phys. Res.* A **735** 102–9

[6] Santisteban J R, Edwards L, Fizpatrick M E, Steuwer A and Withers P J 2002 Engineering applications of Bragg-edge neutron transmission *Appl. Phys.* A **74** s1433–6

[7] Sowards J W, Daniel , Hussey S, Jacobson D L, Ream Stan and Williams Paul 2018 Correlation of neutron-based strain imaging and mechanical behavior of armor steel welds produced with the hybrid laser arc welding process *J. Res. Nat. Inst. Stand. Technol.* **123** 123011

[8] Tremsin A S, Kockelmann W, Paradowska A M, Zhang S-Y, Korsunsky A M, Shinohara T, Feller W B and Lehmann E H 2016 Investigation of microstructure within metal welds by energy resolved neutron imaging *J. Phys.: Conf. Ser.* **746** 012040

[9] Zhu B, Leung N, Kockelmann W, Kabra S, London A J, Gorley M, Whiting M J, Wang Y and Sui T 2022 Revealing the residual stress distribution in laser welded Eurofer97 steel by neutron diffraction and Bragg edge imaging *J. Mater. Sci. Technol.* **114** 249–60

[10] Santisteban J R, Edwards L and Stelmukh V 2006 Characterization of textured materials by TOF transmission *Physica* B **385–6** 636–8

[11] Sato H, Kamiyama T and Kiyanagi Y 2011 A Rietveld-type analysis code for pulsed neutron Bragg-edge transmission imaging and quantitative evaluation of texture and microstructure of a welded α-iron plate *Mater. Trans.* **52** 1294–302

[12] Withers P J, Preuss M, Steuwer A and Pang J W L 2007 Methods for obtaining the strain-free lattice parameter when using diffraction to determine residual stress *J. Appl. Crystallogr.* **40** 891–904

[13] Tremsin A S, Ganguly S, Meco S M, Pardal G R, Shinohara T and Feller W B 2016 Investigation of dissimilar metal welds by energy-resolved neutron imaging *J. Appl. Crystallogr.* **49** 1130–40

[14] Sato H *et al* 2013 Upgrade of Bragg edge analysis techniques of the RITS code for crystalline structural information imaging *Phys. Proc.* **43** 186–95

[15] Arabi-Hashemi A, Maeder X, Figi R, Schreiner C, Griffiths S and Leinenbach C 2020 3D magnetic patterning in additive manufacturing via site-specific *in situ* alloy modification *Appl. Mater. Today* **18** 100512

[16] Facchini L, Vicente N, Lonardelli I, Magalini E, Robotti P and Molinari A 2010 Metastable Austenite in 17–4 precipitation-hardening stainless steel produced by selective laser melting *Adv. Eng. Mater.* **12** 184–8

[17] Bacak M *et al* 2020 Neutron dark-field imaging applied to porosity and deformation-induced phase transitions in additively manufactured steels *Mater. Des.* **195** 109009

[18] Tremsin A S, Gao Y, Dial L C, Grazzi F and Shinohara T 2016 Investigation of microstructure in additive manufactured Inconel 625 by spatially resolved neutron transmission spectroscopy *Sci. Technol. Adv. Mater.* **17** 324–36

[19] Tremsin A S, Gao Y, Makinde A, Bilheux H Z, Bilheux J C, An K, Shinohara T and Oikawa K 2021 Monitoring residual strain relaxation and preferred grain orientation of additively manufactured Inconel 625 by *in situ* neutron imaging *Addit. Manuf.* **46** 102130

[20] Kalentics N, Boillat E, Peyre P, Ćirić-Kostić S, Bogojević N and Logé R E 2017 Tailoring residual stress profile of selective laser melted parts by laser shock peening *Addit. Manuf.* **16** 90–7

[21] Morgano M, Kalentics N, Carminati C, Capek J, Makowska M, Woracek R, Maimaitiyili T, Shinohara T, Loge R and Strobl M 2020 Investigation of the effect of laser shock peening in additively manufactured samples through Bragg edge neutron imaging *Addit. Manuf.* **34** 101201

[22] Busi M, Kalentics N, Morgano M, Griffiths S, Tremsin A S, Shinohara T, Logé R, Leinenbach C and Strobl M 2021 Nondestructive characterization of laser powder bed fusion parts with neutron Bragg edge imaging *Addit. Manuf.* **39** 101848

[23] Busi M *et al* 2022 Bragg edge tomography characterization of additively manufactured 316L steel *Phys. Rev. Mater.* **6** 053602

[24] Sumarli S, Polatidis E, Malamud F, Busi M, Navarre C, Esmaeilzadeh R, Logé R and Strobl M 2022 Neutron Bragg edge imaging for strain characterization in powder bed additive manufacturing environments *J. Mater. Res. Technol.* **21** 4428–38

[25] Ghasemi-Tabasi H, Trtik P, Jhabvala J, Meyer M, Carminati C, Strobl M and Logé R E 2021 Mapping spatial distribution of pores in an additively manufactured gold alloy using neutron microtomography *Appl. Sci.* **11** 1512

[26] Brooks A J, Knapp G L, Yuan J, Lowery C G, Pan M, Cadigan B E, Guo S, Hussey D S and Butler L G 2017 Neutron imaging of laser melted ss316 test objects with spatially resolved small angle neutron scattering *J. Imaging* **3** 58

[27] Brooks A J, Ge J, Kirka M M, Dehoff R R, Bilheux H Z, Kardjilov N, Manke I and Butler L G 2017 Porosity detection in electron beam-melted Ti–6Al–4V using high-resolution neutron imaging and grating-based interferometry *Prog. Addit. Manuf* **2** 125–32

[28] Brooks A J, Yao H, Yuan J, Kio O, Lowery C G, Markötter H, Kardjilov N, Guo S and Butler L G 2018 Early detection of fracture failure in SLM AM tension testing with Talbot–Lau neutron interferometry *Addit. Manuf.* **22** 658–64

[29] Song G *et al* 2017 Characterization of crystallographic structures using Bragg-edge neutron imaging at the spallation neutron source *J. Imaging* **3** 65

[30] Xie Q, Song G, Gorti S, Stoica A D, Radhakrishnan B, Bilheux J C, Kirka M, Dehoff R, Bilheux H Z and An K 2018 Applying neutron transmission physics and 3D statistical full-field model to understand 2D Bragg-edge imaging *J. Appl. Phys.* **123** 074901

[31] Sofinowski K, Wittwer M and Seita M 2022 Encoding data into metal alloys using laser powder bed fusion *Addit. Manuf.* **52** 102683

[32] Sun Z, Tan X, Tor S B and Chua C K 2018 Simultaneously enhanced strength and ductility for 3D-printed stainless steel 316L by selective laser melting *NPG Asia Mater.* **10** 127–36

[33] Sofinowski K A, Raman S, Wang X, Gaskey B and Seita M 2021 Layer-wise engineering of grain orientation (LEGO) in laser powder bed fusion of stainless steel 316L *Addit. Manuf.* **38** 101809

[34] Dehoff R R, Kirka M M, Sames W J, Bilheux H, Tremsin A S, Lowe L E and Babu S S 2015 Site specific control of crystallographic grain orientation through electron beam additive manufacturing *Mater. Sci. Technol.* **31** 931–8

[35] Tremsin A S, Perrodin D, Losko A S, Vogel S C, Bourke M A M, Bizarri G A and Bourret E D 2017 Real-time crystal growth visualization and quantification by energy-resolved neutron imaging *Sci. Rep.* **7** 46275

[36] Strickland J, Tassenberg K, Sheppard G, Nenchev B, Perry S, Li J, Dong H, Burca G, Kelleher J and Irwin S 2020 2D single crystal Bragg-dip mapping by time-of-flight energy-resolved neutron imaging on IMAT@ISIS *Sci. Rep.* **10** 20751

[37] Strickland J, Nenchev B, Tassenberg K, Perry S, Sheppard G, Dong H, Zhang R, Burca G and D'Souza N 2021 On the origin of mosaicity in directionally solidified Ni-base superalloys *Acta Mater.* **217** 117180

[38] Peetermans S and Lehmann E H 2013 Simultaneous neutron transmission and diffraction contrast tomography as a non-destructive 3D method for bulk single crystal quality investigations *J. Appl. Phys.* **114** 124905

[39] Malamud F, Santisteban J R, Gao Y, Shinohara T, Oikawa K and Tremsin A 2022 Non-destructive characterization of the spatial variation of γ/γ' lattice misfit in a single-crystal Ni-based superalloy by energy-resolved neutron imaging *J. Appl. Crystallogr.* **55** 228–39

[40] Sakurai Y, Sato H, Adachi N, Morooka S, Todaka Y and Kamiyama T 2021 Analysis and mapping of detailed inner information of crystalline grain by wavelength-resolved neutron transmission imaging with individual Bragg-dip profile-fitting analysis *Appl. Sci.* **11** 5219

[41] Sato H *et al* 2017 Inverse pole figure mapping of bulk crystalline grains in a polycrystalline steel plate by pulsed neutron Bragg-dip transmission imaging *J. Appl. Crystallogr.* **50** 1601–10

[42] Grazzi F, Cantini F, Morgano M, Busi M and Park J-S 2021 Microstructural characterization of a single crystal copper rod using monochromatic neutron radiography scan and tomography: a test experiment *Appl. Sci.* **11** 7750

[43] Li W, Zhang S Y, Kabra S, Tremsin A, Abbey B, Kirkwood H, Terret D, Ndoye S and McDevitt E T 2014 Characterisation of residual stress due to fillet rolling on bolts made of a nickel base superalloy *Adv. Mater. Res.* **996** 670–5

[44] Staf H, Forssbeck Nyrot E and Larsson P-L 2018 On the usage of a neutron source to determine the density distribution in compacted cemented carbide powder compounds *Powder Metall.* **61** 389–94

[45] Richter T, Keplinger O, Shevchenko N, Wondrak T, Eckert K, Eckert S and Odenbach S 2018 Single bubble rise in GaInSn in a horizontal magnetic field *Int. J. Multiphase Flow* **104** 32–41

[46] Takenaka N, Fujii T, Ono A, Sonoda K, Tazawa S and Nakanii T 1994 Visualization of streak lines in liquid metal by neutron radiography *Nondestruct. Test. Eval.* **11** 107–13

[47] Cha J E, Lim I C, Kim H R, Kim C M, Nam H Y and Saito Y 2005 Measurement of liquid-metal two-phase flow with a dynamic neutron radiography *Transactions of the Korean Nuclear Society Autumn Meeting* **3** *(Busan, Korea, October 27–28)*

[48] Saito Y, Mishima K, Tobita Y, Suzuki T, Matsubayashi M, Lim I C and Cha J E 2005 Application of high frame-rate neutron radiography to liquid-metal two-phase flow research *Nucl. Instrum. Methods Phys. Res.* A **542** 168–74

[49] Sarma M, Ščepanskis M, Jakovičs A, Thomsen K, Nikoluškins R, Vontobel P, Beinerts T, Bojarevičs A and Platacis E 2015 Neutron radiography visualization of solid particles in stirring liquid metal *Phys. Proc.* **69** 457–3

[50] Ščepanskis M, Sarma M, Vontobel P, Trtik P, Thomsen K, Jakovičs A and Beinerts T 2017 Assessment of electromagnetic stirrer agitated liquid metal flows by dynamic neutron radiography *Metall. Mater. Trans.* B **48** 1045–54

[51] Birjukovs M, Zvejnieks P, Lappan T, Sarma M, Heitkam S, Trtik P, Mannes D, Eckert S and Jakovics A 2021 Particle tracking velocimetry in liquid gallium flow about a cylindrical obstacle *Exp. Fluids* **63** 1–19

[52] Baranovskis R, Sarma M, Ščepanskis M, Beinerts T, Gaile A, Eckert S, Räbiger D, Lehmann E H, Thomsen K and Trtik P 2020 Investigation of particle dynamics and solidification in a two-phase system by neutron radiography *Magnetohydrodynamics* **56** 4C3–50

[53] Baake E, Fehling T, Musaeva D and Steinberg T 2017 Neutron radiography for visualization of liquid metal processes: bubbly flow for CO_2 free production of hydrogen and solidification processes in em field *IOP Conf. Ser.: Mater. Sci. Eng.* **228** 012026

[54] Musaeva D, Baake E, Köppen A and Vontobel P 2017 Application of neutron radiography for *in situ* visualization of gallium solidification in travelling magnetic field *Magnetohydrodynamics* **53** 583–93

[55] Derimow N A, Santodonato L J, Mills R and Abbaschian R 2018 In-situ imaging of liquid phase separation in molten alloys using cold neutrons *J. Imaging* **4** 1–14

[56] Derimow N, Santodonato L J, MacDonald B E, Le B, Lavernia E J and Abbaschian R 2019 In-situ imaging of molten high-entropy alloys using cold neutrons *J. Imaging* **5** 1–11

[57] Kihm K *et al* 2013 Neutron imaging of alkali metal heat pipes *Phys. Proc.* **43** 323–30

[58] Birjukovs M, Dzelme V, Jakovics A, Thomsen K and Trtik P 2020 Argon bubble flow in liquid gallium in external magnetic field *Int. J. Appl. Electromagn. Mech.* **63** S51–7

[59] Birjukovs M, Dzelme V, Jakovics A, Thomsen K and Trtik P 2020 Phase boundary dynamics of bubble flow in a thick liquid metal layer under an applied magnetic field *Phys. Rev. Fluids* **5** 61601

[60] Birjukovs M, Trtik P, Kaestner A, Hovind J, Klevs M, Gawryluk D J, Thomsen K and Jakovics A 2021 Resolving gas bubbles ascending in liquid metal from low-SNR neutron radiography images *Appl. Sci.* **11** 43–6

[61] Trtik P and Lehmann E H 2013 Comment on 'Demonstration of achromatic cold-neutron microscope utilizing axisymmetric focusing mirrors' *Appl. Phys. Lett.* **102103** 183508

[62] Mingrone F, Calviani M, Torregrosa Martin C, Aberle O, Bacak M, Chiaveri E, Fornasiere E, Perillo-Marcone A and Vlachoudis V 2019 Development of a neutron imaging station at the n_TOF facility of cern and applications to beam intercepting devices *Instruments* **3** 32

[63] Leung C L A, Marussi S, Atwood R C, Towrie M, Withers P J and Lee P D 2018 *In situ* x-ray imaging of defect and molten pool dynamics in laser additive manufacturing *Nat. Commun.* **9** 1355

[64] Ghasemi-Tabasi H *et al* 2022 Direct observation of crack formation mechanisms with operando laser powder bed fusion x-ray imaging *Addit. Manuf.* **51** 102619

[65] Hocine S, Van Swygenhoven H, Van Petegem S, Chang C S T, Maimaitiyili T, Tinti G, Ferreira Sanchez D, Grolimund D and Casati N 2020 Operando x-ray diffraction during laser 3D printing *Mater. Today* **34** 30–40

[66] Cabeza S, Özcan B, Cormier J, Pirling T, Polenz S, Marquardt F, Hansen T C, López E, Vilalta-Clemente A and Leyens C 2020 Strain monitoring during laser metal deposition of inconel 718 by neutron diffraction *Superalloys* ed S Tin, M Hardy, J Clews, J Cormier, Q Feng, J Marcin, C O'Brien and A Suzuki (Cham: Springer International Publishing) pp 1033–45

Part III

Natural materials and processes

IOP Publishing

Neutron Imaging
From applied materials science to industry
Markus Strobl and Eberhard Lehmann

Chapter 7

Processes in soil and plants

Anders Kaestner, Andrea Carminati, Mohsen Zare and Pascal Benard

Understanding the interactions between water, soil, and plants is highly relevant for terrestrial ecosystems for predicting the consequences of climate change on agriculture, forests, and natural hazards mitigation.

The study of water and transport processes in porous media, such as in soil or simplified in sand packings, were among the first processes studied using neutron imaging, e.g. [1, 2]. This sample and measurement technique combination comes naturally due to the high neutron contrast between water and soil. The method allows quantifying the distribution and amount of water in the bulk of a sample without necessarily being able to resolve the pore space. It is even possible to quantify the amount of water in the sample using single-neutron radiographs, which reduces the time needed for each time step of the process and thus allows faster time-series acquisition compared to computed tomography. This aspect of neutron imaging permits the study of larger samples than reasonable using methods that depend on the ability to resolve the pores to measure the water content.

The sample dimensions and dynamics of soil samples can be adopted to match the capabilities of neutron imaging without losing relevance of the experiment. The representative elementary volume, level of detail, and the time scale of the observed processes can be captured within the field of view with sufficient spatial and temporal resolution offered by the detectors at the flux of most neutron sources.

Early neutron imaging studies mainly focussed on quantifying the water distribution and transport of trace elements in the soil using radiography for 2D images and, to some extent, tomography to obtain the 3D water distribution. More recent experiments aim at specific questions about the interaction between soil and water transport in roots.

This chapter starts with the main actor in many porous media experiments and the component that delivers the most significant contrast in neutron images, water. The following sections then focus on processes in the vadose zone. The vadose zone in the soil contains a mixture of water and air. The processes are in the rhizosphere,

doi:10.1088/978-0-7503-3495-2ch7

© IOP Publishing Ltd 2024

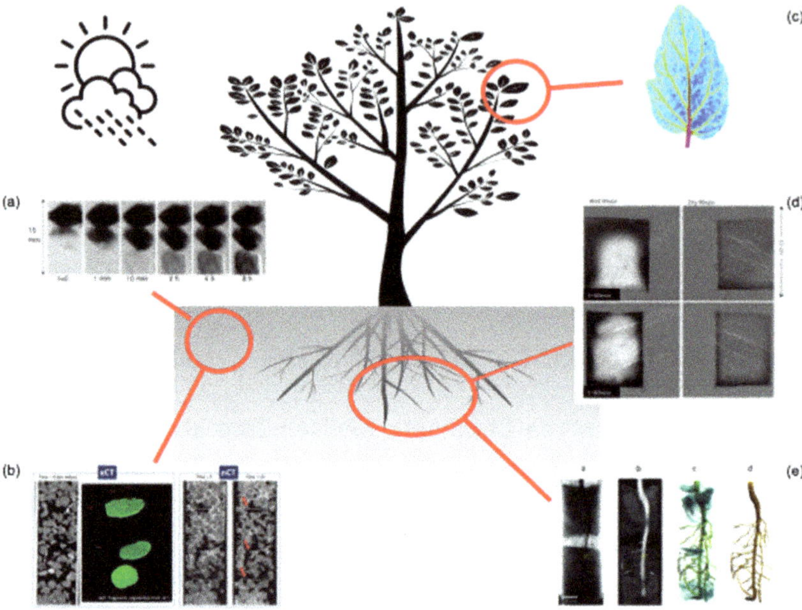

Figure 7.1. An overview of experiment types using neutron imaging to observe water in soil and plants. Soil hydrology (a) and (b) in 2D and 3D. (a) Reprinted from [66], copyright (2007), with permission from Elsevier. (b) Reprinted from [12], copyright (2023), with permission from Elsevier. Evaporation from foliage (c), reprinted from [75], copyright (2014), with permission from Springer Nature. Water transport in rhizosphere and roots (d) and (e). (d) Reprinted from [65] CC BY 4.0. (e) Reprinted from [25] CC BY 4.0.

which is the narrow segment of soil near the roots and represent the interaction between soil and roots and the general water distribution in the plants. Figure 7.1 summarizes imaging results related to water transport in the soil as a pure structural medium in the vadose zone and in the rhizosphere as interaction between plants and soil.

7.1 Water

Water is essential for the existence of life; it acts as a transport fluid for nutrition and trace elements without being nutrition itself. Water is also essential in many technical applications such as chemical reactors and electrochemical systems such as fuel cells; see Part IV of this book.

Water is, furthermore, one of the most studied liquids using neutron imaging. The reason is the high neutron sensitivity to hydrogen. The attenuation coefficient for water in Nature is in the order of 3.2–3.8 cm^{-1}, depending on the used neutron spectrum. The attenuation coefficient values are less for a thermal spectrum and greater for a cold spectrum. An attenuation coefficient of this magnitude contrasts the materials forming many porous media, usually based on minerals and silicon, that often have attenuation coefficients of less than 1 cm^{-1}.

Neutron imaging experiments to study water cycle processes involve all three aggregation forms; solid, liquid, and gas. Early neutron imaging experiments aimed at detecting the presence of water in the observed specimen. Meanwhile, it is possible to quantify the absolute volume of water using a radiograph. However, the quantification requires that the attenuation coefficient of water, Σ_{water}, is known for the spectrum of the imaging beamline. A step wedge with increasing water thickness can be used to determine the attenuation coefficient if it is unknown. Typically, the thickness of a water layer is determined using the following relations between two images; the initial condition, I_{dry}, and the wetted sample, I_{wet}:

$$\begin{cases} I_{\text{dry}} = I_0 e^{-\left(\mathrm{d}_{\text{container}}\Sigma_{\text{container}}+\mathrm{d}_{\text{porous}}\Sigma_{\text{porous}}\right)} \\ I_{\text{wet}} = I_0 e^{-\left(\mathrm{d}_{\text{container}}\Sigma_{\text{container}}+\mathrm{d}_{\text{porous}}\Sigma_{\text{porous}}+\mathrm{d}_{\text{water}}\Sigma_{\text{water}}\right)} \end{cases}$$

Dividing these two images provides the change in transmission caused by water

$$I_{\text{wet}}/I_{\text{dry}} = e^{-\mathrm{d}_{\text{water}}\Sigma_{\text{water}}} \Longrightarrow \mathrm{ad}_{\text{water}} = -\ln\left(\frac{I_{\text{wet}}}{I_{\text{dry}}}\right)/\Sigma_{\text{water}}$$

Here, d_{water} represents the thickness of water in the beam direction. The water volume behind each pixel is calculated using the pixel area multiplied by d_{water}. It is often more relevant to measure changes in water content over time than to make single observations. The changes are usually measured differentially relative to an initial condition of the sample, e.g., when it is saturated or dry. The image time series of water entering or leaving the sample is used to quantify the amount of water over time and to build models explaining the behaviour of the water distribution in the structure [3, 5].

The equation above is simplified to illustrate the principle of obtaining the water content in a sample using neutron radiographs. The primary interaction mechanism between neutrons and water (hydrogen) is scattering. The presence of water introduces high uncertainty in the interpretation of the grey levels in the image due to bias caused by scattered neutrons. This bias has a smooth intensity profile that increases behind regions with a higher density of scattering elements. Therefore, it is essential to remove the bias caused by scattering from the sample and instrument background to avoid misinterpretations of the data. The correction for scattering can be done using methods described in chapter 20 and involves measuring additional reference images to estimate the contribution from scattered neutrons.

The liquid and solid forms of water are trivial to detect using neutrons and represent most neutron imaging experiments. These attenuation coefficients of the two aggregation forms are hard to separate using a white neutron spectrum. There is, however, a subtle difference between the two aggregation forms in their attenuation spectra. This difference indicates that wavelength-resolved imaging is a method for distinguishing between the two forms. Siegwart [82] demonstrated that separating ice and liquid water in fuel cells is possible using the ratio between two images, one made below and the other above a given wavelength threshold. Detecting water vapour is more challenging due to the low number of water molecules per volume unit. Thus, it requires long neutron transmission distances through the vapour to allow detection and quantification. Vapour can, however, be part of either condensation or evaporation processes to and from liquid water.

Table 7.1. Comparing physical properties of D_2O and H_2O.

Property	D_2O	H_2O	Unit
Melting point	3.83	0	°C
Boiling point	101.72	100.0	°C
Density	1.1056	0.9982	g cm^{-3}
Viscosity	1.25	1.005	mPa s
Surface tension	71.93	71.97	mN m^{-1}
Cross section @ 1.8 Å	13.1	72.0	barn
Attenuation coefficient @ 1.8 Å	0.587	5.647	1 cm^{-1}

Therefore, the vapour amount is indirectly measured by measuring changes in the sample's liquid water as very large volumes are required to detect any changes directly.

Neutrons are sensitive to different isotopes of the elements, chapter 1. The isotope sensitivity is, in particular, relevant to water as there is a significant difference in attenuation coefficient between the two isotopes ^1H and ^2H; the latter is also called deuterium. ^1H is the most abundant isotope, with nearly 100% of all H atoms. The abundance of ^2H is only 0.0156%. The significant contrast difference between the isotopes and the in-comparison slight difference in relevant physical properties, table 7.1, makes it attractive to use combinations of natural water, H_2O, and deuterated water or heavy water, D_2O, as tracers in diffusion and transport experiments [3, 6, 9].

Heavy water in high concentrations is toxic to living organisms [10, 11], a fact that must be considered while planning experiments that involve the use of heavy water. The heavy water does, however, not affect organisms severely in low concentrations or during short periods to be washed out by natural water after serving its purpose as a tracer.

7.2 Vadose zone hydrology

The vadose zone is the region between the Earth's surface and the groundwater table. The pores in this region are typically partially saturated with a mixture of water and air; thus, it is also called the unsaturated zone. This zone is vital for terrestrial ecosystems and is called the critical zone in this context. The vadose zone controls energy exchange and matter exchange, e.g., water and carbon between land and the atmosphere. Thus, the vadose zone plays a central role in climate predictions, provides a habitat for plant and microbial growth, and is a filter controlling water quality.

Soil properties and function depend on the soil texture and structure. The texture describes the particle size distribution and how the soil particles arrange into larger clusters, such as soil aggregates. In the upmost part of the vadose zone, the soil is a complex medium with high biological activity, containing a higher organic fraction, and is the location of most roots. Of particular relevance is the soil region around the roots, the rhizosphere. Plants and soil interact in the rhizosphere. The plants take up water and nutrients here while they deposit carbon in the soil. The interest in studying structures and transport processes in the soil is manifold.

Several vadose zone processes and properties vary with water saturation, i.e., the fraction of the pore space saturated with water. The energy state of water, the water potential, which, among other effects, determines water availability to plants, depends on water saturation. The ability of water flow in porous media, the hydraulic conductivity, varies with water saturation. Besides the water flow, the solute transport depends on water saturation due to the dependence on diffusion and dispersion. Gas diffusion is much faster in the air phase than in the liquid one, and thus the diffusion of gas depends on water saturation, as water impedes it. This relation is, for example, essential for oxygen availability to bacteria and plants. Finally, water saturation is vital for the ability of plant roots to extract water from the soil. As the soil is inherently heterogeneous across scales, measuring the spatial distribution of water in soils at varying scales is crucial. Measuring the spatial distribution of soil properties is needed to predict flow and transport processes. Soil organic matter, and its spatial distribution, is an important component of soils, which is important for water retention and biogeochemical processes [12].

The high sensitivity of neutrons to water and organic matter makes neutron imaging an optimal tool to investigate vadose zone processes. It allows for quantifying the spatial and temporal dynamics of water saturation in soils with varying structures. Perfect *et al* review the application of neutron imaging on vadose zone processes [5].

7.2.1 Static quantification of soil water content and soil organic matter

Water distribution is a transient process. However, despite its dynamic nature, it is important to know the equilibrium water distribution in soils at a steady state (for a given volume of water per soil volume). This water distribution is particularly important to characterize water retention, which is the relation between soil volumetric water content, soil matric potential, and spatial heterogeneity. In a recent application, Cheng *et al* applied neutron imaging to calculate water retention curves [13]. The water retention describes the soil's ability to retain the water under different levels of suction or pressure. They quantified the water retention curve based on transmission images and derived the matric potential from the height. Water thickness was measured using a calculation that combined the attenuation coefficient of water and a beam hardening factor based on a calibration procedure using step wedges. The beam hardening is a consequence of the energy dependence of the attenuation in the material. This beam hardening factor compensates, besides beam hardening, also for the effects of scattering from sample and instrument background. Scattering is in fact the main reason for the biased reading [14]. The water content was used to fit the parameters of the van Genuchten model [15]. The results from the imaging experiments were validated by hanging water column experiments. Carminati *et al* used a similar approach to investigate whether and to what extent the water retention curve is impacted by root activity [16]. The results are discussed in the sections below.

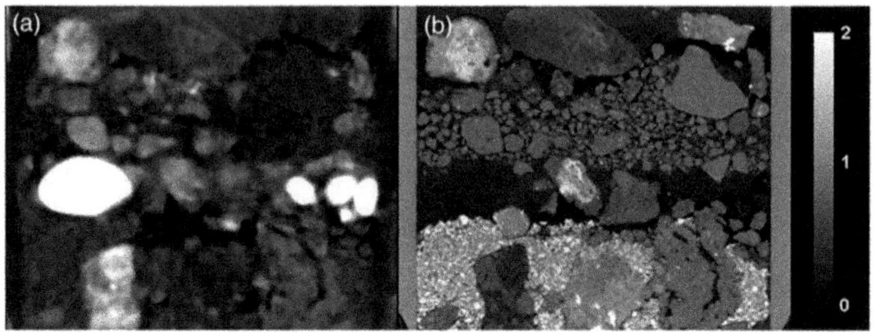

Figure 7.2. Vertical slices of neutron (a) and x-ray (b) tomography showing the distribution of attenuation coefficients (cm^{-1}). Reprinted from [17] John Wiley & Sons. Copyright 2021 The Authors. European Journal of Soil Science published by John Wiley & Sons Ltd on behalf of British Society of Soil Science.

The combination of neutron and x-ray imaging allowed Koestel *et al* to estimate the composition of the soil matrix. It showed the possibility of quantifying soil organic matter distribution in dry samples [17]. Neutron imaging and x-ray CT have complementary attenuation characteristics in soil–water systems, figure 7.2. Combining the two modalities makes it possible to provide more specific information on soil composition. Soils consist primarily of minerals, mainly silicon and aluminum, which are relatively translucent to neutrons contrary to x-rays, and organic matter, which is more strongly attenuated by neutrons than x-rays. Combining x-ray and neutron imaging demonstrated abilities to identify the change in bulk densities and the content of soil organic matter or clay domain. Minerals are crystals that can be probed by neutron diffraction methods and, in particular, observing the Bragg-edges caused by d-spacings in the minerals. Koestel *et al* also report the first neutron time-of-flight imaging experiments to better distinguish between organic matter and clay minerals. They conclude that further development of the analysis is still required to separate the minerals from organic residues in soil materials.

7.2.2 Flow and transport in porous media

Understanding the soil properties and processes that impact the flow of water and transport of solutes is a core topic in vadose zone research due to both the complexity of the topic and its relevance for the transport of pollutants through the soil, e.g., [18, 19]. In addition, the high sensitivity of neutrons imaging to water makes it an ideal tool for investigating water flow and solute transport in soils.

Deinert *et al* used real-time neutron radiography to quantify water content profiles in the sand [20]. Their set-up allowed for a time resolution of 30 ms with a field of view of ca. 400 cm^2 and a pixel size of 0.5 mm. With this set-up, they were able to quantify the variation in moisture content across a wetting front moving at constant velocity.

Soils are heterogeneous porous media that require 3D information to understand the ongoing processes completely. Schaap *et al* [21] and Kaestner *et al* [22] used neutron tomography to obtain the 3D distribution of water in a heterogeneous sand column undergoing repeated drainage-wetting cycles. The column had a rectangular shape with a squared base, a side length of 50 mm, and a volume of 105 cm^3, and it was comprised of a heterogenous arrangement of cubic regions. In all, there were 101 cubes with fine and 49 cubes with coarse sand. The time series of tomograms made it possible to explore the impact of soil structure on water saturation during imbibition and drainage. These experiments also, for the first time, demonstrated the effect of scattering correction [23] on a larger scale beyond method validation experiments. The use of deuterated water increased the penetration depth compared to natural water allowing investigation of how the liquid was distributed in the relatively large column. The high frame rate of the used flat panel detector allowed the acquisition of tomographies within 56 s at a pixel size of 127 μm. Although the flat panel detector was, at the time, the fastest option for rapid image acquisition, about a decade later, it was possible to reach these frame rates using camera-based systems, which offers greater flexibility [24, 25].

A similar approach was later also used by Snehota *et al* [26] to investigate air entrapment in heterogeneous sample packings during infiltration in quasi-saturated conditions. The authors used 2D time-series imaging to capture the relatively fast changes in water content and 3D imaging during quasi-steady state conditions, figure 7.3. Similar to Schaap *et al* [21], the authors used deuterated water, D_2O, to increase the penetration depth to investigate thicker samples—in this case, a cylinder

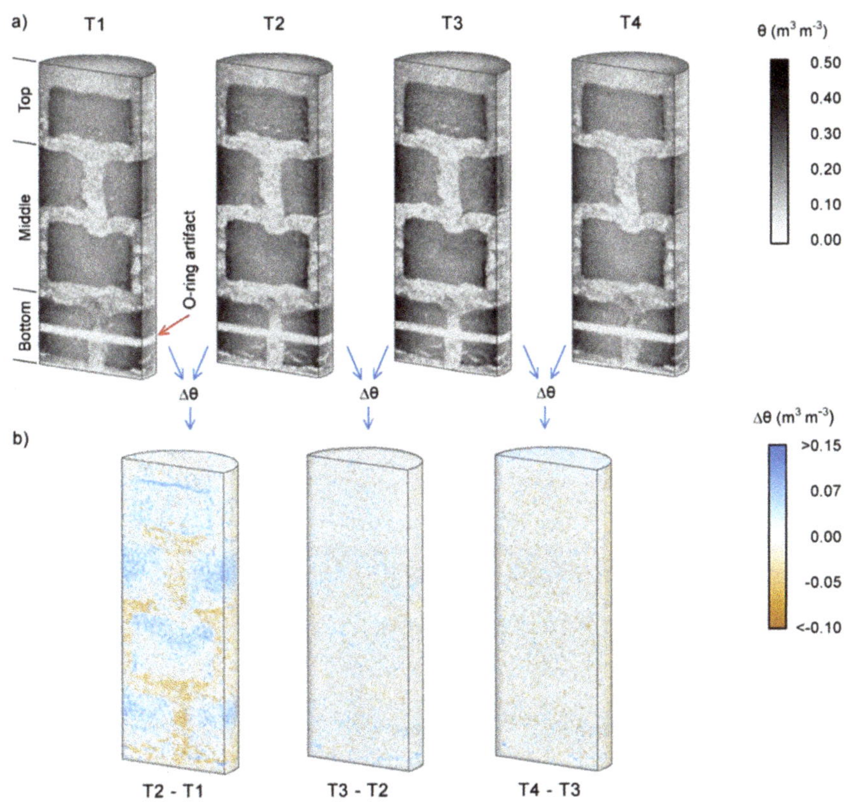

Figure 7.3. Vertical slice of the 3D image illustrating the water content and the temporal changes at different steps of the infiltration and drainage experiment. Reprinted from [26] John Wiley & Sons. Copyright 2015. American Geophysical Union. All Rights Reserved.

with a diameter 34 mm. The water sensitivity of the imaging method allowed detecting the air-entrapment dynamics during near-saturation water infiltration.

7.2.3 Evaporation

Evaporation is a key factor in soil drying. The low vapour pressure of the atmosphere compared to the pressure in the soil controls evaporation, where the gas phase is typically saturated with vapour. In wet soils, evaporation is controlled by atmospheric conditions like solar radiation, temperature, humidity, and wind. But as the soil dries, capillary forces become insufficient to sustain the evaporation demand of the atmosphere and overcome the needed viscous forces [27]. A central question in evaporation research in vadose zone hydrology is to find the critical point when the soil starts to limit the evaporation rate and how this critical point depends on soil properties, such as soil texture.

Shokri *et al* [28] used neutron imaging to quantify the morphology of the drying front and water content dynamics during evaporation in sandy substrates. They found that the drying front showed irregular patterns and isolated liquid clusters. Furthermore, they found that the vapour between the drying surface and the boundary layer limited the evaporation rate. The images helped to prove the theory on the forces controlling evaporation rates from soils.

Zheng *et al* and Benard *et al* [29, 30] extended the work of Shokri *et al* to rhizosphere studies. They investigated how the evaporation rates are affected by plant growth-promoting rhizobacteria and plant mucilage. Both studies showed that biofilms suppress evaporative fluxes and thus maintain soil moisture. Furthermore, the neutron radiographs obtained by Benard *et al*, figure 7.4, showed that the evaporation

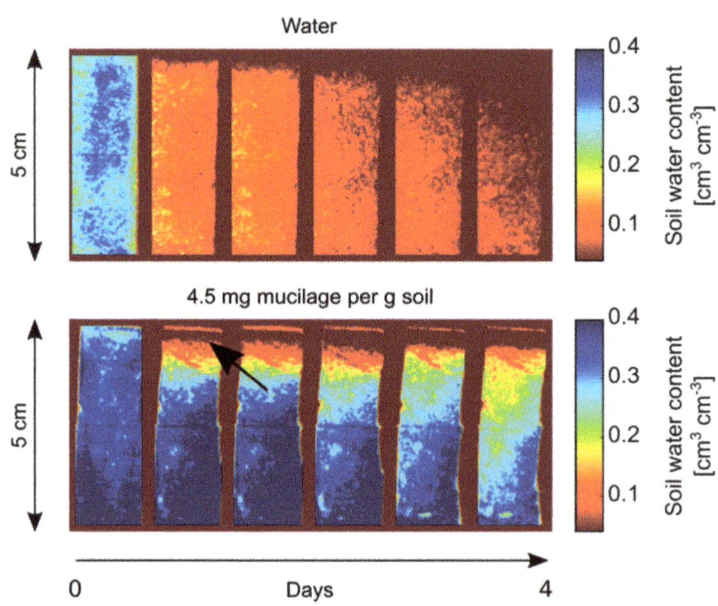

Figure 7.4. Water distribution in soils without and with mucilage during evaporation. Reproduced from [31] CC BY 4.0.

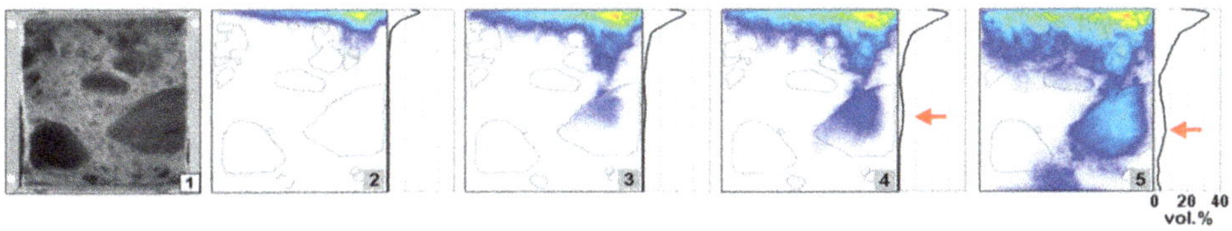

Figure 7.5. Water distribution in heterogeneous soil material, adapted from [36] John Wiley & Sons. Copyright Soil Science Society of America.

suppression was concomitant with forming a thin, dry layer near the soil surface, where plant exudates were deposited as 2D surfaces reaching across the pore space. Another study by Benard *et al* highlighted the potential of soil bacteria to adapt to fluctuations in soil moisture utilizing the same mechanism as illustrated in figure 7.4 [31]. The authors observed accelerated drying of soil near the surface which initiated a period of reduced evaporation governed by vapour diffusion.

7.2.4 Undisturbed and structured soils

Most studies on infiltration and evaporation described above were performed in repacked homogenous soil substrates mainly consisting of sand. Natural soils are, however, inherently heterogeneous and structured. x-ray CT has been instrumental in visualizing the arrangement of soil particles and the liquid phase [32, 34]. In addition, the ability of neutron imaging to quantify water has also been used to successfully study infiltration in structured soils [36] (figure 7.5). Examples include the infiltration into aggregate packing [35] and in mine soils [32]. These examples illustrate the importance of soil structures like macropores and aggregates on preferential flow.

7.3 Rhizosphere water dynamics

Root water uptake and the dynamics are closely linked to the soil water content in the rhizosphere [4 ,16, 37]. When they extract soil water, roots create a gradient in water potential between the bulk soil and the root surface. The induced water flow results in soil moisture depletion, which affects a manifold of physical and biological processes, such as root water uptake, solute and gas diffusion, and microbial activity [30]. These processes are vital for plant growth. Therefore, understanding the rhizosphere's water dynamics is of fundamental importance.

Early observations of reduced soil water content near roots were interpreted as a depletion caused by root water uptake in these regions. Authors of [33, 38–40] used light transmission, MRI, and x-ray tomography and to image root growth and soil rewetting after irrigation of lupine and maize for up to 4 weeks. The authors observed reduced soil water content near older parts of the root system, which was suspected to result from either a quick depletion by root water uptake or increased porosity. Nakanishi *et al* [41] on the other hand, observed a gradual increase in water content across the rhizosphere of soybean with thermal neutron tomography, figure 7.6.

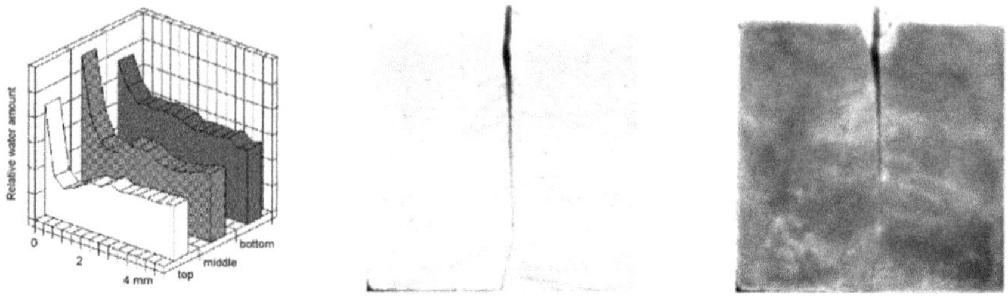

(a) (b) (c)

Figure 7.6. Gradual increase in water content towards the roots surface of soybean resolved by neutron tomography (a), reprinted from [41], copyright (2003), with permission from Springer Nature. The effect of reduced water content near roots was also observed in radiography studies. Panel (b) shows the soil after drying and in (c) the soil was rewetted, reprinted from [16], copyright (2010), with permission from Springer Nature.

In another study, this gradual increase was confirmed for soybean as well as for spring bean and pea roots, using the same technique [9]. Furthermore, the gradient in soil water content appeared less pronounced at greater depths. Similar observations were made by Tumlinson *et al* in the vicinity of corn roots during root water uptake at different depths captured by neutron tomography [3]. At first, these studies appeared to contradict each other as some reported a decrease in soil water content across the rhizosphere while others made a contrary observation. Nevertheless, neutron imaging became a preferred method for investigating rhizosphere water dynamics [7, 8].

Carminati *et al* [16] used neutron radiography to measure the soil moisture distribution in the rhizosphere of lupin. Increased water content during drying and temporal water repellency was observed within the vicinity of roots, figures 7.6(b) and (c). The moisture dynamics alteration was attributed to root-exuded mucilage, a highly polymeric blend of substances released from the root tip.

Esser *et al* [42] combined neutron radiography and tomography to quantify water distribution in the rhizosphere of lupin and maize. The authors observed an increased soil water content in the rhizosphere during and after irrigation, especially near the root tips. This increase in water content is likely to be caused by high water retention of mucilage [43–46]. Locally decreased soil water content was attributed to root water uptake in these regions. Using a combination of x-ray and neutron tomography, Mawodza *et al* observed an increase in the water content of wheat roots in contact with moist soil aggregates [45]. However, moisture gradients within the rhizosphere were not reported. Moradi *et al* [46] quantified the dynamics of water content in the rhizosphere of chickpea, lupin, and maize by neutron tomography using thermal and cold neutrons. The authors measured increased soil water content near the roots at 4, 6, and 8 cm depths of all plant species, figure 7.7. Following Carminati *et al* [16] the authors explained this observation with a modified soil water retention of the rhizosphere. Moradi *et al* [47] showed with a radiography time-series study that there is a significant delay in drying and rewetting the rhizosphere of lupin grown in sandy soil. The dry rhizosphere's increased soil

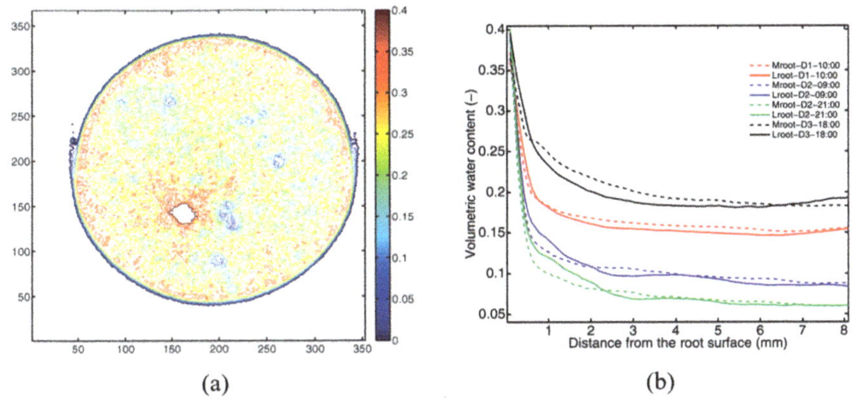

Figure 7.7. Water distribution near the root of chickpea (a). The water distribution in the lower part of the root network (b). Reprinted from [46] John Wiley & Sons. Copyright 2011 The Authors. New Phytologist Copyright 2011 New Phytologist Trust.

water retention and decreased wettability were attributed to the presence of root-exuded mucilage and other polymeric substances released by microorganisms.

Similarly, Ahmed *et al* [48] explained the reduced rhizosphere water content by the wettability of the mucilage in the rhizosphere of lupin and maize grown in sandy soils. A series of experiments using time-series neutron radiography elucidated the impact of dry mucilage on rhizosphere wettability. Kroener *et al* explained the susceptibility of coarse-textured rhizosphere soil to turn water-repellent with the comparably low surface area needed to be covered by dry, low-wettability mucilage to render the soil temporarily non-wettable [49]. The authors acquired time series of neutron radiography of capillary rise experiments in different soil textures and at varying chia mucilage contents. The threshold-like behaviour of the water infiltration near a critical mucilage content observed by Kroener *et al* was later confirmed with similar capillary rise experiments for maize mucilage in a variation of different soil textures and wettabilities [50, 51].

Later, Benard *et al* demonstrated the water retaining effect of maize mucilage in soil [30]. In these experiments, mucilage amended glass beads were packed next to water-wet glass beads. After a period of equilibration, the difference in water content was quantified using neutron radiography.

A problem arising when performing neutron radiography experiments with roots and the rhizosphere is the contradicting requirements of root network morphology and the neutron penetration depth in wet soil material. The roots need large volumes to grow unconstrained, soon reaching the limit of completely blocking all neutrons from penetrating the sample. The compromise to this problem is to grow the plants in flat aluminum containers (i.e., rhizotrons or rhizoboxes). This container shape allows roots to grow unconstrained in two directions, and at the same time, the slab thickness allows sufficient neutron transmission for the images. The degree of transmission from a moisture-saturated soil slab defines the slab thickness. Although adverse, considering the geometry of a naturally accessible soil volume, this

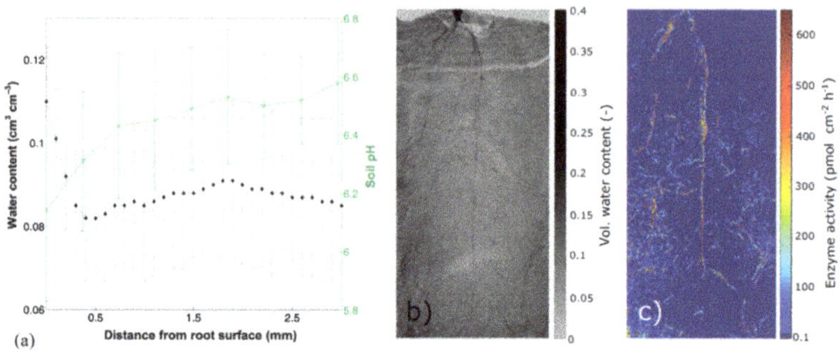

Figure 7.8. Average water content derived from time-series neutron radiography and pH captured with fluorescence imaging as a function of distance from the surface of maize crown roots (a), [53] John Wiley & Sons, copyright 2017 WILEY-VCH Verlag GmbH & Co. KGaA, Weinheim. Volumetric water content rhizobox soil and barley root system imaged with neutron radiography beside phosphatase activity mapped with soil zymography (adapted [55], copyright (2019), with permission from Springer Nature).

limitation offers the opportunity to combine neutron imaging of rhizosphere water with other methods requiring containers of similar dimensions.

Rudolph-Mohr *et al* presented a combination of techniques to capture gradients in soil water content, pH, and oxygen in the rhizosphere of lupin [52]. This combination of time-series neutron radiography and fluorescence imaging was later used to map the rhizosphere of the crown and lateral roots of maize plants [53], figure 7.8(a). The authors observed increased water retention in the vicinity of roots from both lupin and maize. Furthermore, the observed water content gradients became steeper during soil drying. Using a combination of neutron radiography and $^{14}CO_2$ labelling, Holz *et al* [54] showed how the rhizosphere's modified water dynamics help maintain root exudate diffusion in drying soil. They further investigated the impact of increased water content in the barley rhizosphere on phosphatase activity by combining neutron radiography and zymography [55]. In this work, it was found that there is a significant increase in water retention near the roots which maintained a high enzyme activity during soil drying, figure 7.8(b) and (c). The work with the combined methods of $^{14}CO_2$, zymography, and neutron radiography was continued by Bilyera *et al*, who studied different exudation strategies of three maize genotypes [56]. The maize genotypes showed contrasting exudation strategies, explained by different compensation strategies to ensure high microbial activity in the rhizosphere.

7.3.1 Root water uptake and flow in plants

Plant roots are vital in extracting water from the soil and transporting it to the shoot to sustain the plant's transpirational demands. To fulfil this function, plants develop a large complex root architecture consisting of several roots differentiated in diameters, branching order, positions, and anatomy. Bearing this complexity in mind, a key question will be the functioning of different root types and locations in water extraction from soils, particularly under limited water conditions. The answer to this question requires experimental knowledge about where and how fast water is extracted from soils by different root types and locations.

Recent advances in imaging water and root distribution in soil using different advanced imaging techniques such as x-ray computed tomography [57], light transmission imaging [38], magnetic resonance imaging [58], and neutron tomography [46] enabled *in situ* quantification of soil water content changes around the roots of transpiring plants. These studies provided precious information on soil roots and water distribution [16, 59, 60]. However, the attempt to relate the change in soil water content to the location of root water uptake was not plausible due to the quick redistribution of water within soil. Typically any attempt to quantify root water uptake from the change in soil water content requires detailed modelling of water flow within soil and root systems. The method requires accurate knowledge of the soil hydraulic properties in the root zone, particularly at the root–soil interface.

The challenge in quantifying root water uptake by transpiring plants growing in soils was recently overcome with the help of neutron radiography combined with the injection of D_2O. Matsushima *et al* used neutron radiography combined with D_2O injection in soils to visualize water fluxes in the roots and shoots of living plants [61]. They used D_2O as a tracer of normal water within soil and plant due to the large difference in the neutron attenuation coefficient of H_2O versus D_2O for cold neutrons, a factor of 8 according to [23]. This remarkable difference in neutron attenuation makes the flow of D_2O within porous media (i.e., root and soil) distinguishable from existing normal water. Inspired by this work, Zarebanadkouki *et al* introduced the first quantitative technique to locate the spatiotemporal distribution of water fluxes along the root system of plants growing in soils [62, 63] figures 7.9(a1), (a2). The technique consists of monitoring and modelling the transport of deuterated water (D_2O) into roots. D_2O was injected into the soil and its transport within the roots was monitored using time-series neutron radiography. They found root water uptake is not uniform along the root system with varying contributions of different root types and locations in lupin [63] and maize plants [64].

Recently the application of neutron radiography to study the flow of water along the soil–root system has been extended to 3D observation using ultra-fast neutron tomography. This breakthrough technology has been achieved by a drastic acceleration of the image acquisition process, which is necessary to resolve water flow within soils and roots in 3D. To meet the timing requirements to resolve the water flow, Zarebanadkouki *et al* employed the first ultra-fast neutron tomography with a temporal resolution of 30 s per full volume frame [24, 65]. This experiment made it possible to visualize water transport across the root tissue of 12 days old lupin plants, figures 7.9(b1)–(b3). Their contribution showed that the application of an ultra-fast neutron tomography combined with the injection of D_2O and modelling its transport across the root tissue is capable of revealing water fluxes across the root tissue at the cellular level. They found that water flows nonuniformly across the root tissue. Tötzke *et al* extended the application of the ultra-fast neutron tomography combined with injection of D_2O to the entire plant root system and visualized the progression of D_2O front within the root system [25], figures 7.9(c1)–(c6).

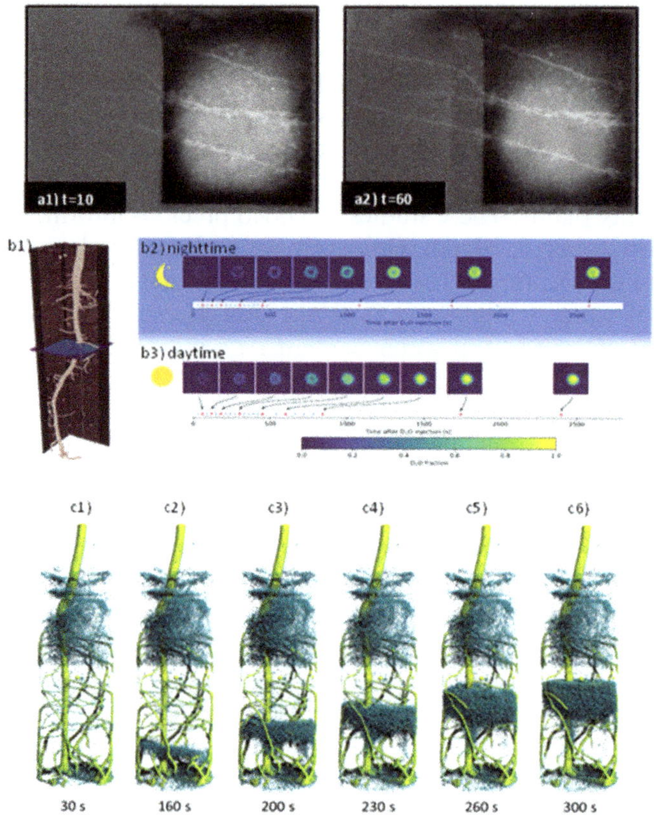

Figure 7.9. (a1)–(a2) Neutron radiography of D_2O (used as a tracer of normal water) transport within the root tissue (adapted from [63] John Wiley & Sons, copyright 2013 The Authors. New Phytologist copyright 2013 New Phytologist Trust). This data shows that root water uptake is not uniform along the root system, with varying contributions of different root types and locations. (b1)–(b3) Ultra-fast neutron tomography of the D_2O transport across the root tissue during day and nighttime (adapted from [65] CC BY 4.0). These results confirmed the capability of ultra-fast neutron tomography to resolve the transport of water across the root tissue at the cellular level. (c1)–(c6) Time-series neutron tomography of the D_2O transport along the root system. The D_2O was injected at the bottom of the soil (adapted from [25] CC BY 4.0).

7.3.2 Root growth and nutrition or pollution transport

There is often remarkable heterogeneity in the spatial distribution of natural soil resources such as water and nutrients or of anthropogenically induced resources such as fertilizers and contaminations [66, 67]. There have been numerous attempts to uncover the feedback within root developments and spatial distribution of resources in soils in either case. Neutron radiography has proven to be an excellent tool to study *in situ* root development within soils and its plasticity in response to soil patchy resources by overcoming the technical challenge of visualizing roots as the hidden half of plants.

Plants may preferentially vary their root proliferation towards soil regions with a higher water and/or nutrient contents [68, 69]. Felderer *et al* used neutron radiography to uncover the role of cluster roots, bottlebrush-like structures of densely haired rootlets [70]. They showed that white lupin plants boost cluster root growth in soil regions with limited water content more than in regions with P limited content. Similarly, in contaminated soils it is fundamental to understand how plant root development responds to soil contaminated zones, particularly in the context of phytoremediation [71]. Phytoremediation is a plant-based approach in which plants are used to extract and remove pollutants or lower their bioavailability in soils. In such a case, preferential root proliferation into soil patches containing elevated concentrations of the target metal might be an important factor in metal acquisition by hyperaccumulator plants. In other words, identifying hyperaccumulator plants possessing root proliferation into soil patches containing a specific pollutant can be ideal for use in phytoremediation plants. Menon *et al* monitored the root growth of lupin plants over a period of 3 weeks and showed that root growth is significantly decreased in the soil zones contaminated with boron (B) or zinc (Zn) [72]. In their study they demonstrated that lupin roots have a mechanism to avoid contaminated soil. Moradi *et al* monitored the root development of *Berkheya coddii* (a Ni hyperaccumulator) and chickpea (a non-hyper Ni accumulator) in soil heterogeneously spiked with Ni [73]. They observed a remarkable morphological change in the root system of *B. coddii* in Ni-contaminated soils (thicker diameter, less branching, lower root length) compared to the control soil with no preferential proliferation of *B. coddii* towards the Ni-spiked regions.

7.3.3 Water dynamics in the shoot system and transpiration

The focus of this chapter has mainly been water flow and solute transport in soil and the root system. The full water management of a plant does, however, also involve transport and flow in the shoot system. Here, the leaves play an important role, in particular, due to the transpiration which acts as the sink of the flow and pulls water from the soil through the roots. This has been observed by studying the root water uptake in day-and-night cycles [62] which is an indication of the shoots' role in the water cycle. Matsushima *et al* used cold neutrons to measure the water thickness in Hedera leaves and observed a logarithmic relation between water thickness and time during dehydration by evaporation from the leave's surface [74]. Defraeye *et al* took these results further using tomography by showing that it is possible to quantify the amount of water in the leaves within between 7% accuracy compared to gravimetric measurements for thin leaves like tomato, while the water content for thicker leaves was less accurate with more than 30% deviations [75]. Defraeye *et al* were also able to show how light changes the water uptake of the leaf. D_2O is, as shown earlier in this chapter, often used as tracer in flow experiments and Matsushima *et al* [76] demonstrated that the flow of D_2O in neutron images can be measured from grey level changes in the images of rose peduncles using optical flow image analysis algorithms. Later they also used D_2O as trace to estimate the water flow velocity Matsushima [78].

The applications of leaf studies using neutron imaging has mainly been to understand the effects of the environment in plants. This includes air as well as soil pollution.

Boron is an essential micronutrient to plants in correct concentrations [83, 84]. Some soils do, however, have toxic levels of boron due to fertilizers and spreading of ash on the fields. The effect of too high boron levels is less efficient crop growth. To remediate soils with toxic levels of boron and make it suitable for crop is to plant boron tolerant plants. Poplar is one such plant species (cite Robinson). Boron is in the view of neutrons a high-contrast element for neutrons and specifically the ^{10}B isotope can be detected at very low concentrations. Initially, the impact of boron was initially neglected in the quantification of water content due to very small amounts in many natural plants [74]. Rees *et al* did, however, use the high boron sensitivity to show that certain species of poplar are accumulating boron and better than many other plants. In these experiments, plants were grown in soils with different concentrations of boron which resulted in the uptake and accumulation by the plant [77]. The highest concentration of boron was found at the tips of the veins in the leaf. Second highest concentration was found in the roots, while the stem appeared to merely serve as transport path between roots and leaves and show lower boron concentrations. Matsushima [61, 78] used a combination of neutron and chlorophyll fluorescence analysis imaging to investigate the impact of air pollution by SO_2 in hibiscus plants. They concluded that there is a severe reduction in transpiration and photosynthesis in plants that were exposed to SO_2 at amounts corresponding to average exposure near a city street.

The stem of the plant has different purposes, providing mechanical stability of the plant and to provide a means for transporting nutrients and water between leaves and roots. Important here is the xylem and phloem which provide a passive mechanism of transport. Matsushima, furthermore, demonstrated the ability to measure flow in the xylem [78]. These results lead to more detailed experiments to study transport in the xylem using combinations of neutron imaging and x-ray imaging or NMR. These studies aimed at specific anatomic questions like the formation and impact of embolism in the vessels in the xylem [79]. In this study it was possible to detect pockets of gas formed in the vessels under different light and irrigation conditions. Ryu studied the impact of ion transport in the stem of olive and laurel plants using synchrotron and neutron imaging [80]. Here, the ions K^+ and Ca^{2+} were of main interest. Both ions are included in the physiochemical control of the sap water flow in the plant. Most above-mentioned studies of the stem were made with stems severed from the plant. Severing the samples is convenient for the experiment preparation, but it has the disadvantage that the plant cannot be studied as a whole system like Matsushima *et al* did in [61]. Malone *et al* used neutron radiography as a validation method for their NMR-based method to measure the water content in *in vivo* plant stems [81]. In this work, NMR was used as primary technique for long-term monitoring. The simultaneous acquisition using the two modalities confirmed the water contents observed with NMR.

References

[1] Wilson N E, Harms A A and Emery J J 1975 A neutron radiographic *in vitro* examination of soils *Can. Geotech. J.* **12** 152–6

[2] Jasti J K, Lindsay J T and Fogler H S 1987 Flow imaging in porous media using neutron radiography *SPE Annual Technical Conference and Exhibition (Dallas, TX)* SPE-16950-MS
[3] Tumlinson L G, Liu H, Silk W K and Hopmans J W 2008 Thermal neutron computed tomography of soil water and plant roots *Soil Sci. Soc. Am. J.* **72** 1234–42
[4] Okuni Y, Furukawa J, Matsubayashi M and Nakanishi T 2002 Water accumulation in the vicinity of a soybean root imbedded in soil revealed by neutron beam *Anal. Sci./Suppl.* **17** i1499–501
[5] Perfect E, Cheng C-L and Lehmann P 2011 Neutron imaging and applications: a reference for the imaging community *Vadose Zone J.* **10** 1336–7
[6] Carminati A, Kaestner A, Hassanein R, Ippisch O, Vontobel P and Flühler H 2007 Infiltration through series of soil aggregates: neutron radiography and modeling *Adv. Water Resour.* **30** 1168–78
[7] Oswald S E, Menon M, Carminati A, Vontobel P, Lehmann E and Schulin R 2008 Quantitative imaging of infiltration, root growth, and root water uptake via neutron radiography *Vadose Zone J.* **7** 1035–47
[8] Zarebanadkouki M, Kroener E, Kaestner A and Carminati A 2014 Visualization of root water uptake: quantification of deuterated water transport in roots using neutron radiography and numerical modeling *Plant Physiol.* **166** 487–99
[9] Nishiyama H 2005 Water gradient profiles at bean plant roots determined by neutron beam analysis *J. Radioanal. Nucl. Chem.* **264** 313–7
[10] Katz J J 1960 The biology of heavy water *Sci. Am.* **203** 106–17
[11] Basov A, Fedulova L, Vasilevskaya E and Dzhimak S 2019 Possible mechanisms of biological effects observed in living systems during $^{2}H/^{1}H$ isotope fractionation and deuterium interactions with other biogenic isotopes *Molecules* **24** 4101
[12] Kim K, Kaestner A, Lucas M and Kravchenko A 2023 Microscale spatiotemporal patterns of water, soil organic carbon, and enzymes in plant litter detritusphere *Geoderma* **438** 116625
[13] Cheng C L *et al* 2012 Average soil water retention curves measured by neutron radiography *Soil Sci. Soc. Am. J.* **76** 1184–91
[14] Hassanein R, Lehmann E and Vontobel P 2005 Methods of scattering corrections for quantitative neutron radiography *Nucl. Instrum. Methods Phys. Res.* A **542** 353–60
[15] van Genuchten M T 1980 A closed-form equation for predicting the hydraulic conductivity of unsaturated soils *Soil Sci. Soc. Am. J.* **44** 892–8
[16] Carminati A *et al* 2010 Dynamics of soil water content in the rhizosphere *Plant Soil* **332** 163–76
[17] Koestel J *et al* 2022 Potential of combined neutron and X-ray imaging to quantify local carbon contents in soil *Eur. J. Soil Sci.* **73** e13178
[18] Flury M, Flühler H, Jury W A and Leuenberger J 1994 Susceptibility of soils to preferential flow of water: a field study *Water Resour. Res.* **30** 1945–54
[19] Flury M 1996 Experimental evidence of transport of pesticides through field soils—a review *J. Environ. Qual.* **25** 25–45
[20] Deinert M R, Parlange J-Y, Steenhuis T, Throop J, Ünlü K and Cady K B 2004 Measurement of fluid contents and wetting front profiles by real-time neutron radiography *J. Hydrol. (Amst)* **290** 192–201
[21] Schaap J D *et al* 2008 Measuring the effect of structural connectivity on the water dynamics in heterogeneous porous media using speedy neutron tomography *Adv. Water Resour.* **31** 1233–41
[22] Kaestner A *et al* 2007 Mapping the 3D water dynamics in heterogeneous sands using thermal neutrons *Chem. Eng. J.* **130** 79–85

[23] Hassanein R 2006 *Correction Methods for the Quantitative Evaluation of Thermal Neutron Tomography* (Swiss Federal Institute of Technology)

[24] Zarebanadkouki M *et al* 2015 On-the-fly neutron tomography of water transport into lupine roots *Phys. Procedia* **69** 292–8

[25] Tötzke C, Kardjilov N, Manke I and Oswald S E 2017 Capturing 3D water flow inrooted soil by ultra-fast neutrontomography *Sci. Rep.* **7** 6192

[26] Snehota M, Jelinkova V, Sobotkova M, Sacha J, Vontobel P and Hovind J 2015 Water and entrapped air redistribution in heterogeneous sand sample: quantitative neutron imaging of the process *Water Resour. Res.* **51** 1359–71

[27] Lehmann P, Assouline S and Or D 2008 Characteristic lengths affecting evaporative drying of porous media *Phys. Rev.* E **77** 056309

[28] Shokri N, Lehmann P, Vontobel P and Or D 2008 Drying front and water content dynamics during evaporation from sand delineated by neutron radiography *Water Resour. Res.* **44** W06418

[29] Zheng W *et al* 2018 Plant growth-promoting rhizobacteria (PGPR) reduce evaporation and increase soil water retention *Water Resour. Res.* **54** 3673–87

[30] Benard P *et al* 2019 Microhydrological niches in soils: how mucilage and eps alter the biophysical properties of the rhizosphere and other biological hotspots *Vadose Zone J.* **18** 1–10

[31] Benard P, Bickel S, Kaestner A, Lehmann P and Carminati A 2023 Extracellular polymeric substances from soil-grown bacteria delay evaporative drying *Adv. Water Resour.* **172** 104364

[32] Helliwell J R *et al* 2013 Applications of x-ray computed tomography for examining biophysical interactions and structural development in soil systems: a review *Eur. J. Soil Sci.* **64** 279–97

[33] Wildenschild D and Sheppard A P 2013 X-ray imaging and analysis techniques for quantifying pore-scale structure and processes in subsurface porous medium systems *Adv. Water Resour.* **51** 217–46

[34] Schlüter S *et al* 2022 Microscale carbon distribution around pores and particulate organic matter varies with soil moisture regime *Nat. Commun.* **13** 2098

[35] Carminati A and Flühler H 2009 Water infiltration and redistribution in soil aggregate packings *Vadose Zone J.* **8** 150–7

[36] Badorreck A, Gerke H H and Vontobel P 2010 Noninvasive observations of flow patterns in locally heterogeneous mine soils using neutron radiation *Vadose Zone J.* **9** 362–72

[37] Carminati A and Vetterlein D 2013 Plasticity of rhizosphere hydraulic properties as a key for efficient utilization of scarce resources *Ann. Bot.* **112** 277–90

[38] Garrigues E, Doussan C and Pierret A 2006 Water uptake by plant roots: I—formation and propagation of a water extraction front in mature root systems as evidenced by 2D Light transmission imaging *Plant Soil* **283** 83–98

[39] Hainsworth J M and Aylmore L A G 1989 Non-uniform soil water extaction by plant roots *Plant Soil* **113** 121–4

[40] Segal E, Kushnir T, Mualem Y and Shani U 2008 Microsensing of water dynamics and root distributions in sandy soils *Vadose Zone J.* **7** 1018–26

[41] Nakanishi T M *et al* 2003 Water movement in a plant sample by neutron beam analysis as well as positron emission tracer imaging system *J. Radioanal. Nucl. Chem.* **255** 149–53

[42] Esser H G, Carminati A, Vontobel P, Lehmann E H and Oswald S E 2010 Neutron radiography and tomography of water distribution in the root zone *J. Plant Nutr. Soil Sci.* **173** 757–64

[43] McCully M E and Boyer J S 1997 The expansion of maize root-cap mucilage during hydration. 3. Changes in water potential and water content *Physiol. Plant.* **99** 169–77
[44] Read D B, Gregory P J and Bell A E 1999 Physical properties of axenic maize root mucilage *Plant Soil* **211** 87–91
[45] Mawodza T, Burca G, Casson S and Menon M 2020 Wheat root system architecture and soil moisture distribution in an aggregated soil using neutron computed tomography *Geoderma* **359** 113988
[46] Moradi A B *et al* 2011 Three-dimensional visualization and quantification of water content in the rhizosphere *New Phytol.* **192** 653–63
[47] Moradi A B *et al* 2012 Is the rhizosphere temporarily water repellent? *Vadose Zone J.* **11** vzj2011.0120
[48] Ahmed M A *et al* 2018 Engineering rhizosphere hydraulics: pathways to improve plant adaptation to drought *Vadose Zone J.* **17** 1–12
[49] Kroener E, Ahmed M A and Carminati A 2015 Roots at the percolation threshold *Phys. Rev.* E **91** 042706
[50] Benard P, Kroener E, Vontobel P, Kaestner A and Carminati A 2016 Water percolation through the root-soil interface *Adv. Water Resour.* **95** 190–8
[51] Ahmed M A, Kroener E, Benard P, Zarebanadkouki M, Kaestner A and Carminati A 2016 Drying of mucilage causes water repellency in the rhizosphere of maize: measurements and modelling *Plant Soil* **407** 161–71
[52] Rudolph-Mohr N, Vontobel P and Oswald S E 2014 A multi-imaging approach to study the root–soil interface *Ann. Bot* **114** 1779–87
[53] Rudolph-Mohr N, Tötzke C, Kardjilov N and Oswald S E 2017 Mapping water, oxygen, and pH dynamics in the rhizosphere of young maize roots *J. Plant Nutr. Soil Sci.* **180** 336–46
[54] Holz M, Zarebanadkouki M, Kaestner A, Kuzyakov Y and Carminati A 2018 Rhizodeposition under drought is controlled by root growth rate and rhizosphere water content *Plant Soil* **423** 429–42
[55] Holz M, Zarebanadkouki M, Carminati A, Hovind J, Kaestner A and Spohn M 2019 Increased water retention in the rhizosphere allows for high phosphatase activity in drying soil *Plant Soil* **443** 259–71
[56] Bilyera N *et al* 2021 Maize genotype-specific exudation strategies: an adaptive mechanism to increase microbial activity in the rhizosphere *Soil Biol. Biochem.* **162** 108426
[57] Pierret A, Kirby M and Moran C 2003 Simultaneous x-ray imaging of plant root growth and water uptake in thin-slab systems *Plant Soil* **255** 361–73
[58] Pohlmeier A *et al* 2008 Changes in soil water content resulting from ricinus root uptake monitored by magnetic resonance imaging *Vadose Zone J.* **7** 1010–7
[59] Warren J M *et al* 2013 Neutron imaging reveals internal plant water dynamics *Plant Soil* **366** 683–93
[60] Dara A, Moradi B A, Vontobel P and Oswald S E 2015 Mapping compensating root water uptake in heterogeneous soil conditions via neutron radiography *Plant Soil* **397** 273–87
[61] Matsushima U, Kardjilov N, Hilger A, Graf W and Herppich W B 2008 Application potential of cold neutron radiography in plant science research *J. Appl. Bot. Food Qual.* **82** 90–8
[62] Zarebanadkouki M, Kim Y X, Moradi A B, Vogel H-J, Kaestner A and Carminati A 2012 Quantification and modeling of local root water uptake using neutron radiography and deuterated water *Vadose Zone J.* **11** vzj2011.0196

[63] Zarebanadkouki M, Kim Y X and Carminati A 2013 Where do roots take up water? Neutron radiography of water flow into the roots of transpiring plants growing in soil *New Phytol.* **199** 1034–44

[64] Ahmed M A, Zarebanadkouki M, Meunier F, Javaux M, Kaestner A and Carminati A 2018 Root type matters: measurement of water uptake by seminal, crown, and lateral roots in maize *J. Exp. Bot.* **69** 1199–206

[65] Zarebanadkouki M, Trtik P, Hayat F, Carminati A and Kaestner A 2019 Root water uptake and its pathways across the root: quantification at the cellular scale *Sci. Rep.* **9** 12979

[66] McKane R B *et al* 2002 Resource-based niches provide a basis for plant species diversity and dominance in arctic tundra *Nature* **415** 68–71

[67] Liu D 2021 Root developmental responses to phosphorus nutrition *J. Integr. Plant Biol.* **63** 1065–90

[68] Bao Y *et al* 2014 Plant roots use a patterning mechanism to position lateral root branches toward available water *Proc. Natl Acad. Sci.* **111** 9319–24

[69] Fromm H 2019 Root plasticity in the pursuit of water *Plants* **8** 236

[70] Felderer B, Vontobel P and Schulin R 2015 Cluster root allocation of white lupin (*Lupinus albus* L.) in soil with heterogeneous phosphorus and water distribution *Soil Sci Plant Nutr* **61** 940–50

[71] Yan A, Wang Y, Tan S N, Mohd Yusof M L, Ghosh S and Chen Z 2020 Phytoremediation: a promising approach for revegetation of heavy metal-polluted land *Front Plant Sci.* **11** 359

[72] Menon M *et al* 2007 Visualization of root growth in heterogeneously contaminated soil using neutron radiography *Eur. J. Soil Sci.* **58** 802–10

[73] Moradi A B, Conesa H M, Robinson B H, Lehmann E, Kaestner A and Schulin R 2009 Root responses to soil Ni heterogeneity in a hyperaccumulator and a non-accumulator species *Environ. Pollut.* **157** 2189–96

[74] Matsushima U, Kawabata Y, Hino M, Geltenbort P and Nicolaï B M 2005 Measurement of changes in water thickness in plant materials using very low-energy neutron radiography *Nucl. Instrum. Methods Phys. Res.* A **542** 76–80

[75] Defraeye T *et al* 2014 Quantitative neutron imaging of water distribution, venation network and sap flow in leaves *Planta* **240** 423–36

[76] Matsushima U, Kardjilov N, Hilger A, Lehmann E H, Kaestner A and Herppich W B 2009 Cold neutron radiography for non-destructive analysis of food water status *5th Int. Technical Symp. on Food Processing, Monitoring Technology in Bioprocesses and Food Quality Management*

[77] Rees R, Robinson B H, Menon M, Lehmann E, Günthardt-Goerg M S and Schulin R 2011 Boron accumulation and toxicity in hybrid poplar (*Populus nigra* × *euramericana*) *Environ. Sci. Technol.* **45** 10538–43

[78] Matsushima U, Herppich W B, Kardjilov N, Graf W, Hilger A and Manke I 2009 Estimation of water flow velocity in small plants using cold neutron imaging with D_2O tracer *Nucl. Instrum. Methods Phys. Res.* A **605** 146–9

[79] Tötzke C *et al* 2013 Visualization of embolism formation in the xylem of liana stems using neutron radiography *Ann. Bot.* **111** 723–30

[80] Ryu J, Ahn S, Kim S-G, Kim T and Lee S J 2014 Interactive ion-mediated sap flow regulation in olive and laurel stems: physicochemical characteristics of water transport via the pit structure *PLoS One* **9** e98484

[81] Malone M W *et al* 2016 *In vivo* observation of tree drought response with low-field NMR and neutron imaging *Front. Plant Sci.* **7** 564

[82] Siegwart M *et al* 2019 Distinction between super-cooled water and ice with high duty cycle time-of-flight neutron imaging *Rev. Sci. Intrum.* **90** 103705
[83] Howe P D 2013 A review of boron effects in the environment *Biol. Trace Elem. Res.* **66** 153–66
[84] Brdar-Jokanoviá M 2020 Boron toxicity and deficiency in agricultural *Int. J. Mol. Sci.* **21** 1424

IOP Publishing

Neutron Imaging
From applied materials science to industry
Markus Strobl and Eberhard Lehmann

Chapter 8

Neutron imaging and wood

David Mannes, Peter Niemz and Walter Sonderegger

Wood is a natural, renewable material and potential carbon sink, which plays an important role as a building and engineering material. This is due to its advantageous properties such as being lightweight and having good mechanical properties at the same time. In addition, wood shows good insulating properties, easy machinability and high availability. Due to these properties, it is probably the one material which has accompanied mankind since its earliest beginning. Therefore, wood also represents a very important material when it comes to cultural heritage objects.

It is a porous, anisotropic material, which is structurally organized on several hierarchical levels ranging from the molecular level (organization of macromolecules, such as cellulose, microfibrils, etc) over the microscopic level (organization of different cell types to tissues) to the macroscopic level (tree rings, knots, etc).

Non-destructive testing methods play an important role for the appraisal of different wood properties. In a very pragmatic approach this permits for example an improved classification and grading of timber and hence an optimized and more efficient utilization in wood industry. Non-destructive evaluation can also help for a better understanding of internal processes and mechanisms allowing improvement of the utilization but also optimization of established and development of new wood-based products [1].

Many non-destructive evaluation methods are already standard in the wood industry (e.g. in-line inspection with x-ray and/or optically in the production line of sawmills, etc). Compared to these widely used methods, neutron imaging is limited to a niche existence regarding wood analysis. This is due to the limited availability of neutron sources and the fact that these sources represent in general large-scale facilities, which do not allow for high-throughput in-process evaluation of industrial production. Nevertheless, neutron imaging has proved to be particularly well suited to study certain questions. This is mainly due to the high sensitivity for hydrogen and hence water; neutron imaging allows localizing and quantifying even small changes in the moisture content of reasonably large wood samples. In this chapter we give an overview on the possibilities and limitations of the method when it comes to investigations of wood and show in which areas of application neutron imaging can be used.

doi:10.1088/978-0-7503-3495-2ch8

© IOP Publishing Ltd 2024

8.1 Neutron imaging and wood—theoretical considerations

Wood is a complex, organic and naturally grown material. No two pieces of wood are identical, which often makes generalizations of experimental results complicated. Wood consists to a major extent of a number of hydrocarbon compounds, where cellulose, lignin and the group of hemi-celluloses are the most prominent. While this appears to be a very complex conglomerate of different components, the situation is much simplified when viewing it from the neutron imaging perspective. For the interaction behaviour of incident neutrons the actual elements are more relevant than the actual chemical compounds present in a material. For wood, this simplifies the situation considerably because the elemental composition of wood is more or less identical for most of the wood species, only differing in a few trace elements [2]. When considering wood from the neutron perspective the elemental composition can be reduced to carbon, oxygen and hydrogen, which make up for 50%, 44% and 6%, respectively. All other elements with nitrogen being the most dominant can be neglected for the considerations regarding the neutron attenuation coefficient of wood.

As for many other examples, hydrogen plays a dominant role for the interaction of neutrons with the material. For wood, hydrogen, which only contributes with $6\%_{\mathrm{weight}}$ to the wood substance is responsible for ca. 90% of the overall attenuation of the incident neutrons [3] (see table 8.1).

Mannes *et al* [3] and Reda *et al* [4] compared the theoretically expected attenuation coefficients of wood with experimentally determined ones.

In order to allow for quantitative neutron imaging it proved to be necessary to correct the attenuation coefficients for scattering effects to obtain reliable results [3]; this is due to the fact that the interaction of neutrons and hydrogen occurs almost

Table 8.1. Theoretical microscopic cross-sections and attenuation coefficients for a simplified physical model of wood (density 0.6 g cm^{-3}) at the imaging beamlines NEUTRA (thermal spectrum) and ICON (cold spectrum); only carbon, oxygen and hydrogen are taken into account as constituents. Reproduced with permission from [3], copyright De Gruyter.

		C	O	H	Wood
Material density, ρ	(g cm^{-3})				0.6
Relative fraction	(%)	50	44	6	
Number density, n	(cm^{-3})	1.51E+22	9.94E+21	2.17E+22	4.67E+22
Atomic weight, A	(g mol^{-1})	12	16	1	
Thermal neutron spectrum (NEUTRA)					
Microscopic cross-section, σ	(cm^2)	4.93E−24	4.00E−24	4.70E−23	
Attenuation coefficient, Σ	(cm^{-1})	0.07	0.04	1.02	1.13
Relative share of Σ	(%)	6	4	90	
Mass attenuation, Σ/ρ	(cm^2 g^{-1})				1.88
Cold neutron spectrum (ICON)					
Microscopic cross-section, σ	(cm^2)	5.29E−24	4.23E−24	6.02E−23	
Attenuation coefficient, Σ	(cm^{-1})	0.08	0.04	1.31	1.42
Relative share of Σ	(%)	6	3	91	
Mass attenuation, Σ/ρ	(cm^2 g^{-1})				2.37

exclusively as scattering. Multiple scattering within the sample, but also from the experimental set-up can therefore falsify the results [5, 6]. Mannes *et al* [3] also investigated the influence of different (European and tropical) wood species, the content of extracts and the density of wood (using densified wood specimens). They showed that the content of extracts and also the presence of different trace elements (which can be presumed for the tropical wood species) only play a minor role. They also showed that the neutron attenuation coefficient of dry wood can be assumed to be solely dependent on the density of the sample but not the exact elemental composition taking all trace elements into consideration.

8.2 Wood structure

The high attenuation coefficients on the one hand and the general limited spatial resolution of neutron detector systems on the other hand are limiting the applicability of neutron imaging methods for the investigation of wooden structure. The high attenuation coefficients limit the sample size to very few centimetres at best [7]. The spatial resolution (i.e. in the order of few tens of micrometres), which can be achieved with standard neutron imaging detectors, allows only for a very limited study of the anatomical wood features; here only relatively large structures such as large vessels in ring-porous wood or resin ducts can be clearly visualized and studied [8]. Most smaller anatomical features such as tracheids, parenchyma cells or pits with diameters down to micrometre range remain in general below the resolution limit given by the used detector systems. Using more recent detector developments such as the Paul Scherrer Institute (PSI) neutron microscope [9, 10] with its spatial resolution of ca. 5 μm, might allow us to overcome these limitations.

A macroscopic wood feature, which can more easily be studied is tree rings. Due to secondary growth, woody plants gather cell layer over cell layer over time. These growth zones can show different densities, which is particularly pronounced for trees in the temperate climate zone. Here, the production of new cells is interrupted in winter and restarts in spring with the beginning of a new vegetation period. Wood produced in spring, the so-called early wood, shows lower density due to larger cell diameters necessary for efficient transport of water, while the wood produced towards the end of the growing season in late summer and autumn, the so-called late wood, shows much higher densities. Depending on the site and changing climatic and growing conditions the produced annual tree rings differ in size. Hence, the sequence of these annual rings yields information on these conditions and is typical for specific tree species and geographical regions. In dendrochronology this information is used for the dating of wooden objects or the reconstruction of climatic conditions. The sequence of low-density early wood and high-density late wood results in a typical density profile in the wood, which can be investigated by a variety of methods; a standard method, which is widespread and frequently used is x-ray microdensitometry. To assess how well neutron imaging-based approaches would work and complement this type of investigation, Mannes *et al* [11] and Keunecke *et al* [12] compared the results of x-ray methods with results acquired with thermal and cold neutrons, respectively.

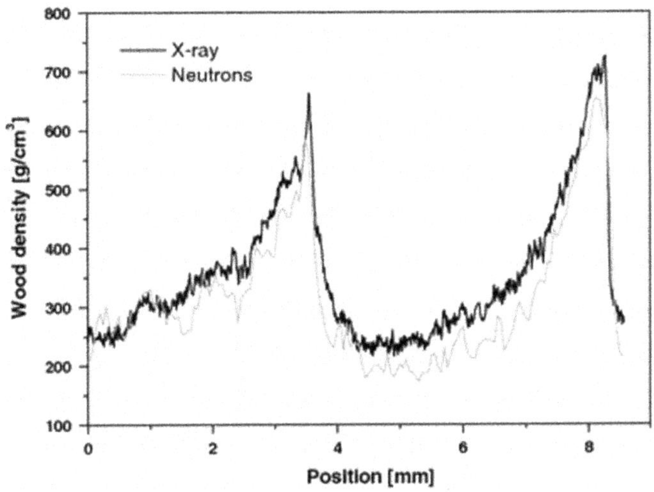

Figure 8.1. Density profile of spruce samples determined with neutron and x-ray imaging. Reprinted from [11], copyright (2007), with permission from Springer Nature.

For the study of density variations in tropical wood species, De Ridder *et al* [13] compared x-ray tomography, neutron tomography and drilling resistance measurements to gravimetrically determined wood densities. They found significant correlation between the density profile determined using neutron tomography and the gravimetrically determined one; only x-ray densitometry showed slightly higher values.

Cherubini *et al* [14, 15] used neutron imaging as one among many other methods in the attempt to get more precise dendrochronological results for a better dating of the Santorini eruption which had occurred during antiquity.

Overall, it could be stated that both approaches, using either x-ray or neutrons can yield equivalent results (cf figure 8.1). In the case of x-rays the attenuation can be attributed to the interaction with carbon and oxygen, while the neutron attenuation of wood is linked to the hydrogen content in wood (cf table 8.1). Higher spatial resolution and better availability nevertheless make standard x-ray techniques the method of choice for most dendrochronological questions. An exception might only be found for example for more complex cultural heritage objects where the wood is hidden behind metal layers. That these hidden wood features can be made visible was shown, e.g. by Mannes *et al* [16] and Masalles *et al* [17], although the ring sequences were in these cases not used for dendrochronological analyses.

8.3 Wood and adhesives

Adhesives play an important role when it comes to the production and utilization of wood-based materials. One important parameter is the penetration depth of the adhesive. As wood ahesives contain in general hydrogen, it was very early tried to use the high sensitivity of neutrons to investigate this topic. Niemz *et al* [18] showed that neutron imaging can be used to analyze the rough structures of glue joints and the penetration behaviour of the glue into the wood. Figures 8.2 and 8.3 show results

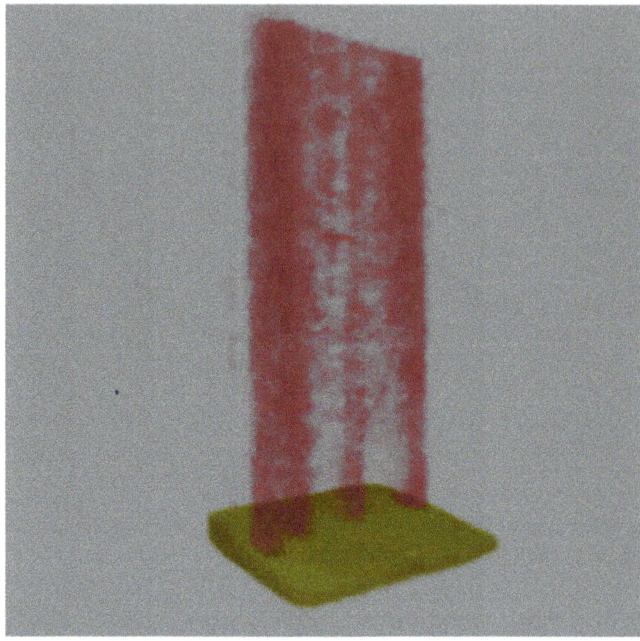

Figure 8.2. Bond lines tested (UF resin) with neutron tomography. Reprinted from [18], copyright (2004), with permission from Springer Nature.

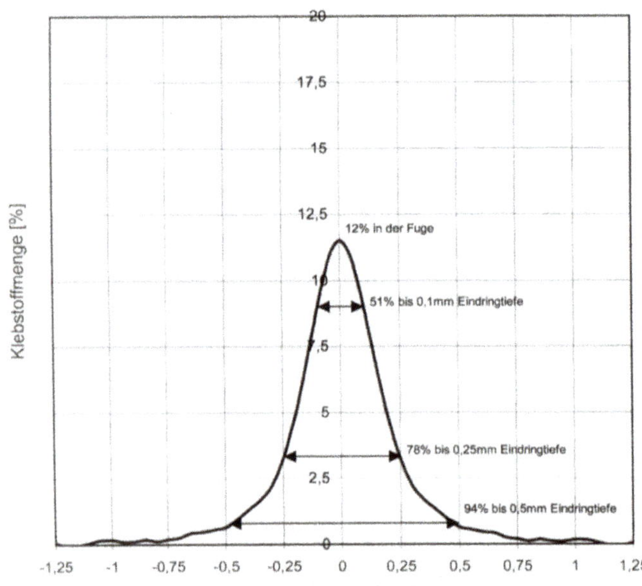

Figure 8.3. Relative distribution of the adhesive (PVAc) vertical to the bond line determined by neutron imaging. Reprinted from [18], copyright (2004), with permission from Springer Nature.

from [18]. Mannes *et al* [19] also presented an approach to quantify the penetration depth of the adhesive into the wood structure.

Nevertheless, neutron imaging could not be established as a standard method for the study of the penetration depth of adhesive into wood as the contrast between the adhesives and the hydrogenous wood is not good enough to precisely discriminate the two materials.

8.4 Wood and water

The relation between wood and water is paramount when it comes to wood as a building or engineering material. In a tree, the main function of the woody tissues beside the structural stability is the transport of water. Hence, it can be seen as an array of successive capillary tubes, for liquid water transport in the direction of the stem axis. Wood is furthermore a hygroscopic material, which can absorb and release moisture from the atmosphere depending on the actual climatic conditions. Together with the moisture content, also many physical wood properties change as they are dependent on it. In the following paragraph a quick overview on the general dependence of wood properties and moisture content is given.

8.4.1 Determination of moisture content

The parameter for assessing the moisture content of wood is the wood moisture ω, which is calculated according to:

$$\omega = \frac{m_\omega - m_0}{m_0} \quad (\%) \tag{8.1}$$

ω—moisture content in %
m_ω—mass wet wood with moisture ω
m_0—mass oven dry wood ($\omega = 0\%$)

Wood is a porous material with a very large inner surface [20]. Moisture is absorbed in the cell wall by sorption from the water present in the air (up to the fibre saturation point (FSP), i.e. when the cell wall is fully saturated) and capillary forces, which is the case for example for liquid water, (e.g. when the wood is immersed in the water) or the application of adhesives with a proportion of water (40%–50%) in the adhesive (e.g. UF (**U**rea **F**ormaldehyde), MF (**M**elamine **F**ormaldehyde), PVAc (**P**oly**v**inyl **Ac**etate), PF (**P**henol **F**ormaldehyde) resins).

The transport of moisture within the capillary-porous system occurs for free water, above the fibre saturation point through capillary forces; below fibre saturation moisture transport happens through diffusion [21–23].

The equilibrium moisture content, is the wood moisture content, which establishes under certain climatic conditions, i.e. temperature and relative humidity. Usually isothermal conditions are assumed (wood moisture content as a function of the relative humidity and temperature of the surrounding climate). Local differences in relative humidity and temperature in heated rooms also lead to local effects such as cracking (e.g. wood floor) or delamination of adhesive joints (e.g. furniture).

Due to the high relevance of moisture content and transport for the utilization of wood as a building and engineering material a broad variety of methods is used and has been used for its determination [21, 24–29].

The moisture content influences almost all properties of wood, especially within the hygroscopic area up to fibre saturation point [22, 24, 25, 30, 31]. If there is a change in moisture content within the hygroscopic area (up to FSP), the wood swells (during absorption from moisture) or shrinks (when it releases moisture). In the case of large cross-sections (e.g. in laminated timber beams), moisture profiles form over the cross-section, which lead to tension and, in extreme cases, cracks [28].

8.4.2 Methods for determination of moisture content

As a consequence of the importance of the moisture content on the wood properties, a multitude of different measuring approaches is used [21, 25–27, 31]:

- Gravimetric method (oven drying) (basic method), see equation (8.1);
- measurement of the electrical resistance or dielectric properties;
- radiometric methods x-ray, gamma radiation [32, 33], or neutrons [7, 34];
- nuclear magnetic resonance (NMR) spectroscopy [35];
- selected other methods like optical (spectroscopy measurement from moisture content with near-infrared (NIR) spectroscopy) [26].

The gravimetric (oven drying) method is in general used as reference for all other methods. In many cases this method is sufficient if only the overall moisture content of entire samples or objects is of interest for the respective application or scientific question. Nevertheless, there were also attempts to apply more sophisticated methods even for the determination of the overall moisture content. Tzscherlich [34] tested a method based on the counting of scattered neutrons using a so-called Troxler probe to ascertain the moisture content of wood chips.

The focus of interest in the last years changes towards investigations of local moisture contents and transport. Using conventional methods, it is often difficult to carry out non-destructive local moisture content measurements with high spatial resolution. For small and medium sized samples and simplified models, neutron imaging methods can prove to be a very powerful tool.

8.4.3 Neutron imaging as a non-destructive testing method for the localization and quantification of wood moisture contents

The high sensitivity for hydrogen and the importance of the moisture content for the utilization of wood, makes the interaction between wood and water the predominant topic when it comes to investigations of wood properties using neutron imaging methods.

8.4.3.1 Determination of local moisture distribution using neutron imaging

Local moisture distribution was one of the first questions researchers tried to tackle using neutron imaging techniques. Nakanishi and Matsubayashi [36] showed that neutron imaging allows to determine zones with different moisture contents in green, freshly cut stem disks. In these first experiments, evacuated aluminum cassettes with x-ray-film and Gd-converter foil were used as a detector system. As neutron imaging is a non-destructive technique it also allows studying processes and comparing and observing identical samples over a certain time period. To make use of this, Nakanishi *et al* [37, 38] used a similar set-up using stem disks to determine their moisture content and the effect of a drying process. Here, identical samples were measured repeatedly in intervals of a few hours. Even though the study had to revert to x-ray film, it was still possible to determine radial moisture content profiles as well as a 2D mapping of the local moisture distribution over the whole sample and to observe the development of the results over time.

A similar approach, acquiring repeatedly neutron radiographs of samples in an interval of several minutes was used by Islam *et al* [39]. Here, samples of different types of wood plastic composites (WPCs) were radiographed in a dry reference state and subsequently immerged in water for periods between 5 and 20 min, after which they were imaged again. As a detector system, neutron sensitive imaging plates were used. The comparison of the results allowed obtaining clear tendencies on the water absorption behaviour of the different WPC samples over time.

The examples presented so far, relied on removing the samples between the measurements. For more dynamic and directional processes, it is favourable that samples remain untouched between the different points of measurement. Lehmann [40] demonstrated in a feasibility experiment how digital detectors can be used to measure and visualize the directional water uptake in beech wood samples.

8.4.3.2 Water uptake due to capillary forces

The utilization of digital detector systems hence opens up new possibilities when it comes to different moisture-related dynamic processes in wood, e.g. the uptake of liquid water by capillary forces.

First feasibility studies [40] were extended, and more wood species and wood-based materials were tested. Lehmann *et al* [41, 42] observed the uptake of water through capillary forces in different types of wood and wood-based material (cf figure 8.4). Dry samples of two hardwood species (chestnut (*Castanea sativa*), beech (*Fagus sylvatica*)), two softwood species (Norway spruce (*Picea abies*), pine (*Pinus sylvestris*)) and two wood-based materials (particle board, MDF) were placed in a shallow water basin. The water uptake in longitudinal direction was observed over a time period of 300 min using neutron radiography and compared to gravimetrically determined water uptake. For the evaluation, the resulting images were referenced to the initial dry state; in this step the images were divided by the reference image resulting in a difference image showing only the changes between the individual images in the series and the reference state, which was presumed to be mainly the difference in water content. It was stated that differences in the uptake behaviour, such as lower water absorption in chestnut in comparison to the other tested wood species, can reliably be determined. Nevertheless, it is already pointed out that the swelling of wood can be problematic. Swelling and shrinkage of wood occurs whenever the moisture content of wood changes below the FSP. Hence, dimensions

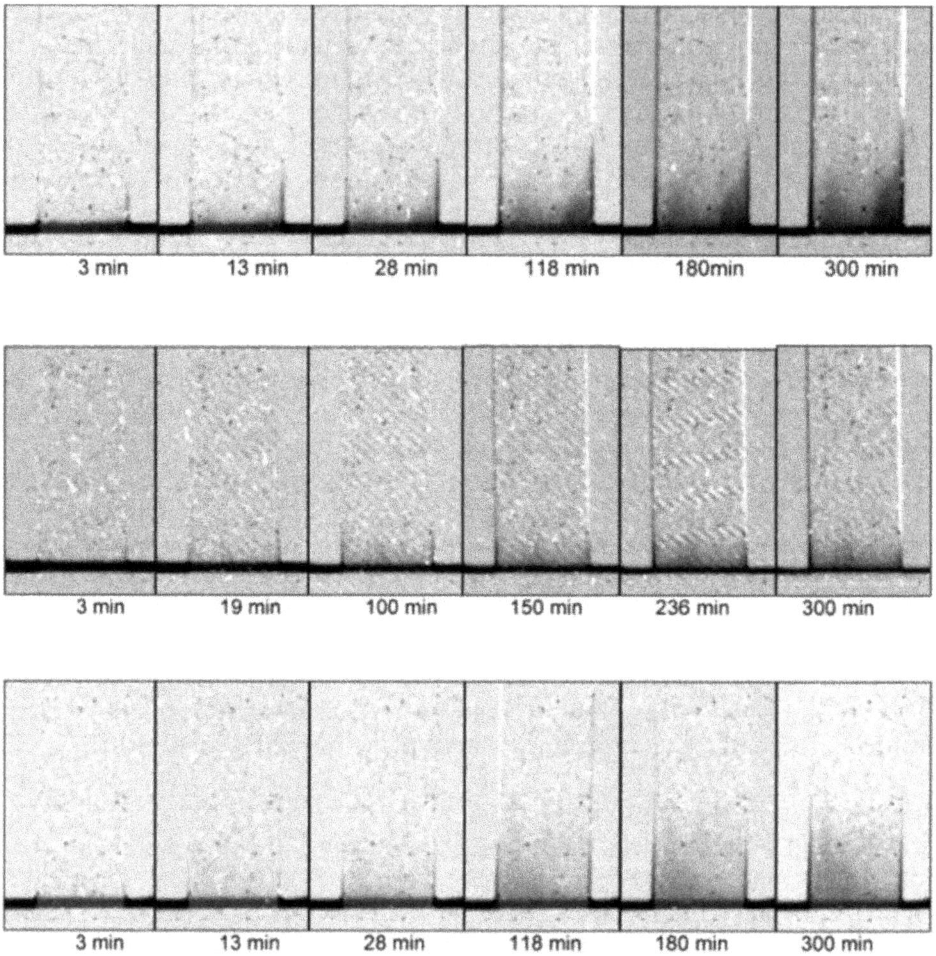

Figure 8.4. Water uptake of Norway spruce (top), chestnut (middle) and scots pine (down) depending on time. Reprinted from [42], copyright (2001), with permission from Springer Nature.

and geometry of the samples do not match perfectly anymore during the referencing step in the image processing. In the image regions, where swelling and shrinkage occurs, evaluation is hence very challenging.

The uptake of water by capillary force remains an important topic and is investigated in various other studies. Sedighi-Gilani and collaborators used neutron radiography to study the liquid water transport in softwood [43, 44]; using neutron imaging the team describes and quantifies the differences in the way water is transported in the wood tissue with respect to the different organizational directions in wood (longitudinal, radial, tangential), early and late wood as well as the differences between sapwood and heartwood, see figure 8.5.

Not only the uptake of liquid water is studied; Desmarais *et al* [45] compared the transport behaviour of a polar and non-polar liquid dipping softwood samples in

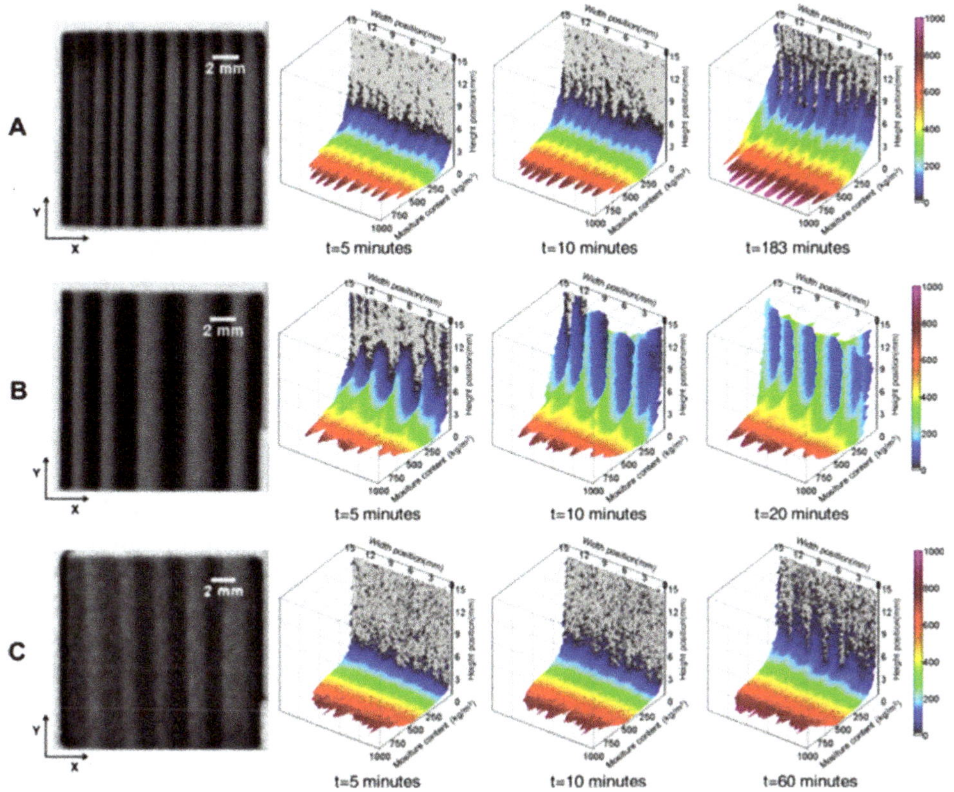

Figure 8.5. Moisture content spatial distribution after 5, 10 min and at the end of the experiment in (A) spruce, (B) pine and (C) fir during water uptake in longitudinal direction, from dry state. Reprinted from [43], copyright (2012), with permission from Springer Nature.

water, respectively, and decane in order to study the moisture capacity of the wood cell walls.

Besides plain pieces of wood, it was also investigated how liquid water penetrates in joint timber. Niemz *et al* [46] showed this on the example of different types of wooden corner connections. The team compared connections with blunt and mitre joints which had been exposed to alternating climate over a couple of months in order to induce artificial aging. The alternating climate causes the wood to swell and shrink. Wood shows considerable differences in the swelling/shrinkage behaviour in its three orthotropic directions (longitudinal direction (max. swelling for spruce 0.0%–0.4%) and transversal direction (max. swelling for spruce: radial 3.5%, tangential 8.5%)). This induces stress and strain in the corner joints; due to hindered swelling and gradual plastic deformation, the joints open up over time. To what extent this is taking place depends very much on the type of joint (blunt or mitre) and the type of adhesive used. Neutron imaging was used in this context to assess the differences in the water uptake behaviour of the tested joint types, with respect to amount, speed and penetration depth of the penetrating water. The study showed that the artificial aging had much higher impact on the square joints (blunt insert) in comparison to the mitre connection. Here, the joints open up much less, which led to smaller amounts of water penetrating the corner joints.

Sonderegger *et al* [47] investigated how defects influence the barrier effect of surface coatings. Two different coating systems were tested and two different types of defects were applied to the surface (cut with saw and microtome blade, respectively). The penetration of the water was inspected over a time of 16 h. The differences between coating and width of the failure are only marginal.

8.4.3.3 Investigation on the hygroscopic behaviour of wood

When it comes to the properties of wood, much depends on its hygroscopic behaviour. Almost all physical properties of wood are somehow depending on the moisture content of wood, especially in the range from dry wood to the fibre saturation point. As here already small changes in the moisture content can induce considerable changes in the behaviour of wood, neutron imaging proves to be a very useful tool for investigating these changes. The high sensitivity for hydrogen and hence water allows assessing small changes in the water content quantitatively.

One of the first studies focusing on the investigation of moisture changes in the hygroscopic range was carried out by Mannes *et al* [48]. Oven dry samples of Norway spruce (*Picea abies*) and European Beech (*Fagus sylvatica*) were exposed to a differential climate with dry conditions on the one side of the samples and more humid conditions (ca. 27 °C/86% rH) on the other side of the samples. All other sides were insulated, allowing only moisture content change along one direction. Neutron imaging was used to assess and visualize the transport of moisture in longitudinal direction over several hours. The results subsequently allowed determining and quantifying the diffusion processes in the wood samples for the longitudinal direction of the two species, see figure 8.6.

Rosner *et al* [49] used neutron imaging to study the water movement within annual rings of Norway spruce (*Picea abies*) samples during the drying process. Small saturated samples of spruce sapwood were dried over a period of 22 h at ambient conditions. The drying process was assessed by continuous imaging of the sample over the whole experiment duration. The water content was additionally monitored by a balance, on which the samples were positioned throughout the experiment.

In a slightly different approach, Lanvermann *et al* [50] studied how moisture content changes vary between early wood and late wood regions. Samples of Norway spruce (picea abies) were subjected to an adsorption/desorption regime with relative humidity (rH) values varying between 0% rH and 95% rH.

Zhou *et al* [51] also studied the hygroscopic behaviour of spruce during adsorption and desorption. In this study, neutron imaging was combined with local temperature measurements. The samples were insulated on four sides allowing moisture transport in one direction and thermocouples were inserted in the sample at different heights of the sample. This allowed correlating the temperature changes in the sample to the moisture content changes, which were assessed simultaneously using neutron imaging.

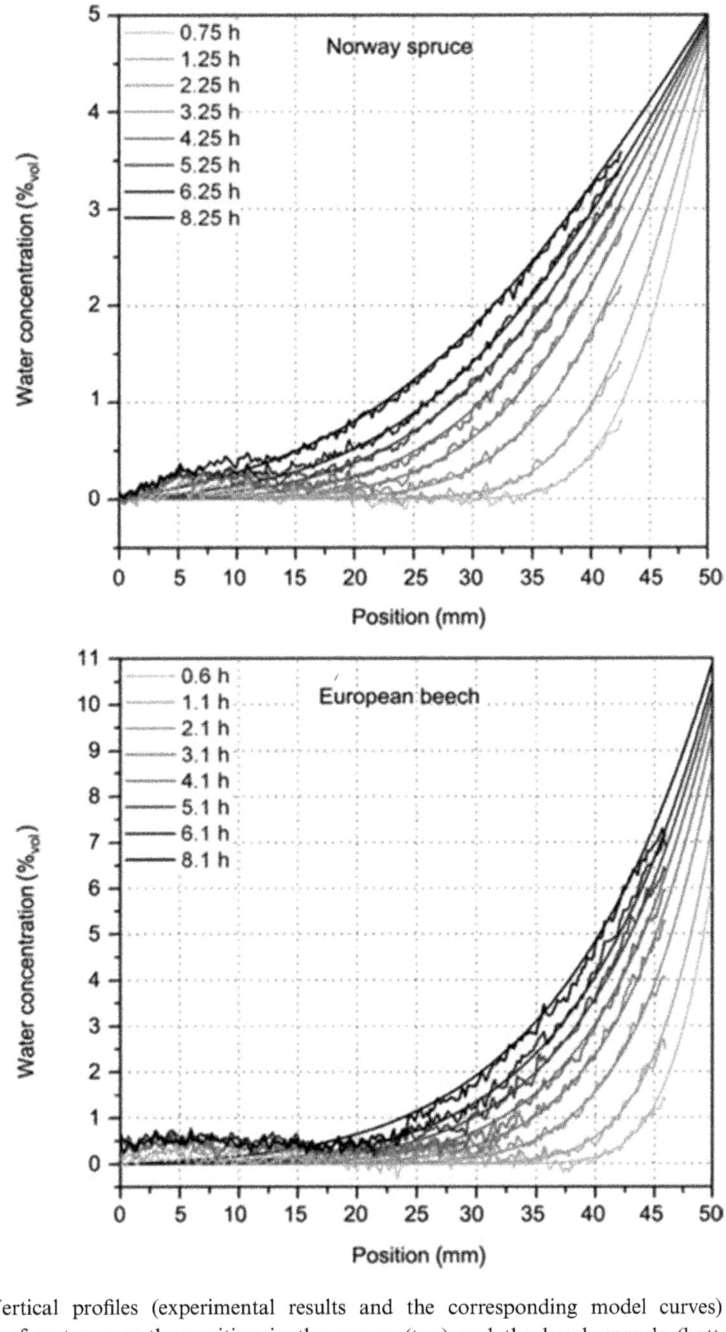

Figure 8.6. Vertical profiles (experimental results and the corresponding model curves) of volumetric concentration of water over the position in the spruce (top) and the beech sample (bottom) during the diffusion experiment; the left hand side represents the specimen's lower part (toward the silica gel), the right hand panel is the upper part open to the wet climate (ca. 27 °C at 86% rH). Reproduced with permission from [48], copyright De Gruyter.

The investigations on the hygroscopic behaviour of wood using neutron imaging were not restricted to plain wood. Very quickly the method was also extended and used to study wood-based materials as well as the influence of adhesives and coatings on the hygroscopic influences. Sonderegger *et al* [52] used neutron imaging to assess and quantify the diffusion behaviour in multilayer boards. Multilayer wood samples with different types of adhesives used for bonding were studied. The samples were insulated on four sides allowing only one direction for the moisture transport and subjected to differential climate with dry climatic conditions on the one side and high relative humidity on the other side. The diffusion process was observed over a period of 2.5 months, repeating the neutron imaging experiments in intervals of a few days to several weeks, depending on the time available at the facility. This allowed determining the diffusion coefficients for the different varieties tested.

In a similar approach Mannes *et al* [53] tested two-layer samples bonded with different historically relevant types of adhesive such as bone, hide or fish glue. Neutron imaging was again used to determine the water transport processes through the adhesive layer. This allowed determining the diffusion coefficients for the different types of adhesive. The properties of such types of adhesives differ considerably from the modern synthetic adhesives used nowadays in building, production of wood-based materials or carpentry. The relevance is given by the necessity of such information for the best way to preserve, consolidate and conservate the large amount of wooden cultural heritage objects.

Sonderegger *et al* [47] studied to what extent surface coating with and without artificial surface affects the water diffusion in spruce wood samples (*Picea abies*). In this study two coating systems were compared in a rather similar approach used in Sonderegger *et al* [52]. Samples were exposed to a differential climate for 10 days and inspected by neutron imaging at the beginning of the experiment, after 4 days and after 10 days. The results showed clear differences between the investigated coating systems: one system (acrylate/polyurethane) still showed clear decelaration of the diffusion, when compared to an untreated reference sample. The other coating system (silica-based) showed an initial decelaration, but had already reached the level of the reference at the end of the experiment.

In another study with cultural heritage relevance, Lämmlein *et al* [54] investigated how different types of varnishing and coating procedures influence the hygroscopic behaviour of wood used for violin making. These results were correlated with the change of the vibrational properties of the wood samples. Violin varnishes represent a multilayer system. Playing the instruments induces substantial wear. In order to better understand how the individual layers influence the sorption behaviour and vibrational properties of the wood the experiments were carried out on samples with different treatments and varnish systems. The overall set-up was similar to a couple of experiments described above; the samples were insulated, allowing only moisture transport in one direction, in this case through the surface coated with the different pretretments, respectively, varnish systems. The two 'open' sides were exposed to the same climatic conditions, hence simulating a region inside the body of the violin with one varnished (or 'worn off') surface on the outside and one untreated surface on the inside of the presumed violin. Neutron imaging allowed quantitatively determining the dynamics of the moisture content changes in high spatial and temporal resolution (figure 8.7). This allowed comparison of the impact of the individual

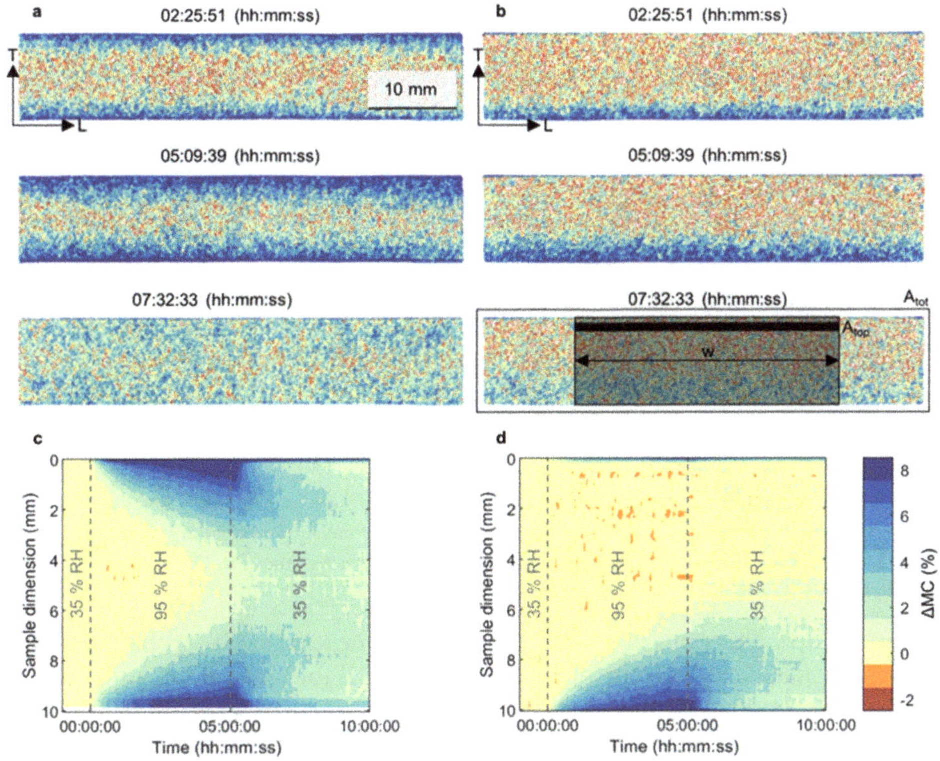

Figure 8.7. Spatial changes in moisture content (MC) compared to the initial equilibrated wood MC at 35% RH for (a) an unvarnished control and (b) a sample treated with a multilayer varnish system with four layers of alcohol varnish on the top surface after 2.5 h (top) and 5 h (middle) at 95% RH and after 2.5 h of desorption (bottom) (humidity was decreased to 35% RH after 5 h). Change in MC profiles from the top (0 mm) to bottom surface (10 mm), averaged over the width $w = 30$ mm for (c) the unvarnished reference sample and (d) the sample with complete varnish system. The averaging along w leads to a reduced noise for the results and allows a more detailed visualization of the changes over time. (L: longitudinal and T: tangential wood direction). Reprinted from [54] CC BY 4.0.

treatments and varnish systems and subsequently correlatation of the results to the change of vibrational properties tested with laser vibrometer modal analysis.

8.4.3.4 Moisture content changes induced by thermal changes

Changes in the wood moisture content can not only be induced by changes in the relative humidity in the ambient air but also by thermal changes. Sedighi-Gilani *et al* [55, 56] investigated the moisture displacement induced by steep temperature gradients. Samples of European beech (*Fagus sylvatica*), which had been conditioned at 20 °C/80% rH were exposed to one-sided heating. The samples were placed on a foil heated to 150 ° C, respectively 250 °C. Neutron imaging allowed visualizing and quantifying the drying front and moisture content changes. The experiment showed a zone with increased moisture content moving ahead of the actual drying front (figure 8.8). The study included also a comparison of the heat induced moisture content movement for the

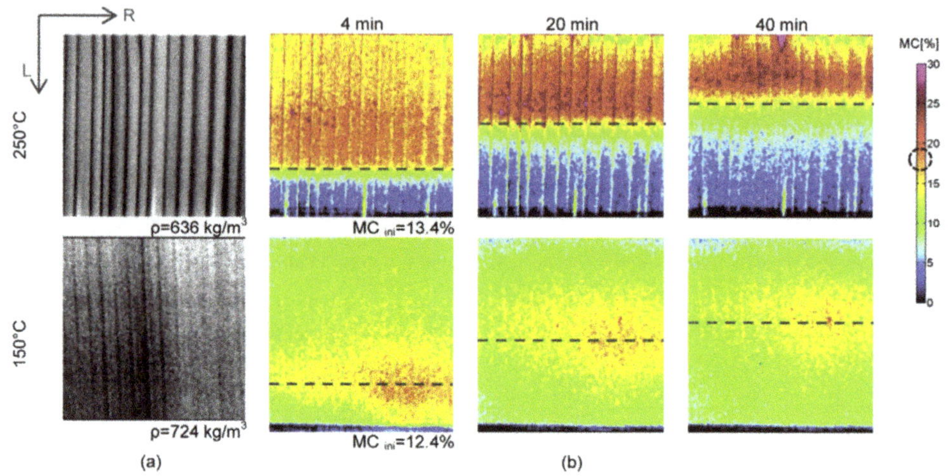

Figure 8.8. Raw neutron image of the samples (ROI: 40 × 40 mm^2) for heating in the longitudinal direction (250 °C on the top row and 150 °C on the bottom row), (b) plots of moisture content after 4, 20 and 40 min of heating. Moisture front (15% moisture content) is highlighted by dashed lines for 250 °C heating. Reprinted from [56], copyright (2014), with permission from Elsevier.

three orthotropic directions of wood. A very similar experiment by Abbasion *et al* [57] was used to validate a hygrothermo-mechanical model for wood using a poromechanical approach. In contrast to the previous study in this experiment softwood samples (Norway spruce (*Picea abies*))) were used, which had been conditioned at ambient temperature and 50% rH, respectively 80% rH. Temperature was regulated to 150 °C, respectively 250 °C. As described in Sedighi-Gilani *et al* [56] the moisture content distribution can be visualized and quantified over time. This allowed validating the developed model which is largely in agreement with the experimentally obtained data.

Heat and the water vapour movement induced by it, play an important rule, when it comes to wood processing and the production of wood-based materials. The production of wood-based composites such as particle boards, fibre boards or plywood comprises in general heat and pressure. Solbrig *et al* [58] used thermal and cold neutrons to study water vapour movement using a simplified model set-up of the hot pressing process. Fibre board and particle board samples were subjected to one-sided heat and vapour exposure. The neutron imaging experiments were carried out independently at the cold imaging beamline ANTARES at FRM-II in Garching (D), while the experiments with thermal neutrons were conducted at the NEUTRA beamline at PSI (CH). The results of both facilities are in good agreement. Neutron imaging allowed assessment of the wavefront-like water vapour movement and its visualization.

8.4.3.5 Correction of swelling and shrinkage

The problem with studies of hygroscopic properties using neutron imaging is the swelling and shrinkage behaviour of wood. When studying this phenomenon with neutron imaging the images of the time series are in general referenced to an initial starting state. When the moisture content changes during the experiment, the dimensions of the sample change accordingly. By referencing the images the

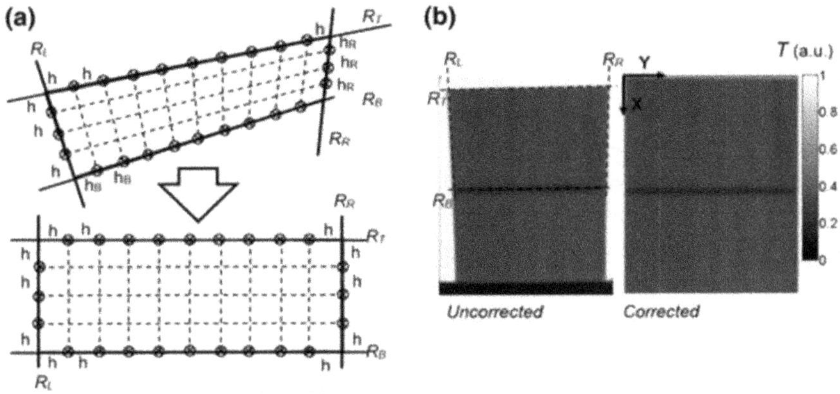

Figure 8.9. Deformation correction in neutron images using an equidistant grid. (a) Image referencing method. (b) Neutron transmission images before (uncorrected) and after (corrected) deformation correction. Reprinted from [53], copyright (2014), with permission from Elsevier.

dimensional changes induce uncertainties and errors to some extent. There are different approaches to cope with this problem. When only dealing with small moisture content changes (and especially if the main interest is in the transport in longitudinal direction and the image is viewing the tangential surface tangential–longitudinal plane), swelling/shrinkage can be neglected [48]. To compensate for the deformation, e.g. if the interest is on the cross-section of a wood sample radial–tangential plane or if higher moisture content changes had to be studied, different approaches were used. One possibility is to register and scale back the images in a way that the sample outline in every image fits the reference state again [43, 53], see figure 8.9.

Lanvermann *et al* [50] combined the neutron imaging with digital image correlation. On this study, the sample deformation was observed and registered with two additional digital cameras. This deformation was then used to correct for the dimensional changes in the different images.

Sanabria *et al* [59] used an approach solely using neutron images. The adaptive correction algorithm also allows compensating for varying deformations over the sample area. This also allows studying specimens which show varying extent of deformation in different areas of the sample, e.g. regions with late and early wood and inhomogeneous moisture content change.

Couceiro *et al* [60] propose using a combination of neutron and x-ray imaging to determine the moisture content in the images. Here, not a reference image is used but complementary image information from another modality, namely x-ray.

8.5 Wood modification

Neutron imaging can be used to evaluate modified wood, and also to monitor some modification processes *in situ*. Wood can be differently modified. Two well-established modifications are thermal treatment and acetylation. Both promise an improvement of the durability for outdoor applications.

Plaza *et al* [61] show differences in the attenuation coefficient of unmodified, acetylated and thermally degraded wood as well as of unmodified and acetylated WPCs.

Sonderegger *et al* [62] investigated swelling and shrinkage as well as moisture content changes of spruce and beech wood during different thermal treatments (with different relative humidity cycles and temperatures in the range of 70 °C–150 °C) by means of neutron radiation. The advantage of this method is the high sensitivity to hydrogen and the fact that some metals like aluminium, which was used for the equipment (a pressurised testing chamber with an integrated specimen holder), are

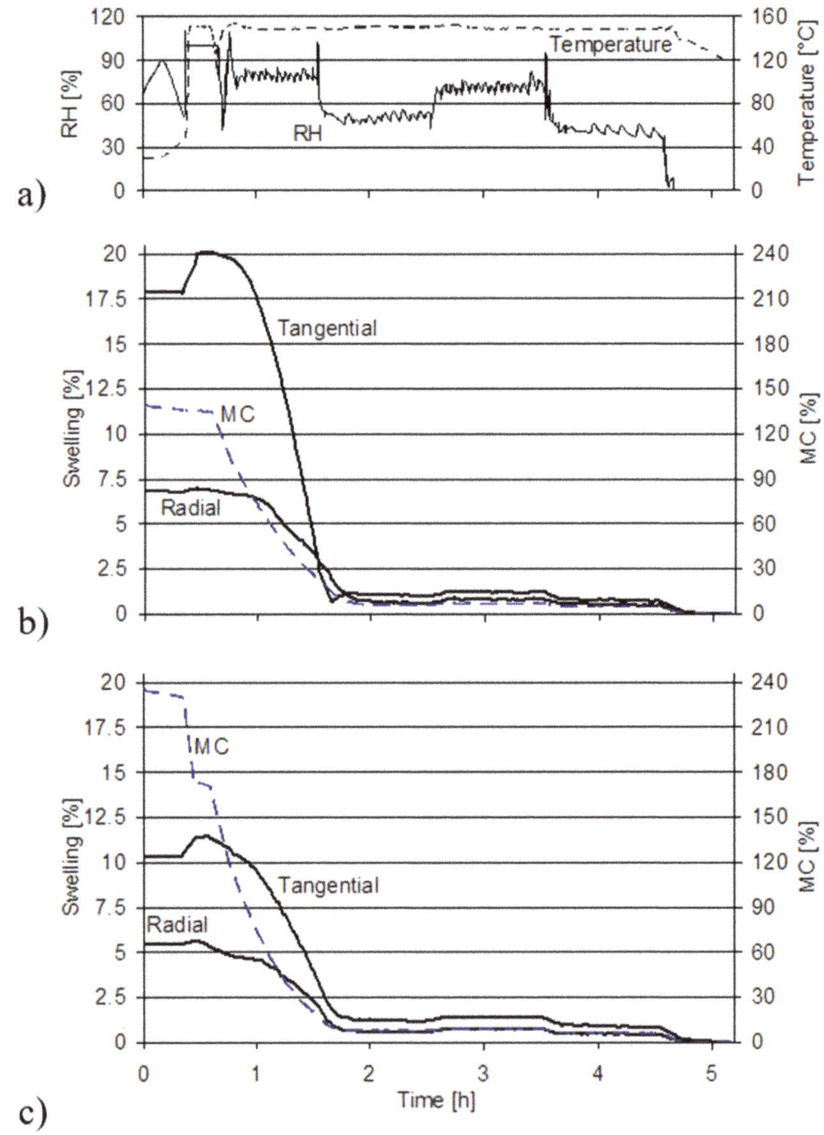

Figure 8.10. Dimensional and MC changes of beech (b) and spruce (c) wood during cycle 4 at 150 °C. Reproduced with permissions from [62], copyright De Gruyter.

practically transparent to neutrons. This allowed making visible the specimens in the equipment by neutron imaging during the treatment with different steam pressures and temperatures and to determine the dimensional and moisture content changes by means of the images with a specific evaluation software. Figure 8.10 shows a cycle with different relative humidities at a temperature of 150 °C.

In a different approach for wood modification, Gilani and Schwarze [63] investigated how different white rot fungi influence the hygroscopic properties of Norway spruce (*Picea abies*) and Sycamore (*Acer pesudoplatanus*). These wood species are traditionally used in making of musical instruments, violins in particular. Incubation with rot fungi modifies the properties of wood and improves its acoustic properties [64]. Neutron imaging was used in this study to visualize and quantify to what extent the hygroscopic properties are modified by this treatment.

It is debatable if pyrolysis of wood can be counted as wood modification, although it definitely modifies the chemical structure and composition of wood. As clean energy production plays an ever-growing role, a better understanding of combustion processes of biomass such as wood, is gaining in importance. In order to gain this better understanding, Ossler *et al* [65, 66] used neutron imaging to *in situ* monitor the changes in the hydrogen content of different wood species during pyrolysis. For the experiments a vacuum furnace was placed in a cold neutron imaging beamline. Samples of European and American beech (*Fagus sylvatica* and *Fagus grandifolia*), poplar (*Populus trichocarpa*) and loblolly pine (*Pinus taeda*) as well as pelletized biomass samples were vacuumpyrolyzed from room temperature to 400 °C [66], respectively 1000 °C [65]. Neutron imaging allowed obtaining a real time assessment of the hydrogen loss in the different samples over time. Besides the radiography mode, neutron tomography was also carried out before and after pyrolyzation. This allowed obtaining insight on the varying loss for the different types of tissue (such as xylem and phloem) happening during pyrolyzation.

8.6 Wooden cultural heritage

Wood is one of the oldest materials which has been used by mankind for every sort of purpose, starting from the manufacturing of tools to its use as a building material, but also for the manufacturing of religious or artistic objects. For the study of wooden cultural heritage objects the utilization of non-destructive testing methods plays an important role. In this context neutron imaging methods were also applied in a multitude of projects. The big challenge, when it comes to cultural heritage objects on the one hand is the fact that the integrity of the objects must not be altered (or destroyed in the worst case) and on the other hand the hydrogen content. The high sensitivity for hydrogen is in general one of the strengths of neutron imaging studies, e.g. when it comes to the detection of small amounts of hydrogenous materials (such as water, adhesives, etc). The size of many wooden cultural heritage objects though makes it difficult to use neutron imaging, as the objects can often not be penetrated by neutrons. Hence, neutron imaging is restricted to relatively small objects still allowing high enough transmission of neutrons for decent results. One area, where this is the case is for many wooden musical instruments. In particular, many wood wind instruments have a size which still allows the utilization of neutron imaging methods. Festa *et al* [67] studied a set of seven historical wind instruments using neutron and x-ray tomography in order to gain more knowledge on the construction and materials and material combinations used in the manufacturing process. Kirsch and Mannes [68] tested the applicability of neutron imaging for different cases. Relatively large instruments, such as violins pose a problem for investigations using neutron tomography as the transmission can drop to the noise level of the detector, making the reconstruction difficult and artefact prone. On smaller instruments featuring metal parts, such as keyed recorders, neutron imaging can be used very successfully. In contrast to x-ray, wood and metal parts can be penetrated to a sufficient degree resulting in satisfying results, yielding information on the working mechanisms and

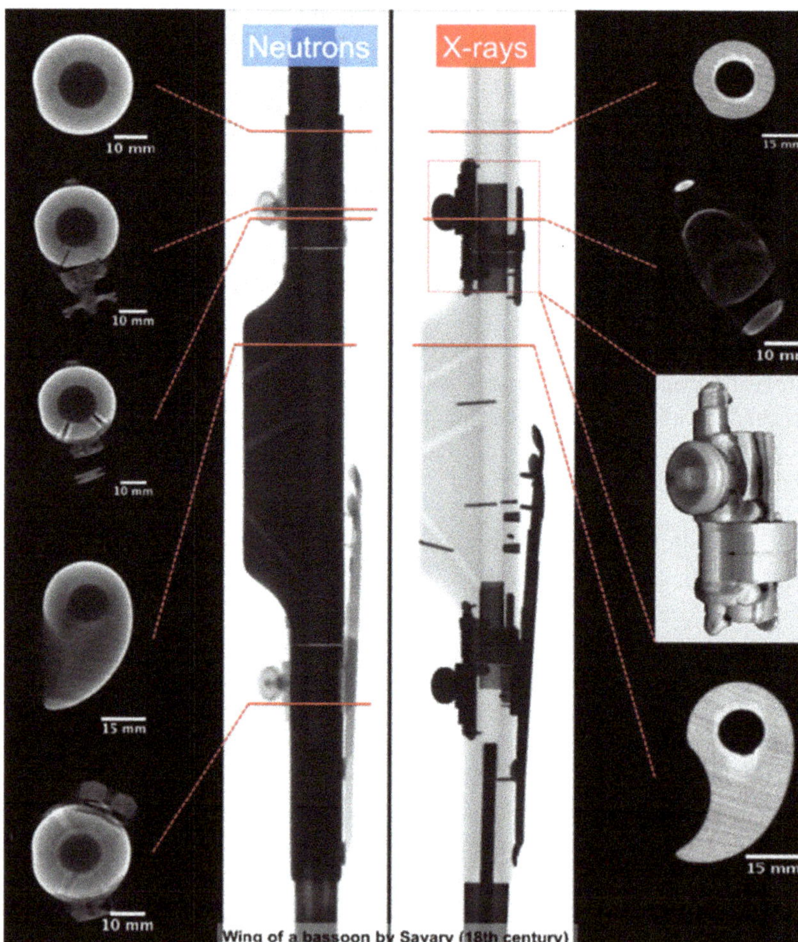

Figure 8.11. Results of the neutron and x-ray investigation of the Savary bassoon; the centre (white background) shows projections with the respective radiation (left neutrons; right x-rays); in the periphery (black background) slices and 3D visualization of the reconstructed CT-data are shown. Reprinted from [69], copyright (2017), with permission from Springer Nature.

dimensions of the key system as well as on the dimensions and shape of the wood body. A similar result is presented by Festa *et al* [69] on the example of a bassoon part (figure 8.11). The presented object well points out some of the possibilities and limitations of neutron tomography, when it comes to neutron imaging. The parts, where the size of the object limits the transmission only yield limited information; smaller parts with a combination of wood and metal allow on the other hand detailed insight on both types of material, which would be much harder to obtain with other methods such as x-ray imaging.

In some cases the mix of materials present in cultural heritage object makes it hard or impossible to obtain an integral insight into the object using only one method. In such cases it can be beneficial and necessary to use and combine two or more methods. One such example is an ancient short sword found in lake Zug (Switzerland). The object has a well preserved wooden hilt abundandly decorated with small metallic inlays. Using only x-ray or neutron tomography would have revealed only part of the information. By using both modalities it was possible to combine the result and hence to obtain complete information on the object and its wooden hilt [16]. The project is discussed in more detail in chapter 15.

One way to overcome the problem of lacking transmission is the utilization of higher energies. Osterloh *et al* [70–73] used fast neutrons to prove the feasibility to study larger wooden objects. Osterloh *et al* [73] and Osterloh and Nusser [74] used fast neutrons to study the penetration behaviour of carbolineum, in a couple of ecclesiastical wooden artefacts. The objects had been treated in the early 20th century to stop biological degradation. The objective of the study is to gain more insight into how the chemical has penetrated the bulk of the material in order to find possible solutions to clean the objects again from it, as the carbolineum which is in the meantime hazardous and harmful to health and the environment, has started discolouring the pieces of art.

Conservation and the application of conservants and consolidants as well as other topics touching cultural heritage objects are discussed in more detail in chapter 17.

The high sensitivity for hydrogen plays the paramount role, when it comes to the investigation of wood-related topics with neutron-based techniques. On the one hand wood is a hydrogen-containing material and most of the interaction with neutrons results from interaction with hydrogen in the wood components. On the other hand, wood is a hygroscopic material and most of its properties depend on the moisture content. The hydrogen content in the wood components limits the size of the wooden samples which can reasonably be investigated using neutrons. The high sensitivity for hydrogen and hence water allows at the same time studying even smallest changes in the moisture content with high spatial resolution, which would be more difficult or even impossible using other non-destructive testing methods. Neutron imaging therefore allows measuring and studying, e.g. changes in the moisture concentrations within tree rings or at the boundaries between materials with different diffusion resistance (adhesive joints, coatings).

Since wood swells and shrinks when there is a change in humidity, the change in volume must be corrected in the edge areas or in zones with local differences in humidity such as glue joints or coatings. Suggestions for this are presented.

Neutron imaging (radiography and tomography) offers good complementary options to the classic methods of humidity measurement such as electrical resistance measurement, dielectric humidity measurement or NIR spectroscopy. This is particularly true, when it comes to investigations of spatially resolved, dynamic, time-dependent moisture-related phenomena.

References

[1] Bucur V 2003 Neutron imaging *Nondestructive Characterization and Imaging of Wood* (Berlin: Springer) pp 281–98

[2] Rowell R 2005 *Handbook of Wood Chemistry and Wood Composites* (Boca Raton, FL: Taylor and Francis)

[3] Mannes D, Josic L, Lehmann E and Niemz P 2009 Neutron attenuation coefficients for non-invasive quantification of wood properties *Holzforschung* **63** 472–8

[4] Reda K, Said S A, Fayez-Hassan M, Tartor B A and Bashter I I 2018 Neutron attenuation coefficients for different types of Egyptian wood *World Appl. Sci. J.* **36** 674–9

[5] Hassanein R, Lehmann E and Vontobel P 2005 Methods of scattering corrections for quantitative neutron radiography *Nucl. Instrum. Meth.* A **542** 353–60

[6] Boillat P, Carminati C, Schmid F, Grünzweig C, Hovind J, Kaestner A and Lehmann E H 2018 Chasing quantitative biases in neutron imaging with scintillator-camera detectors: a practical method with black body grids *Opt. Express* **26** 15769–84

[7] Mannes D C 2009 *Non-destructive Testing of Wood by Means of Neutron Imaging in Comparison with Similar Methods* (Zürich: Diss. ETH Zürich) p 172

[8] Mannes D, Niemz P and Lehmann E 2009 Tomographic investigations of wood from macroscopic to microscopic scale *Wood Res.* **54** 33–44

[9] Trtik P and Lehmann E H 2016 Progress in high-resolution neutron imaging at the Paul Scherrer institute—the neutron microscope project *J. Phys. Conf. Ser.* **746** 012004

[10] Trtik P, Hovind J, Grünzweig C, Bollhalder A, Thominet V, Mannes D C, Kaestner A and Lehmann E H 2015 Improving the spatial resolution of neutron imaging at Paul Scherrer Institute—the neutron microscope project *Phys. Proc.* **69** 169–76

[11] Mannes D, Lehmann E, Cherubini P and Niemz P 2007 Neutron Imaging versus standard x-ray densitometry as method to measure tree-ring wood density *Trees Struct. Funct.* **21** 605–12

[12] Keunecke D, Mannes D, Evans R, Lehmann E and Niemz P 2010 Silviscan vs. neutron imaging to generate radial softwood density profiles *Wood Res.* **55** 49–60

[13] De Ridder M *et al* 2011 High-resolution proxies for wood density variations in *Terminalia superba Annal. Bot.* **107** 293–302

[14] Cherubini P, Humbel T, Beeckman H, Gärtner H, Mannes D, Pearson C, Schoch W, Tognetti R and Ley-Yadan S 2013 Olive tree-ring problematic dating: A comparative analysis on Santorini (Greece) *PLoS One* **8** e54730

[15] Cherubini P, Humbel T, Beeckman H, Gärtner H, Mannes D, Pearson C, Schoch W, Tognetti R and Lev-Yadun S 2014 The olive-branch dating of the Santorini eruption *Antiquity* **88** 267–73

[16] Mannes D, Schmid F, Frey J, Schmidt-Ott K and Lehmann E 2015 Combined neutron and x-ray imaging for non-invasive investigations of cultural heritage objects *Phys. Proc.* **69** 653–60

[17] Masalles A, Lehmann E and Mannes D 2015 Non-destructive investigation of 'The Violinist' a lead sculpture by Pablo Gargallo, using the neutron imaging facility NEUTRA in the Paul Scherrer Institute *Phys. Proc.* **69** 636–45

[18] Niemz P, Mannes D, Lehmann E, Vontobel P and Haase S 2004 Untersuchungen zur Verteilung des Klebstoffes im Bereich der Leimfuge mittels Neutronenradiographie und Mikroskopie *Holz als Roh- und Werkstoff* **62** 424–32

[19] Mannes D, Niemz P, Kläusler O and Lehmann E 2005 Non-invasive study of adhesive penetration into wood by means of neutron imaging methods *Proc. of the 14th Int. Symp. on Nondestructive Testing of Wood* ed F W Bröker (Aachen: Shaker) pp 61–72

[20] Plötze M and Niemz P 2011 Porosity and pore size distribution of different wood types as determined by mercury intrusion porosimetry *Europ. J. Wood Prod.* **69** 649–57

[21] Niemz P and Sonderegger W 2017 *Physik des Holzes und der Holzwerkstoffe* (Leipzig: Fachbuchverlag im Carl Hanser Verlag) p 580

[22] Siau J F 1995 *Wood: Influence of Moisture on Physical Properties* (Keene, NY: Department of Wood Science and Forest Products, Virginia Polytechnic Institute and State University) p 227

[23] Skaar C 1988 *Wood-Water Relations* (Berlin: Springer)

[24] Kollmann F 1955 *Technologie des Holzes und der Holzwerkstoffe* vol 2 (Heidelberg: Springer)

[25] Kollmann F and Côté Jr W A 1968 *Principles of Wood Science and Technology* **vol 1** (Berlin: Springer) p 592

[26] Niemz P and Sander D (eds) 1990 *Prozessmesstechnik in der Holzindustrie* (Leipzig: Fachbuchverlag) p 288

[27] Pellerin R F and Ross R J 2002 *Nondestructive Evaluation of Wood* (Madison, WI: Forest Products Society)

[28] Yin Q and Liu H-H 2021 Drying stress and strain of wood: a review *Appl. Sci.* **11** 5023

[29] Niemz P, Teischinger A and Sandberg D 2023 *Springer Handbook of Wood Science and Technology* (Cham: Springer Nature Switzerland AG)

[30] Kollmann F 1936 *Technologie des Holzes* (Berlin: Springer)

[31] Kollmann F 1951 *Technologie des Holzes und der Holzwerkstoffe* vol 1 (Berlin: Springer)

[32] Couceiro J, Hansson L, Sehlstedt-Persson M, Vickberg T and Sandberg D 2020 The conditioning regime in industrial drying of Scots pine sawn timber studied by x-ray computed tomography: a case-study *Eur. J. Wood Prod.* **78** 673–82

[33] Couceiro J, Lindgren O, Hansson L, Söderström O and Sandberg D 2019 Real-time wood moisture-content determination using dual-energy x-ray computed tomography scanning *Wood Mater. Sci. Eng.* **14** 437–44

[34] Tzscherlich S 1989 Vergleich verschiedener Verfahren der Feuchte- und Dichtemessung von Hackschnitzeln mit Gamma- und Neutronenstrahlung *PhD Thesis* Dresden: TU Dresden (in German)

[35] Arends T, Barakat A J and Piel L 2018 Moisture transport in pine wood during one-sided heating studied by NMR *Exp. Therm. Fluid Sci.* **99** 259–71

[36] Nakanishi T M and Matsubayashi M 1997 Nondestructive water imaging by neutron beam analysis in living plants *J. Plant Physiol.* **151** 442–5

[37] Nakanishi T M, Karakama I and Sakur T 1998 Moisture imaging of a camphor tree by neutron beam *Radioisotopes* **47** 387–91

[38] Nakanishi T M, Okano T, Karakama I, Ishihara T and Matsubayashi M 1998 Three-dimensional imaging of moisture in wood disk by neutron beam during drying process *Holzforschung* **52** 673–6

[39] Islam M N, Khan M A, Alam M K, Zaman M A and Matsubayashi M 2003 Study of water absorption behavior in wood plastic composites by using neutron radiography techniques *Polym.-Plast. Technol. Eng.* **42** 925–34

[40] Lehmann E H 2000 Neutron imaging *Neutron Scattering in Novel Materials: Proceedings of the Eighth Summer School on Neutron Scattering* ed A Furrer (Singapore: World Scientific) pp 22–36

[41] Lehmann E, Vontobel P and Niemz P 2000 Investigation of moisture distribution in wooden structures by neutron radiography *Sci. Tech. Rep. 1999* vol 6 (Villigen: PSI) pp 53–5

[42] Lehmann E, Vontobel P, Scherrer P and Niemz P 2001 Anwendung der Methode der Neutronenradiographie zur Analyse von Holzeigenschaften *Holz als Roh- und Werkstoff* **59** 463–71

[43] Sedighi-Gilani M, Griffa M, Mannes D, Lehmann E, Carmeliet J and Derome D 2012 Visualization and quantification of liquid water transport in softwood by means of neutron radiography *Int. J. Heat Mass Transfer* **55** 6211–21

[44] Sedighi-Gilani M, Vontobel P, Lehmann E, Derome D and Carmeliet J 2014 Liquid uptake in Scots pine sapwood and hardwood visualized and quantified by neutron radiography *Mater. Struct.* **47** 1083–96

[45] Desmarais G, Sedighi Gilani M, Vontobel P, Carmeliet J and Derome D 2016 Transport of polar and nonpolar liquids in softwood imaged by neutron radiography *Transp. Porous Media* **113** 383–404

[46] Niemz P, Lehmann E, Vontobel P, Haller P and Hanschke S 2002 Untersuchungen zur Anwendung der Neutronenradiographie zur Beurteilung des Eindringens von Wasser in Eckverbindungen aus Holz *Holz als Roh- und Werkstoff* **60** 118–26

[47] Sonderegger W, Glaunsinger M, Mannes D, Volkmer T and Niemz P 2015 Investigations into the influence of two different wood coatings on water diffusion determined by means of neutron imaging *Eur. J. Wood Prod.* **73** 793–9

[48] Mannes D, Sonderegger W, Hering S, Lehmann E and Niemz P 2009 Non-destructive determination and quantification of diffusion processes in wood by means of neutron imaging *Holzforschung* **63** 589–96

[49] Rosner S, Riegler M, Vontobel P, Mannes D, Lehmann E, Karlsson B and Hansmann C 2012 Within-ring movement of free water in dehydrating Norway spruce sapwood visualized by neutron radiography *Holzforschung* **66** 751–6

[50] Lanvermann C, Sanabria S J, Mannes D and Niemz P 2014 Combination of neutron imaging (NI) and digital image correlation (DIC) to determine intra-ring moisture variation in Norway spruce *Holzforschung* **68** 113–22

[51] Zhou X, Carmeliet J and Derome D 2021 Assessment of moisture risk of wooden beam embedded in internally insulated masonry walls with 2D and 3D models *Build. Environ.* **193** 1–15

[52] Sonderegger W, Hering S, Mannes D, Vontobel P, Lehmann E and Niemz P 2010 Quantitative determination of bound water diffusion in multilayer boards by means of neutron imaging *Eur. J. Wood Prod.* **68** 341–50

[53] Mannes D, Sanabria S, Funk M, Wimmer R, Kránitz K and Niemz P 2014 Water vapour diffusion through historically relevant glutin-based wood adhesives with sorption measurements and neutron radiography *Wood Sci. Technol.* **48** 591–609

[54] Lämmlein S L, Mannes D, Van Damme B, Schwarze F W and Burgert I 2019 The influence of multi-layered varnishes on moisture protection and vibrational properties of violin wood *Sci. Rep.* **9** 1–9

[55] Sedighi Gilani M, Vontobel P, Lehmann E, Carmeliet J and Derome D 2014 Moisture migration in wood under heating measured by thermal neutron radiography *Exp. Heat Transfer* **27** 160–79

[56] Sedighi Gilani M, Abbasion S, Lehmann E, Carmeliet J and Derome D 2014 Neutron imaging of moisture displacement due to steep temperature gradients in hardwood *Int. J. Therm. Sci.* **81** 1–12

[57] Abbasion S, Carmeliet J, Sedighi Gilani M, Vontobel P and Derome D 2015 A hygro-thermo-mechanical model for wood: part A. Poroelastic formulation and validation with neutron imaging *Holzforschung* **69** 825–37

[58] Solbrig K, Frühwald K, Ressel J B, Mannes D, Schillinger B and Schulz M 2015 Radiometric investigation of water vapour movement in wood-based composites by means of cold and thermal neutrons *Phys. Proc.* **69** 583–92

[59] Sanabria S, Lanvermann C, Michel F, Mannes D C and Niemz P 2015 Adaptive neutron radiography correlation for simultaneous imaging of moisture transport and deformation in hygroscopic materials *Exp. Mech.* **55** 403–15

[60] Couceiro J, Hansson L, Mannes D, Niemz P and Sandberg D 2022 Estimation of the moisture content in wood by combination of neutron and x-ray imaging *Proc. of the 22nd Int. Nondestructive Testing and Evaluation of Wood Symp., General Technical Report FPL-GTR-290*
Wang X and Ross R J (eds) (Madison, WI: U.S. Department of Agriculture, Forest Service, Forest Products Laboratory) pp 40–7

[61] Plaza N Z, Ibach E R, Hasburgh L E and Taylor M 2021 Using thermal neutron imaging in forest products research *Environ. Sci. Proc.* **3** 1–7

[62] Sonderegger W, Mannes D, Kaester A, Hovind J and Lehmann E 2015 On-line monitoring of hygroscopicity and dimensional changes of wood during thermal modification by means of neutron imaging methods *Holzforschung* **69** 87–95

[63] Sedighi Gilani M and Schwarze F W 2015 Hygric properties of Norway spruce and sycamore after incubation with two white rot fungi *Holzforschung* **69** 77–86

[64] Schwarze F W, Spycher M and Fink S 2008 Superior wood for violins—wood decay fungi as a substitute for cold climate *New Phytol.* **179** 1095–104

[65] Ossler F, Finney C E, Warren J M, Bilheux J C, Zhang Y, Mills R A, Santodonato L J and Bilheux H Z 2021 Dynamics of hydrogen loss and structural changes in pyrolyzing biomass utilizing neutron imaging *Carbon* **176** 511–29

[66] Ossler F, Santodonato L J, Warren J M, Finney C E, Bilheux J C, Mills R A, Skorpenske H D and Bilheux H Z 2019 *In situ* monitoring of hydrogen loss during pyrolysis of wood by neutron imaging *Proc. Combust. Inst.* **37** 1273–80

[67] Festa G, Tardino G, Pontecorvo L, Mannes D C, Senesi R, Gorini G and Andreani C 2014 Neutrons and music: Imaging investigation of ancient wind musical instruments *Nucl. Instrum. Methods Phys. Res., Sect.* B **336** 63–9

[68] Kirsch S and Mannes D 2015 X-ray CT and neutron imaging for musical instruments—a comparative study *Science* **55** 188–96

[69] Festa G, Mannes D and Scherillo A 2017 Neutrons unveil secrets of musical instruments *Neutron Methods for Archaeology and Cultural Heritage. Neutron Scattering Applications and Techniques* ed N Kardjilov and G Festa (Cham: Springer) pp 53–66

[70] Osterloh K R, Bücherl T, Hasenstab A, Rädel C, Zscherpel U, Meinel D, Weidemann G, Goebbels J and Ewert U 2007 Fast neutron radiography and tomography of wood as compared to photon based technologies *Proc. of the DIR 2007—Int. Symp. on Digital Industrial Radiology and Computed Tomography* pp 25–7

[71] Osterloh K, Rädel C, Zscherpel U, Meinel D, Ewert U, Bücherl T and Hasenstab A 2008 Fast neutron radiography and tomography of wood *Insight—Non-Destructive Testing and Condition Monitoring* **50** 307–11

[72] Osterloh K, Bücherl T, Jechow M, Fratzscher D, Wrobel N, Tannert T, Hasenstab A, Zscherpeli U and Ewert U 2010 Radiological examination of wood with neutrons, different perspectives *Proc. of the Int. Workshop on Fast Neutron Detectors and Applications* pp 7–11

[73] Osterloh K *et al* 2015 Computed tomography with x-rays and fast neutrons for restoration of wooden artwork *Phys. Proc.* **69** 472–7

[74] Osterloh K R and Nusser A 2014 X-ray and neutron radiological methods to support the conservation of wooden artworks soaked with a polluting impregnant 'Carbolineum' *Proc. of the 11th European Conf. on Non-Destructive Testing (ECNDT 2014)*

Chapter 9

Geology

Anders Kaestner, Alessandro Tengattini and Veerle Cnudde

Within the realm of geology, the investigation of geomaterials holds immense significance. Here, different aspects are relevant, e.g., material composition, the behaviour as a porous medium exposed to liquids, or as structures in mechanical systems. Geomaterials refer to the solid materials that constitute the Earth's crust, including rocks, minerals, sediments, and soils. Processes in the soil are a specialization of these processes and were covered already in chapter 7. Understanding the properties and behaviour of geomaterials is crucial for various aspects of geology and engineering, such as resource-exploration, natural hazard assessment, environmental studies, and engineering projects.

There are manifold research questions related to the study of geomaterials. The type of question affects the choice of method to probe the material. Neutron imaging techniques are mainly used for studies related to (a) characterizing samples in terms of material composition, porosity, and fracture analysis, (b) fluid flow and transport processes inside the stone [1], and (c) chemo-thermo-mechanical properties of and processes within rocks [2]. The study of geomaterials counts to the earliest applications of neutron imaging. Quantitative experiments first became efficient by introducing digital imaging systems at the high flux neutron sources; this allowed tomography and time-series acquisition of radiographs. In early experiments, geologists used neutron scattering [3] and later radiography on films to detect water or resign distribution in pieces of rocks [4–6] and some 20 years later, with the availability of efficient digital cameras, also neutron tomography [7–9]. The earliest study of a dynamic process in geomaterials dates to the 1980s [6]. Here, neutron radiography was used to capture water moving in rock pores. Fast tomography time series was first introduced using amorphous silicon flat panels [10, 11], which provided the highest frame rates at the time. This technology was replaced by more flexible camera-based systems [12, 13]. The well-known properties of neutrons with greater sensitivity to some light elements, such as hydrogen, and lower sensitivity to others, such as most metals, compared to the equivalent x-ray technique and its

doi:10.1088/978-0-7503-3495-2ch9

© IOP Publishing Ltd 2024

isotope sensitivity, offered significant advantages in studying geomaterials. Additionally, neutron computed tomography (CT) can be easily combined with x-ray CT, allowing the simultaneous acquisition of complementary datasets [14–16]. This method further helped the characterization of the 3D structure of the geological material and understanding of complex processes, e.g., [17–21].

Today, a large number of studies in the domain of geology and rock mechanics adopt neutron imaging to cast light on a wide range of processes, from understanding and prediction of the effect of strain localization on fluid flow for reservoir exploitation [22, 23] to pollution remediation studies [24], storage of nuclear waste [19, 20, 25] or CO_2, conservation and restoration strategies against weathering [26, 27], to mention but a few.

9.1 Characterization of geomaterials

Enhancing our understanding of the studied materials is typically the first crucial step. This knowledge plays a significant role in planning experiments on geomaterials and in the context of studying fluid dynamics, geomechanics, and chemical reactions within these materials. In this context, three essential aspects of material characterization come into focus: (a) Identification of materials present and their spatial distribution within a given specimen; (b) analysis of sample texture; and (c) evaluation of voids within the material, enabling the resolution of distinctive features such as cracks and the determination of pore network properties. These characterizations collectively contribute to a comprehensive understanding of the 3D composition, structure, and properties of the materials under investigation, facilitating more informed experimental design and analysis.

Geomaterials can often be well characterized through their chemical composition. Chemical analysis is, however, often destructive and does not resolve the spatial distribution of the components in the sample. Imaging can characterize the spatial distribution of different components comprising the specimens. Features that can be described are clusters or layers of varying chemical compositions. The choice of imaging modality decides what information is revealed from the identified regions. X-rays and neutrons are two modalities often used to analyze geomaterials with sensitivities, for example, biased towards density and hydrogen content, respectively. Here, the phrasing *biased towards* is worth paying attention to as many factors contribute to the local attenuation coefficient, e.g., porosity, material mixtures, amount of residual pore water, or even bound water. This complexity can hinder classification if the decision is solely based on the attenuation coefficient from a single modality. An approach that has been proposed is to combine images obtained using neutrons and x-rays to exploit the complementary nature of the information that the two modalities provide by Vontobel *et al* [28], Chiang *et al* [29], and Martell *et al* [30]. The authors tabulate calculated attenuation coefficients for selected geomaterials using neutrons and x-rays. This combination aims to provide material distribution maps of the specimens, allowing for a less ambiguous classification of the specimen composition.

The materials presented in figure 9.1 can roughly be categorized into three classes. (a) The first class includes minerals with significant variations in density and

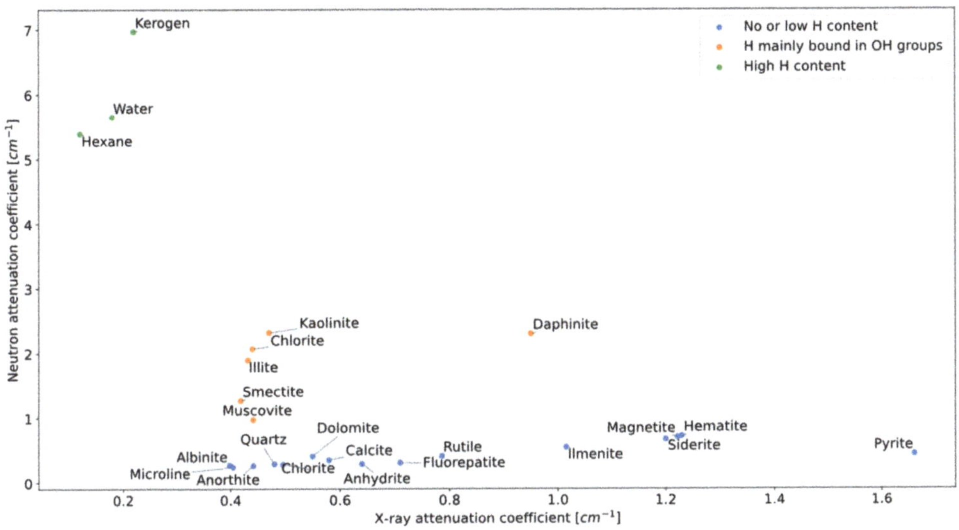

Figure 9.1. Neutron and x-ray attenuation coefficients for a selection of materials relevant for imaging experiments with geomaterials.

generally no hydrogen content. These minerals have a widespread distribution in their x-ray attenuation due to substantial density variations. Neutrons, on the other hand, are less sensitive to materials in this category as many of these materials are represented by different combinations of the same elements, which have very similar neutron attenuation. The second class of minerals is the opposite of the first. The materials in this class all include OH groups that directly impact the neutron attenuation (b). The OH groups add little to the density of these materials. Thus, the contrast variations for x-rays are relatively small. The third class contains water, low-density organic solids, and liquid materials (c).

These categories could naturally be extended by adding minerals containing, for example, boron or lithium, two light elements with high neutron attenuation. In environment protection studies, it would also be relevant to add heavy metals like cadmium, e.g., [24].

Observing these materials is often the purpose of many imaging experiments. Here, the high hydrogen content makes the neutrons the ideal probe. The attenuation coefficient pairs shown in figure 9.1 are calculated using the attenuation coefficient tables for x-rays and neutrons. These tables are specified for a single energy. It is, however, important to note that the beam is not monochromatic in most experiments, and the measured image represents the weighted average attenuation over the neutron spectrum. Therefore, the measured attenuation coefficients will deviate from those in the plot due to different source spectra and natural variations in material density and composition.

Furthermore, some values in figure 9.1 are averages for mineral families with similar composition and properties. Still, these values indicate which material combinations are likely to provide contrast differences that can be used to classify

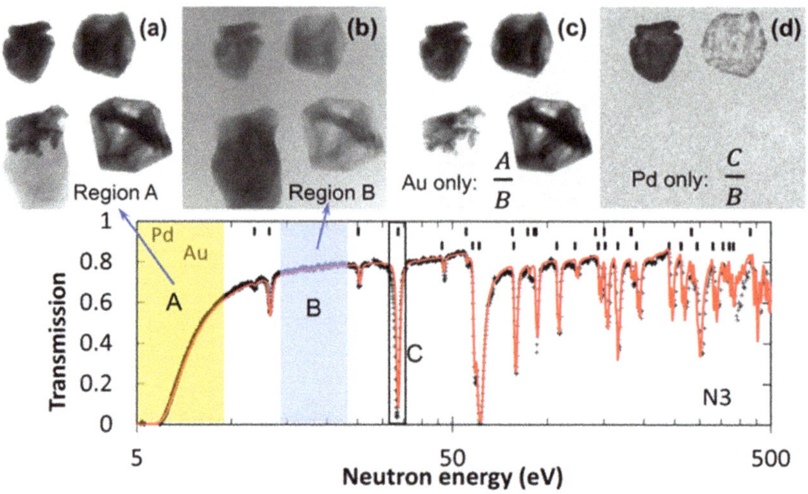

Figure 9.2. Identifying regions containing Au and Pd using neutron resonance imaging. Reproduced from [31]. CC BY 4.0.

regions in the samples. Some materials are still hard to separate even though the full neutron spectrum combined with x-rays is used; haematite (Fe_2O_3) and magnetite ($Fe^{3+}{}_2Fe^{2+}O_4$) are examples of such minerals. In such cases, it has been suggested to use wavelength-resolved imaging like in [31], which studies the attenuation at different wavelengths and investigates the bulk crystallinity of the gold single crystals, figure 9.2. The method can identify the gold by using the resonance of gold in spectrum segment 'C' in the figure. This is done by computing the ratio between images from different parts of the neutron spectrum.

There are reports of content mapping using radiography and tomography; whether 2D or 3D data is needed depends on the target application and the complexity of the studied specimens. Measurements using radiography mostly require slab-shaped specimens where the thickness is known, as the measured values must be divided by the thickness to obtain the attenuation coefficient, assuming a homogeneous distribution along the irradiation direction. Tomography has the advantage that the reconstructed voxel values are already given as attenuation coefficients, and there is no need to do any further corrections, i.e., the material distribution does not have to be homogeneous as for radiography. The quantitative measurements of the attenuation coefficient for a material are susceptible to biases caused by sample and background scattering. These biases must be corrected to avoid great errors in the quantification. The need for correction applies to radiography and tomography alike. Methods to correct scattering are described in chapter 20. A further effect that makes it hard to compare the attenuation coefficient between instruments is the shape of the neutron spectrum for instruments with a cold spectrum which have greater variations in spectrum shape. Polychromaticity is particularly relevant for characterizing crystalline materials [31]. Each crystalline material has a spectral fingerprint thanks to diffraction in the crystalline texture. This can be used in the characterization process.

Besides the ability to separate and identify different materials, there is also a more fundamental planning aspect of knowing the attenuation coefficients of the involved materials. Namely, they set an upper limit on the sample dimensions in the beam direction that can be used without attenuating the radiation to levels below the noise level of the experiment.

Although neutron imaging is a very powerful tool, it is not able to perform element mapping directly. Achieving element mapping requires calibration measurements of isolated compounds. A method with low reliability as it is difficult to account for mixtures of two or more compounds. A solution is the complementary use of Prompt Gamma Activation Analysis (PGAA) that enhances the imaging technique [32] with information about the present elements. The combination allows a coarse spatial element mapping to support the analysis of the neutron tomography data.

The sample meso- and macro-texture and morphology provide valuable information for continued experiments as this allows categorizing the samples into groups with similar expected behaviour [33]. Later, this information can also be used to model the observed processes more accurately. Particularly relevant information is how voids are distributed and interconnected in the specimens. Here, we are talking about two types of voids: pores and cracks. Pores are small voids between the solid components that form the geomaterial. For pores, two highly relevant pieces of information to characterize are the statistical distribution of their size and their connectivity. The sample dimensions required to obtain a statistically representable sample (generally named representable elementary volume, REV) in turn requires a large field of view (FOV) with the consequence that the pixel sizes are sometimes larger than pores in the pore network. The result is that not all pores cannot be resolved. However, it is still possible to measure the porosity of the specimen using neutron imaging even when not all pores can be resolved. This can be done using indirect methods like comparing the oven-dry specimen to the fully saturated specimen [34]. It is, furthermore, common to perform a complementary high-resolution x-ray tomography in a separate scan to provide the missing pore network information. Cracks are a different type of void; they are often caused by mechanical actions and separate the specimen's regions by a thin sheet-like gap that is narrow in one direction and extends widely in the other directions. The number of cracks and their dimensions are relevant for chemo-hydro-thermo-mechanical studies which will be covered in the following section on dynamic processes. In order to highlight the presence of fractures inside materials, often water or hydrogen-rich materials within the fissures have been used as tracers to aid their detection and establish their connectivity (e.g., [4]).

9.2 Dynamic processes in geomaterials

The characterization and visualization of processes in geomaterials is very important as these processes are linked to a broad range of scientific questions involving for instance different aspects of hydrology, including solute transport, the mechanics of the materials, and their chemical interaction with components in the solute transport. This section on processes starts with studies of the hydrology of the rocks, which is the foundation for more complex experiments. The complexity will be increased with other geological processes like mechanical loading, thermal changes, and chemical reactions.

Progress in these experiments has been achieved thanks to new camera technologies with short read-out times providing near-gapless acquisition of the images used in the time series. Combining such cameras with continuously turning rotational stages results in on-the-fly tomography [12, 13, 35]. These detectors, in combination with improved experiment infrastructure [36] allow *in situ* 2D and 3D time-lapse neutron imaging, also known as 4D neutron imaging of fluid and pollutant percolation in rocks down to speeds of 1.5 s per tomography [37–39].

9.2.1 Rock hydrology and solute transport

Understanding fluid flow in rocks is crucial in geology to quantify many natural processes such as groundwater flow and naturally triggered seismicity, but also to study displacement of contaminants [24], the eligibility of subsurface waste storage [40], geothermal energy usage [41], oil and gas recovery and artificially induced seismicity, to mention but a few.

As neutrons are particularly sensitive to the presence of hydrogen, neutron imaging is very applicable to any geological study where water is involved. In this context, the method can provide important additional insights into the study of permeability evolution and subsurface transport of pollutants in the rock. Here, two key parameters that control the variability of fluid flow and the movement of dissolved chemical species are (a) the local hydraulic conductivity, and (b) the local sorption properties of the dissolved chemical species by the solid matrix. These parameters can be constrained through tomography imaging of rock samples subjected to fluid injection under constrained flow rate and pressure. Several of these works focus on the roles of surfactants (used in hydraulic fracturing fluids in the oil and gas industry), fissure aperture and surface roughness, and water salinity [42, 43].

The properties of neutron imaging are ideal for exploring fluid localization in porous materials due to the high but variable sensitivity of neutrons to significant differences in attenuation coefficient between the different hydrogen isotopes, chapter 7. Therefore, neutron imaging was used for fundamental research related to the study of fluid flow in porous media [44, 45] or to study the evolution of permeability in porous media undergoing loading, as detailed below, as well as very applied research dealing with, for instance, the testing of conservation products inside building stones [1, 46] or to quantify stone weathering [27, 47, 48].

Neutron imaging can, besides the qualitative visualization of the distribution of hydrogen-rich liquids, also be used to quantify the hydrogen-rich liquid content inside the material as a function of time. A development that was promoted by methods to increase the accuracy of the water content quantification. One approach to quantifying the water content is to use calibration methods based on the neutron transmission through different thicknesses of water [49]. This approach is rather empirical and only corrects for the directly observed effect, it requires new calibration for each liquid, and does not consider sample shape or instrument background. The neutron scattering causes the nonlinearity observed in the step wedge calibrations. Therefore, methods for scattering corrections were introduced to allow a generalized physics-based approach for quantitative neutron radiography and tomography that considers the sum of the scattering contributions [50, 51] and later by [52, 53] to provide more precise estimations of the liquid content from variations in image intensity. These methods are covered in more detail in chapter 20.

A central neutron imaging application in hydrogeology is the study of spontaneous and flow-driven imbibition in fractured rocks [54–56]. The relevance is mainly to understand the evolution of permeability in reservoir rocks for aquifer exploitation or where water or brine is injected to extract hydrocarbons, also called fracking [57, 58], and for CO_2 sequestration. A further application for water injection is geothermal energy extraction [41, 59]. Typically, these experiments observe the rise height of water or brine in a natural or prepared fracture as a function of time. The early part of this process can be described using Washburn's equation [60], which states that the height of the water column is proportional to $t^{1/2}$.

$$L = \sqrt{\frac{\gamma r t \cos\phi}{2\eta}}$$

It is often relevant to determine the wetting angle, ϕ, between liquid and solid material, as it, with the pore radius, r, controls the imbibition properties of the material. The remaining parameters describe the liquid in terms of dynamic viscosity, η, and surface tension, γ. This model is valid for early uptake when other factors like gravity and diffusion are less relevant. The neutron imaging experiments initially used mainly time-series radiography [61, 62]. The need for tomography using either neutrons or x-rays soon was evident as it allows for better structural descriptions for heterogeneous specimens. The surface roughness, the tortuosity of the porous material, and the fracture impact the rise velocity sufficiently to deviate from the expected $t^{1/2}$ proportionality. A model based on the fractal dimension of the pore network proposed by Cai *et al* [63] was applied in the modelling [64–66]. The typical model system with a fractured rock is shown in figure 9.3.

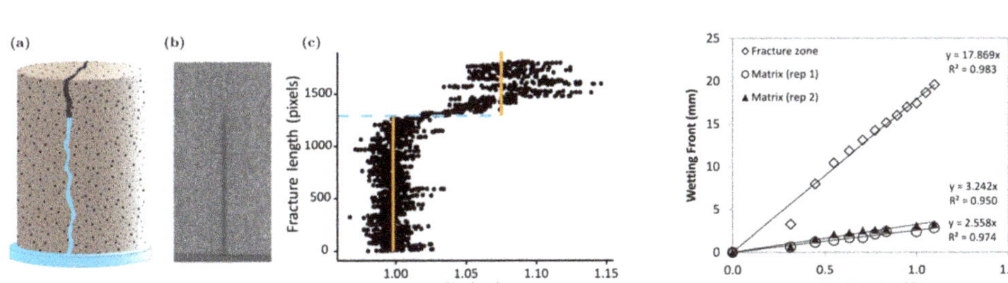

Figure 9.3. Model system for spontaneous imbibition (a). The level of the rise can be seen as a darker (opaque) region in the normalized image (b) with a vertical intensity profile at the fracture (c), reprinted from [54], copyright (2015), with permission from Elsevier. The time series provide plots of the wetting front (d), reproduced from [66] CC BY 4.0.

This model was still incomplete, and was studied by Zambrano *et al* [55] who used a combination of neutron radiography time series to capture the rising process complemented by neutron and x-ray tomography of the initial and final states of the sample. The x-ray 3D data was used in a lattice-Boltzmann simulation which showed that unresolved pores have an impact on the imbibition in the rock sample. This conclusion was based on the fact that the neutron tomography showed water in regions that the simulation could not predict. This leads to the hypothesis that the imbibition/diffusion of liquid in the rock matrix also plays a relevant role in the rise velocity in the fracture. The effect of this bi-porosity system was shown by Cheng *et al* [54] and Brabazon *et al* [66], see figure 9.3, who measured the wetting front in fracture and matrix. McFarlane *et al* [67] concluded that the model of Cai *et al* is insufficient to describe spontaneous imbibition in the fracture [67] which caused imprecise measurements of the wetting angles from the imbibition experiment. Siddiqui *et al*'s [68] complementary measurements using SAXS and SEM-FIB conclude that the slower rise in the fracture is caused by configurational diffusion into the rock matrix, figure 9.4. This means that not only the pore sizes in the network but also the molecule sizes of the liquids involved play a role in the imbibition process. The consequence is particularly relevant in systems with different liquids where, for example, water is added to push hydrocarbons out of the matrix.

Besides the intrinsic interest in the spatial distribution of fluids (e.g., to determine the water retention curves of porous media [35, 45]), the water content determination can help assess the connected porosity [55, 69].

Physical, chemical, and biological weathering has a profound impact on the Earth's landscape and on its building infrastructure. Rock and masonry are constantly damaged and disaggregated by chemical reactions, water infiltration and temperature changes [70–72]. Strengthening efforts to protect and safeguard the world's cultural and natural heritage is one of the United Nations' Targets for Sustainable Cities and Communities in the *2030 Agenda for Sustainable Development*.

Fluids are a major driver of rock weathering: they trigger, among others, dissolution, precipitation, frost damage, and salt weathering. The key to manipulating weathering lies in understanding and controlling fluid flow within the rocks' internal pore structure, thereby influencing the related pore-scale processes. Neutron radiography is often used to study fluid migration inside building stones for the conservation of cultural heritage. Already in 2008, Cnudde *et al* [73] employed rapid neutron radiography to observe and quantify water exchange within a stone, to quantify the water exchange between a stone and the atmosphere, while Xue *et al* [74] quantified water exchange between different materials, such as fresh and old cement mortars. In the latter, the influence of the initial moisture conditions in mortar on water exchange at the interfaces was studied. This approach allows them to obtain the dynamic water content profiles inside materials as a function of time as well as the desorption coefficient. In addition, neutron radiography also allows investigating visually and quantitatively the influence of cracks on water redistribution inside materials, such as stone, cement, and mortars.

As the fluid flow in a porous material is influenced by the pore structure, the impact of different pore structures in fluid flow or reactive transport is often studied.

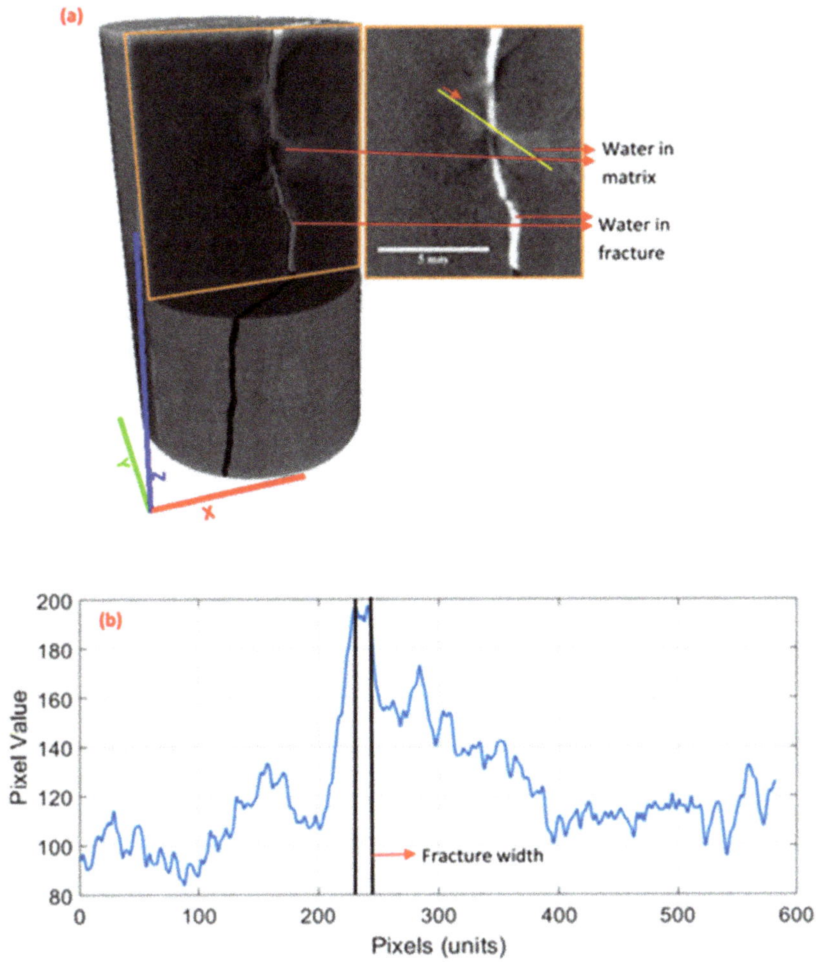

Figure 9.4. Demonstration that imbibed water is not constrained to the fracture, but also enters the stone matrix. Reproduced from [68] CC BY 4.0. The yellow line in the rendered volume (a) indicates the location of the profile plotted in (b).

For instance, when focusing on clays which are common in engineered barrier systems for the underground disposal of radioactive waste (e.g., [19, 20]), their alteration processes need to be known. Also, here, the porosity inside clays is a key parameter for, and a key outcome of, reactive transport (such as dissolution and precipitation reactions). In this case, the water content derived from neutron imaging can be used as a proxy for porosity in saturated samples. These datasets provide valuable input to better model fluid flow behaviour, constrain reactive transport models [40] or validate current models [75].

The neutron imaging datasets are crucial to increase our understanding of the internal processes inside stones. They are needed to validate numerical modelling research on fluid flow or crystallization in porous media, as these data are indispensable for identifying phenomena to be included in models and for model validation. They can also be used to validate alternative lower-resolution approaches often used in on-site research [76].

Stone weathering is a highly relevant topic for the preservation of old buildings. Many of these buildings are built using sedimentary rocks, such as sandstone and limestone. The porous nature of these stones makes them sensitive to moisture variations caused by the climate. Stones in the building foundation furthermore interact with moisture from the soil. An important aspect of the climate-driven moisture variations is the material damage caused by solute transport where, for instance, in the past sulfur from polluted rain created an acidic environment in the pores which eroded the stones. On the other hand, 2D, 3D, and 4D information on saline transport, salt crystal distributions, and fracturing due to salt crystallization is essential for understanding the coupling between transport and mechanical processes related to salt weathering [26, 27, 48, 77]. This knowledge is required to select appropriate materials and conservation methods for our building infrastructure. Often neutron imaging data is complemented by x-ray imaging data [78–80] as detailed 3D structural information is needed to understand the pore-scale dynamics better. When combining x-rays and neutron imaging, it becomes possible to combine the higher spatial resolution description of the microstructure, which is more easily achievable with x-ray micro-tomography and the higher hydrogen sensitivity of neutron imaging when studying processes such as liquid transport, salt crystal distributions, and fracturing processes due to (salt) precipitation. For example, x-ray CT can characterize the microstructure while neutron radiographies provide insight into its effect on fluid flow [19, 55, 64, 81]. In other studies, like [82, 83], this combination was also used to characterize the capillary bridges in a given material state. Due to its non-destructiveness, repeated wetting–drying cycles and repeated uptake and drying of saline solutions are possible. With the ability to follow these processes, it is possible to quantify salt accumulation inside the stone and to monitor potential induced fractures. Further examples of the coupled use of neutrons and x-rays are also reported in the next section, which focuses on coupled processes.

In the past, neutron imaging was also used to study the effect of conservation products as the damage from weathering can be reduced by impregnation, for instance by stone consolidants, which produces a layer to protect and restore the stability of the stones [1]. The penetration depth of these consolidants is important for the efficiency of the treatment and has been studied both as an isolated process [1, 46, 84] and in terms of its effect on the water or brine uptake in the impregnated stones.

When combining x-rays and neutron imaging, it becomes possible to combine the higher spatial resolution achievable with x-ray micro-tomography and the higher hydrogen sensitivity of neutron imaging when studying processes such as liquid transport, salt crystal distributions, and fracturing processes due to (salt) precipitation. For example, x-ray CT can characterize the microstructure while neutron radiographies provide insight into its effect on fluid low [55, 85]. In other studies, like [82, 86], this combination was also used to characterize the capillary bridges in a given material state. Due to its non-destructiveness, repeated wetting–drying cycles and repeated uptake and drying of saline solutions are possible. With the ability to follow these processes, it is possible to quantify salt accumulation inside the stone and to monitor potential induced fractures. Further examples of the coupled use of neutrons and x-rays are also reported in the next section, which focuses on coupled processes.

9.2.2 Mechanics

Geomechanics is a field that explores the mechanical behaviour of geological materials and their interaction with the surrounding environment. By bringing subsurface processes into a laboratory setting, neutron tomography or a combination of x-ray and neutron tomography can immediately visualize these processes. One of the advantages of neutron imaging is its high penetration into metals which allows the construction and use of intricate pressure vessels and the imposition of extreme boundary conditions for geophysics and geomechanics [87, 88]. As such, subsurface processes can be brought to the laboratory environment and immediately visualized using neutron tomography or a combination of x-ray and neutron tomography. For very small samples, generally a few mm, the more powerful x-ray sources available can penetrate these cells. Nonetheless, only neutron tomography is applicable for geomaterials where the REVs are too large and the thick metallic walls are required to withstand the applied pressures [89]. Pure mechanical experiments are otherwise less relevant with neutrons as the structural information is also well represented in x-ray imaging and mostly with higher resolution.

With acquisition of time or load series of a loaded sample it is possible to quantify the strain field of a loaded sample with sub-pixel accuracy [90] a technique that has been improved to work *in situ* with the ability to obtain volume sequences with higher frame rates [91].

Tengattini *et al* [92] provide a detailed review with a comprehensive exploration of geomechanics applications utilizing neutron imaging. The following section will address their combined aspects as mechanics and hydraulics often intertwine in various applications.

9.2.3 Hydro-thermo-chemo-mechanically coupled processes

Besides the intrinsic interest in the flow of fluids within porous media, numerous complex couplings can occur between fluid flow and the evolution of the underlying microstructure. A notable case is hydraulic fracturing (or hydro-fracking), which is particularly interesting to the oil, gas and geothermal industries. In this case, high-pressure fluid injected into rocks opens fractures, which can be followed in *operando* through rapid neutron imaging to reveal the essential role of natural shale heterogeneities, such as its bedding planes. Extreme stress conditions can be easily achieved in *operando* thanks to the high metal penetrations of neutrons [58, 68]. Converse coupling is also of great interest to the oil, gas, and water reservoir industries.

While the permeability of homogeneous media is relatively well known, in fact, as the samples are loaded, strain typically localizes in more or less narrow areas. These are known to severely alter the fluid flow, which is best captured by neutron imaging.

As mentioned in the previous sections, in fact, the isotope sensitivity of neutrons makes it possible to study the fluid permeability by repeated alternating flushes of D_2O and H_2O while subsequent steps of mechanical loading induce different degrees of deformation in the rock [23, 93]. This reveals the effect that localized mechanical deformation has on the flow pathways, depending on the nature of the localization (e.g., dilatant or compacting) and of the rock (e.g., its porosity and connectivity).

Clay rocks have, thanks to their low permeability, been identified to provide a suitable hosting environment for the long-term storage of nuclear waste, which is particularly relevant for the nuclear industry. There are several aspects to study in these rocks and the permeability of the clay–cement interface in the long term [40]. The hydro-mechanical couplings can also be even more profound on a shorter time scale. In these clay rocks [19, 20], the mere presence of water can induce profound changes to the microstructure. These complex hydro-mechanical couplings are best studied using neutron imaging to track the fluid distribution, and x-ray imaging, to detect changes in the rock microstructure. Combining the modalities reveals a coupled relation between water penetration and fracture formation. The fracture initiation is driven by the loss of suction within the—usually partially saturated—rock in contact with water. Then the fracture propagation drives the deeper penetration of the water through the rock. This relation results in greater damage to the rocks as the two interactions enhance each other.

Another relevant coupling for engineering is the hydro-thermo-mechanical couplings in porous media, for example, to study the fire resistance of construction materials. In this case, the migration moisture within samples heated at high temperatures can be tracked with neutron imaging [94–97]. At the same time, simultaneous x-ray tomographies can reveal details about the opening fissures and, thus, about the role they play in preferential drying [21]. Chapter 3 covers the application of neutron imaging in civil engineering construction materials and processes.

An alternative notable example of coupled interactions is the behaviour of the wet sand, in particular bentonites, which can be used as a mould for metal casts. Here the amount of needed material depends on the extent and spatial distribution of its dehydration [98, 99]. These initial studies were completed with rapid time series allowing one to follow crack formation in the drying sand [100, 101] and show that the water was pushed by the heating front, figure 9.5. The sand structure broke near

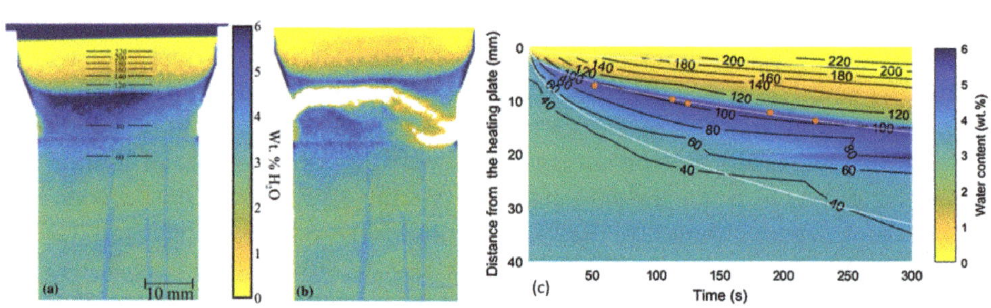

Figure 9.5. Tensile experiment in heated wet sand. Initial and final sample state (a) and (b) shows the moisture distribution in the sample. Monitoring over time resulted in the plot in (c). Reproduced from [100] CC BY 4.0.

the interface between dry and wet sand. Some other studies analyze through neutron imaging the coupled relationship between water content and shrinkage/swelling in clays [102, 103].

Today, our climate is affected by the large amount of CO_2 produced mainly from the combustion of fossil fuels. CO_2 is a greenhouse gas that causes an increase in the global temperature with devastating consequences. Therefore, on the one hand, the production of CO_2 must be reduced, but it is also important to prevent the produced CO_2 from being released into the atmosphere. Carbon capture and storage [104] is a solution that involves the storage of CO_2 underground, such as in sandstone formations at depths below 1 km underground. The concept of this technique is to pump CO_2 gas or a CO_2-rich brine into these sandstone formations. In these types of studies, one can examine for instance if CO_2 reacts with the host rock, which might be calcium-rich sandstone. Due to this reaction solid, stable minerals like calcium carbonate can be produced, visible by neutron imaging [8, 105]. Problems to be addressed in this process are pore network blocking for instance by precipitates formed by the reaction and so-called wormholes caused by the acidity of the CO_2-rich fluid [106] which reduce the efficiency of the sequestration and can also affect the mechanical stability of the site in severe cases.

9.3 Outlook

Geological samples have been studied using neutron imaging since the initial development of the method. Today, the primary method is still based on the use of transmission images in the form of radiography and tomography. This is often done in the form of a time series of 2D or 3D images to be able to capture the changes in liquid distribution and also structural changes in the specimens. The availability of infrastructure and instruments suited for performing dynamic experiments, as well as simultaneous/complementary x-ray imaging has opened new possibilities for the use of neutron imaging for applications in geology. The evolution of imaging instrumentation and a better understanding of relevant scales and materials have moved the experiment character away from feasibility studies into full-scale scientific experiments addressing specific questions directly impacting understanding processes relevant to geologists.

Currently, most experiments still use the fundamental functionality of the imaging technique, transmission-based imaging with full neutron spectrum provided by the beamline. Techniques like wavelength-resolved neutron imaging and dark-field imaging have the capacity to add new aspects to the investigation. With wavelength-resolved imaging, it is possible to provide information about the crystalline texture using diffraction. Dark-field imaging will be able to explore correlation lengths under, for example, different loading states or degrees of liquid saturation.

References

[1] Cnudde V *et al* 2007 Determination of the impregnation depth of siloxanes and ethylsilicates in porous material by neutron radiography *J. Cult. Herit.* **8** 331–8

[2] Wilson J E, Goodwin L B and Lewis C J 2003 Deformation bands in nonwelded ignimbrites: petrophysical controls on fault-zone deformation and evidence of preferential fluid flow *Geology* **31** 837–40

[3] Gardner W and Kirkham D 1952 Determination of soil moisture by neutron scattering *Soil Sci.* **73** 391–402

[4] Subramanian R V and Burkhart D 1973 Determination by neutron radiography of the location of polymeric resins injected in rock fissures *Nucl. Technol.* **17** 184–8

[5] Kettle R J, Clark M A and D G 1987 Developments in the use of thermal neutron radiography for studying mass transfer in a partially frozen soil *Neutron Radiography* ed J P Barton, G Farny, J-L Person and R. H. (Dordrecht: Springer) pp 271–9

[6] Jasti J K, Lindsay J T and Fogler H S 1987 Flow imaging in porous media using neutron radiography *Paper presented at the SPE Annual Technical Conference and Exhibition (Dallas, TX)*

[7] Winkler B *et al* 2002 Neutron imaging and neutron tomography as non-destructive tools to study bulk-rock samples *Eur. J. Mineral.* **14** 349–54

[8] Wilding M, Lesher C E and Shields K 2005 Applications of neutron computed tomography in the geosciences *Nucl. Instrum. Methods Phys. Res.* A **542** 290–5

[9] Carlson W D 2006 Three-dimensional imaging of earth and planetary materials *Earth Planet. Sci. Lett.* **249** 133–47

[10] Dierick M, Vlassenbroeck J, Masschaele B, Cnudde V, Van Hoorebeke L and Hillenbach A 2005 High-speed neutron tomography of dynamic processes *Nucl. Instrum. Methods Phys. Res.* A **542** 296–301

[11] Kaestner A *et al* 2005 Mapping the three dimensional water dynamics in heterogeneous sands using thermal neutrons *4th World Congress in Industrial Process Tomography*

[12] Zarebanadkouki M *et al* 2015 On-the-fly neutron tomography of water transport into lupine roots *Phys. Procedia.* **69** 292–8

[13] Tötzke C, Kardjilov N, Manke I and Oswald S E 2017 Capturing 3D water flow in rooted soil by ultra-fast neutron tomography *Sci. Rep.* **7** 6192

[14] Kaestner A P, Mannes D and Lehmann E 2016 Combined neutron and x-ray imaging—imaging with a second opinion *6th Conf. on Industrial Computed Tomography (iCT) 2016*; J Kastner (ndt.net)

[15] Tengattini A, Atkins D, Giroud B, Andò E, Beaucour J and Viggiani G 2017 NeXT-grenoble, a novel facility for neutron and x-ray tomography in grenoble *3rd Int. Conf. on Tomography of Materials and Structures, ICTMS2017 (Lund, Sweden)* p 163

[16] LaManna J M, Hussey D S, Baltic E and Jacobson D L 2017 Neutron and x-ray tomography (NeXT) system for simultaneous, dual modality tomography *Rev. Sci. Instrum.* **88**

[17] Wyrzykowski M, Ghourchian S, Münch B, Griffa M, Kaestner A and Lura P 2021 Plastic shrinkage of mortars cured with a paraffin-based compound—bimodal neutron/x-ray tomography study *Cem. Concr. Res.* **140** 106289

[18] Tengattini A, Lenoir N, Andò E and Viggiani G 2021 Neutron imaging for geomechanics: a review *Geomech. Energy Environ.* **27** 100206

[19] Stavropoulou E *et al* 2020 Dynamics of water absorption in callovo-oxfordian claystone revealed with multimodal x-ray and neutron tomography *Front. Earth Sci. (Lausanne)* **8** 6

[20] Stavropoulou E *et al* 2018 Liquid water uptake in unconfined Callovo Oxfordian clay-rock studied with neutron and x-ray imaging *Acta Geotech.* **14** 19–33

[21] Cheikh Sleiman H, Tengattini A, Briffaut M, Huet B and Dal Pont S 2021 Simultaneous x-ray and neutron 4D tomographic study of drying-driven hydro-mechanical behavior of cement-based materials at moderate temperatures *Cem. Concr. Res.* **147** 106503
[22] Tudisco E *et al* 2019 Fast 4-D imaging of fluid flow in rock by high-speed neutron tomography *J. Geophys. Res.-Solid Earth* **124** 3557–69
[23] Etxegarai M, Tudisco E, Tengattini A, Viggiani G, Kardjilov N and Hall S A 2021 Characterisation of single-phase fluid-flow heterogeneity due to localised deformation in a porous rock using rapid neutron tomography *J. Imaging* **7** 275
[24] Cordonnier B *et al* 2019 Neutron imaging of cadmium sorption and transport in porous rocks *Front. Earth Sci. (Lausanne)* **7** 306
[25] Shafizadeh A, Gimmi T, van Loon L R, Kaestner A P, Mäder U K and Churakov S V 2020 Time-resolved porosity changes at cement-clay interfaces derived from neutron imaging *Cem. Concr. Res.* **127** 105924
[26] Derluyn H *et al* 2013 Characterizing saline uptake and salt distributions in porous limestone with neutron radiography and x-ray micro-tomography *J. Build. Phys.* **36** SI 353–74
[27] Dewanckele J *et al* 2014 Neutron radiography and x-ray computed tomography for quantifying weathering and water uptake processes inside porous limestone used as building material *Mater. Charact.* **88** 86–99
[28] Vontobel P, Lehmann E and Carlson W D 2005 Comparison of x-ray and neutron tomography investigations of geological materials *IEEE Trans. Nucl. Sci.* **52** 338–41
[29] Chiang W S *et al* 2018 Simultaneous neutron and x-ray imaging of 3D structure of organic matter and fracture in shales *Petrophysics* **59** 153–61
[30] Martell J *et al* 2022 The scale of a martian hydrothermal system explored using combined neutron and x-ray tomography *Sci. Adv.* **8** eabn3044
[31] Tremsin A S, Rakovan J, Shinohara T, Kockelmann W, Losko A S and Vogel S C 2017 Non-destructive study of bulk crystallinity and elemental composition of natural gold single crystal samples by energy-resolved neutron imaging *Sci. Rep.* **7** 40759
[32] Kis Z, Szentmiklósi L, Schulze R and Abraham E 2017 Prompt gamma activation imaging (PGAI) *Neutron Methods for Archaeology and Cultural Heritage* ed N Kardjilov and G Festa (Cham: Springer) pp 303–20
[33] Dingwell D B *et al* 2016 Eruptive shearing of tube pumice: pure and simple *Solid Earth* **7** 1383–93
[34] de Beer F C and Middleton M F 2006 Neutron radiography imaging, porosity and permeability in porous rocks *S. Afr. J. Geol.* **109** 541–50
[35] Kaestner A *et al* 2007 Mapping the 3D water dynamics in heterogeneous sands using thermal neutrons *Chem. Eng. J.* **130** 79–85
[36] Syed A *et al* 2021 A portable triaxial cell for beamline imaging of rocks under triaxial state of stress *Meas. Sci. Technol.* **32** 95403
[37] Tudisco E *et al* 2019 Fast 4-D imaging of fluid flow in rock by high-speed neutron tomography *J. Geophys. Res. Solid Earth* **124** 3557–69
[38] Jailin C, Etxegarai M, Tudisco E, Hall S A and Roux S 2018 Fast tracking of fluid invasion using time-resolved neutron tomography *Transp. Porous Media* **124** 117–35
[39] Tötzke C, Kardjilov N, Lenoir N, Manke I, Oswald S E and Tengattini A 2019 What comes NeXT?—high-speed neutron tomography at ILL *Opt. Express* **27** 28640
[40] Shafizadeh A, Gimmi T, Van Loon L R, Kaestner A P, Mäder U K and Churakov S V 2020 Time-resolved porosity changes at cement-clay interfaces derived from neutron imaging *Cem. Concr. Res.* **127** 105924

[41] Bingham P *et al* 2015 Neutron radiography of fluid flow for geothermal energy research *Phys. Procedia* **69** 464–71

[42] Das S, Adeoye J, Dhiman I, Bilheux H Z and Ellis B R 2019 Imbibition of mixed-charge surfactant fluids in shale fractures *Energy Fuels* **33** 2839–47

[43] Brabazon J W *et al* 2019 Rock fracture sorptivity as related to aperture width and surface roughness *Vadose Zone J.* **18** 1–10

[44] Perfect E *et al* 2014 Neutron imaging of hydrogen-rich fluids in geomaterials and engineered porous media: a review *Earth Sci. Rev.* **129** 120–35

[45] Papafotiou A *et al* 2008 From the pore scale to the lab scale: 3-D lab experiment and numerical simulation of drainage in heterogeneous porous media *Adv Water Resour* **31** 1253–68

[46] Graziani G *et al* 2018 Neutron radiography as a tool for assessing penetration depth and distribution of a phosphate consolidant for limestone *Constr. Build. Mater.* **187** 238–47

[47] Masschaele B *et al* 2004 High-speed thermal neutron tomography for the visualization of water repellents, consolidants and water uptake in sand and lime stones *Radiat. Phys. Chem.* **71** 807–8

[48] Barone G *et al* 2014 Neutron radiography for the characterization of porous structure in degraded building stones *J. Instrum.* **9** C05024 C05024

[49] Kang M *et al* 2013 Water calibration measurements for neutron radiography: application to water content quantification in porous media *' Nucl. Instrum. Methods Phys. Res.* A **708** 24–31

[50] Hassanein R, Lehmann E and Vontobel P 2005 Methods of scattering corrections for quantitative neutron radiography *Nucl. Instrum. Methods Phys. Res.* A **542** 353–60

[51] Hassanein R, de Beer F, Kardjilov N and Lehmann E 2006 Scattering correction algorithm for neutron radiography and tomography tested at facilities with different beam characteristics *Phys. B Condens. Matter.* **385–6**

[52] Boillat P *et al* 2018 Chasing quantitative biases in neutron imaging with scintillator-camera detectors: a practical method with black body grids *Opt. Express* **26** 15769–84

[53] Carminati C *et al* 2019 Implementation and assessment of the black body bias correction in quantitative neutron imaging *PLoS One* **14** e0210300

[54] Cheng C-L *et al* 2015 Rapid imbibition of water in fractures within unsaturated sedimentary rock *Adv. Water Resour.* **77** 82–9

[55] Zambrano M, Hameed F, Kaestner A, Mancini L and Tondi E 2019 Implementation of dynamic neutron radiography and integrated x-ray and neutron tomography in porous carbonate reservoir rocks *Front. Earth Sci. (Lausanne)* **7** 329

[56] Lewis H *et al* 2023 Interactions between imbibition and pressure-driven flow in a microporous deformed limestone *Transp. Porous Media* **146** 559–85

[57] Roshankhah S, Rubino V, Marshall J P, Rosakis A J and Andrade J E 2019 Monitoring laboratory hydraulic fractures in pre-fractured shale using high-resolution optical and neutron imaging techniques *53rd U.S. Rock Mechanics/Geomechanics Symp. (New York)*

[58] Roshankhah S *et al* 2018 Neutron imaging: a new possibility for laboratory observation of hydraulic fractures in shale? *Geotech. Lett.* **8** 316–23

[59] Bingham P, Polsky Y and Anovitz L 2013 Neutron imaging for geothermal energy systems *Proc. Image Processing: Machine Vision Applications VI*; P R Bingham and E Y Lam 8661 86610K

[60] Washburn E W 1921 The dynamics of capillary flow *Phys. Rev.* **17** 273–83

[61] Lunati I, Vontobel P, Kinzelbach W and Lehmann E 2003 Laboratory visualization of two-phase flow in a natural fracture by neutron tomography *Groundwater in Fractured Rocks* (Prague: UN- ESCO IHP-VI)

[62] Middleton M F, Murdoch U, Li K and de Beer F 2005 Spontaneous imbibition studies of Australian reservoir rocks with neutron radiography *Paper presented at the SPE Western Regional Meeting (Irvine, CA)* SPE-93634-MS

[63] Jian-Chao C, Bo-Ming Y, Mao-Fei M and Liang L 2010 Capillary rise in a single tortuous capillary *Chin. Phys. Lett.* **27** 054701

[64] Wu Y, Zhao Y and Li P 2019 Effect of the heterogeneity on sorptivity in sandstones with high and low permeability in water imbibition process *Processes* **7** 260

[65] Perfect E, Brabazon J W and Gates C H 2020 Forward prediction of early-time spontaneous imbibition of water in unsaturated rock fractures *Vadose Zone J.* **19** e20056

[66] Brabazon J W, Perfect J S E, Gates C H, Dhiman I, Bilheux H Z, Bilheux J-C and McKay L D 2019 Spontaneous imbibition of a wetting fluid into a fracture with opposing fractal surfaces: theory and experimental validation *Fractals* **27** 1940001

[67] McFarlane J *et al* 2021 Effect of fluid properties on contact angles in the eagle ford shale measured with spontaneous imbibition *ACS Omega* **6** 32618–30

[68] Siddiqui M A Q, Salvemini F, Ramandi H L, Fitzgerald P and Roshan H 2021 Configurational diffusion transport of water and oil in dual continuum shales *Sci. Rep.* **11** 2152

[69] Degueldre C *et al* 1996 Porosity and pathway determination in crystalline rock by positron emission tomography and neutron radiography *Earth Planet. Sci. Lett.* **140** 213–25

[70] Abd A E-G E and Milczarek J J 2004 Neutron radiography study of water absorption in porous building materials: anomalous diffusion analysis *J. Phys. D: Appl. Phys.* **37** 2305–13

[71] El Abd A, Czachor A and Milczarek J 2009 Neutron radiography determination of water diffusivity in fired clay brick *Appl. Radiat. Isot.* **67** 556–9

[72] Kis Z, Sciarretta F and Szentmiklósi L 2017 Water uptake experiments of historic construction materials from Venice by neutron imaging and PGAI methods *Mater. Struct.* **50** 159

[73] Cnudde V *et al* 2008 High-speed neutron radiography for monitoring the water absorption by capillarity in porous materials *Nucl. Instrum. Methods Phys. Res.* B **266** 155–63

[74] Xue S, Zhang P, Lehmann E H, Hovind J and Wittmann F H 2021 Neutron radiography of water exchange across the interface between old and fresh mortar *Measurement (Lond.)* **183** 109882

[75] Bultreys T, De Boever W and Cnudde V 2016 Imaging and image-based fluid transport modeling at the pore scale in geological materials: a practical introduction to the current state-of-the-art *Earth Sci. Rev.* **155** 93–128

[76] Blanche J *et al* 2020 Dynamic fluid ingress detection in geomaterials using k-band frequency modulated continuous wave radar *IEEE Access* **8** 111027–41

[77] Derluyn H 2012 Salt transport and crystallization in porous limestone—neutron-x-ray imaging and poromechanical modeling *PhD Thesis* ETH Zurich

[78] Cnudde V, Masschaele B, Dierick M, Vlassenbroeck J, van Hoorebeke L and Jacobs P 2006 Recent progress in x-ray CT as a geosciences tool *Appl. Geochem.* **21** 826–32

[79] Cnudde V and Boone M N 2013 High-resolution x-ray computed tomography in geosciences: a review of the current technology and applications *Earth Sci. Rev.* **123** 1–17

[80] Withers P J *et al* 2021 X-ray computed tomography *Nat. Rev. Methods Primers* **1** 18

[81] Mascle M *et al* 2023 Investigation of salt precipitation dynamic in porous media by x-ray and neutron dual-modality imaging *Sci. Technol. Energy Transit.* **78** 11

[82] Ryu J, Ahn S, Kim S-G, Kim T and Lee S J 2014 Interactive ion-mediated sap flow regulation in olive and laurel stems: physicochemical characteristics of water transport via the pit structure *PLoS One* **9** e98484

[83] Kim F H, Penumadu D, Kardjilov N and Manke I 2016 High-resolution x-ray and neutron computed tomography of partially saturated granular materials subjected to projectile penetration *Int. J. Impact Eng.* **89** 72–82

[84] Hameed F, Schillinger B, Rohatsch A, Zawisky M and Rauch H 2009 Investigations of stone consolidants by neutron imaging *Nucl. Instrum. Methods Phys. Res.* A **605** 150–3

[85] Zhao Y *et al* 2017 Effects of microstructure on water imbibition in sandstones using x-ray computed tomography and neutron radiography *J. Geophys. Res.-Solid Earth* **122** 4963–81

[86] Kim F H, Penumadu D, Gregor J, Kardjilov N and Manke I 2013 High-resolution neutron and x-ray imaging of granular materials *J. Geotech. Geoenviron. Eng.* **139** 715–23

[87] Gabet T, Vu X H, Malecot Y and Daudeville L 2006 A new experimental technique for the analysis of concrete under high triaxial loading *J. Phys. IV (Proc.)* **134** 635–40

[88] Shimada M 2000 *Mechanical Behaviour of Rocks Under High Pressure Conditions* 1st edn (Routledge)

[89] Yehya M, Andò E, Dufour F and Tengattini A 2018 Fluid-flow measurements in low permeability media with high pressure gradients using neutron imaging: application to concrete *Nucl. Instrum Methods Phys. Res.* A **890** 35–42

[90] Hall S A 2013 Characterization of fluid flow in a shear band in porous rock using neutron radiography *Geophys. Res. Lett.* **40** 2613–8

[91] Tudisco E, Hall S A, Charalampidou E M, Kardjilov A, Hilger N and Sone H 2015 Full-field measurements of strain localisation in sandstone by neutron tomography and 3D-volumetric digital image correlation *Proc. of the 10th World Conf. on Neutron Radiography (WCNR-10)* ; E H Lehmann, A P Kaestner and D Mannes *Physics Procedia* **69** 509–15

[92] Tengattini A, Lenoir N, Ando E and Viggiani G 2021 Neutron imaging for geomechanics: a review *Geomech. Energy Environ.* **27** 100206

[93] Tudisco E, Hall S A, Charalampidou E M, Kardjilov N, Hilger A and Sone H 2015 Full-field measurements of strain localisation in sandstone by neutron tomography and 3D-volumetric digital image correlation *Phys. Procedia* **69** 509–15

[94] Zeilinger A and Huebner R 1976 Moisture transport in a concrete of the SNR-300 investigated by neutron transmission *Kerntechnik* **18** 119–25

[95] Toropovs N *et al* 2015 Real-time measurements of temperature, pressure and moisture profiles in high-performance concrete exposed to high temperatures during neutron radiography imaging *Cem. Concr. Res.* **68** 166–73

[96] Dauti D, Tengattini A, Dal Pont S, Toropovs N, Briffaut M and Weber B 2018 Analysis of moisture migration in concrete at high temperature through *in situ* neutron tomography *Cem. Concr. Res.* **111** 41–55

[97] Tengattini A, Dal Pont S, Cheikh Sleiman H, Kisuka F and Briffaut M 2020 Quantification of evolving moisture profiles in concrete samples subjected to temperature gradient by means of rapid neutron tomography: influence of boundary conditions, hygro-thermal loading history and spalling mitigation additives *Strain* **56** e12371

[98] Jordan G, Eulenkamp C and Schmahl W W 2010 Neutron radiography study of the dehydration kinetics of smectites in moulding sands *Acta Crystallogr. A-Found. Adv.* **66** S297–8

[99] Schillinger B, Calzada E, Eulenkamp C, Jordan G and Schmahl W W 2011 Dehydration of moulding sand in simulated casting process examined with neutron radiography *Nucl. Instrum. Methods Phys. Res.* A **651** 312–4

[100] Schiebel K, Jordan G, Kaestner A, Schillinger B, Boehnke S and Schmahl W W 2018 Neutron radiographic study of the effect of heat-driven water transport on the tensile strength of bentonite-bonded moulding sand *Transp. Porous Media* **121** 369–87

[101] Schiebel K *et al* 2018 Effects of heat and cyclic reuse on the properties of Bentonite-bonded sand *Eur. J. Mineral.* **30** 1115–25

[102] Żołądek J, Milczarek J J and Fijał-Kirejczyk I 2008 Dynamic neutron radiography studies of water migration in beds of natural zeolite *Nukleonika* **53** 113–9

[103] Fijał-Kirejczyk I, Milczarek J J, Banaszak J, Żołądek J and Trzciński A 2010 Drying of kaolin clay cylinders: dynamic neutron radiography studies *Defect Diffus. Forum* **297–301** 508–12

[104] Kelemen P, Benson S M, Pilorgé H, Psarras P and Wilcox J 2019 An overview of the status and challenges of CO_2 storage in minerals and geological formations *Front. Clim.* **1** 9

[105] Peng S 2021 Advanced understanding of gas flow and the Klinkenberg effect in nanoporous rocks *J. Pet. Sci. Eng.* **206** 109047

[106] Selvadurai A P S, Couture C-B and Rezaei Niya S M 2017 Permeability of wormholes created by CO_2-acidized water flow through stressed carbonate rocks *Phys. Fluids* **29** 096604

Part IV

Energy research

Chapter 10

Fuel cells and electrolysers

Pierre Boillat and Markus Strobl

As of 2020, the energy landscape in many countries has started a shift towards a higher share of renewable energies, as a means to address climate change and other issues related to the energy supply. The output of renewable sources such as solar, wind and hydro power being typically in the form of electrical energy, this trend goes in pair with an increasing share of electrification of the transportation system. Nevertheless, a purely electrical energy system is unrealistic, as the matching between fluctuating energy supply and consumption requires large storage capacities. Hydrogen as an energy carrier provides an interesting solution to these needs, due to its low cost of storage. Furthermore, hydrogen is a commodity already used nowadays, in particular in chemistry. The vast majority of the hydrogen produced today stems from fossil sources, and producing hydrogen from renewable sources in times of excess production offsets the corresponding CO_2 emissions. Hydrogen also provides new opportunities for zero emission vehicles. The key components of a hydrogen-based energy system are electrolyzers, which allow producing hydrogen from electrical energy, and fuel cells, which allow the use of hydrogen in stationary and mobile applications. In the following chapter, we will show how neutron imaging can be used to give insight into the inner workings of such devices, in particular in the presence of two-phase flow phenomena.

10.1 Polymer electrolyte fuel cells (PEFCs)

10.1.1 PEFC basics

Polymer electrolyte fuel cells (PEFCs) convert the chemical energy contained in a fuel (e.g. hydrogen) to electrical energy. One of their major applications is the use—in combination with electrical motors—in mobility applications. Like battery electric vehicles, fuel cell electric vehicles result in zero emissions of CO_2 and other pollutants at the point of use, the only product besides electricity and heat being water. While the higher complexity of fuel cell systems make them more expensive compared to batteries for small vehicles, the lower marginal cost of extending the

doi:10.1088/978-0-7503-3495-2ch10 © IOP Publishing Ltd 2024

storage capacity make them a very suitable alternative for larger vehicles and when large autonomy range and fast refilling are required. PEFC-based vehicles are technically mature and available as commercial products, but a large scale use of this technology still relies on system cost reduction. Two major and complementary directions can be pursued to reach this goal for a system of a given power output: reduce the amount of costly components (and in particular of the precious metal used for electrocatalysis) or increase the power output per cell area. The latter approach can highly benefit from a clear understanding of the performance limiting processes occurring in the fuel cell, including those related with the presence of liquid water.

A typical PEFC structure is shown in figure 10.1. The electrochemical reaction used to produce electrical energy is composed of two half reactions, the hydrogen oxidation reaction (HOR) on the anode side—where the hydrogen fuel is supplied—and the oxygen reduction reaction (ORR) on the cathode side. The oxygen used for the latter is typically taken from the ambient air, although some specialty designs also use pure oxygen stored in a separate tank. The operating temperature of PEFCs (typically below 80 °C) is both an asset and an issue. It is low enough so that PEFCs can start up in a very short time and do not require special materials resisting high temperatures. On the other side, the water produced as a result of the electrochemical reaction is usually

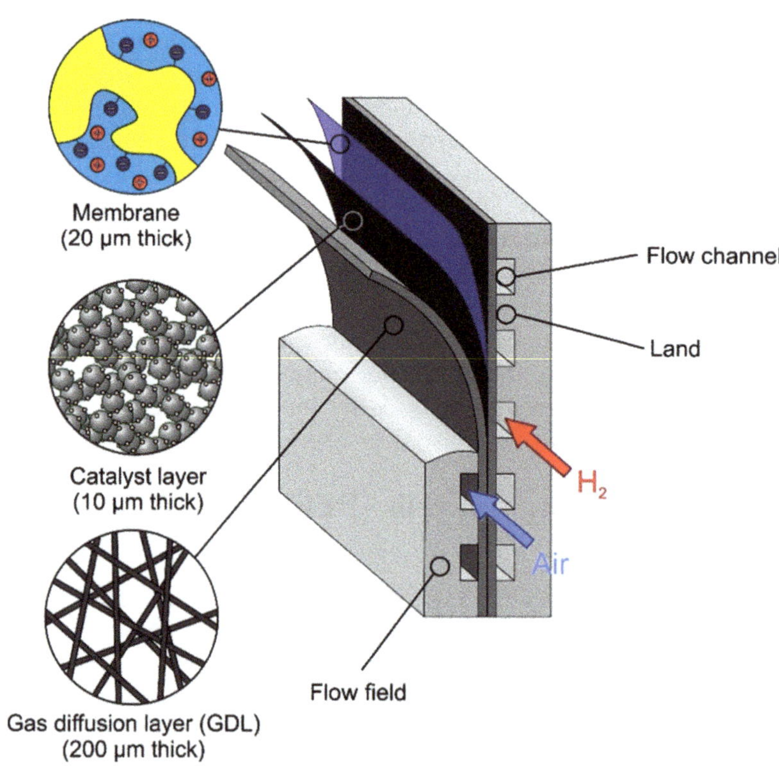

Figure 10.1. Key elements composing the structure of a polymer electrolyte fuel cell (PEFC).

liquid at these temperatures. Because the supply of reactant gases and the removal of liquid water share the same pathways, two-phase flow occurs which renders both the operation of PEFCs and the understanding of the performance limitations very complex. In the following paragraphs, we will review how this two-phase flow phenomena affects the operation within different components of the fuel cell.

10.1.2 Water in gas flow channels

In fuel cell designs with a high power density—a prerequisite for mobility applications—the reactant gases need to be actively circulated through flow channels, whose patterns constitute the so-called *flow field.* To prevent an unacceptably large pressure drop across the cell, flow field of automotive scale (several hundreds of cm^2) cannot be composed of a single serpentine channel and need to use several channels in parallel. A classical design for a technical size fuel cell use mostly straight parallel channels, with the notable exception of the '3D mesh' flow field used in the Toyota Mirai, one of the first commercially available fuel cell automobiles. In general, water accumulation in flow channels can be detrimental to performance and durability, as the blocking of channels can result in air of hydrogen starvation—the latter being in particular detrimental to performance. A key characteristic of neutron imaging for the study of two-phase flow patterns in fuel cell flow fields is the ability of neutrons to penetrate through a variety of materials, including the stainless steel sheets commonly used for the flow fields. This is combined with a very high sensitivity to liquid water, with detection limits in the range of 5–10 μm of water thickness. Depending on the instrument, the beam diameter can be as large as 400 mm (as in the NEUTRA instrument), and fuel cells of full technical size can be imaged in a single exposure. In the so-called through-plane measurement configuration, where the neutrons are transmitted through the plane of the cell, the averaged quantity of water over different layers is measured. This means that different cells in a stack cannot be imaged separately. For this reason, radiography is usually performed on a single cell instead of a stack. The distinction between anode and cathode flow fields is not direct, as these contributions are also superposed in the through-plane image. However, several strategies can be used to distinguish anode and cathode flow field water from through-plane images. In some cases, the design of the flow fields on the different sides is different enough to make the distinction. When this is not the case, a possible strategy is to measure time sequences and observe the direction of movement of water, as anode and cathode gas flows are usually fed in opposite directions. Finally, a more complex analysis strategy consists in interrupting the cell operation at a given point, and successively purging the anode and gas flow channels [1] (figure 10.2).

If a direct visualization of water in separate sides (anode and cathode) or between different cells of a stack is sought, other possibilities include neutron tomography, resulting in full 3D imaging, and in-plane imaging. Tomography is scarcely used and has been so far limited to proof-of-concept studies [4, 5], because of the strong constraints it imposes on the measurement set-up, and the limited resolution obtained for cells of a technically relevant size. In-plane imaging, where the neutron

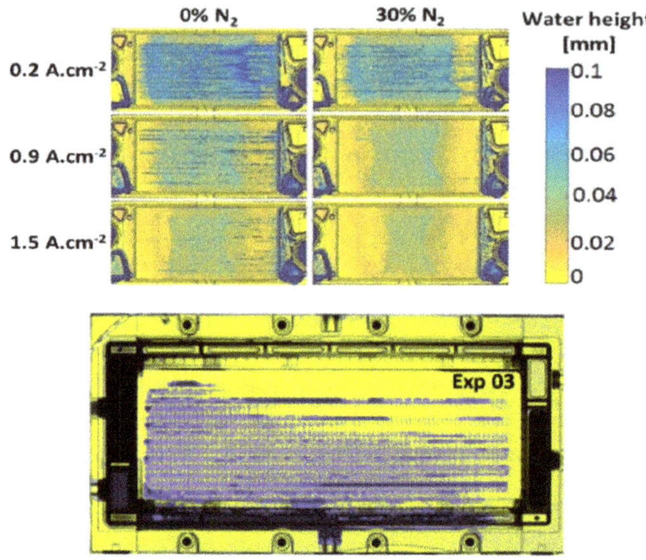

Figure 10.2. Neutron radiographies of single fuel cells of full technical size. Measurements realized with through-plane radiography. (a) Cell with a 300 cm^2 active area having parallel channels in the same orientation on the anode and cathode side (Autostack-core project, image reproduced from [2]). In this case, the water in the straight flow channels was identified as being mostly on the anode side from the direction of movement in time series. (b) Cell with a 140 cm^2 active area with anode and cathode channels having perpendicular orientations [3]. In this case, the anode (horizontal) and cathode (vertical) water accumulations can be deduced from their orientation. Reprinted from [3], copyright (2020), with permission from Elsevier.

beam axis is parallel to the cell membrane, allows imaging fuel cells with a much lower degree of set-up complexity, as the cell does not need to be rotated. One of the important limitations of such configuration is that the size of the cell in the beam axis direction is usually limited to 20–30 mm to obtain a sufficient transmission through all the structural materials and the water. Nevertheless, this imaging configuration is very well suited for the imaging of elongated cells, where the dimension along the primary direction of the gas flow—which is the direction with the strongest heterogeneity—is the full technical dimension, and the size in the perpendicular direction is reduced. A further difficulty is introduced by the very high aspect ratio of fuel cells when seen from the side: the dimension along the gas flow is typically 100 mm or more, while the total cell thickness is typically between 1 and 2 mm. To keep the ability of imaging the whole cell without scanning and having a sufficiently high resolution to distinguish the different layers of the cell, anisotropic setups have been developed [6, 7]. One particular possibility is to use a detector surface that is tilted instead of the usual configuration perpendicular to the neutron beam. This results in magnification and therefore a higher resolution only in the direction where it is needed. An example is shown in figure 10.3. in this study [8], the water accumulation in cathode flow channels was studied when using an interdigitated flow field design. This design allows a better supply of oxygen to the active sites, but is usually deemed unsuitable for PEFCs because of water accumulation in

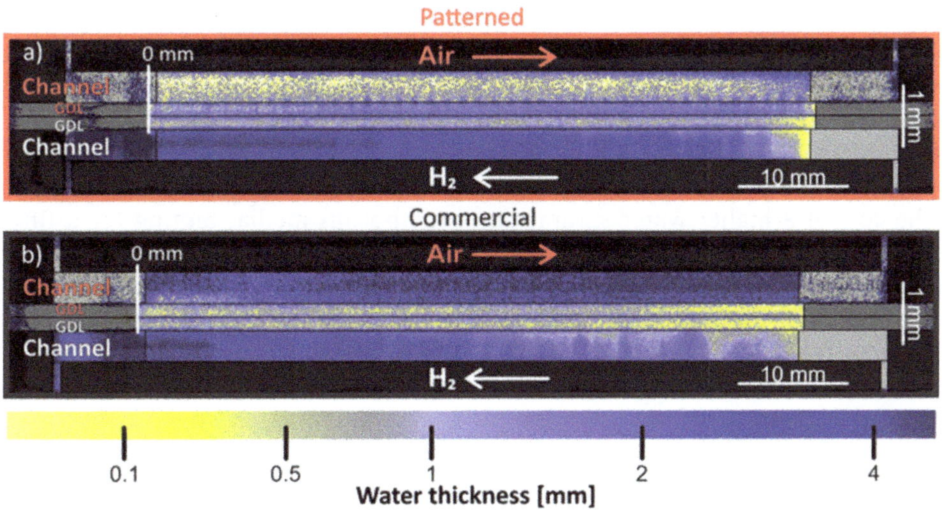

Figure 10.3. Neutron imaging of the water distribution in PEFCs with interdigitated cathode flow field comparing the use of a special GDL with patterned wettability (a) with a standard GDL (b). Water accumulation in the dead-ended cathode channels (top side of the cell) is greatly reduced by the use of a patterned GDL. Due to the anisotropic resolution enhancement, the horizontal and vertical scales are different. Reproduced from [8], copyright 2019 The Authors. Published on behalf of The Electrochemical Society by IOP Publishing Limited).

the cathode flow field. This example, based on pressure loss measurements and neutron imaging, demonstrated that using a particular type of porous media with hydrophilic regions [9] for discharging the water effectively prevented this issue. The resolution of 20–30 μm typically obtained with the in-plane imaging configuration allows not only distinguishing water between the anode and cathode flow channels, but also in the porous media used in the cell, which is the topic of the following paragraph.

10.1.3 Water in porous media

The so-called gas diffusion layer (GDL) in fuel cells fulfils numerous functions, including the fine distribution of gaseous reactants in the regions between the flow channels, and the electrical and thermal contact in the regions under the channels. Because water has to be removed through these porous layers, two-phase flow also plays an important role in this component. In most designs, the transport of oxygen to the active sites occurs by diffusion through this layer. Water accumulation in the GDL pores results in a reduction of the effective diffusivity, and the correspondingly reduced oxygen concentration at the active sites induces performance losses. The most adequate method for imaging water in the porous media of operating fuel cells depends on the constraints that are acceptable on the cell design. X-ray tomography, with either lab sources or synchrotron instruments, allows resolving the pore space and water distribution in three dimensions [10] and is the method of choice if the cell can be made small enough and with materials transparent to x-rays. Neutron imaging, on the other hand, allows imaging cells of larger dimensions and with less constraint on the materials choice. This can be used for in-plane imaging of the water content in porous media in elongated cells such as in the example (figure 10.3) of the interdigitated flow field cell already presented in the previous paragraph. In this particular example, it could be identified that the better water discharging capability between inlet and outlet channels thanks to the hydrophilic regions comes at the cost of a higher water saturation in the porous media, hinting for a further improvement potential of this material. This high-resolution in-plane imaging configuration is not the only possibility for analyzing water in fuel cell GDLs. First, depending on the application and operating conditions, there may be no water present in the flow channels, in which case the remaining contrast is from GDL water. A first example of such a situation is the higher current density operation points (1.5 $A{\cdot}cm^{-2}$) of the through-plane measurement example shown in the previous paragraph (figure 10.2). A further example is given here for a fuel cell using a particular operation scheme called evaporative cooling, where liquid water is fed to the cell for both humidification and cooling by evaporation [11, 12]. In the particular evaporative cooling scheme developed at PSI, the porous media with hydrophilic regions is used to spread over a larger area water fed by a single supply channel. Besides the observation of the cell cooling by evaporation, the effectiveness of the water spreading could be observed *in situ* using neutron imaging figure 10.4.

This example also illustrates one of the key features of neutron imaging, namely the ability to measure water distribution in real operating conditions, including in cell with a complex additional instrumentation: the particular cell used for analyzing the evaporative cooling capabilities included a sophisticated control of the thermal boundaries, with heat flux control through thermoelectric elements and feedback sensors. Finally, an interesting case is the situation where the water content in the porous media needs to be measured when water is also present in the gas flow channels. Besides the possibility already mentioned previously of successively purging the gas flow channels [1], investigations were conducted on the use of neutron dark-field imaging (DFI) to analyze the water distribution specifically in the porous media. The base idea is that DFI, whose contrast is not based on attenuation but on neutron scattering at ultra-low angles, is only sensitive to water structures with sizes of a few micrometers and less, which are only found in porous media. Because DFI is based on the coherent scattering cross-section, this approach is much more effective when using heavy water, which means *in situ* experiments also need to be conducted using deuterium gas. Very promising *ex situ* proof-of-concept measurements were realized [13] showing that heavy water in porous media can easily be distinguished even when superposed with a channel full of water. First *in situ* experiments, however, are much more challenging to analyze due to the lower fraction of volume occupied by water in porous media during operation [14] (figure 10.5).

10.1.4 Distinction of liquid water and ice

The ability to start-up at temperatures below 0 °C is an important characteristic of the PEFC technology. It is well known [15, 16] that the polymer electrolyte

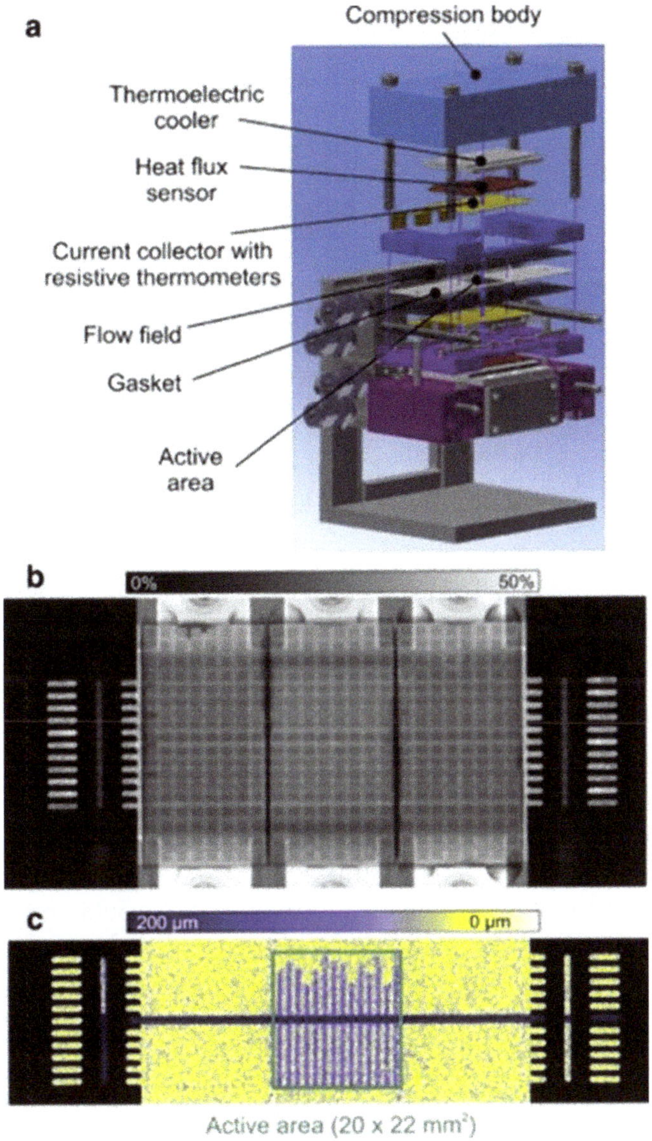

Figure 10.4. Neutron imaging of a laboratory cell for studying evaporative cooling. (a) Buildup of the instrumented cell, showing the different elements in the neutron path including heat flux sensors and thermos-electric coolers. (b) Neutron image of the cell structure. (c) Neutron image of the water distribution during injection through the central channel. The hydrophilic channels in the porous media are able to spread the water nearly homogeneously over the active area. Reprinted from [52], copyright (2017), with permission from Elsevier.

membranes are sufficiently conductive at low temperatures to allow fuel cell operation, but the product water accumulating in porous layers in the form of ice is a concern. The usual PEFC freeze start strategy is to have the stack heating up sufficiently fast to reach a temperature of 0 °C before ice blocks the pores and shuts

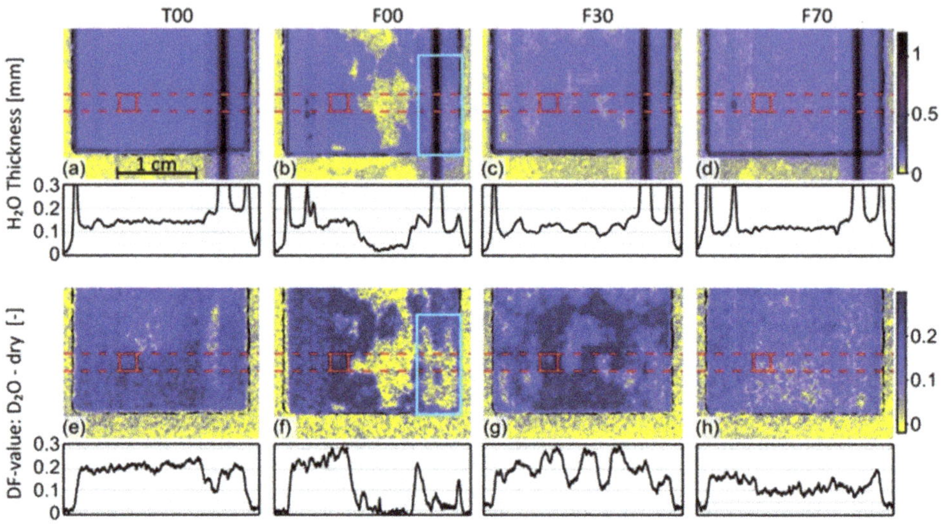

Figure 10.5. Comparison of neutron attenuation and dark-field contrast of water in porous media. The top row (a)–(d) shows the liquid water distribution for various porous media samples. Besides water in the sample itself, accumulations on the border and the filled injection channel are clearly visible. The bottom row (e)–(h) shows the distribution (using heavy water) in the porous material only and was obtained from the dark-field contrast using neutron grating interferometry. Reproduced from [13], copyright 2019 The Author(s). Published on behalf of The Electrochemical Society by IOP Publishing Limited.

off the supply of reactants [17–19]. A complicating (but also helpful) aspect of freeze-starts is that electrochemically produced water can remain in super-cooled state within the porous layers and even be drained as a liquid to the flow channels [20–22]. In a publication using dynamic high-resolution imaging, Oberholzer *et al* [22] demonstrated that PEFC kept at temperature of −10 °C could operate isothermally for tens of minutes, with the water being discharged in liquid form to the flow channels. It was shown that freezing occurs randomly, but can also be triggered by a mechanical shock, identified as a sudden drop of the cell voltage, coupled with an increase of the quantity of water in the central region of the cell. The explanation is that, when freezing occurs, the liquid water discharge is stopped, and ice rapidly accumulated in the catalyst layer.

While the super-cooled state is helpful to extend the time fuel cells can operated below 0 °C, it is also of concern for the durability, because the physical expansion occurring during freezing can result in mechanical damage [23, 24]. In this context, a better understanding of the freezing behaviour in operating PEFCs is of high interest. One key characteristic of the neutron interaction with materials containing hydrogen is the large inelastic scattering cross-section, which results in the fact that the attenuation of the neutron beam is not only a function of the density of hydrogen atoms but also of their motions. The attenuation for long neutron wavelengths depends in particular of molecular diffusion, which is significant in the liquid state and nearly nonexistent in the frozen state. As a result, the wavelength dependence of the attenuation cross-section is a function of the aggregate state, and the latter can

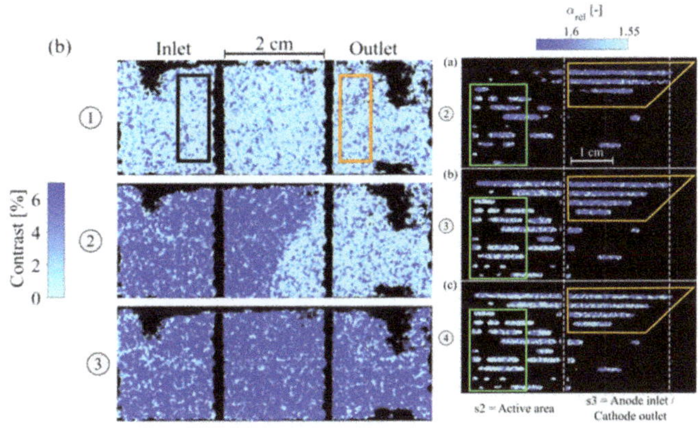

Figure 10.6. Use of wavelength resolved neutron imaging to identify water and ice in fuel cells. Left: water (blue)/ice (light blue) in a stack of GDLs subject to different temperatures. The centre image used a temperature gradient to produce water on one side and ice on the other side, clearly showing the shape of the boundary. Right: *Operando* measurement of freezing in a fuel cell. In the centre image, the water accumulations in one part (green rectangle) are already frozen, while the water accumulations in the other parts (orange trapeze) are still liquid. Reproduced from [27], copyright 2019 The Author(s). Published on behalf of The Electrochemical Society by IOP Publishing Limited.

be detected based on measurement at different neutron wavelengths (or energies) [25, 26]. Based on this, the apparition of local freezing effects—meaning that the water network is not completely connected and the freezing does not propagate to the whole cell when initiated—could be identified [27, 28]. While further methodological improvements are necessary to improve the sensitivity, these results indicate that fuel cell designs promoting the separation of water in separated clusters may make them more resilient to freezing conditions (figure 10.6).

10.2 Polymer electrolyte water electrolyzers (PEWEs)

10.2.1 PEWE basics

The function of water electrolyzers is to produce hydrogen using water as a feedstock and electricity as a power source. Among the different available technologies, the polymer electrolyte water electrolyzer (PEWE) is based on a membrane technology similar to the PEFC described previously. This technology is characterized by a very dynamic response, making it a highly suitable candidate for the storage of excess energy from fluctuating renewable sources. Importantly, PEWEs and other water electrolysis technologies have already been in use for several decades, but for a different purpose. The major historical application is the production of hydrogen for specialist applications requiring high purity, and for which energy efficiency is not a major concern. In consequence, currently available PEWE devices tend to be over-engineered in terms of structure, and not optimized for efficiency.

Function-wise, a PEWE can be seen as a fuel cell operated in the reverse direction, and, as such, its structure is similar to the fuel cell structure described

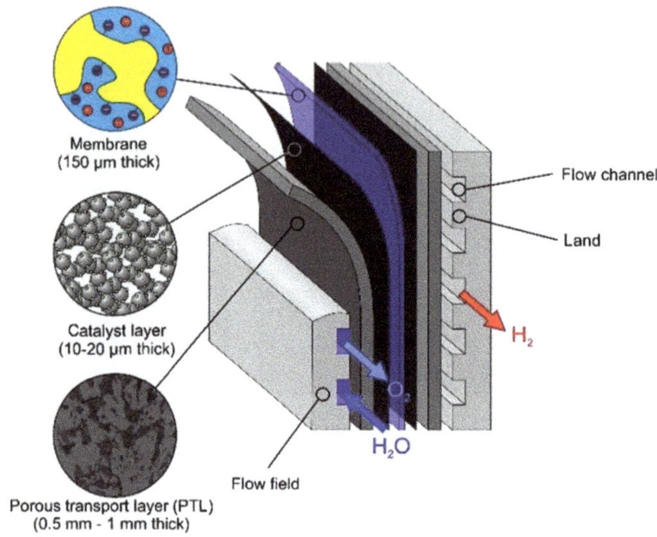

Figure 10.7. Base structural elements of a PEWE. The figure illustrates the case where water is fed in the anode only, but some designs use water feeding in both the anode and cathode side.

in figure 10.7, but with the electrochemical reaction in the reverse direction. However, the materials used in PEWEs need to respond to different constraints for different reasons. Firstly, the electrochemical potential on the anode side (where O_2 is evolved) is very high, and carbon-based materials are not suitable as they would oxidize to CO_2 very rapidly. Second, it is usually desirable to produce hydrogen at a relatively high pressure (typically 30–50 bar, and up to 300 bar for high-pressure electrolysis), which sets higher constraints on the material's robustness. Finally, the platinum-based catalysts used for oxygen reduction in fuel cells are very poor for oxygen evolution, and iridium-based catalysts are used instead. Concerning the other structural elements, porous media, when used, are typically made of sintered titanium and called *porous transport layers* (PTLs). For cost reasons, some designs do not make use of porous media but use a coarse titanium mesh.

10.2.2 Water/gas distribution in flow channels

Similarly to fuel cells, water electrolyzers need their reactant—in this case, water—to be actively circulated. Water is consumed on the anode side and a logical configuration is to feed it in the anode compartment, though designs also exist where water is circulated both on the anode and cathode side. The production of O_2 on the anode side and of H_2 on the cathode side result, as in fuel cells, in the formation of two-phase flow patterns where the liquid reactant and gaseous products compete for the same pathways. To date, there is little published research about the visualization of two-phase flow pattern in the supply channels of electrolyzers using neutron imaging. Selamet *et al* [29] used a combination of neutron imaging and optical visualization to analyze the two-phase flow patterns in an

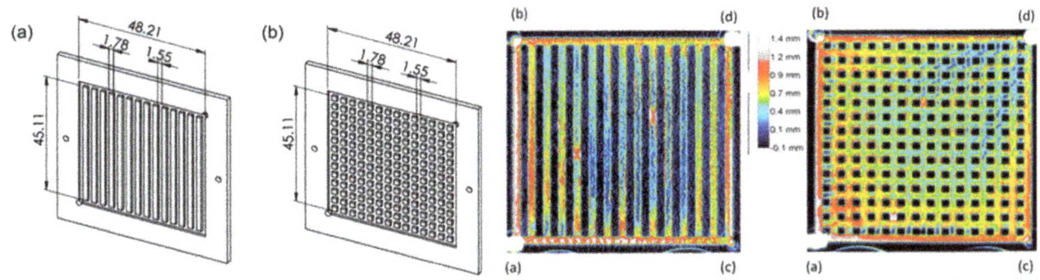

Figure 10.8. Comparison of the two-phase flow pattern in operating electrolyzers using two different types of flow fields. The blue regions indicate a lower water thickness, meaning a higher gas accumulation. Reprinted with permission from [30]. Copyright (2020) American Chemical Society.

electrolyzer cell using a coarse mesh flow field. More recently, Minnaar *et al* [30] used visualization by neutron radiography to compare the water content in two different types of water supply flow fields (see figure 10.8).

They observed that a so-called 'pin type' design resulted in a lower amount of gas accumulation (shown in blue in figure 10.8) than a design using parallel channels. As illustrated by this result, but also by the fact that the literature on this topic is still very scarce, the visualization of water/gas distribution in the flow fields of operating electrolyzers still has an important unused potential in helping the optimization of this component.

10.2.3 Water/gas distribution in porous media

In designs including flow channels and porous media, these pathways include first the transport across the PTL then along the flow fields. Two-phase flow patterns in porous media result in different patterns of interaction, depending on the fluid velocity and characteristics. For oxygen and water in the anode side of PEWE, theory predicts a capillary fingering transport mode, although the possibility of being in the transition region towards viscous fingering transport mode is also considered. First measurements realized with neutron imaging [31] did show that the profile of water/gas fraction remains constant over a wide range of current densities —meaning, a wide range of oxygen flow rates. These results, completed by further measurements combined to a theoretical analysis [32], confirmed that the main transport mode is capillary fingering, as the independence to gas flow indicates that viscous forces do not play an important role, at least at the temperature of 50 °C. It must be mentioned that other researchers [33] reported a slight increase of gas accumulation in the region near the electrode for higher current densities, in particular at higher temperatures such as 80 °C (figure 10.9).

While these results shared an important light on the fundamentals of two-phase flow in PEWE porous media, an important unresolved question is whether the significant fraction of porous media occupied by gas is detrimental to PEWE performance. Unlike in fuel cells, the transport of the reactant through the porous media does not rely on diffusion but on convection. Due to the incompressibility of water, the reactant activity is the same in all locations where liquid water is in

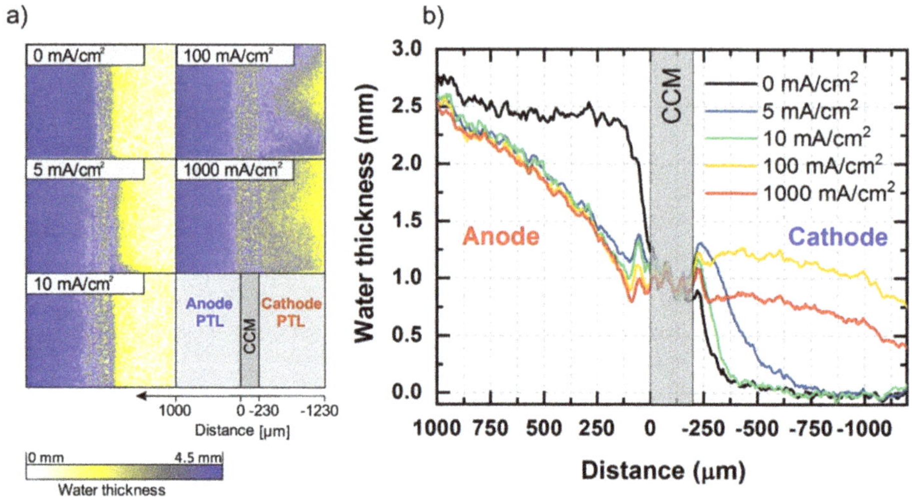

Figure 10.9. Water distribution in the porous transport layers of an operating electrolyzer over a large range of current densities. (a) Neutron radiographs with the water thickness along the neutron beam represented in false colour. (b) Corresponding quantified profiles of water thickness. Reproduced from [32], copyright The Author(s). Published by IOP Publishing Ltd. CC BY 4.0.

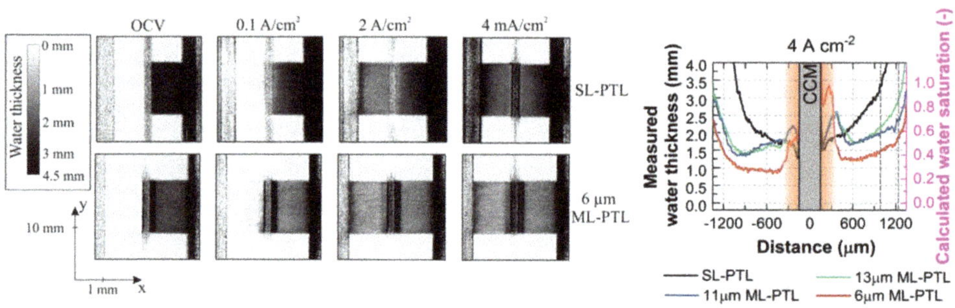

Figure 10.10. Left: neutron radiographies showing the *operando* water distribution in PTLs with and without a micro porous layer. Right: quantitative water profiles extracted from neutron radiographies for four different materials (one PTL without MPL and three PTLs with MPL having different pore sizes). Reproduced from [35], copyright The Author(s). Published by IOP Publishing Ltd. CC BY 4.0.

contact with the catalyst particles. Therefore, the water supply can either be disrupted or not, with no intermediate situation—except potentially on the very local scale in the catalyst layer, where some particles not directly in contact with liquid water may still be supplied by vapour or through the ionomer. The imaging results showed previously indicate that the water supply is not disrupted, as this would result in a dramatic drop of water fraction near the catalyst layer.

This was confirmed by measurements with further materials. The used materials, which include a *micro porous layer* (MPL) [34] were previously shown to improve the performance of electrolyzers. As demonstrated with neutron imaging (see figure 10.10), including an MPL strongly affects the water gas distribution, not only in the region covered by the MPL but further away within the bulk of the material [35]. When using an MPL—in particular, those having the smallest pore sizes—a very low gas saturation is observed in the MPL region, but in a significantly increased gas saturation is seen in the bulk. Even so, the water supply is not disrupted, as exemplified by the high water saturation in the MPL despite constant gas production. The high water saturation in the MPL region is not necessarily the underlying cause of performance improvement. First, MPL with different pore sizes result in different water saturation while the performance improvement due to the present of an MPL is essentially independent of the pore size [34]. Second, there are many other possible explanations as to why the presence of an MPL improves performance, such as a reduced length of pathways inside the catalyst layer [36].

In summary, the observation of two-phase flow in electrolyzer porous media using neutron imaging helps elucidating the transport mechanisms. An important result is that, in sintered titanium PTLs, there is no disruption of water transport through the bulk of the material. If any performance loss is related to the occupation of pore space by the evolved gas, it has to be sought on the very local scale inside the catalyst layer. On the other side, this result shows that the use of sintered titanium PTL might be replaced by simpler, cheaper to produce structure—without going to the extent of the coarse meshes, which result in a lower performance. Neutron imaging will be a valuable asset to identify the best compromise among the different possible material structures.

10.2.4 Cation contamination

Cation contamination is recognized as one of the major issues for PEWE durability. Once fed into the electrolyzer, the positively charge ions can replace the protons in the membrane and in the ionomer used to provide ionic conductivity to the catalyst layers. This results in a lower proton conductivity of these materials, and overall in a decrease of performance. The membrane-electrodes assembly can be regenerated by immersion into an acid solution, but the corresponding operation requires a full disassembly of the electrolyzer stack. Methods for *in situ* identification and regeneration of cation contamination are therefore of high interest.

It was shown [37] that the polarization curve of an electrolysis cell (measurement of the cell voltage as a function of the current density) contaminated by cations exhibits a characteristic hysteresis which does not appear in a pristine cell. The voltage was shown to be much higher—equivalent to a lower performance—when measuring with increasing currents than when measuring with decreasing currents. This is explained by the fact that the cation contaminants, when operated at high current density, migrate towards the cathode catalyst layer under the effect of the electric field. As their impact is lower than when they are in the membrane or in the anode catalyst layer, the performance is better after the cell has been operated at high current density, which results in the hysteresis. This phenomenon could be visualized with the help of neutrons, using Gd^{3+} ions as a model contaminant. Although their mobility in the membrane material is not quantitatively the same as the typical cation contaminants (e.g. Fe^{3+}), their general behaviour is similar. Due to

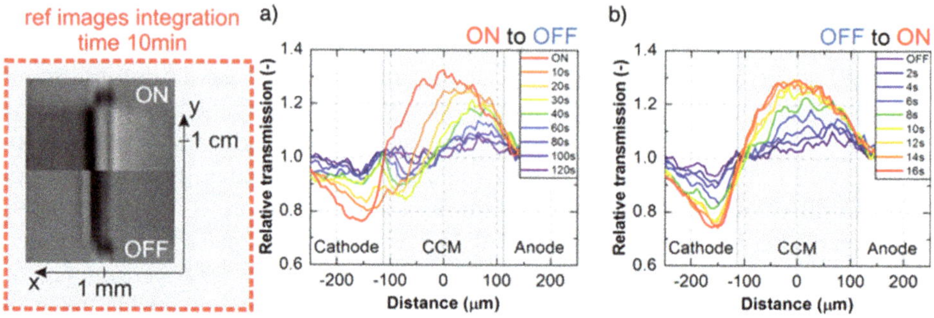

Figure 10.11. (Left) Neutron radiograph of an electrolysis cell contaminated with Gd^{3+} ions (upper half during operation and lower half without operation). (Right) Corresponding transmission profiles relative to the transmission without operation. Reproduced from [38], copyright The Author(s). Published by IOP Publishing Ltd. CC BY 4.0.

the exceptionally large cross-section of Gd for neutron absorption, the distribution of Gd ions can be visualized even at very low concentrations.

As shown in figure 10.11, the transient distribution of ions during on–off cycles could be visualized. When the current is started, the ions migrate towards the cathode catalyst layer, and the equilibrium profile is established within a little more than 10 seconds. When stopping the electrolyzer, the ions diffuse back to the membrane, a process that lasts approximately one minute. The use of these visualization experiments for the validation of a physical model of cation transport [38] allow a dependent understanding of the behaviour of cation contaminants. This can be used as a basis for the development of early identification strategies (based on transient voltage measurements), but also for the development of *in situ* regeneration methods. As an example, promising results were obtained using water acidification by saturation with CO_2 [37].

10.3 Solide oxide cells (SOC)

10.3.1 SOC basics

Solid oxide fuel and electrolysis cells are viable alternatives to PEFCs and PEWEs. Solid oxide fuel cells (SOFCs) have been considered the most promising fuel cell technology for stationary applications based on their low operating costs and extremely high electrical efficiencies reaching up to 60% and total efficiencies even towards 80%, if in addition the produced heat can be utilized in so-called combined heat and power (CHP) units [39]. While SOFCs and PEFCs, also known as proton-exchange membrane fuel cells (PEMFCs), share their basic operating principles (compare section 10.2.1), they yet have many differences as well. For example, PEFCs are bound to hydrogen as fuel, but SOFCs can operate additionally with hydrocarbon fuels like methane, natural gas and propane to produce electricity, which also implies less sensitivity to the purity of the fuel. Furthermore, SOFCs do not require expensive catalyst materials like platinum or ruthenium. However, SOFCs operate at elevated temperatures up to around 800 °C, implying slower start-up and fuel cell stacks that are bulkier than those of PEFCs, which makes them less suited for portable applications like in transportation. On the other hand, they are favourable for distributed generation and electric utility purposes.

In analogy to the case of PEFCs and PEWEs, solid oxide electrolysis cells can also be realised, and it is even possible to combine both functions, fuel and electricity production, in a single cell, which is referred to as a reversible solid oxide fuel cell (RSOFC) [40].

SOFCs are composed of at least three layers with sub-millimeter thicknesses: a porous anode, a gas tight electrolyte and a porous cathode. Oxygen is delivered on the cathode side during operation. A thin porous layer of some tens of micrometers formed typically by LSM–YSZ (strontium-doped lanthanum manganate—yttria-stabilized zirconia) or LSCF (lanthanum strontium cobalt ferrite) constitutes the oxygen electrode, i.e. the cathode. After reduction at such a cathode, oxygen ions are conducted towards the anode, e.g. Ni-YSZ, through an about 10 μm thick dense electrolyte (typically YSZ). There the oxygen ions react with the fuel, e.g. hydrogen, in which case the oxidation produces water, in the form of stream, and free electrons. The latter generate an electric current when travelling to the cathode through an external electrical circuit. For achieving efficient ion conduction, this process requires very high temperatures. Higher temperatures facilitate better ion conductivity in YSZ but mechanical and chemical stability of the cell decreases with temperature [41]. The optimal operation temperature of Ni-YSZ based SOFC around 800 °C is a compromise between these competing effects.

Today numerous materials for electrodes and electrolyte, different cell support options (anode-, cathode- or electrolyte-supported cells) and different cell designs (e.g. planar or tubular cells) are considered. However, nickel–zirconia cermet (ceramic–metallic composite) has been one of the most popular anode and anode support materials in decades [42] and currently, for anodes as well as anode supports, a cermet of Ni and YSZ constitutes the state-of-the-art in planar SOFCs. Due to the fact that Ni-YSZ electrodes and supports perform well in both SOFC and SOEC [40] they appear even well-qualified for RSOFCs. However, the technology also still requires significant research efforts to optimize the performance, and in particular improve the durability and mechanical integrity of these fuel cells. Especially, the high operation temperatures (around 800 °C) cause issues in SOFCs which can be related to corrosion and, due to mismatching thermal expansion coefficients of different SOFC components, i.e. materials, to thermal stresses. It could be shown [43] that mainly processes taking place in the Ni-YSZ cermet of corresponding solid oxide cells cause severe degradations limiting the lifetime and long-term performance. Direct insights into these processes under operation conditions are thus indispensable for the understanding and improvement of this technology.

10.3.2 Redox behaviour of NiO-YSZ SOC anodes

The brittle ceramic layers of SOFCs are subject not only to stresses due to temperature gradients, thermal cycling and thermal expansion coefficient mismatches, but also the phase transition occurring in the Ni-YSZ cermet during reduction and re-oxidation of NiO and Ni, respectively, as well as Ni particle coarsening [43] during long-term operation of SOFC change the microstructure and dimensions of the material and lead to stresses in the constraint fuel cell and fuel cell stacks. Understanding the microstructure evolution under stress in operation conditions is crucial for improving the integrity and thus the durability of SOCs.

The phase and microstructural transitions during reduction and re-oxidation of NiO-YSZ composites have been researched for decades using *ex situ* diffraction and microscopy techniques such as XRD, TEM and EDS. More recent studies also utilized *in situ* microscopy techniques such as environmental TEM [44] and controlled atmosphere high-temperature scanning probe microscopy (CAHT-SPM) [45] to investigate morphological changes in NiO-YSZ during reduction. On the other hand, macroscopic techniques such as impedance spectroscopy, dilatometry, thermo-gravimetry have also been employed. However, all these methods meet severe limitations with respect to capturing the process on the required length and time scales to provide a full picture of the local evolutions in the bulk material under realistic conditions.

Therefore, neutron imaging, and in particular diffraction contrast neutron imaging (compare chapter 2), was chosen to enable true *in situ* and spatially resolved observations of bulk NiO-YSZ reduction and oxidation under realistic operation conditions and variations of such [46]. While the total attenuation of Ni and NiO are very similar when a broad spectrum is applied, their Bragg edge patterns differ significantly (figure 10.12) and with Ni being a strong coherent scatterer, wavelength resolved neutron imaging enables imaging phase fractions with temporal resolution in *in situ* experiments. In fact, the various measurements of this investigation were the first examples of time-resolved phase mapping with diffraction contrast neutron imaging [47] and required a tailored sample environment to enable temperatures of up to 900 °C in hydrogen atmosphere and the possibility to apply inhomogeneous strain fields to the sample [48].

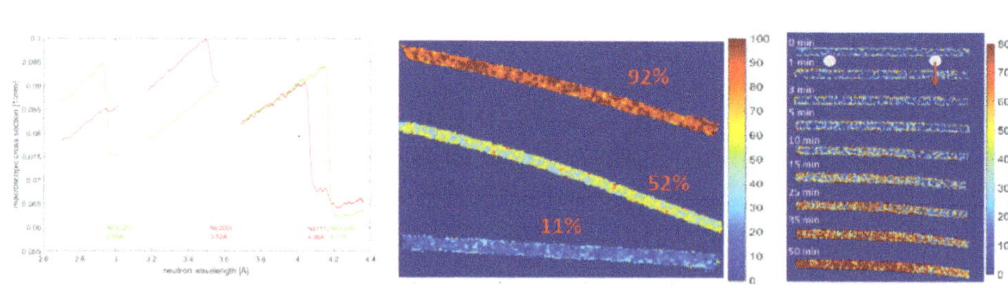

Figure 10.12. The most pronounced Bragg edges of Ni and NiO measured by ToF transmission of Ni-YSZ and NiO-YSZ samples (left hand side); Mapping the reduced phase content in anode/support samples of 900 μm thickness through local Bragg edge height analysis (mid). Reproduced from [49]. Copyright IOP Publishing Ltd. All rights reserved. Time-resolved phase transformation study at high temperature in hydrogen atmosphere of a NiO-YSZ sample under load condition. Reproduced from [47] with permission from the International Union of Crystallography.

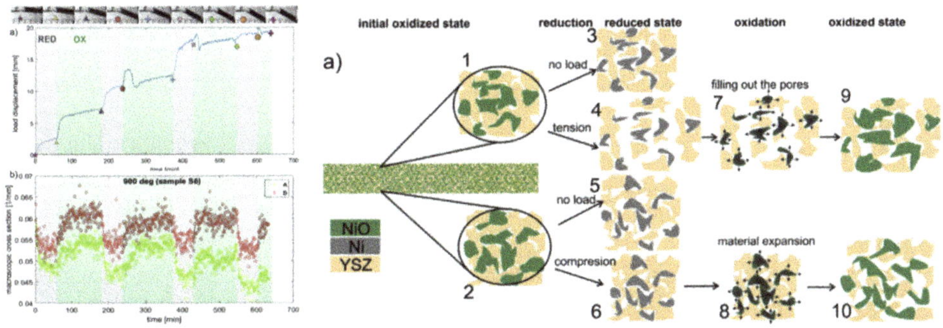

Figure 10.13. Monochromatic imaging of continuous redox cycling under applied load providing phase transformation and deformation behavior enabling to derive a model of microstructural evolution of NiO/Ni-YSZ andode/support. Reprinted from [50], Copyright (2017), with permission from Elsevier.

Initial measurements revealed that both, tensile and compressive stresses, just like increased temperatures lead to an accelerated NiO reduction [49]. First time-resolved *in situ* studies subsequently allowed following local reduction rates and tying them to local stress levels as well as to the applied temperatures of up to 1070 °C [47]. It was further found that samples measured in an approach of interrupting the process, instead of applying temporal resolution in a continuous process, displayed a bias towards equalizing differences of regions of different stress. This raises some doubts concerning the reliability of *ex situ* studies. *In situ* time-resolved monochromatic neutron imaging studies during repetitive redox cycling found that accelerated creep in NiO/Ni-YSZ is not only found during reduction but also during oxidation and that regions with less applied mechanical stress tended to increased cracking, potentially due to lower phase transformation rates and corresponding creep [50]. These works led to a detailed model of the microstructural evolution during redox processes in NiO-YSZ and thus a better understanding of the response and damage mechanisms of the material in operation and were complemented by tomographic 3D phase imaging illustrating the context of redox behaviour and structural damage in the volume of NiO-YSZ material (figure 10.13) [51].

References

[1] Boillat P, Iranzo A and Biesdorf J 2015 *J. Electrochem. Soc.* **162** F531

[2] Cochet M, Scheuble D, Hannesen U and Boillat P 2018 Experimental study of the water management in the Autostack-Core Evo2 fuel cell with neutron radiography, in, p 38, Paul Scherrer Institut (PSI) *Annual Report 2017—Electrochemistry Laboratory*

[3] Iranzo A, Gregorio J M, Boillat P and Rosa F 2020 *Int. J. Hydrogen Energy* **45** 12432

[4] Manke I, Hartnig C, Grünerbel M, Kaczerowski J, Lehnert W, Kardjilov N, Hilger A, Banhart J, Treimer W and Strobl M 2007 *Appl. Phys. Lett.* **90** 184101

[5] Takenaka N, Asano H, Sugimoto K, Murakawa H, Hashimoto N, Shindo N, Mochiki K and Yasuda R 2011 *Nucl. Instrum. Methods Phys. Res.* A **651** 277

[6] Boillat P, Frei G, Lehmann E H, Scherer G G and Wokaun A 2010 *Electrochem. Solid-State Lett.* **13** B25

[7] Boillat P, Kramer D, Seyfang B C, Frei G, Lehmann E, Scherer G G, Wokaun A, Ichikawa Y, Tasaki Y and Shinohara K 2008 *Electrochem. Commun.* **10** 546
[8] Manzi-Orezzoli V, Siegwart M, Cochet M, Schmidt T J and Boillat P 2019 *J. Electrochem. Soc.* **167** 054503
[9] Forner-Cuenca A, Biesdorf J, Gubler L, Kristiansen P M, Schmidt T J and Boillat P 2015 *Adv. Mater.* **27** 6317
[10] Xu H, Bührer M, Marone F, Schmidt T J, Büchi F N and Eller J 2021 *J. Electrochem. Soc.* **168** 074505
[11] Cochet M, Forner-Cuenca A, Manzi-Orezzoli V, Siegwart M, Scheuble D and Boillat P 2020 *J. Electrochem. Soc.* **167** 084518
[12] Striednig M, Cochet M, Boillat P, Schmidt T J and Büchi F N 2022 *J. Power Sources* **517** 230706
[13] Siegwart M, Harti R P, Manzi-Orezzoli V, Valsecchi J, Strobl M, Grünzweig C, Schmidt T J and Boillat P 2019 *J. Electrochem. Soc.* **166** F149
[14] Siegwart M, Manzi-Orezzoli V, Valsecchi J, Harti R P, Kagias M, Strobl M, Grünzweig C, Schmidt T J and Boillat P 2020 *J. Electrochem. Soc.* **167** 064509
[15] Kusoglu A and Weber A Z 2017 *Chem. Rev.* **117** 987
[16] Thompson E L, Capehart T W, Fuller T J and Jorne J 2006 *J. Electrochem. Soc.* **153** A2351
[17] Schießwohl E, von Unwerth T, Seyfried F and Brüggemann D 2009 *J. Power Sources* **193** 107
[18] Thompson E L, Jorne J, Gu W and Gasteiger H A 2008 *J. Electrochem. Soc.* **155** B625
[19] Thompson E L, Jorne J, Gu W and Gasteiger H A 2008 *J. Electrochem. Soc.* **155** B887
[20] Ishikawa Y, Hamada H, Uehara M and Shiozawa M 2008 *J. Power Sources* **179** 547
[21] Ishikawa Y, Shiozawa M, Kondo M and Ito K 2014 *Int. J. Heat Mass Transfer* **74** 215
[22] Oberholzer P, Boillat P, Siegrist R, Perego R, Kästner A, Lehmann E, Scherer G G and Wokaun A 2011 *J. Electrochem. Soc.* **159** B235
[23] Hwang G S, Kim H, Lujan R, Mukundan R, Spernjak D, Borup R L, Kaviany M, Kim M H and Weber A Z 2013 *Electrochim. Acta* **95** 29
[24] Palecki S, Gorelkov S, Wartmann J and Heinzel A 2017 *J. Power Sources* **372** 204
[25] Biesdorf J, Oberholzer P, Bernauer F, Kaestner A, Vontobel P, Lehmann E H, Schmidt T J and Boillat P 2014 *Phys. Rev. Lett.* **112** 248301
[26] Siegwart M, Woracek R, Márquez Damián J I, Tremsin A S, Manzi-Orezzoli V, Strobl M, Schmidt T J and Boillat P 2019 *Rev. Sci. Instrum.* **90** 103705
[27] Siegwart M, Huang F, Cochet M, Schmidt T J, Zhang J and Boillat P 2020 *J. Electrochem. Soc.* **167** 064510
[28] Stahl P, Biesdorf J, Boillat P and Friedrich K A 2016 *J. Electrochem. Soc.* **163** F1535
[29] Selamet O F, Pasaogullari U, Spernjak D, Hussey D S, Jacobson D L and Mat M D 2013 *Int. J. Hydrogen Energy* **38** 5823
[30] Minnaar C, De Beer F and Bessarabov D 2020 *Energy Fuels* **34** 1014
[31] Seweryn J, Biesdorf J, Schmidt T J and Boillat P 2016 *J. Electrochem. Soc.* **163** F3009
[32] Zlobinski M, Schuler T, Büchi F N, Schmidt T J and Boillat P 2020 *J. Electrochem. Soc.* **167** 084509
[33] Lee C H, Lee J K, Zhao B, Fahy K F, LaManna J M, Baltic E, Hussey D S, Jacobson D L, Schulz V P and Bazylak A 2020 *J. Power Sources* **446** 227312
[34] Schuler T, Ciccone J M, Krentscher B, Marone F, Peter C, Schmidt T J and Büchi F N 2020 *Adv. Energy Mater.* **10** 1903216
[35] Zlobinski M, Schuler T, Büchi F N, Schmidt T J and Boillat P 2021 *J. Electrochem. Soc.* **168** 014505

[36] Schuler T, Schmidt T J and Büchi F N 2019 *J. Electrochem. Soc.* **166** F555
[37] Babic U, Zlobinski M, Schmidt T J, Boillat P and Gubler L 2019 *J. Electrochem. Soc.* **166** F610
[38] Zlobinski M, Babic U, Fikry M, Gubler L, Schmidt T J and Boillat P 2020 *J. Electrochem. Soc.* **167** 144509
[39] Stambouli A B and Traversa E 2002 *Renew. Sustain. Energy Rev.* **6** 433
[40] Minh N Q and Mogensen M B 2013 *Interface Mag.* **22** 55
[41] Möbius H-H 2003 *High Temperature and Solid Oxide Fuel Cells* ed S C Singhal and K Kendall (Amsterdam: Elsevier) p 23
[42] Atkinson A, Barnett S, Gorte R J, Irvine J T S, McEvoy A J, Mogensen M, Singhal S C and Vohs J 2004 *Nat. Mater.* **3** 17
[43] Hauch A, Ebbesen S D, Jensen S H and Mogensen M 2008 *J. Electrochem. Soc.* **155** B1184
[44] Simonsen S B, Agersted K, Hansen K V, Jacobsen T, Wagner J B, Hansen T W and Kuhn L T 2015 *Appl. Catal.,* A **489** 147
[45] Hansen K V, Jacobsen T, Thyden K, Wu Y and Mogensen M B 2014 *J. Solid State Electrochem.* **18** 1869
[46] Makowska M G, Strobl M, Lauridsen E M, Frandsen H L, Tremsin A S, Kardjilov N, Manke I, Kelleher J F and Theil Kuhn L 2015 *J. Appl. Crystallogr.* **48** 401
[47] Makowska M G, Strobl M, Lauridsen E M, Kabra S, Kockelmann W, Tremsin A, Frandsen H L and Theil Kuhn L 2016 *J. Appl. Crystallogr.* **49** 1674
[48] Makowska M G *et al* 2015 *Rev. Sci. Instrum.* **86** 125109
[49] Makowska M G, Strobl M, Mejdal Lauridsen E, Frandsen H L, Tremsin A, Shinohara T and Kuhn L T 2015 *ECS Trans.* **68** 1103
[50] Makowska M G, Kuhn L T, Frandsen H L, Lauridsen E M, De Angelis S, Cleemann L N, Morgano M, Trtik P and Strobl M 2017 *J. Power Sources* **340** 167
[51] Makowska M G *et al* 2018 *Phys.* B **551** 24
[52] Boillat P, Lehmann E H, Trtik P and Cochet M 2017 Neutron imaging of fuel cells – Recent trends and future prospects *Curr. Opin. Electrochem.* **5** 3–10

IOP Publishing

Neutron Imaging
From applied materials science to industry
Markus Strobl and Eberhard Lehmann

Chapter 11

Batteries

Pierre Boillat, Sigita Trabesinger, Michael Schulz and Markus Strobl

In 2019 the Nobel Prize in Chemistry was awarded to three outstanding scientists for their contributions to the development of the lithium-ion battery (LIB). Stanley Whittingham demonstrated the first working lithium metal battery, John Goodenough improved its energy density by introducing lithium cobalt oxide, while Akira Yoshino made the rechargeable battery safer making it a true LIB, e.g. lithium could be present inside the battery only in its ionic form, eliminating highly reactive lithium metal. All these developments led to the practical rechargeable LIB, which we use today. Nobody can imagine anymore our lives without devices powered by rechargeable batteries, applications of which span from portable devices to the automotive industry to large-scale stationary battery storage for storing energy from renewable sources. Rechargeable non-aqueous LIBs are and will be for the foreseeable future among the most advanced energy storage systems, with the continuously growing need for high gravimetric and volumetric energy and power densities—the key factors, which made LIBs a success story. The other important requirements for future rechargeable batteries are sustainability, low cost and enhanced safety. At the same time, the relative importance of these can change depending on the application, and only the demand for high safety is a firmly remaining precondition. For example, in the case of portable electronics, the price is less important than energy density, while in the automotive sector, the lifetime and costs become decisive.

The main focus of the research in the rechargeable battery field is focused on development of battery active materials for both positive and negative electrodes, as battery properties are strongly dependent on the materials used. At present, specific capacities of battery materials, for both positive and negative electrodes, are reaching their limits and new chemistries and battery architectures are emerging.

On the positive electrode side, for a long time lithium cobalt oxide was the main cathode material, while lithium iron phosphate had a much smaller market share. Due to the dominance of lithium cobalt oxide, prices of raw materials for battery production have sky-rocketed, which drove new cheaper cathode material

 © IOP Publishing Ltd 2024

development, at the same time with higher specific capacity in some cases. The reduction of cobalt amount in cathode active materials led to a whole family of new materials, so-called NMCs,—lithium–nickel–manganese–cobalt oxides,—where different ratios of transition metals are used, while reducing cobalt amount. For example, denomination of NMC111 means that the material contains one third of each of the transition metals. Further developments led to so-called nickel-rich materials, where cobalt amount went down to 5%. Further developments are focused on cobalt-free, and even nickel-free materials with high potentials and specific capacities. At the same time, engineering efforts led to wider applications of lithium iron phosphate, where the future, it is projected, belongs to manganese-doped lithium iron phosphate, possessing higher average voltage as compared with current lithium iron phosphate.

On the negative side, graphite still remains the active material of choice. The development of new anode materials is on the rise and taming the lithium metal anode is considered feasible with newly developed technologies since the introduction of LIBs, at which point it was abandoned with invention of safer Li-ion batteries with graphite anode. However, as lithium metal has a very high capacity (3860 mAh g^{-1}) and a low redox potential (−3.04 V versus standard hydrogen electrode (SHE)), now the race for higher energy density pushes the new efforts towards its commercialization, of course without sacrificing the safety. Other negative electrodes with potential to increase energy density are based on conversion and alloying materials, where silicon is having the front seat. Here the major issues are caused by silicon volume expansion upon the lithiation, where volume of fully lithiated silicon becomes four times larger. This is not manageable within commercial cells and therefore the graphite keeps its main place in Li-ion batteries. However, even if the use of silicon itself is unviable, researcher developed graphite anodes with silicon as capacity-enhancing additive, and graphite in high energy battery cells currently contains 3%–5% of silicon.

For improved safety, the solid state inorganic electrolytes become frontrunners in rechargeable battery research, while a sustainability is sought via developing new more-efficient recycling pathways, researching on alternative metal batteries, such sodium or potassium, as well as formulating advanced Li-ion battery materials without cobalt and even nickel.

All these achievements are impossible without the appropriate analytical techniques, especially *operando* characterization methodologies, which provide information during cycling of the cell under realistic conditions. Analytical techniques based on neutrons play a pivotal role in deciphering LIB materials' working mechanisms due to their sensitivity to light elements.

11.1 Visualization of lithium distribution in different battery technologies

The very high neutron absorption cross-section of the ^{6}Li isotope allows *operando* imaging of the lithium distribution (natural lithium contains 7.5% of ^{6}Li). Depending on the context, the visualization of this distribution can yield different types of information, as described in the following sections.

11.1.1 Lithium plating

In order to improve the energy density, alternatives for graphite negative electrodes are sought. One possibility is to use metallic lithium. Such a configuration does not pose any problem for the discharge process, where lithium is stripped from the electrode—as a matter of fact, most primary lithium batteries use metallic lithium as a negative electrode. The charging process is more challenging, as lithium plating does not occur homogeneously and leads to the formation of dendrites that can protrude through the separator and result in a short circuit.

The visualization of dendrite growth in the separator region was reported by Song *et al* [1], as shown in figure 11.1. They used a ^{7}Li enriched metallic lithium as negative electrode in order to better visualize the deposited lithium with natural composition. The set-up used in this experiment had a limited resolution (about 100 μm), which did not allow seeing the morphology of the dendrites. Nevertheless, it allows a quantification of the amount of dendritic lithium in the separator region. A similar measurement with charge/discharge cycles would in principle allow visualizing how much of the dendritic lithium remains in the separator. However,

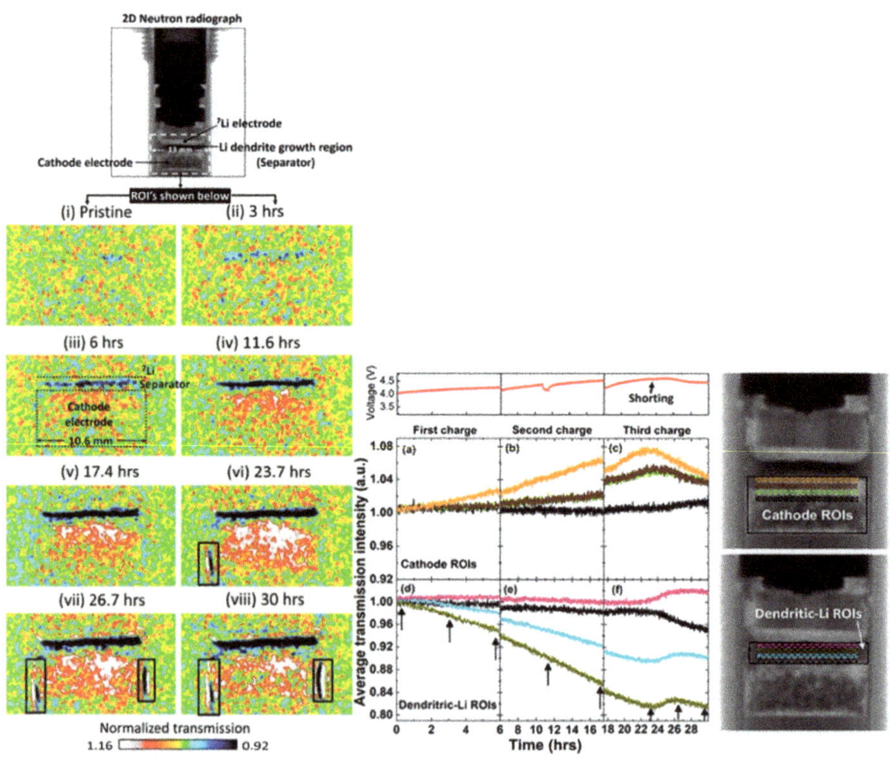

Figure 11.1. Left: visualization of dendrite formation (as evidence by the reduction of transmission in the separator region). The shortcut of the battery reportedly occurred after 24 h (panel vi). Right: temporal profiles of transmission showing the progressive growth of dendritic lithium. Reproduced with permission from [1]. Copyright (2019) American Chemical Society.

it must be noted that the use of ^{7}Li for the negative electrode would complicate the analysis: lithium stripped from this electrode would be included in the electrolyte and positive electrode, resulting in an unknown isotopic composition.

A further proposed cell configuration to improve the energy density is to exclude the negative electrode from the initial cell state and use the copper current collector for lithium deposition. In this way the negative electrode is being formed during the first charge. This type of cell is often referred as 'anode-free' or 'anode-less' and is considered as the battery development generation after lithium metal cells.

Based on this 'anode-less' cell configuration, a modified coin-cell for tracking lithium deposition and stripping *operando* by neutron imaging at high resolution was designed (figure 11.2). In addition, the ^{6}Li lithium metal and ^{6}Li conductive salt were used to facilitate special resolution. A set of carbonate electrolytes with differences in solvent composition were imaged while cycling, selected images are shown in figure 11.3. To allow quantification of equivalent thickness of lithium deposits, the uncycled cell image was used to exclude absorption by ^{6}Li present in electrolyte.

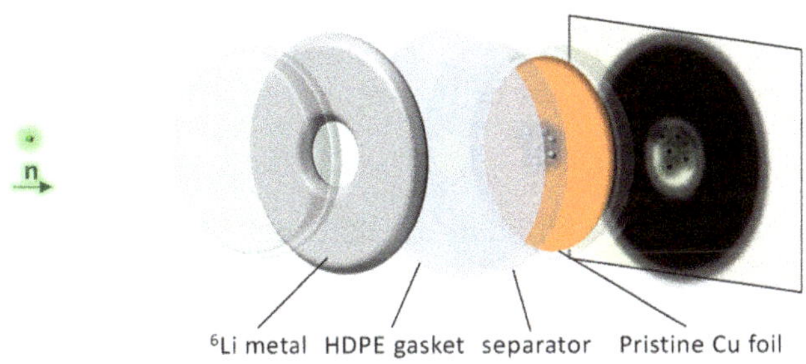

Figure 11.2. Modified coin-cell design for *operando* imaging of lithium distribution and its deposition dynamics.

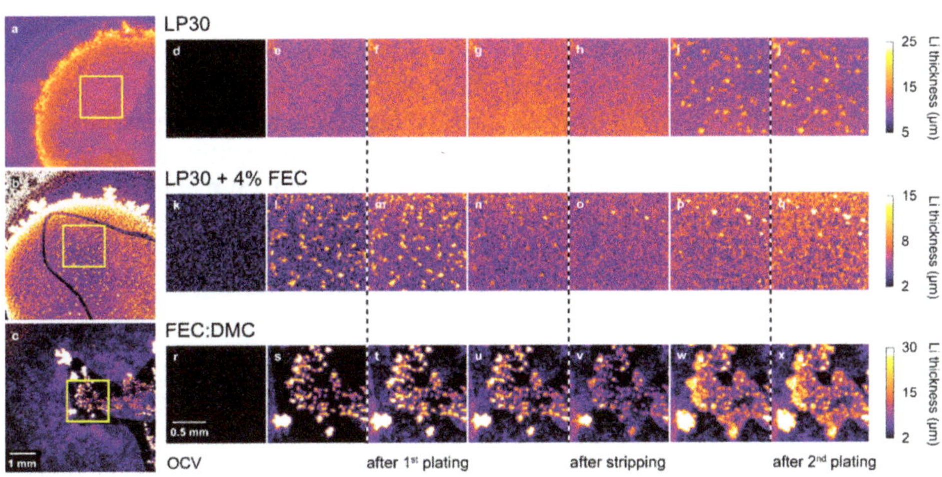

Figure 11.3. Visualization of lithium plating morphology on copper with different electrolytes, colour scale in indicates the equivalent thickness of the deposits and its evolution over the cycling time.

The outer bright ring visible in overall images in figure 11.3 is a set-up-defined artefact due to the concentric test cell's configuration, however, from the central part within this ring, the information on lithium deposition from electrolyte can be extracted.

This experiment allowed not only visualizing lateral lithium deposit distribution but also determining lithium cycling reversibility and quantifying area-dependent inactive lithium (so-called 'dead lithium') formation in each of the tested electrolytes. It has shown that addition of a small amount of fluoroethylene (FEC) to LP30 standard electrolyte (1M $LiPF_6$ in ethylene carbonate and dimethyl carbonate mixture EC:DMC (1:1)) results in less 'dead lithium' formation, however, replacing EC fully by FEC leads to formation of large lithium agglomerates with little reversibility during cycling.

This experiment using *operando* neutron imaging proved to be invaluable for studying dynamic lateral lithium distribution. Due to the use of an isotopically enriched ^{6}Li system, detection limits for Li deposits could be pushed down to the low micrometre range, allowing observation also of thin deposits in the centre of electrodes. Short acquisition times and a high spatial resolution of around 3 μm per pixel allowed precise tracking of metal amounts over the entire experiment. Further developments in the test cell set-up are needed to avoid the set-up defined artefacts.

11.1.2 Lithium distribution in lithium–air batteries

A proposed way of improving the energy density of Li-ion batteries is to avoid intercalation materials on the cathode side and use the redox couple of lithium and oxygen in a corresponding lithium–air battery configuration [2]. While this type of battery theoretically features a very high energy density, it is also associated with numerous challenges which have so far prevented its application. These challenges include the need of a catalyst for the Li/O_2 electrochemistry, but also the impact of the distribution of the $Li–O_2$ reaction products. Nanda *et al* [3] studied this distribution using neutron tomography (see figure 11.4, left) and observed a non-uniform product distribution across the electrode thickness.

The lithium–air battery configuration, to realize its maximum potential in terms of energy density, would require the use of metallic Li as a negative electrode. Using a combination of x-ray and neutron computed tomography, Sun *et al* [4] studied the structural modifications and the distribution of lithium in four different battery samples. They concluded that the irreversible structural changes in the metal Li electrode played an important role in the capacity fading during cycling.

11.1.3 Lithium distribution in intercalation materials

Different processes can limit the transport of lithium during lithiation and delithiation of intercalation materials. Some of them, such as the solid diffusion of lithium to or from the core of electrode particles, result in limitations of charge or discharge rates, but not in heterogeneities on the macroscopic scale. Other limitations include the transport of lithium ions in the electrolyte, and the electric

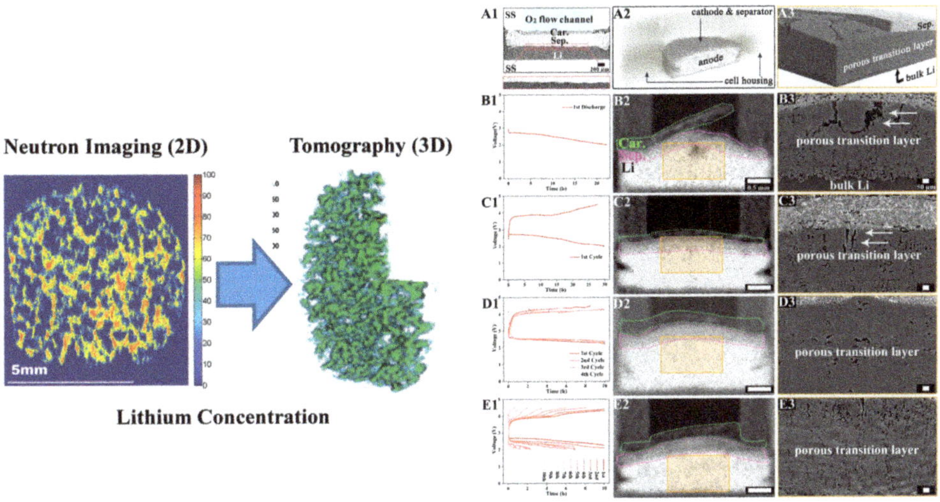

Figure 11.4. Visualization of lithium distribution in lithium–air batteries with neutron imaging. Left: visualization of the discharge products distribution on the cathode side. Reprinted with permission from [3]. Copyright (2012) American Chemical Society. Right: visualization of the shape of the Li metal counter electrode. Reprinted with permission from [4]. Copyright (2019) American Chemical Society.

conductivity of the electrode. When such transport limitations become dominant, they can result in a distribution of state of charge, in particular across the thickness of the electrode. Several authors have reported neutron radiography measurement aimed at visualizing the heterogeneity of lithium distribution across electrodes. Siegel *et al* [5, 6] imaged a pouch-type battery from the side with the goal of measuring the lithium distribution, though their analysis was strongly complicated by the dimensional changes occurring in the battery during charge and discharge. Ziesche *et al* [7] used tomographic imaging with both x-rays and neutrons on primary $LiMnO_2$ cells, showing the complementarity of the methods.

As shown in figure 11.5, while the x-rays are very useful to identify mechanical damage such as cracks, neutrons can identify regions with lower electrolyte content, or different amounts of lithium removal. Ziesche *et al* also demonstrated a virtual unrolling technique in order to perform quantitative analysis from tomography data.

The impact of transport across the electrode is best visible when working with thicker electrodes—not only because of the lower constraints on imaging resolution, but also because the heterogeneities are stronger due to the need to transport the lithium across larger distances. This is evidenced, for example, by the work of Nie *et al* [8], where very thick sintered electrodes (approximately 1 mm) were imaged with neutrons.

As shown in figure 11.6, for the fastest discharges used in this study, only the regions very close to the separator are lithiated, respectively, delithiated. This emphasizes the fact that, for these samples, the ionic transport in the electrolyte is the limiting process, and not the electronic transport.

Most of the research using neutron imaging on batteries which is reported in the literature is conducted using unmodified commercial samples. One of the reasons for

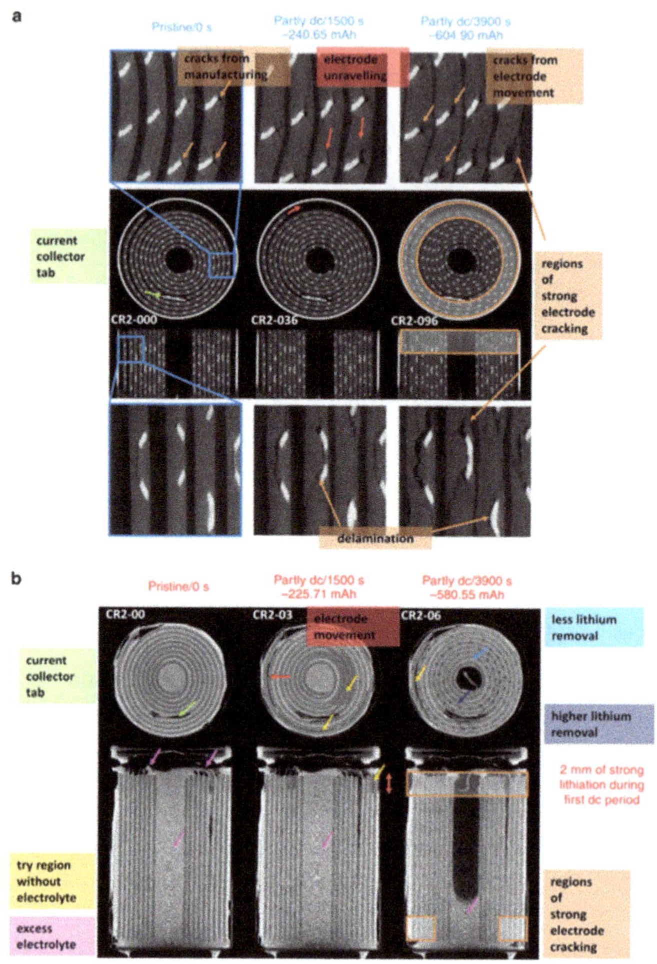

Figure 11.5. Complementary tomography with x-ray and neutrons of a primary $LiMnO_2$ cell. Reproduced from [7]. CC BY 4.0.

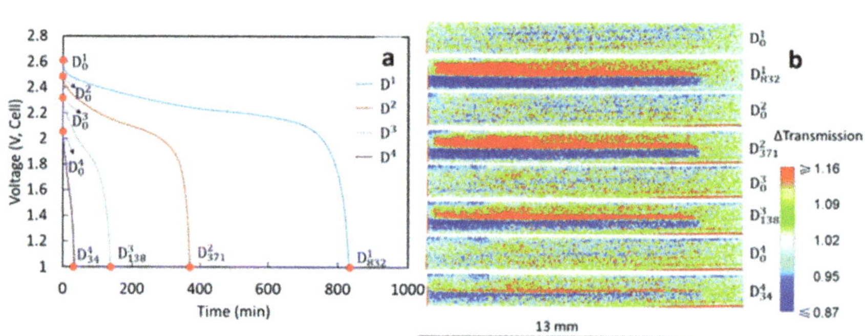

Figure 11.6. Neutron imaging of a $Li_4Ti_5O_{12}/LiCoO_2$ cell with thick sintered electrodes discharged at different C rates. Reproduced from [8] with permission from the Royal Society of Chemistry.

this choice is that the direct imaging of such unmodified samples leaves no doubts about the relevance of the obtained results for real-world applications. A further reason is that using commercially available samples without the need to modify them is experimentally much less demanding. While both are valid reasons, the use of conventional battery geometries also implies drawbacks: non-planar designs such as spiral configurations require tomographic imaging and virtual unrolling. Even for planar designs such as coin cells, the dimensions of the cells are usually not optimal for obtaining the best contrast.

In this context, it is interesting to consider the possibilities offered by customized cell housings. The compatibility of neutron imaging with usual cell construction materials means that such a custom environment can be made representative of a real-world battery. The customization bears the benefits that the electrode general geometry can be chosen to fit the needs of imaging in an optimized way. As an example, we present here a study performed with a custom designed cell having a square geometry of 5×20 mm^2. The dimension of 5 mm in the beam direction is optimal in order to obtain sufficient contrast while keeping a good transmission of the neutron beam. This optimal geometry was combined with the anisotropic enhancement methods already mentioned for fuel cell and electrolysis research (chapter 10). Additionally, the contrast was enhanced using ^{6}Li isotopically enriched electrolyte and electrode materials, and the transmission was improved by using a deuterated electrolyte. As a result of these optimizations, a good image quality and contrast-to-noise ratio could be obtained for a pixel size of 6 μm and an effective resolution of approximately 20 μm, while keeping a reasonably low integrated exposure time of 2.5 min.

As illustrated in figure 11.7, this optimized set-up allows the measurement of a state-of-charge profile across electrodes of 150 μm during a relatively fast charge

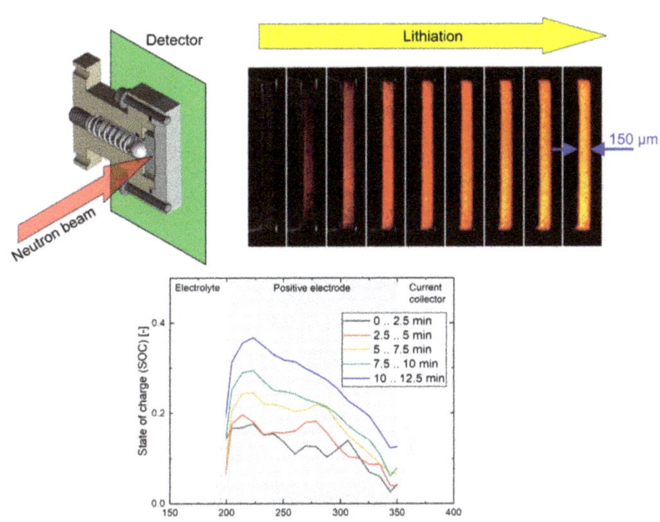

Figure 11.7. High-resolution anisotropic neutron imaging of a Li-ion cell with a customized geometry. The state-of-charge is obtained by interpolation between the image of the fully charged and fully discharged cell.

(charge rate of 1 C, corresponding to a nominal charging time of 1 h). At this rate, the portion on the side of the electrolyte charges much faster than the portion near the current collector. Similarly to the previous example with sintered electrodes, this result indicates that the transport of lithium in the electrolyte is limiting the charging process.

11.1.4 Li concentration in electrolytes

The visualization of the Li^+ ion concentration distribution in the electrolyte gives insight into the transport properties and is a valuable information to validate overall Li transport models of Li-ion batteries. Within the same cell, as described in figure 11.7, and thanks to the isotopically enriched lithium salt and the use of deuterated electrolytes, the impact of charging at a 1 C rate on the Li^+ ion concentration across a thick separator could be visualized.

As shown in figure 11.8, a gradient of Li^+ concentration is quickly established after starting the charging process, and remains stable during the charge. It must be mentioned that the apparition of such a gradient is a consequence of the very thick separator used here. Commercial Li-ion batteries use much thinner separators, having a typical thickness of 20 μm. Nevertheless, the observed gradient is a valuable input for battery modelling validation. Moreover, this observation emphasizes how the use of thick separators in laboratory research may impact the results, in particular for high rate cycling.

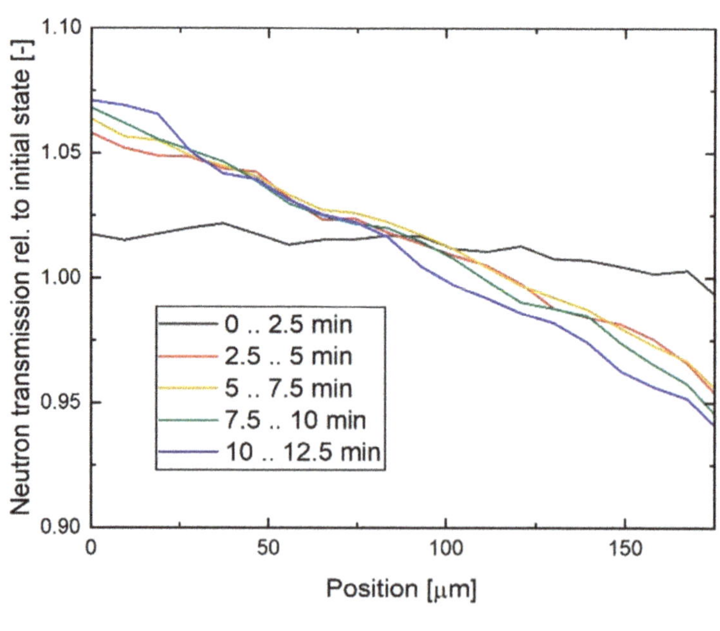

Figure 11.8. Change of neutron transmission due to a gradient of Li^+ ion concentration at charging rates of 1 C. The measurements were realized with the same cell as illustrated in figure 11.7. The separator thickness is approximately 175 μm.

11.2 Mapping lithiation phases

Powder diffraction is typically used to evaluate the lithiation phases of the crystalline electrode materials as it is sensitive to the crystalline phase transitions occurring during lithiation and delithiation. Diffraction can be applied *operando* [9] and in contrast to neutron imaging is not sensitive to other processes in the battery during charge and discharge such as the formation of the solid–electrolyte interface, which contains Li and hydrogen and thus contributes significantly to attenuation in neutron imaging. Thus, diffraction is very well suited to observing the state of charge and the corresponding performance of an electrode during cycling. However, *operando* neutron diffraction does not provide local insights, but averages over all electrode material in the neutron beam. Here neutron imaging can make a difference as has been outlined in the previous section. However, it cannot easily evaluate the lithiation phases and thus the actual intercalation, when only the conventional attenuation signal is available for analyses. Therefore, several attempts have been reported to utilize diffraction contrast (compare chapter 2) in order to resolve the local lithiation state independent of other contributions to attenuation.

An initial study of a commercial Li-ion battery established the feasibility to resolve lithiation phases in a battery by Bragg edge imaging [10]. However, the spatial resolution was limited and similar to diffraction, although the full-field illumination provides information from several regions of interest simultaneously, and the exposure times were not compatible with *operando* measurements. In particular, also the rolled cylindrical geometry of the 18650 Li-ion battery did not allow addressing individual electrodes individually.

However, another study set out to investigate the heterogeneity of lithiation in an ice-templated carbon electrode at different states of charge utilized diffraction contrast tomography (compare with chapter 2) to obtain the local lithiation states [11]. In order to achieve a resolution of about 320 μm, tomographic scans of 2 days were required. A 0.66 mm thick electrode at 33% SOC was reconstructed for the wavelength-resolved data of the time-of-flight (ToF) tomography and the distribution of C, LiC_{12} and LiC_6 in the volume was estimated from the corresponding Bragg edges per voxel (figure 11.9).

In order to study lithium intercalation and the corresponding evolution of lithiation phases in a thick carbon electrode of 400 μm thickness a concurrent diffraction and diffraction contrast imaging approach was undertaken at the diffractometer SENJU at JPARC [12], which was retrofitted with an MCP-type ToF imaging detector (compare chapter 19). The 90 deg detectors of SENJU were sufficient to resolve the integral phase evolution of graphite, intermediate phase (including LiC_{30}, LiC_{24}, LiC_{18}), LiC_{12} and LiC_6 *operando* during C/35 cycling utilizing a wavelength range from 2 to 6.4 Å (figure 11.10).

Analysing the Bragg edges in the corresponding transmission image stacks yielded different maps to identify the lithiation stages (figure 11.11). While figures 11.11(a) and (b) display the voltage evolution and the total attenuation over time and electrode thickness, figure 11.11(c) displays the position shift of the C (112) edge at 2.31 Å to the LiC_{12}(302) Bragg edge at 2.33 Å and back and thus the

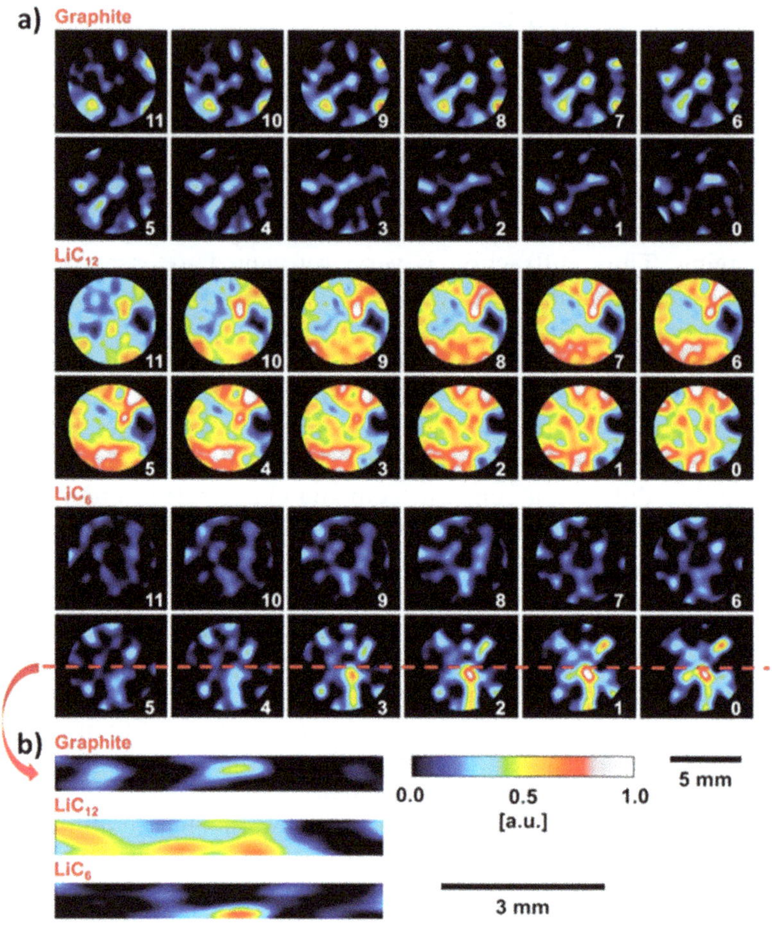

Figure 11.9. Pixel layer wise reconstructed cross sections of distribution of graphite, LiC_{12} and LiC_6 phase in a graphite electrode at 33% SOC. Reproduced from [11]. CC BY 4.0.

transformation via intermediate phases. The edge in this region vanishes with the advent of LiC_6. Figure 11.11(d) displays the appearance and disappearance of the LiC_6(303) edge at 2.36 Å, which coincides with the vanished LiC_{12} edge in figure 11.11(c). Finally, figure 11.11(e) displays the height of an edge overlapping for C(110) and the weaker C_{12}(300) at 2.46 Å. This resembles well the analyses of figure 11.11(c) and correlates as well with the apperance of LiC_6 in the central time frame. However, it has to be noted that the data is yet relatively noisy, which might limit the spatial resolution capability for the respective phases with regards to the thickness of the electrode. However, the measurement efficiently illustrates the principle and feasibility of *operando* diffraction contrast resolution of lithiation phases. The approach can be further improved by utilizing deuterated electrolyte to improve transmission as well as a contrast-optimized geometry and potentially a higher flux, as the source at that time was operated at limited power.

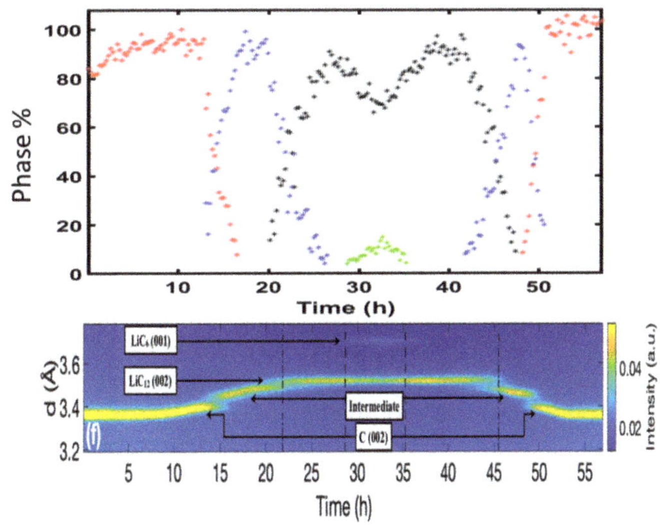

Figure 11.10. *Operando* diffraction pattern and extracted integral lithiation phase evolution in the carbon electrode. Reproduced from [12]. CC BY 4.0.

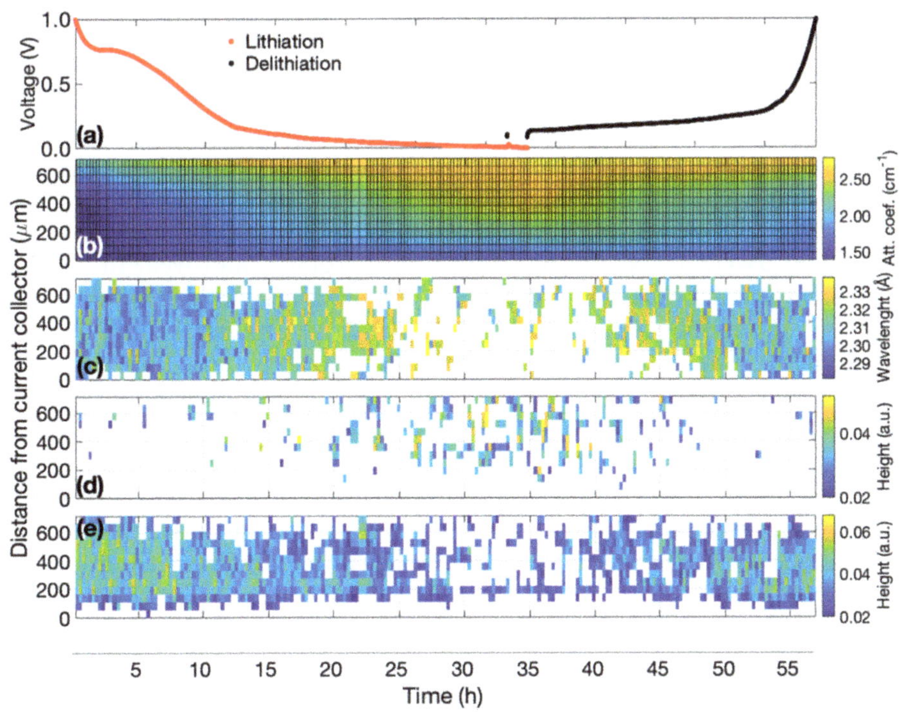

Figure 11.11. (a) Voltage profile of lithiation and delithiation cycle of the pristine half-cell; (b) attenuation map; (c) map of Bragg edge position for two adjacent edges of C and LiC_{12}; (d) map of a LiC_6 Bragg edge height. (e) Map of Bragg edge height for two overlapping edges of C and LiC_{12} [12]. Reproduced from [12]. CC BY 4.0.

11.3 Distribution and properties of the electrolyte

11.3.1 Dynamics of electrolyte filling of Li-ion batteries

The optimization of the production process of LIBs is a key to higher cell performance as well as lower production costs. Neutron imaging has been used in a series of studies aiming at the visualization and optimization of the process of filling new cells with liquid electrolyte. Typically, both in pouch cells and hard case cells dry electrode stacks are assembled and subsequently inserted into the case. As a next step, electrolyte is dosed into the case under low pressure conditions before the cell case is sealed. The process of wetting the electrodes with electrolyte relies on capillary forces, which can be time consuming and may be hindered by remaining gas in the pores. It is important to ensure completed wetting before the formation of the cell, since otherwise an inhomogeneous solid–electrolyte interface may be formed on the cell anode, which may affect the cell performance. A direct observation of the wetting process is limited by the visually intransparent cell housing as well as the required vacuum chamber to maintain low pressure. In this context, the excellent contrast of neutrons for the hydrogen-containing liquid electrolyte may be used for the *in situ* visualization of the dynamics of the wetting of the cell stack with electrolyte.

Knoche *et al* [13] used a custom-made vacuum chamber with aluminum windows for a first *in situ* visualization of the wetting process of pouch cells. The cell was placed inside the vacuum chamber and electrolyte could be dosed automatically into the cell while neutron transmission images were acquired regularly. The pressure inside the vacuum chamber could be electronically controlled and the influence of different pressure profiles on the wetting speed was evaluated. The sealing of the cell was simulated with a pneumatically driven sealing bar on the top of the pouch. It was shown that for a fast wetting it is essential to ensure that remaining gas in the cell is not trapped by surrounding liquid. If an excess amount of liquid on top of the cell stack was avoided, the wetting process was strongly improved. In a following study by Habedank *et al* [14], the effect of different electrode porosities was investigated. It was shown that significantly shorter wetting times could be achieved by increasing the porosity from 30% to 40%. An even larger speed-up of the wetting process by more than an order of magnitude was achieved by laser structuring the electrode material, thereby showing the great potential for optimization of the cell production process.

In a similar study it was shown by Weydanz *et al* [15] that for hard case prismatic cells the wetting process is rather homogeneous from the sides towards the centre of the cell. Application of low pressure before dosing was found to speed up the process by a factor of two.

The influence of the cell format on the wetting process was studied by Günter *et al* [16] based on a comparison of pouch cells with a z-folded stack and hard case cells with a flat wound roll. After dosing, the electrolyte was found to accumulate in the free space in the cell housing and the wetting process was observed to depend on the possible axes of electrolyte penetration into the cell stack. While the electrolyte can penetrate from four sides into the folded stack, a rolled stack only offers two sides for the electrolyte uptake, resulting in a slower wetting process.

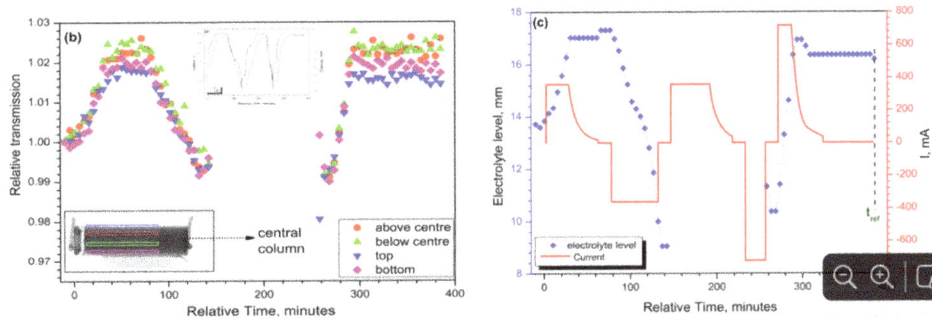

Figure 11.12. Changing transmission signature in different active regions of commercial battery during cycling (inset) illustrating a 'breathing' of the active battery (left); on the right hand side the corresponding filling state of the central (inactive) column of the battery with electrolyte, displaying the naturally complementary filling behaviour. Reproduced from [17]. CC BY 4.0.

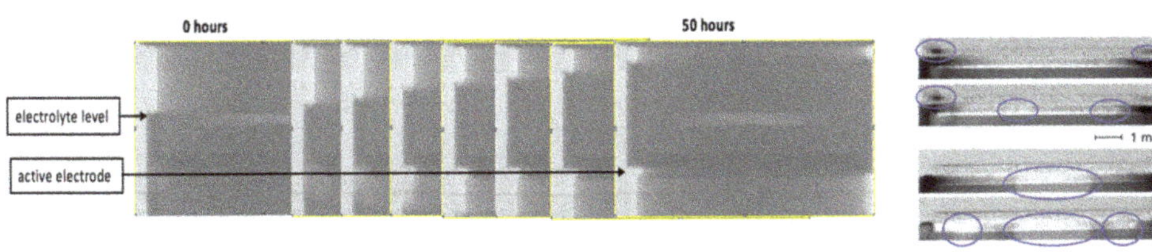

Figure 11.13. Left hand side: redistribution of electrolyte away from the active volume observed during cell operation; right hand side: local redistributions in difference images of a half-cell during operation displaying pockets (dark areas) taking up electrolyte, and bright areas evolving where gas bubbles are formed in the electrolyte. Reproduced from [12]. CC BY 4.0.

Also, the distribution and redistribution of electrolyte during battery cycling is of high relevance and can be observed with *operando* neutron imaging. An example is given by observations published by Nazer *et al* [17] which exposed a commercial cylindrical battery to neutron imaging during cycling and the redistribution of electrolyte between the active volume and the inactive hollow central column of the battery was observed (figure 11.12). The filling state is found to (anti)correlate directly with the cell voltage (inset left hand side figure 11.12) in the central column and the active battery volume, respectively.

Another measurement by Lacatusu *et al* [12] with custom designed half-cells to study the (de)lithiation of a thick carbon electrode (compare with section 11.2) was able to observe a split of the electrolyte filling at the active layer of the cell lifting the upper part of the electrolyte against gravity away from the active volume (figure 11.13, left). Similar measurements also demonstrate the observation of gas bubbles forming in the electrolyte during operation (figure 11.13, right).

11.3.2 Physico-chemical properties of electrolytes

While the distribution of the electrolyte plays a significant role, as explained in the previous section, the performance of LIBs is also closely related to the physico-chemical properties of the electrolyte. In particular, the degradation of capacity over time can be related to degradation processes in the electrolyte. On the other hand, the loss of capacity at low temperatures is usually larger than explained by simple activation processes and probably implies a partial change of the electrolyte aggregate state. To understand these limitations, the development of *in situ* analysis methods are highly demanded. Classical vibrational spectroscopy methods (e.g. Raman or Fourier transform infrared spectroscopy) are limited by the required optical access. Some studies have been reported using specialist set-ups such as the insertion of hollow core optical fibres in pouch cells to extract electrolyte during operation [18], or the use of custom designed cells with an optical window [19]. However, when the use of an unmodified battery design is desired, photon-based vibrational spectroscopy methods can only be used in post mortem studies.

In this context, neutron spectroscopy has an interesting potential for probing the characteristics of electrolytes in ways that are not accessible by other methods. It is well known that the scattering of neutrons by hydrogen atoms has an important inelastic component, which renders this interaction dependent on the motion of hydrogen atoms. The longer-range motions, such as translational diffusion modes, are typically accessed by quasi-elastic neutron scattering (QENS), which can be used, for example, for probing the molecular motions in solid polymer electrolytes [20]. Inelastic neutron scattering (INS), involving larger energy transfers (a few tens or hundreds of meV), in contrast, can be used to probe the more localized motions corresponding to the molecular vibration spectrum. INS is traditionally used for measuring vibrational spectra in crystals, but its use has also been demonstrated to analyze the vibrational spectrum of hydrogen in storage materials [21]. An interesting aspect of the inelastic interaction of neutrons with hydrogen atoms is that its magnitude is sufficient to appear as a prominent feature of the transmission spectrum. In consequence, the information on hydrogen atom motions usually obtained with QENS or INS can also be accessed in wavelength-resolved neutron imaging experiments. This so-called *spectroscopic neutron imaging* methodology has the advantage over traditional QENS/INS that spatially resolved information is obtained.

As an example, we present here the results obtained from ToF neutron imaging on several battery solvent samples [22], including one sample that had suffered from partial solidification of the electrolyte. The hydrogen cross-section shown in figure 11.14 corresponds to solvents typically used in LIBes electrolytes. The case of the EC:DEC solvent, which suffered partial solidification and sedimentation, as seen from the images, allows demonstration of the potential of spectroscopic neutron imaging to analyze electrolyte changes in a spatially resolved fashion.

In the lower section of the sample, where the sedimentation occurred, the spectrum (normalized to the cross-section of hydrogen in polyethylene) is much flatter and approaches that of its ethylene carbonate (EC) component. There are two possible reasons for this: first, EC is solid at the temperature of 17 °C at which the experiments were conducted. Second, partial solidification and sedimentation possibly results in the enrichment of the sedimentation region by the component having the highest melting point (EC in this case). The simulation of the neutron transmission spectra in

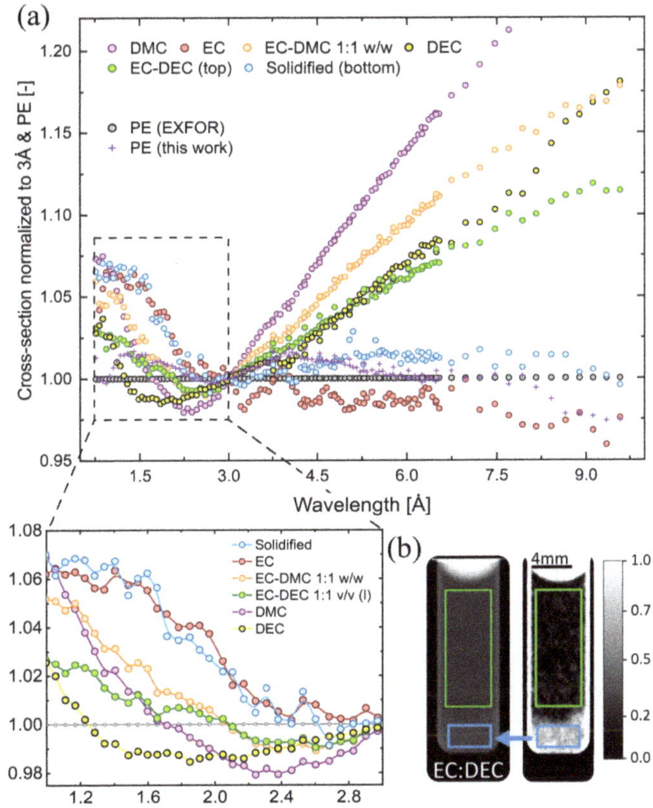

Figure 11.14. Attenuation cross-section of H atoms in different battery solvents normalized to the cross-section of H in polyethylene and to their value at 3 Å. One sample (EC:DEC) has a sedimented region at the bottom, showing a different cross-section behaviour. Reproduced from [22]. Open access CC BY 4.0.

different cases allows obtaining more insight. According to the simulations, the solidification alone is not sufficient to explain the large change in cross-section for the solidified region. However, assuming an important enrichment in the EC component (85% instead of 50% in the original mixture), the change of neutron cross-section can be explained. The model used for this is only based on the type of groups (CH_2 or CH_3) containing the hydrogen atoms. In consequence, neutron spectroscopic imaging is expected to be suitable to probe composition changes in electrolyte as a function of degradation, but also partial solidification in low-temperature conditions.

The major interest of neutron spectroscopic imaging for the research on battery electrolytes is its fully noninvasive aspect. Neutron-based experiments do not require optical access and are even compatible with hard metal casings as neutrons have a mean free path of approximately 10 mm in stainless steel and 100 mm in aluminum. The most important limitation is the neutron flux, which limits the accessible spatial resolution. However, upcoming facilities with high flux and ToF capabilities such as the European Spallation Source (ESS) are expected to significantly push this limitation.

References

[1] Song B, Dhiman I, Carothers J C, Veith G M, Liu J, Bilheux H Z and Huq A 2019 *ACS Energy Lett.* **4** 2402
[2] Abraham K M and Jiang Z 1996 *J. Electrochem. Soc.* **143** 1
[3] Nanda J, Bilheux H, Voisin S, Veith G M, Archibald R, Walker L, Allu S, Dudney N J and Pannala S 2012 *J. Phys. Chem.* C **116** 8401
[4] Sun F *et al* 2019 *ACS Energy Lett.* **4** 306
[5] Siegel J B, Lin X, Stefanopoulou A G, Hussey D S, Jacobson D L and Gorsich D 2011 *J. Electrochem. Soc.* **158** A523
[6] Siegel J B, Stefanopoulou A G, Hagans P, Ding Y and Gorsich D 2013 *J. Electrochem. Soc.* **160** A1031
[7] Ziesche R F *et al* 2020 *Nat. Commun.* **11** 777
[8] Nie Z, Ong S, Hussey D S, LaManna J M, Jacobson D L and Koenig G M 2020 *Mol. Syst. Design Eng.* **5** 245
[9] Vitoux L, Reichardt M, Sallard S, Novák P, Sheptyakov D and Villevieille C 2018 *Front. Energy Res.* **6** 76
[10] Kino K, Yonemura M, Ishikawa Y and Kamiyama T 2016 *Solid State Ionics* **288** 257
[11] Ziesche R F, Tremsin A S, Huang C, Tan C, Grant P S, Storm M, Brett D J L, Shearing P R and Kockelmann W 2020 *J. Imaging* **6** 136
[12] Lăcătuşu M-E *et al* 2021 A multimodal operando neutron study of the phase evolution in a graphite electrode arXiv:2104.03564v1 [cond-mat.mtrl-sci]
[13] Knoche T, Zinth V, Schulz M, Schnell J, Gilles R and Reinhart G 2016 *J. Power Sources* **331** 267
[14] Habedank J B, Günter F J, Billot N, Gilles R, Neuwirth T, Reinhart G and Zaeh M F 2019 *Int. J. Adv. Manuf. Technol.* **102** 2769
[15] Weydanz W J, Reisenweber H, Gottschalk A, Schulz M, Knoche T, Reinhart G, Masuch M, Franke J and Gilles R 2018 *J. Power Sources* **380** 126
[16] Günter F J, Rössler S, Schulz M, Braunwarth W, Gilles R and Reinhart G 2020 *Energy Technol.* **8** 1801108
[17] Nazer N S, Strobl M, Kaestner A, Vie P J S and Yartys V A 2022 *Electrochim. Acta* **427** 140793
[18] Miele E, Dose W M, Manyakin I, Frosz M H, Ruff Z, De Volder M F L, Grey C P, Baumberg J J and Euser T G 2022 *Nat. Commun.* **13** 1651
[19] Zhou Y, Doerrer C, Kasemchainan J, Bruce P G, Pasta M and Hardwick L J 2020 *Batteries & Supercaps* **3** 647
[20] Sinha K, Wang W, Winey K I and Maranas J K 2012 *Macromolecules* **45** 4354
[21] Borgschulte A, Terreni J, Billeter E, Daemen L, Cheng Y, Pandey A, Łodziana Z, Hemley R J and Ramirez-Cuesta A J 2020 *Proc. Natl Acad. Sci.* **117** 4021
[22] Carreón Ruiz R *et al* 2023 *Mater. Today Adv.* **19** 100405

IOP Publishing

Neutron Imaging
From applied materials science to industry
Markus Strobl and Eberhard Lehmann

Chapter 12

Hydrogen economy

Andreas Borgschulte and Pavel Trtik

Providing sufficient energy to everybody is a prerequisite for civilized human life: the relationship between human development index and per capita primary energy consumption suggests that at least 4 kW per person consumption is required for people to achieve a high level of development sustaining food, health, education and wealth [1, 2]. This sums up to nearly 30 TW, if all seven billion people of the Earth were to demand this, twice the current world power consumption (15 TW in 2018 [3]). Energy production distribution and conversion is thus big business: global net imports for fossil fuels reached around 1.9 trillion $ in 2015 (according to the World Trade Organization (WTO) 2017) [4]. The corresponding conversion machines belong to the largest human-made infrastructures ever built. The currently largest oil refinery, the Jamnagar Refinery, India, has a capacity of 1.24 million barrels of oil per day, and covers over 30 km^2 of land. To sustain the world with food, nitrogen fertilization is obligatory. The world energy consumption of the Haber–Bosch process, in which the fertilizer precursor NH_3 is produced, is about 1%–2%. A typical ammonia production plant produces up to 3000 t NH_3 per day [5] in reactors of up to 20 m length [6] (see also figure 12.1). The production yield and efficiency hinge on details of atomistic reactions on the catalysts' surfaces as well as on macroscopic transport phenomena in the reactor [5]. Accordingly, a multitude of concepts, techniques, and analysis/characterization methods covering length scales from nm to several meters are used to support and improve current technology (figure 12.1). Ammonia can be stored safely by absorption in ionic salts. This application is a good example, where neutron imaging can help improving the engineering of the storage device [7].

The situation is aggravated because of the urgent need to transform the current energy scenario based on fossil fuels to a renewable one [8]. In this respect, ammonia production is a good example. Currently, various new ideas for novel, 'greener' synthesis routes are proposed [9] with mostly very low technology readiness levels (TRLs). Only after successful upscaling and repetitive optimization of process steps on the microscopic as

doi:10.1088/978-0-7503-3495-2ch12

© IOP Publishing Ltd 2024

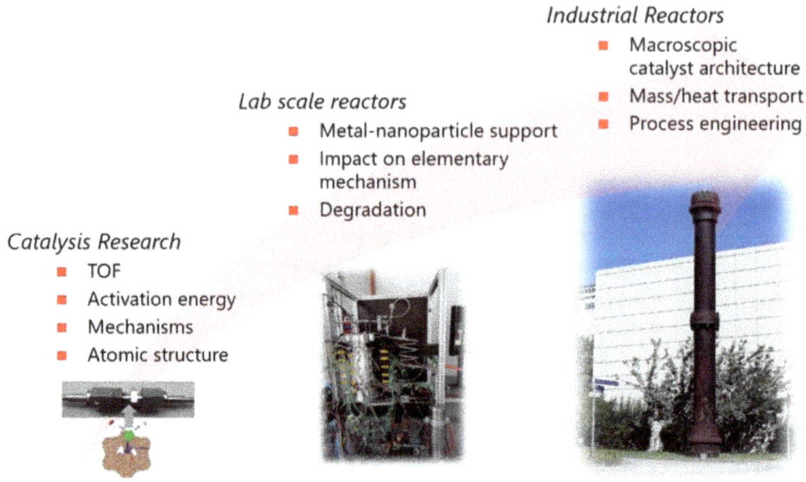

Figure 12.1. Research and development levels of chemical energy conversion: from catalysis research focusing on microscopic phenomena, lab-scale reactors laying the basis of the industrial catalysts and process management, to the set-up and operation of large-scale industrial reactors. The length scales vary over orders of magnitude from nm to several meters. Right image: this Ammoniak Reaktor BASF image has been obtained by the authors from the Wikimedia website where it was made available under a CC BY-SA 3.0 licence. It is included within this book on that basis. It is attributed to Drahkrub.

well as macroscopic level requiring the above-mentioned various methods, can the overall success and the CO_2 neutrality of a novel route be reliably assessed.

In this chapter, we cannot review the full body of modern catalysis and process technology for energy applications; but we want to outline how neutrons can bridge the wide range of length scales and different scientific questions for energy research. Figure 12.2 illustrates the outline along the example of CO_2 methanation on Ni-catalysts. This process is an important ingredient in the power-to-gas scenario as a seasonal storage concept for solar energy, which will be discussed in detail later. The figure depicts that neutron analysis techniques can contribute to R&D on the microscopic, mesoscopic, and macroscopic levels: inelastic neutron scattering reveals reaction mechanisms [10, 11], neutron imaging visualizes diffusion processes in catalyst pellets [12], and neutrons quantify mass transport in lab-scale reactors [13], respectively. The explanation of the results will nevertheless need some reminder on the background of the processes and corresponding scientific questions to be addressed by neutrons. We will tackle:

- hydrogen storage (section 12.1),
- catalyzed reactions such as CO_2 reduction (section 12.2).

The fundamental properties of the neutron, i.e., its (rest) mass, energy (wavelength), and spin are utilized in a number of analysis methods such as (spin-dependent) neutron powder diffraction (NPD), small angle neutron scattering (SANS) [108] (please link), inelastic neutron scattering (INS) [15], quasi-elastic neutron scattering [16], and neutron radiography and reflectometry.

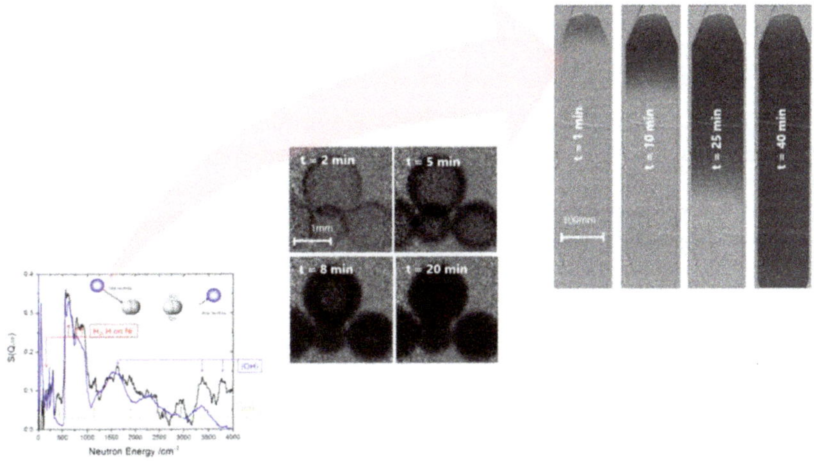

Figure 12.2. Neutrons for energy research. Left panel: inelastic neutron scattering reveals reaction mechanisms. Reproduced from [14]. CC BY 4.0. Middle panel: neutron imaging visualizes diffusion processes in catalyst pellets. Reprinted with permission from [12]. Copyright (2018) American Chemical Society. Right panel: neutrons quantify mass transport in lab-scale reactors. Reproduced from [13] with permission from the Royal Society of Chemistry.

The main focus will be laid on neutron imaging and related methods, while neutron diffraction and spectroscopy are only briefly covered, as excellent reviews and books exist on the matter.

12.1 Hydrogen storage

Hydrogen has the potential of being the energy vector of the future, replacing the fossil fuels used today by storing the in principle sufficiently available solar energy. However, hydrogen is a gas with very low density at room temperature. Thus, one of the challenges is to improve the relatively low gravimetric and volumetric energy density of high-pressure gas cylinders. The common conception is that consumers will not accept diminished performance compared to their fossil fuel powered vehicles, in particular cars, and that any replacement must therefore at least match the cruising range, refuelling time, durability, price and safety. This is a great challenge, which still awaits a solution [17]. Many ideas and systems based on various physical and chemical interactions have been tested, which might be divided into the following categories [18]:

- physical containment (i.e., compression);
- liquefaction;
- physisorption (e.g., adsorption of H_2 onto the surface of highly porous materials) chemical bonding (e.g., metal hydrides).

In this chapter, we will focus on the storage of hydrogen in matter, i.e., by physisorption and chemical bonding only (figure 12.3). Here, neutron methods are particularly suited to support research and development. For the other methods, we refer to literature [17, 19–21].

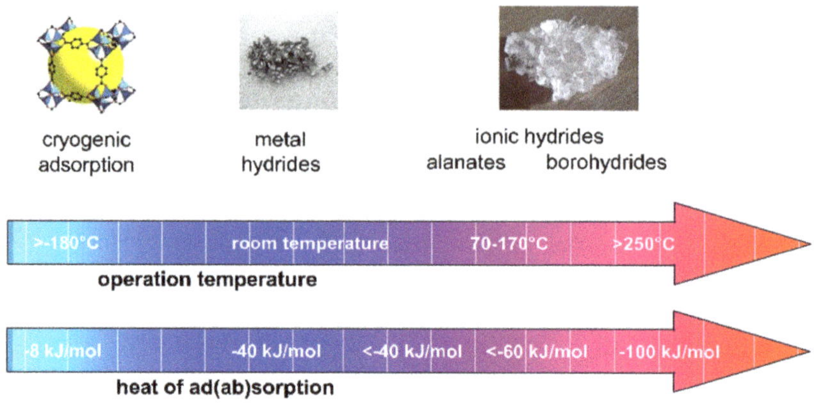

Figure 12.3. Operation temperature for an ambient pressure device using the depicted materials and the corresponding enthalpy range. Reproduced from [19] with permission from the Royal Society of Chemistry.

12.1.1 Hydrogen storage in metal hydrides

Metal hydrides based on transition metals and alloys are used in specific applications as storage materials. These metal hydrides are well established and serve as fast and reliable hydrogen storage materials in (stationary) hydrogen reservoirs [22], and as hydride electrode materials in rechargeable nickel metal hydride batteries [23–26]. Hydrogen absorption from the gas phase as well as from the electrolyte is possible (see, e.g., reference [26]). However, the maximum storage capacity in these alloys is typically less than 2 mass% [20, 27]. Complex hydrides exceed the gravimetric hydrogen density of transition metal hydrides by one order of magnitude, but still suffer from various drawbacks such as sluggish kinetics at low temperatures [20, 28, 107]. The physics and chemistry in complex hydrides differs significantly from that of intermetallic hydrides (figure 12.4).

Eventually, the hydrogen storage material will be placed in a macroscopic container. Although the basic properties such as equilibrium pressure and heat of ab/desorption are defined by the intrinsic properties, kinetics of a macroscopic material is controlled by complex phenomena on the microscale as well as on the macroscale by elementary reactions as well as by mass and heat transport. Here, imaging techniques are particularly important. Neutron imaging and nuclear magnetic resonance imaging are practically the only methods, which can visualize hydrogen directly with neutron imaging having some advantages over MRI because of the dense metallic systems typically used. We highlight two aspects of neutron imaging in the next two sections.

Despite of its relevance in science and technology, the qualitative and quantitative determination of hydrogen in matter is challenging [29]. The standard approaches are volumetric or gravimetric [30, 31] techniques to follow the hydrogenation reaction and to record pressure–concentration (p–c) isotherms. Neither technique can deliver spatial information, as is possible by neutron imaging. The interaction of neutrons with matter (scattering and absorption) is isotope-selective. The interaction of neutrons with many types of materials is relatively weak, while neutrons have a large (scattering) cross-section

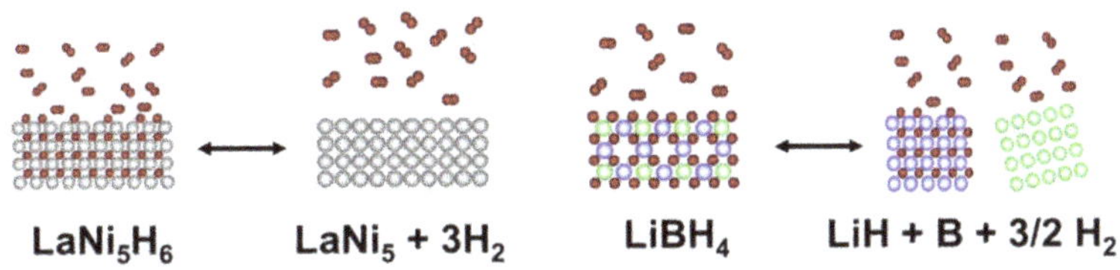

Figure 12.4. Sketch of hydrogen sorption in (intermetallic) metal hydrides such as $LaNi_5H_6$ (left) and in complex hydrides such as $LiBH_4$ (right).

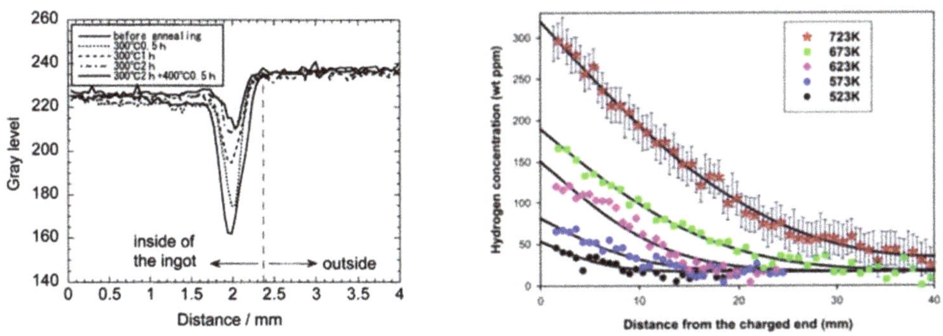

Figure 12.5. Left panel: hydrogen profiling in a Mg–Ni ingot by neutron imaging as charged and after annealing. Reprinted from [32], Copyright (2000), with permission from Elsevier. Right panel: depth profile of hydrogen in Zr–2.5% Nb annealed at various temperatures obtained from neutron imaging. Reprinted from [35], Copyright (2021), with permission from Elsevier.

with hydrogen. Thus, the scattering of neutrons by a sample can be traced back to the absolute amount of hydrogen in it. Realized as an imaging method, the time dependent distribution of hydrogen can be determined with a spatial resolution in the order of several tens micrometers and [32], and even the 3D distribution of hydrogen in Ti-alloys using neutron tomography [18]. As a demonstration of the technique, Sakaguchi *et al* followed the diffusion of hydrogen into Mg–Ni alloy powders and ingots [18, 33]. The method is compatible to high pressures and temperatures in an operating chemical reactor [34].

Figure 12.5 shows grey level line scans of neutron images taken perpendicular to the surface of a Mg_2Ni ingot after exposure to hydrogen and subsequent annealing at elevated temperatures. At 300 °C, an averaged diffusion length of 500 μm over the annealing time of 2 h is calculated from a bulk diffusion coefficient of 3×10^{-11} m^2 s^{-1}. The absence of a clear shift of the peak indicates that hydrogen desorbs faster from the surface than it diffuses into the ingot [32].

In general, investigation of diffusion phenomena by neutron imaging only makes sense with macroscopic object and diffusion lengths of tens of micrometers.

Shukla *et al* studied the hydrogen diffusivity in 40 mm long Zr–2.5% Nb alloy pressure tubes [35]. Due to the high diffusivity of hydrogen in such alloys, a several mm large hydrogen gradient in the tube is formed and observed by neutron imaging (figure 12.5). Modelling the concentration profiles yields the temperature dependent diffusion coefficient with high accuracy.

A great advantage of neutron imaging is that it is a **quantitative** method with a concentration range from the ppm level [36] to high concentration. However, although diffusion of hydrogen is generally fast compared to other light elements, the effective diffusion lengths observed in a typical materials is only a few tens of micrometers impeding the use of neutron imaging for many pertinent questions regarding transport phenomena in hydrogen storage materials and membranes.

A critical phenomenon impeding hydrogen sorption kinetics in complex hydrides is the formation/decomposition of various phases and subsequent segregation during cycling [28]. Most hydrogen storage materials such as Ti-doped $NaAlH_4$ consist of agglomerates of micrometer sized particles. It was found by electron microscopy methods that the distribution of elements changes on the microscale during cycling [37], which is below the spatial resolution of typical neutron imaging. In the case of the reactive hydride composites such as $LiBH_4$–MgH_2, the segregation of compounds takes place on a macroscopic scale, because $LiBH_4$ becomes liquid at operation temperatures [38]. Here, the *operando* possibility of neutron imaging [39] is very insightful [40, 41].

The use of *operando* neutron imaging for recording pressure–composition isotherms requires a specific measurement protocol. In the following, we describe a typical measurement procedure using gravimetric techniques as a general blueprint for pcT-measurements (figure 12.6). The absorption/desorption isotherms are performed by adding/removing an aliquot of H_2, allowing equilibrium to be re-established, and measuring the pressure and mass change. The adding/release of hydrogen is supplied using a pressure controller similar to that of a Sieverts apparatus. Two serially connected high-pressure valves are used to add/remove a chosen amount of hydrogen from the

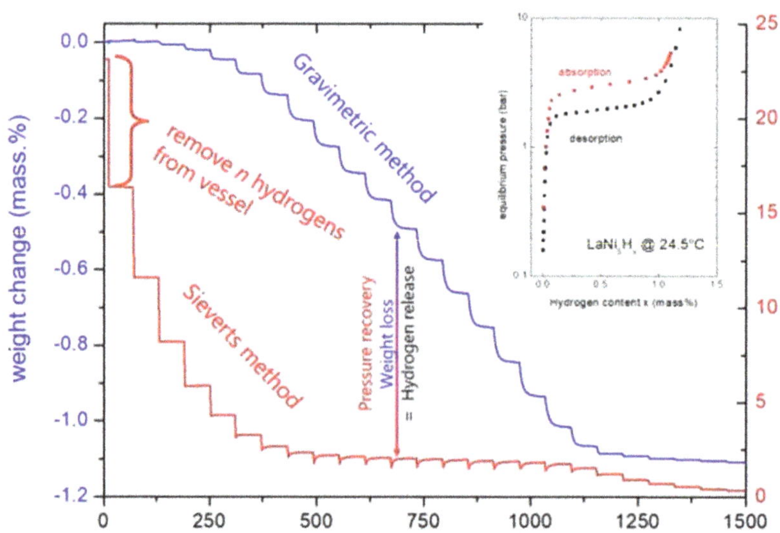

Figure 12.6. Hydrogen desorption from $LaNi_5H_6$ measured gravimetrically (weight change in blue) and volumetrically (pressure dependence in red) using a magnetic suspension balance (Rubotherm, Bochum, Germany). From this, a pressure–composition isotherm is derived (inset, shown for absorption and desorption).

gravimetric chamber. Outside the switching cycles the gravimetric chamber is closed and the recorded pressure change is due to ab/desorption of hydrogen from the sample. Figure 12.6 presents a typical measurement of hydrogen desorption from $LaNi_5H_{5.5}$ showing the instantaneous pressure steps induced by hydrogen release via the pressure controller and subsequent pressure increase due to hydrogen desorption from the sample. Simultaneously, the weight change is recorded, which gives the mass after subtraction of the buoyancy contribution. From this, a pressure–composition isotherm is derived.

To derive p–c isotherms from neutron images, a similar stepwise pressure increase and decrease for absorption is applied, and the amount of hydrogen extracted from the neutron attenuation. To accelerate measurement time and to minimize measurement errors, Nikolic *et al* [42] proposed a combinatorial approach, in which a multitude of samples and/or parameters are measured simultaneously. The method is inspired by the optical hydrogenography approach by Gremaud *et al* [43]. Here, thin film gradients are exposed to hydrogen and analyzed simultaneously by optical detection. However, the transferability of the thin film results to a bulk sample is questionable; unfortunately, bulk samples cannot be probed by the special optical detection. Neutrons, on the other hand, are perfect for bulk samples; instead of a gradient thin film sample, neutron scattering can be applied on individual bulk samples placed in individual reactors (see figure 12.7).

Figure 12.8 presents the pressure–composition isotherms derived from the neutron images shown in figure 12.7 at the thermal neutron beamline NEUTRA [44] at the Swiss Neutron Spallation Source (SINQ), PSI (Switzerland) [45]. The shape of the pcTs and measured plateau pressures of $LaNi_5H_x$ and $LaNi_4CoH_x$ match literature data very well for higher temperature, while the low-temperature data exhibit steps typical for non-equilibrium phenomena, i.e., the measurement steps did not reach equilibrium. Typically, the measurement of a full isotherm of a metal hydride such as $LaNi_5H_x$ takes 12 h and more, see also figure 12.6. It is important to note that the time to reach equilibrium is a materials parameter (kinetics). This is a general challenge of

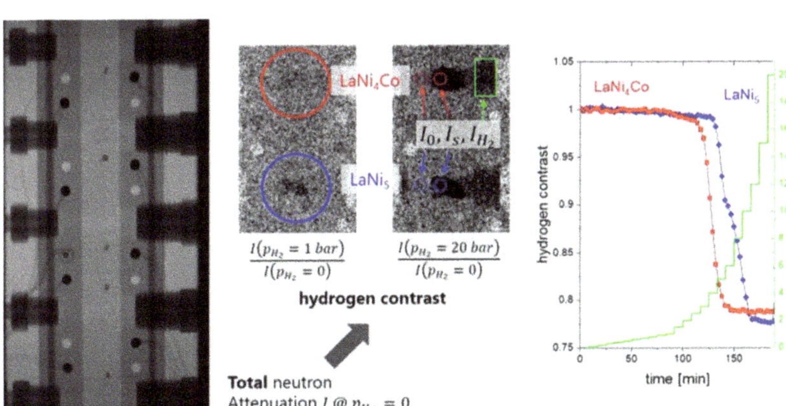

Figure 12.7. Left image: neutron image of a multi-reactor assembly at zero hydrogen pressure. Middle images are the neutron hydrogen contrast images of two reactors containing a $LaNi_5$ and $LaNi_4Co$ sample, respectively.

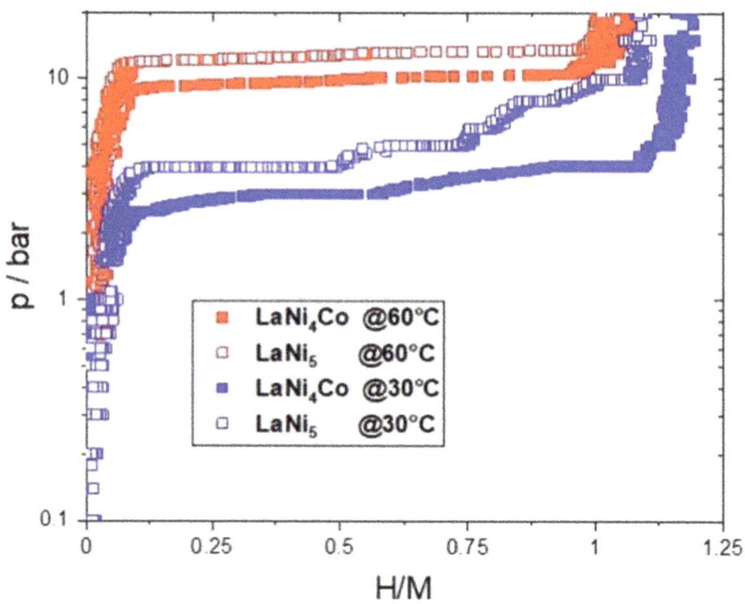

Figure 12.8. Absorption pressure–composition isotherms of $LaNi_5$ and $LaNi_4Co$ derived by neutron imaging.

measuring equilibrium thermodynamics with equipment in large-scale facilities: required long measurements times quickly exceed the granted beam time.

An important advantage of neutron imaging is that the detection of hydrogen is not related to its chemical state: neutrons will be scattered by bulk hydrogen chemically bound in any form as well as hydrogen adsorbed to the surface. In contrast, volumetric and gravimetric methods detect only the amount of exchanged hydrogen is detectable.

12.1.2 Upscaling of metal hydride storage tanks

The design and construction of a large-scale hydrogen storage in metal hydrides is an art of its own. The external conditions set the demanding benchmarks: referring to hydrogen storage in cars, the following performance parameters have to be met [17, 19, 28]:

- storage of approximately 5 kg hydrogen;
- (re-) fuelling time around 2 (5) min;
- desorption pressure around room temperature (at least below 80 °C) above 1 bar.

These benchmarks must of course be met by the intrinsic material properties. This means that currently only intermetallic hydrides are suitable. With including a minimum gravimetric density, which we omitted here, there is currently no material fulfilling all requirements, because the gravimetric hydrogen storage density of intermetallic hydrides is too low [28]. If the temperature condition is relaxed, which has its cause in the use of ambient heat or low-temperature waste heat from fuel cells, more materials options exist (lightweight complex hydrides). However, the energy efficiency of storage is then seriously deteriorated (extra heat generation is needed), and the corresponding complex tank design diminishes the effective gravimetric density [46, 47]. Still, the demonstrated prototypes yield information on the general set-up, limits and remaining technical challenges to overcome.

The main difference of a realistic prototype to a lab-scale hydride reactor is the heat evolution/release during fuelling. From the performance parameters we estimate the heat of hydride formation to be of the order of −30 to −40 kJ (mol $H_2)^{-1}$ [28]. Exchanging 5 kg of hydrogen results in the evolution/release of more than 75 MJ of heat corresponding to a heating/cooling power in the MW-range, if exchanged within minutes. Sophisticated heat management is thus required, less critical for intermetallic hydrides [48], but crucial for lightweight complex hydrides such as $NaAlH_4$ [46]. Corresponding scientific/technical questions include:

- what is the optimum shape and size of a single storage bed? (e.g., rod/disk shaped, additional heat conducting elements) what is the form of the hydride material? (e.g., powder, pellets);
- what is the property determining shape and form? (e.g., heat conductivity, hydrogen transport);
- does degradation take place during cycling? (e.g., segregation, local compaction of the fill, cracks);
- should a specific process during fuelling be introduced? (e.g., temperature/pressure ramps).

Obviously, a method allowing a view into a tank during hydrogen cycling is of greatest help [34, 49–53]. Figure 12.9 is an illustrative example of the neutron

Figure 12.9. Sagittal slices through the tomographic reconstructions of a $LaNi_5$ filled cylinder. (a) The total attenuation coefficients of (a) the empty bed, (b) the bed filled with 94.5 l of D, and (c) the bed after desorbing 60 l of D. Subtracting the empty bed attenuation coefficient yields the D content of the bed: (d) after the absorption of 100 l and (e) after desorption of 60 l of D_2. The grey/dark rod in the middle is a porous Al filter tube for hydrogen gas distribution. Reprinted from [52], Copyright (2010), with permission from Elsevier.

tomography (cross-section) images of a $LaNi_5$ filled cylinder during D_2 cycling. The uneven distribution of hydrogen in the metal hydride as a consequence of the discharging process is clearly visible.

With complex p-metal hydrides such as $LiBH_4$ being liquid at operation temperatures, macroscopic segregation phenomena can occur. Karimi *et al* [41] observed an uneven hydrogen distribution in the reactive hydride composite Li-doped MgH_2–$2LiBH_4$ over various centimeters using neutron tomography.

They allocate the finding to the motion of the liquid $LiBH_4$ induced by wetting of the surrounding solid matrix framework. From this, the authors were able to attribute the degradation of the hydrogen capacity in this material over the cycling procedure to this non-reversible phase separation (see discussion above) as previously reported by Jepsen [54].

Currently, there are only a few hydrogen storage tanks based on metal hydrides, on one hand due to the lack of proper materials, and due to the currently not commercially viable hydrogen storage on the other. It is foreseeable that this will change in the future. In particular for large-scale tanks, neutron imaging will be the only technique to analyze the hydrogen distribution.

One of the few commercial applications of metal hydrides is the metal hydride battery [55]. The simplified operating principle is shuffling of hydrogen from a metal hydride to an oxide electrode. With the high contrast for neutrons, neutron tomography and radiography is the ideal tool to visualize this phenomenon [56].

12.1.3 Hydrogen storage in porous media

The main drawbacks of hydrogen storage in metal hydrides are the kinetic barriers associated with dissociation and diffusion of atomic hydrogen in the case of intermetallic hydrides [28, 55], and formation of hydrogen-containing pseudo anions in the case of complex hydrides [28, 56]. It appears to be obvious to store hydrogen in molecular form by physisorption to high-surface-area materials. Porous media with exceptionally high surface area have been developed [33], among them are zeolites [57], carbon nanostructures [58], and metal–organic frameworks (MOFs) [59–61]. As the binding takes place exclusively at the surface, the hydrogen storage capacity at cryogenic temperatures (i.e., at a surface coverage of one) is directly related to the surface area [58, 60]. Both properties are usually measured volumetrically and/or gravimetrically similar to the method described in section 12.2.2 using hydrogen and a different probe molecule (N_2, He), respectively [33, 62, 63]. There are deviations from the general linear trend, if the heat of physisorption differs significantly from the van der Waals standard value and if multilayer adsorption occurs [62]. Vibrational spectroscopy such as infrared spectroscopy (IR) is ideal to study adsorption phenomena. However, IR is possible only in case of an induced polarization in the hydrogen molecule (e.g., in zeolites) [64], and in general optical methods including Raman have relatively low intensity of the sought hydrogen vibration due to the strong light absorption by the porous medium. Thus neutron vibrational spectroscopy, i.e., INS, has become the standard method for hydrogen adsorption in porous media (carbon: [65]; zeolites: [66], MOFs: [59, 67], and many more).

In principle, neutron imaging should be ideal for probing large-scale hydrogen storage tanks based on porous media, similar to the experiments on metal hydride tanks described in section 12.1.2. However, the process management of hydrogen storage tanks based on porous media is much simpler than that of metal hydride tanks. Thus currently, technical questions focus more on the materials aspects of hydrogen storage in porous media (see above) than on up-scaling, where strong impact from neutron imaging is expected in the future.

12.1.4 Hydrogen in thin films

Hydride thin films are the small sisters of hydride bulk materials and hydride storage tanks. Due to the small dimension, various phenomena such as heat and mass transport can be more easily measured, controlled, and modelled. A great variety of thin film preparation and characterization methods exist, which facilitate the preparation of many new materials and modifications thereof. Hydride thin films are thus often taken as model systems to explore new materials, or to scrutinize specific effects observed in larger, less defined systems earlier on. Apart from their use as model systems, there are applications, which rely on the use of thin films such as hydrogen sensors [68].

In many cases, the hydrogen content in such thin films is derived indirectly, by, e.g., means of optical and electrical methods [43, 69]. The verification and/or calibration of such indirect methods by a direct method is crucial. The small dimension of the sample is challenging for many direct methods, including volumetric and gravimetric methods. An exception may be the mass-based evaluation of hydrogen contents in thin films by vibrating quartz microbalance. However, nuclear methods [70] such as Rutherford backscattering [71] and resonant nuclear reaction analysis [72] are advantageous over volumetric and gravimetric methods. In principle, also neutron imaging is usable. Ott *et al* quantified the amount of water in amorphous aluminum oxide films with a thickness of $\geqslant 50\ \mu m$ by neutron imaging [73]. However, most thin films are at least two orders of magnitude thinner. For this dimension, neutron reflectometry has become the standard solution [68, 74–76].

With neutron applications limited by beam time, neutron reflectometry on thin films is often used as calibration to calibrate simpler methods such as optical spectroscopy [68].

12.2 Catalysis

With the advent of a worldwide renewable energy scenario, the production of hydrocarbons from CO_2 and renewable hydrogen, also called power to X, is receiving increasing attention [77–81]. Many of the underlying hydrogenation reactions have long been known, e.g., the Sabatier reaction ('power to gas')

$$4H_2 + CO_2 \rightarrow CH_4 + 2H_2O \qquad (12.1)$$

was discovered by Paul Sabatier and Jean-Baptiste Senderens as early as 1897 [82]. The Fischer–Tropsch reaction to produce liquid fuels from CO and hydrogen [83, 84]

$$(n + 2)H_2 + nCO \rightarrow C_nH_{2n+2} + nH_2O \quad (12.2)$$

was utilized by Nazi-Germany to produce synthetic fuels in the 1940s, and later by South Africa, in both cases to circumvent the oil embargo [85]. However, the challenge of the future lies in the development of a clean and energy-efficient process to produce a sustainable fuel. In this respect it is telling that although of pivotal importance in heterogeneous hydrogenation reactions, the amount of hydrogen on catalysts during reaction is seldom known. This impedes the determination of reaction mechanisms and development of novel catalysts on one hand, and complicates the optimization of large-scale reactors on the other. With the high selectivity for hydrogen, neutron methods can support both the study of phenomena on the atomic scale as well as on a macroscopic scale.

12.2.1 Elementary reactions on hydrogenation catalysts

Hydrogenation reactions such as the Sabatier reaction 1.5 and ammonia synthesis reaction [9]

$$3/2H_2 + 1/2N_2 \rightarrow NH_3 \quad (12.3)$$

take place on catalysts surfaces, i.e., the gaseous reactants are adsorbed, dissociate, recombine, and eventually the products desorb from the active surface. This so-called heterogeneous catalysis takes place in reactors at elevated temperatures and pressures (see also next section). The overall goal of research and development in catalysis is to find a process and corresponding catalyst with highest conversion yields at mildest conditions [86]. Modern catalysts are complex materials, consisting of various material classes (oxides, metals) structured on length scales from nano- to millimeters. Thus, their characterization is an art of its own, and a particular challenge is the *in situ* detection and quantification of chemical species adsorbed on them. Most methods are based on the interaction of photons and/or electrons.

The challenge of these experimental methods is the determination of the exact amount and dynamics of hydrogen in/on the material during reaction [29]. As hydrogen has only one electron, its detection by analytical tools based on the interaction with electrons is challenging: core-level spectroscopies such as x-ray photoelectron spectroscopy [87] and x-ray absorption [88, 89] cannot be used as a quantitative method for hydrogen, and hydrogen is nearly invisible for x-ray diffraction [90, 91]. A reliable probe must thus make use of the interaction with the proton, which is possible by nuclear magnetic resonance [19, 92, 93] and neutron techniques.

Given the much higher incoherent scattering cross-section of hydrogen (80.26 barn) relative to that of carbon (0.001 barn) and oxygen (0.0008 barn), INS spectra reflect the hydrogen partial phonon density of states of adsorbed hydrogen and hydrogen-containing molecules, which can be directly compared to the diffuse reflectance infrared Fourier transform spectroscopy (DRIFTS) spectra. Despite this promising advantage, INS can only be used post-mortem, because the measurement temperature has to be cryogenic temperatures (around 20 K) due the high scattering background at higher temperatures [11, 94]. To conserve the reaction state as well as possible, the reactor is quenched in liquid nitrogen and then further cooled down to 15 K [14].

DRIFTS spectra contain in most cases more details than the corresponding INS spectra. However, INS reveals some interesting aspects: under methanation conditions, chemisorbed hydrogen on metallic nickel co-exists with NiO [14]. These observations support a redox reaction mechanism of CO_2 reduction on Ni. In general, INS is complementary to DRIFTS. Parker and Lennon [11, 95] give an overview of the various systems relevant in catalysis for energy.

The INS measurements provide an indication that the amount of hydrogen on the catalyst depends on reaction conditions. To prove this, a true *operando* technique is required. Neutron radiography cannot distinguish (easily) between different states of hydrogen on the catalyst, but gives the quantitative amount. This was exploited to follow and quantify hydrogen-containing species in Cu/ZnO catalysts *operando* during methanol synthesis. In figure 12.10, the steady-state amount of hydrogen-containing intermediates on the catalyst is compared to the reaction yields of CO and methanol. Not surprisingly, hydrogen amount and methanol yield are closely related, as expected from simple considerations of the likely reaction mechanism. Additional time-resolved measurements indicate that these intermediates, despite being indispensable within the course of the reaction, slow down the overall reaction steps.

The outstanding catalytic properties of Cu/ZnO for methanol synthesis have ling been a matter of discussion. Various studies emphasized that the peculiar nanostructure of the Cu/ZnO catalyst formed during reduction is the origin of the reactive sites [96–98]. An alternative explanation is that ZnO and/or the ZnCu interface generated upon reduction is a hydrogen reservoir supplying the active sites bound to the transient intermediate with atomic hydrogen. To support this hypothesis, the concentration of H*

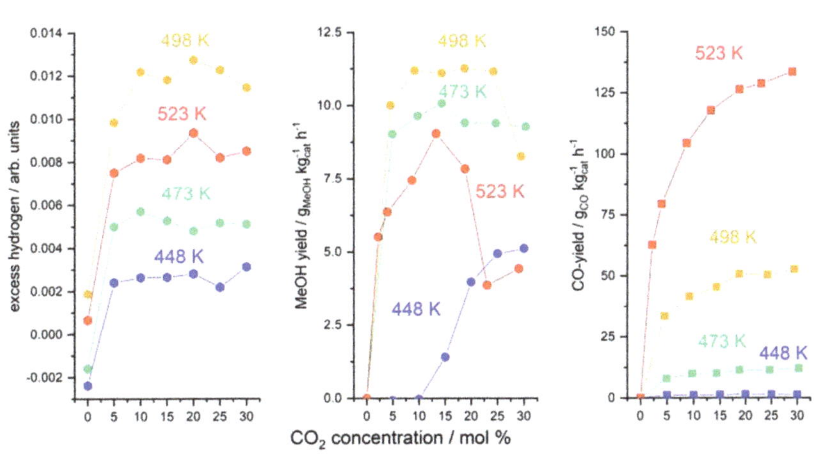

Figure 12.10. Left graph: amount of hydrogen-containing adsorbates ('excess hydrogen') on Cu/ZnO catalyst at 12 bar as a function of the CO_2 concentration and temperature derived from neutron radiography. Simultaneously, the MeOH-yield (middle graph) and CO-yield was determined by IR-gas analysis (right graph). Reproduced from [10] with permission from the Royal Society of Chemistry.

under methanol synthesis conditions should be measured, which is not possible by neutron imaging due to interference with other hydrogen-containing species.

However, probing the amount of hydrogen in the catalyst under pure hydrogen atmosphere gives some indication. Terreni *et al* [10] found an increase of the maximum hydrogen content with temperature. Such a behaviour is untypical for hydrogen chemisorption on surfaces, but typical in bulk absorption of hydrogen at low concentrations (solid solution) supporting the hydrogen spillover effect proposed above.

To further increase the accuracy of uptake experiments, the group combined neutron scattering with hydrogen–deuterium exchange [10, 12]. As discussed above, absolute hydrogen uptake measurements are error prone, because the amount of irreversible hydrogen adsorbed prior the measurement is usually unknown. In particular, hydroxyl groups and similar surface species are very stable, and can only be removed by heating at very high temperatures. For neutron HD experiments, the hydrogen flow to the sample, which has been equilibrated at a given temperature and pressure, is abruptly exchanged for deuterium at the same conditions and later back to hydrogen again. Deuterium has a considerably lower neutron attenuation coefficient $\sigma(_{tot}(D) = 7.64$ barn, compared to $\sigma_{tot}(H) = 82.35$ barn) and thus the neutron contrast is directly proportional to the total amount of hydrogen in the sample. This technique can be extended to a combinatorial method to measure hydrogen uptake in catalysts [42].

12.2.2 Mass transport in catalysts and reactors

An advantage of the *operando* neutron technique is the time-resolved imaging possibility, giving insights on mass transport in catalysts. General scientific questions related to mass transport in heterogeneous catalysis are as follows:

- mass transport on the atomistic level. This is usually probed by spectroscopic methods as discussed above as the spatial resolution of neutron imaging is in the micrometer range.
- mass transport on the mesoscale, e.g., in catalyst pellets. Here diffusion processes influence the kinetics [99], which can be influenced by the shape of the sample and the process management. Temporally- and spatially-resolved neutron imaging is ideal to probe this and can contribute to an optimization of catalysts.
- mass transport on the macroscale, e.g., the mass transport in reactors. The large penetration depth of neutrons allows the study of large reactor up to meter size. The performance of such reactors hinge on geometry and process management, i.e., external and heat mass transport. The distribution of hydrogen-containing reactants and products in the reactor is a consequence of these parameters, and its measurement will help optimization of reactor geometry as well as operation parameters.

There are numerous hydrogenation reactions and correspondent processes [86]; giving an overview of all goes beyond the scope of this chapter. Instead, we demonstrate the use of neutron imaging on specific reactions. One of the simplest catalytic process relevant for hydrogen storage is the conversion of ortho-(o-H_2) to parahydrogen (p-H_2) by paramagnetic materials [100]. The spin-state of mainly o-H_2 at room temperature is quenched during liquefaction resulting in an enrichment of at liquid hydrogen temperatures thermodynamically unstable o-H_2 [101]. The heat of conversion is higher than the heat of liquefaction, i.e., liquid o-H_2 is unstable, and a catalytic conversion process is included in the liquefaction process to convert o-H_2 into p-H_2 [102]. Due to the orthogonal spin moment, o-H_2 and p-H_2 have very different scattering cross-sections [103]. This fact allows visualization of the distribution of o-H_2 and p-H_2 on a ortho–para catalyst (figure 12.11, from [104]). The self-poisoning of the catalyst by the generation of the sought product p-H_2 leads to large inactive areas. The geometrical optimization of the catalyst improving mass transport inspired by the neutron imaging can enhance the conversion rates while minimizing the amount of material needed [104]. It is worth noting that the observed changes are on a macroscopic length scale perfectly suited to be followed by neutron imaging.

A more complex reaction to be mentioned here is the sorption enhanced methanation reaction relevant in the power-to-X scenarios. The concept of sorption enhanced catalysis relies on the use of an appropriate sorbent [105], which serves as an effective sink for specific products of the desired reaction, thereby pushing the reaction in the direction of choice.

To minimize diffusion path length, the sorbent is the support of the catalytically active metal nanoparticle, then called sorption catalyst. The model system for the methanation of CO_2 (equation (1.5)) is nickel nanoparticles dispersed within a zeolite [106]. However, unlike in ordinary catalysis, the reaction yield depends on the state of the sorbent [106]. The effect on catalytic activity was investigated extensively by means of spectroscopy similar to the examples described above [106], first ignoring the effect of microstructure (assuming an equal distribution of water everywhere in the sorption catalyst). In reality, catalysts are macroscopic objects, usually mm sized pellets or beads.

In zeolite based catalysts, Terreni *et al* could show by neutron imaging for this technique that macroscopic diffusion controls the reaction rate in catalyst pellets [12], a very well-known constraint in technical catalysts [99]. For this, the authors related the spatially resolved water uptake in single catalyst beads with the simultaneously measured reaction conversion of CO_2. To calibrate the system,

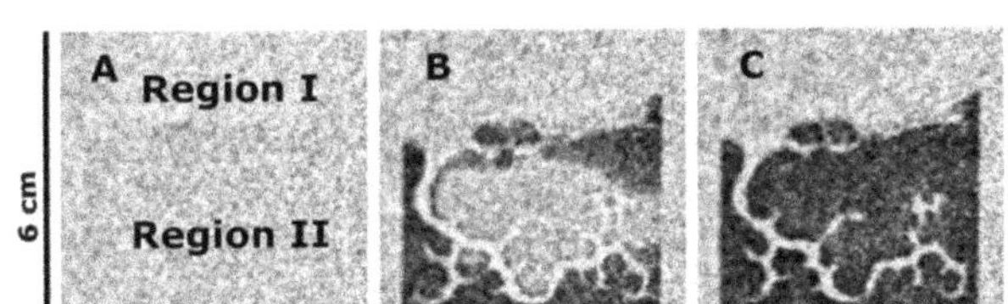

Figure 12.11. Neutron imaging of an ortho–para catalyst during the H_2 loading process at 15 K: no H_2 loaded (A) and adsorption onto the catalyst (B, C). Note the length scale of the images. Reprinted with permission from [104]. Copyright (2019) American Chemical Society.

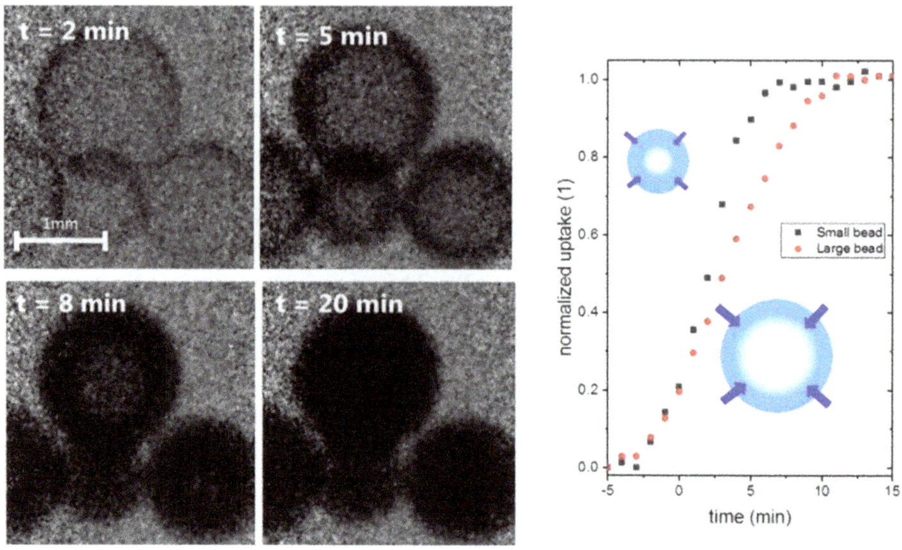

Figure 12.12. Left panel: time series of high-resolution neutron images of the adsorption of water from humidified hydrogen gas. Temperature was 50 °C, water partial pressure 6.5 mbar, total flow 100 ml min^{-1}. Right panel: the normalized uptake of the individual catalyst grains as derived from the log of the integrated neutron attenuation. Reprinted with permission from [12]. Copyright (2018) American Chemical Society.

the authors visualized the water uptake in the beads exposed to water steam (figure 12.12). Apart from being a proof of concept, the measurements demonstrate another important application: the quantification of adsorption of water or basically any hydrogen-containing molecule in adsorbents.

Before a practical realization of a large-scale reactor is possible, critical parameters must be determined to optimize the geometry of the design. An important difference to fixed bed reactors without sorption functions is that the amount of material (sorption catalyst) is significantly higher in sorption reactors, as one product (the water) remains in the bed. The reactor performance is thus linked to the absolute water adsorption capacity. However, this requires a careful definition of the dimensions of the reactor aligned to the planned operation conditions (for the related steam reforming sorption catalysis, see reference [105]). Due to the high neutron attenuation coefficient of hydrogen, the absorbed water in the sorption catalyst gives a high contrast in neutron radiography, making it an ideal tool to follow the formation and distribution of water *operando* in a large-scale reactor, from which guidelines for a design of large-scale reactors can be drawn.

The set-up shown in figure 12.13 may be seen as a general scheme for this kind of experiment. The reactor is placed in the neutron beam, and the attenuation spatially recorded. In parallel, the reactants and products are measured by IR-gas analysis. Even without sophisticated evaluation, the neutron images provide a qualitative understanding of the reaction dynamics: a reaction front establishes. Once the front arrives at the reactor outlet, the sorption catalyst is saturated everywhere. The reaction yield is almost 100% methane, while it immediately drops when reaching

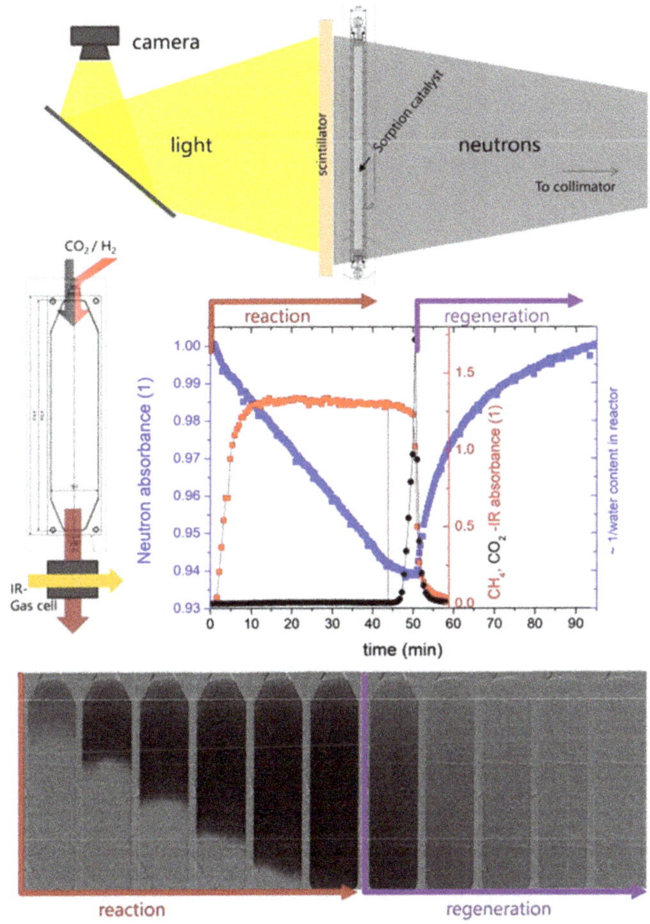

Figure 12.13. Top: experimental set-up used to measure the water content in sorption reactors at instrument NEUTRA at PSI. Middle left: drawing of the reactor with dimensions, and the gas fluxes and gas analysis attached to it. Middle right: experimental data as received from an evaluation of the neutron radiography (changes of neutron absorbance relative to dry reactor in blue) from a time series of pictures (selection shown below) and the corresponding Fourier transform infrared signal of the CO_2 (black dots) and CH_4 (red dots) leaving the reactor. Hydrogen flux is kept constant, during the reaction and regeneration phase the CO_2 flux is switched on and off, respectively. Reproduced from [13] with permission from the Royal Society of Chemistry.

full saturation. With these results, the dimensioning of a reactor and corresponding process management (reaction/drying cycles) are relatively straightforward.

The peculiar dynamics is a consequence of special sorption enhanced process. Most hydrogenation reactions take place in plug-flow reactors operating in a steady-state distribution of concentrations and local temperatures [86]. Still, valuable information can be drawn from such distribution analyses to improve the overall reactor performance. With neutron imaging providing this information, interesting results are to be expected in the future.

References

[1] Dale B E and Ong R G 2012 Energy, wealth, and human development: why and how biomass pretreatment research must improve *Biotechnol. Progr.* **28** 893–8

[2] Martínez D M and Ebenhack B W 2008 Understanding the role of energy consumption in human development through the use of saturation phenomena *Energy Policy* **36** 1430–5

[3] International Energy Agency 2021 Net Zero by 2050(IEA, Paris) https://www.iea.org/reports/net-zero-by-2050

[4] Schmidt J, Gruber K, Klingler M, Klöckl C, Ramirez Camargo L, Regner P, Turkovska O, Wehrle S and Wetterlund E 2019 A new perspective on global renewable energy systems: why trade in energy carriers matters *Energy Environ. Sci.* **12** 2022–9

[5] Rouwenhorst K H R, Krzywda P M, Benes N E, Mul G and Lefferts L 2021 Ammonia production technologies *Techno-Economic Challenges of Green Ammonia as an Energy Vector* ed A Valera-Medina and R Banares-Alcantara (New York: Academic) ch 4, pp 41–83

[6] Suhan M B K, Hemal M N R, Choudhury M A A S, Mazumder M A A and Islam M 2022 Optimal design of ammonia synthesis reactor for a process industry *J. King Saud Univ.—Eng. Sci.* **34** 23–30

[7] Karabanova A, Berdiyeva P, Helfen L, Tengattini A, Bücherl T, Makowska M G, Deledda S and Blanchard D 2021 Neutron radiography for local modelling of thermochemical heat storage reactors: case study on $SrCl_2$-NH_3 *Int. J. Heat Mass Transfer* **178** 121287

[8] COP26 2021 *COP26 Presidency Outcomes—The Climate Pact*

[9] Smith C, Hill A K and Torrente-Murciano L 2020 Current and future role of Haber–Bosch ammonia in a carbon-free energy landscape *Energy Environ. Sci.* **13** 331–44

[10] Terreni J, Billeter E, Sambalova O, Liu X, Trottmann M, Sterzi A, Geerlings H, Trtik P, Kaestner A and Borgschulte A 2020 Hydrogen in methanol catalysts by neutron imaging *Phys. Chem. Chem. Phys.* **22** 22979–88

[11] Parker S F and Lennon D 2021 Net zero and catalysis: how neutrons can help *Physchem.* **1** 95–120

[12] Terreni J, Trottmann M, Delmelle R, Heel A, Trtik P, Lehmann E H and Borgschulte A 2018 Observing chemical reactions by time-resolved high-resolution neutron imaging *J. Phys. Chem.* C **122** 23574–81

[13] Borgschulte A, Delmelle R, Duarte R B, Heel A, Boillat P and Lehmann E 2016 Water distribution in a sorption enhanced methanation reactor by time resolved neutron imaging *Phys. Chem. Chem. Phys.* **18** 17217–23

[14] Terreni J, Sambalova O, Borgschulte A, Rudić S, Parker S F and Ramirez-Cuesta A J 2020 Volatile hydrogen intermediates of CO_2 methanation by inelastic neutron scattering *Catalysts* **10** 433

[15] Ramirez-Cuesta A J and Mitchell P C H 2013 Neutrons and neutron spectroscopy *Local Structural Characterisation* (New York: Wiley) pp 173–224

[16] Kruteva M 2021 Dynamics studied by quasielastic neutron scattering (QENS) *Adsorption* **27** 875–89

[17] Yang J, Sudik A, Wolverton C and Siegel D J 2010 High capacity hydrogen storage materials: attributes for automotive applications and techniques for materials discovery *Chem. Soc. Rev.* **39** 656

[18] Yang L, He L, Huang D, Wang Y, Song Q, Zhao L, Shen X, Tian Z and Wang H 2020 Three-dimensional hydrogen distribution and quantitative determination of titanium alloys via neutron tomography *Analyst* **145** 4156–63

[19] Felderhoff M, Weidenthaler C, von Helmolt R and Eberle U 2007 Hydrogen storage: the remaining scientific and technological challenges *Phys. Chem. Chem. Phys.* **9** 2643
[20] Züttel A 2003 Materials for hydrogen storage *Mater. Today* **6** 24
[21] Elberry A M, Thakur J, Santasalo-Aarnio A and Larmi M 2021 Large-scale compressed hydrogen storage as part of renewable electricity storage systems *Int. J. Hydrogen Energy* **46** 15671–90
[22] Schlapbach L and Züttel A 2001 Hydrogen-storage materials for mobile applications *Nature* **414** 353
[23] Willems J J G 1984 Metal hydrides electrodes stability of $LaNi_5$-related compounds *Philips J. Res.* **39** 1
[24] Notten P H L, Ouwerkerk M, Ledovskikh A, Senoh H and Iwakura C 2003 Hydride-forming electrode materials seen from a kinetic perspective *J. Alloys Compd.* **356–7** 759
[25] Feng F and Northwood D O 2004 Hydrogen diffusion in the anode of Ni/MH secondary batteries *J. Power Sources* **136** 346–50
[26] Züttel A, Meli F and Schlapbach L 1994 Electrochemical and surface properties of $Zr(V_xNi_{1-x})_2$ alloys as hydrogen-absorbing electrodes in alkaline electrolyte *J. Alloys Compd.* **203** 235–41
[27] Zhang Y, Li C and Yuan Z *et al* 2022 Research progress of TiFe-based hydrogen storage alloys *J. Iron. Steel Res. Int.* **29** 537–51
[28] Lai Q, Sun Y, Wang T, Modi P, Cazorla C, Demirci U B, Ares Fernandez J R, Leardini F and Aguey-Zinsou K-F 2019 How to design hydrogen storage materials? fundamentals, synthesis, and storage tanks *Adv. Sustain. Systems* **3** 1900043
[29] Wei T Y, Lim K L, Tsenga Y S and Chan S L I 2017 A review on the characterization of hydrogen in hydrogen storage materials *Renew. Sustain. Energy Rev.* **79** 1122–33
[30] Bielmann M, Kato S, Mauron P, Borgschulte A and Züttel A 2009 Characterization of hydrogen storage materials by means of pressure concentration isotherms based on the mass flow method *Rev. Sci. Instrum.* **80** 083901
[31] Blach T P and Gray E M 2007 Sieverts apparatus and methodology for accurate determination of hydrogen uptake by light-atom hosts *J. All. Compds* **446** 692
[32] Sakaguchi H, Kohzai A, Hatakeyama K, Fujine S, Yoneda K, Kanda K and Esaka T 2000 Visualization of hydrogen in hydrogen storage alloys usingneutron radiography *Int. J. Hydrogen Energy* **25** 1205–8
[33] Samantaray S S, Putnam S T and Stadie N P 2021 Volumetrics of hydrogen storage by physical adsorption *Inorganics* **9** 45
[34] Heubner F, Hilger A, Kardjilov N, Manke I, Kieback B, Gondek J, Banhart and Röntzsch L 2018 In-operando stress measurement and neutron imaging of metal hydride composites for solid-state hydrogen storage *J. Power Sources* **397** 262–70
[35] Shukla S, Singh P, Roy T, Kashyap Y S, Shukla M and Singh R N 2021 Investigation hydrogen diffusivity in Zr–2.5% Nb alloy pressure tube material using metallography and neutron radiography *J. Nucl. Mater.* **544** 152679
[36] Buitrago N L *et al* 2018 Determination of very low concentrations of hydrogen in zirconium alloys by neutron imaging *J. Nucl. Mater.* **503** 98–109
[37] Léon A, Kircher O, Rösner H, Décamps B, Leroy E, Fichtner M and Percheron-Guégan A 2006 SEM and TEM characterization of sodium alanate doped with $TiCl_3$ or small Ti clusters ($Ti_13{\cdot}_6THF$) *J. Alloys Compd.* **414** 190–203

[38] Roedern E, Hansen B R S, Ley M B and Jensen T R 2015 Effect of eutectic melting, reactive hydride composites, and nanoconfinement on decomposition and reversibility of $LiBH_4$–KBH_4 *J. Phys. Chem.* C **119** 25818–25

[39] Börries S 2017 Neutron imaging of metal hydride systems *PhD Thesis* University of Hamburg

[40] Börries S, Metz O, Pranzas P K, von Colbe J M B, Bücherl T, Dornheim M, Klassen T and Schreyer A 2016 Optimization and comprehensive characterization of metal hydride based hydrogen storage systems using *in situ* neutron radiography *J. Power Sources* **328** 567–77

[41] Karimi F *et al* 2021 Characterization of $LiBH_4$–MgH_2 reactive hydride composite system with scattering and imaging methods using neutron and synchrotron radiation *Adv. Eng. Mater.* **23** 2100294

[42] Nikolic M, Longo F, Billeter E, Cesarini A, Trtik P and Borgschulte A 2022 Combinatorial neutron imaging methods for hydrogenation catalysts *Phys. Chem. Chem. Phys.* **24** 27394

[43] Gremaud R, Broedersz C P, Borsa D M, Borgschulte A, Mauron P, Schreuders H, Rector B and Griessen R 2007 Hydrogenography: an optical combinatorial method to find new light-weight hydrogen-storage materials *Adv. Mater.* **19** 2813

[44] Lehmann E H, Vontobel P and Kardjilov N 2004 Hydrogen distribution measurements by neutrons *Appl. Radiat. Isot.* **61** 503–9

[45] Blau B *et al* 2009 The Swiss spallation neutron source SINQ at Paul Scherrer Institut *Neutron News* **20** 5

[46] Na Ranong C *et al* 2009 Concept, design and manufacture of a prototype hydrogen storage tank based on sodium alanate *Chem. Eng. Technol.* **32** 1154–63

[47] von Colbe J M B, Metz O, Lozano G A, Pranzas P K, Schmitz H W, Beckmann F, Schreyer A, Klassen T and Dornheim M 2012 Behavior of scaled-up sodium alanate hydrogen storage tanks during sorption *Int. J. Hydrogen Energy* **37** 2807–11

[48] Bevan A I, Züttel A, Book D and Harris I R 2011 Performance of a metal hydride store on the 'Ross Barlow' hydrogen powered canal boat *Faraday Discuss.* **151** 353–67

[49] Baruj A, Borzone E M, Ardito M, Marín J, Rivas S, Roldán F, Sánchez F A and Meyer G 2015 Neutron radiography analysis of a hydride-based hydrogen storage system *Int. J. Hydrogen Energy* **40** 16913–20

[50] Gondek Ł, Selvaraj N B, Czub J, Figiel H, Chapelle D, Kardjilov N, Hilger A and Manke I 2011 Imaging of an operating LaNi4.8Al0.2–based hydrogen storage container *Int. J. Hydrogen Energy* **36** 9751–7

[51] Herbrig K, Pohlmann C, Gondek H, Figiel N, Kardjilov A, Hilger I, Manke J, Banhart B, Kieback and Röntzsch L 2015 Investigations of the structural stability of metal hydride composites by *in situ* neutron imaging *J. Power Sources* **293** 109–18

[52] Jacobson D L, Hussey D S, Baltic E, Udovic T J, Rush J J and Bowman R C 2010 Neutron imaging studies of metal-hydride storage beds *Int. J. Hydrogen Energy* **35** 12837–45

[53] Wood B M, Ham K, Hussey D S, Jacobson D L, Faridani A, Kaestner A, Vajo J J, Liu P, Dobbins T A and Butler L G 2014 Real-time observation of hydrogen absorption by $LaNi_5$ with quasi-dynamic neutron tomography *Nucl. Instrum. Methods Phys. Res.* B **324** 95–101

[54] Jepsen J 2013 *PhD Thesis* Helmut-Schmidt Universität Hamburg

[55] Borgschulte A, Jain A, Ramirez-Cuesta A J, Martelli P, Remhof A, Friedrichs O, Gremaud R and Züttel A 2011 Mobility and dynamics in the complex hydrides $LiAlH_4$ and $LiBH_4$ *Faraday Discuss.* **151** 213–30

[56] Paskevicius M, Jepsen L H, Schouwink P, Černý R, Ravnsbæk D B, Filinchuk Y, Dornheim M, Besenbacher F and Jensen T R 2017 Metal borohydrides and derivatives—synthesis, structure and properties *Chem. Soc. Rev.* **46** 1565–634

[57] Langmi H W, Walton A, Al-Mamouri M M, Johnson S R, Book D, Speight J D, Edwards P P, Gameson I, Anderson P A and Harris I R 2003 Hydrogen adsorption in zeolites A, X, Y and RHO *J. Alloys Compd.* **356–7** 710–5

[58] Züttel A, Sudan P, Mauron P, Kiyobayashi T, Emmenegger C and Schlapbach L 2002 Hydrogen storage in carbon nanostructures *Int. J. Hydrogen Energy* **27** 203–12

[59] Rosi N L, Eckert J, Eddaoudi M, Vodak D T, Kim J, O'Keeffe M and Yaghi O M 2003 Hydrogen storage in microporous metal-organic frameworks *Science* **300** 1127–9

[60] Hirscher M and Panella B 2007 Hydrogen storage in metal–organic frameworks *Scr. Mater.* **56** 809–12

[61] Ahmed A, Liu Y, Purewal J, Tran L D, Wong-Foy A G, Veenstra M, Matzger A J and Siegel D J 2017 Balancing gravimetric and volumetric hydrogen density in MOFs *Energy Environ. Sci.* **10** 2459–71

[62] Ahmed A, Seth S, Purewal J, Wong-Foy A G, Veenstra M, Matzger A J and Siegel D J 2019 Exceptional hydrogen storage achieved by screening nearly half a million metal-organic frameworks *Nat. Commun.* **10** 1568

[63] Broom D P and Hirscher M 2016 Irreproducibility in hydrogen storage material research *Energy Environ. Sci.* **9** 3368–80

[64] Lamberti C, Zecchina A, Groppo E and Bordiga S 2010 Probing the surfaces of heterogeneous catalysts by *in situ* IR spectroscopy *Chem. Soc. Rev.* **39** 4951–5001

[65] Georgiev P A, Ross D K, Albers P and Ramirez-Cuesta A J 2006 The rotational and translational dynamics of molecular hydrogen physisorbed in activated carbon: a direct probe of microporosity and hydrogen storage performance *Carbon* **44** 2724–38

[66] Ramirez-Cuesta A J, Mitchell P C H, Ross D K, Georgiev P A, Anderson P A, Langmi H W and Book D 2007 Dihydrogen in cation-substituted zeolites X—an inelastic neutron scattering study *J. Mater. Chem.* **17** 2533–9

[67] Liu Y, Brown C M, Neumann D A, Peterson V K and Kepert C J 2007 Inelastic neutron scattering of H-2 adsorbed in HKUST-1 *J. Alloys Compd.* **446** 385–8

[68] Boelsma C, Bannenberg L J, van Setten M J, Steinke N-J, van Well A A and Dam B 2017 Hafnium—an optical hydrogen sensor spanning six orders in pressure *Nat. Commun.* **8** 15718

[69] Huang W, Xiao X, Steichen P, Droulias S A, Brischetto M, Wolff M, Li X and Hjörvarsson B 2021 Combined light and electron scattering for exploring proximity effects on hydrogen absorption in vanadium *Energies* **14** 8251

[70] Barbour J C and Doyle B L 1995 *Handbook of Modern Ion Beam Materials Analysis* ed J R Tesmer and M Nastasi (Pittsburgh, PA: Materials Research Society)

[71] Lanford W A, Trautvetter H P, Ziegler J F and Keller J 1976 New precision technique for measuring the concentration versus depth of hydrogen in solids *Appl. Phys. Lett.* **28** 566–8

[72] Wilde M and Fukutani K 2014 Hydrogen detection near surfaces and shallow interfaces with resonant nuclear reaction analysis *Surf. Sci. Rep.* **69** 196–295

[73] Ott N, Cancellieri C, Trtik P and Schmutz P 2020 High-resolution neutron imaging: a new approach to characterize water in anodic aluminum oxides *Mater. Today Adv.* **8** 100121

[74] Mâaza M, Farnoux B and Samuel F 1993 Study of the hydrogen diffusion in superlattices by grazing angle neutron reflectometry *Phys. Lett.* A **181** 245–50

[75] Callori S J, Rehm C, Causer G L, Kostylev M and Klose F 2016 Hydrogen absorption in metal thin films and heterostructures investigated *in situ* with neutron and x-ray scattering *Metals* **6** 125

[76] Kalisvaart P, Luber E, Fritzsche H and Mitlin D 2011 Effect of alloying magnesium with chromium and vanadium on hydrogenation kinetics studied with neutron reflectometry *Chem. Commun.* **47** 4294–6

[77] Olah G A, Goeppert A and Prakash G K S 2009 *Beyond Oil and Gas: The Methanol Economy* (Heidelberg: Wiley-VCH Verlag GmbH & Co. KGaA)

[78] Patterson B *et al* 2019 Renewable CO_2 recycling and synthetic fuel production in a marine environment *Proc. Natl Acad. Sci.* **116** 12212–9

[79] Götz M, Lefebvre J, Mörs F, Koch A M, Graf F, Bajohr S, Reimert R and Kolb T 2016 Renewable power-to-gas: a technological and economic review *Renew. Energy* **85** 1371–90

[80] Marques Mota F and Kim D H 2019 From CO_2 methanation to ambitious long-chain hydrocarbons: alternative fuels paving the path to sustainability *Chem. Soc. Rev.* **48** 205–59

[81] Dieterich V, Buttler A, Hanel A, Spliethoff H and Fendt S 2020 Power-to-liquid via synthesis of methanol, DME or Fischer–Tropsch-fuels: a review *Energy Environ. Sci.* **13** 3207–52

[82] Sabatier P and Senderens J B 1901 Hydrogenation of CO over nickel to produce methane *J. Soc. Chem. Ind* **21** 504–6

[83] Fischer F and Tropsch H 1923 Über die Herstellung synthetischer Ölgemische (Synthol) durch Aufbau aus Kohlenoxyd und Wasserstoff *Brennstoff-Chem.* **4** 276–85

[84] Schulz H 1999 Short history and present trends of Fischer–Tropsch synthesis *Appl. Catal.,* A **186** 3–12

[85] Kaneko T, Derbyshire F, Makino E, Gray D and Tamura M 2001 Coal liquefaction *Ullmann's Encyclopedia of Industrial Chemistry* (New York: Wiley)

[86] Ertl G, Knözinger H, Schüth F and Weitkamp J 2008 *Handbook of Heterogeneous Catalysis* (New York: Wiley-VCH Verlag GmbH & Co. KGaA)

[87] Nguyen L, Tao F F, Tang Y, Dou J and Bao X-J 2019 Understanding catalyst surfaces during catalysis through near ambient pressure x-ray photoelectron spectroscopy *Chem. Rev.* **119** 6822–905

[88] Newton M A and Dent A J 2013 *In situ Characterization of Heterogeneous Catalysts* ed J A Rodriguez, J C Hanson and P J Chupas (New York: Wiley) pp 75–119

[89] van Bokhoven J A and Lamberti C 2017 *XAFS Techniques for Catalysts, Nanomaterials, and Surfaces* ed Y Iwasawa, K Asakura and M Tada (Cham: Springer International Publishing) pp 299–316

[90] Cheng H, Lu C, Liu J, Yan Y, Han X, Jin H, Wang Y, Liu Y and Wu C 2017 Synchrotron radiation x-ray powder diffraction techniques applied in hydrogen storage materials—a review *Prog. Nat. Sci.: Mater. Int.* **27** 66–73

[91] Buchter F, Lodziana Z, Remhof A, Mauron P, Friedrichs O, Borgschulte A, Züttel A, Filinchuk Y and Palatinus L 2011 Experimental charge density of $LiBD_4$ from maximum entropy method *Phys. Rev.* B **83** 064107

[92] Spencer M S 1998 ^{1}H and ^{13}C nuclear magnetic resonance investigations of the Cu/Zn/Al oxide methanol-synthesis catalyst *Catal. Lett.* **50** 37

[93] Bouchard L S, Burt S R, Anwar M S, Kovtunov K V, Koptyug I V and Pines A 2008 NMR imaging of catalytic hydrogenation in microreactors with the use of para-hydrogen *Science* **319** 442

[94] Mitchell P C H, Parker S F, Ramirez-Cuesta A J and Tomkinson J 2005 *Vibrational Spectroscopy with Neutrons With Applications in Chemistry, Biology, Materials Science and* ed J L Catalysis, Finney and D L Worcester (Singapore: World Scientific Publishing)
[95] Parker S F and Lennon D 2016 Applications of neutron scattering to heterogeneous catalysis *J. Phys.: Conf. Ser.* **746** 012066
[96] Behrens M 2012 The active site of methanol synthesis over Cu/ZnO/Al_2O_3 industrial catalysts *Science* **759** 893
[97] Tisseraud C, Comminges C, Pronier S, Pouilloux Y, A and Le Valant 2016 The Cu–ZnO synergy in methanol synthesis. Part 3: impact of the composition of a selective Cu@ZnOx core–shell catalyst on methanol rate explained by experimental studies and a concentric spheres model *J. Catal.* **343** 106–14
[98] van den Berg R, Prieto G, Korpershoek G, van der Wal L I, van Bunningen A J, Laegsgaard-Joergensen S, de Jongh P E and de Jong K P 2016 Structure sensitivity of Cu and CuZn catalysts relevant to industrial methanol synthesis *Nat. Commun.* **7** 13057
[99] Thiele E W 1939 Relation between catalytic activity and size of particle *Ind. Eng. Chem.* **31** 916–20
[100] Matsumoto M and Espenson J H 2005 Kinetics of the interconversion of parahydrogen and orthohydrogen catalyzed by paramagnetic complex ions *J. Am. Chem. Soc.* **127** 11447–53
[101] Farkas A 1935 *Orthohydrogen, Parahydrogen, and Heavy Hydrogen* (Cambridge: Cambridge University Press)
[102] Grilly E R 1953 The liquefaction and storage of partially converted liquid hydrogen *Rev. Sci. Instrum.* **24** 1–4
[103] MacFarlane R E 1994 *New thermal neutron scattering files for ENDF/B-VI release 2* LA-12639-MS Los Alamos National Lab., NM
[104] Romanelli G, Minniti T, Škoro G, Krzystyniak M, Taylor J, Fornalski D and Fernandez-Alonso F 2019 Visualization of the catalyzed nuclear-spin conversion of molecular hydrogen using energy-selective neutron imaging *J. Phys. Chem.* C **123** 11745–51
[105] Carvill B T, Hufton J R, Anand M and Sircar S 1996 Sorption-enhanced reaction process *AIChE J.* **42** 2765–72
[106] Borgschulte A, Gallandat N, Probst B, Suter R, Callini E, Ferri D, Arroyo Y, Erni R, Geerlings H and Züttel A 2013 Sorption enhanced CO_2 methanation *Phys. Chem. Chem. Phys.* **15** 9620–5
[107] Züttel A, Borgschulte A and Orimo S I 2007 Tetrahydroborates as new hydrogen storage materials *Scr. Mater.* **56** 823
[108] Sartori S, Knudsen K D, Zhao-Karger Z, Bardaji E G, Muller J, Fichtner M and Hauback B C 2010 Nanoconfined magnesium borohydride for hydrogen storage applications investigated by SANS and SAXS *J. Phys. Chem.* C **114** 18785–9

Part V

Hard and soft condensed matter

Chapter 13

Magnetism

Markus Strobl and Michael Schulz

The magnetic moment of the neutron paired with its outstanding transmission characteristics makes it a unique tool to probe bulk magnetism of condensed matter. The study of fundamental magnetic phenomena and material features has long been a key application field in neutron science, in particular neutron scattering [1, 2]. The interaction of neutrons with magnetic fields is also exploited in neutron imaging, even though with a comparably young history (compare section 2.2), despite some early attempts and considerations to utilize neutrons to image magnetic domains [3–10]. While neutron scattering methods are able to investigate magnetic structures up to some 100 nm, full-field transmission neutron imaging has the potential for investigating bulk magnetic domain structures with spatial resolution up to the macroscopic range of components and devices, which are not amenable with any conventional method. The need for domain characterization to establish the link between fundamental physical properties and macroscopic behaviour such as magnetization has fostered advances of conventional domain observation techniques as well as the development of novel methods. A wide range of methods is available today including the Bitter technique, numerous electron microscopy techniques as well as magneto-optical Kerr microscopy, but also more recent methods such as spin-polarized tunnelling and magnetic force microscopy as well as dedicated methods of x-ray spectroscopy [11, 12]. However, all of these methods allow only investigations of surfaces, very small samples or thin films and do not provide insights into bulk domain structures of magnetic materials.

Polarized neutron imaging (sections 2.2 and 19.6) was initially proposed to provide access to magnetic domain structures in the bulk of materials, but it was finally the dark-field contrast imaging modality of neutron grating interferometry (compare sections 2.2 and 19.7) that enabled first visualizations and studies of networks of bulk magnetic domains [13]. Figure 13.1 shows a comparison of a dark-field contrast neutron image and a Faraday image of a high permeability steel lamination [14]. It becomes obvious that the neutron image covers a larger field of view, however, at lower spatial resolution. The neutron dark-field image (DFI)

doi:10.1088/978-0-7503-3495-2ch13 © IOP Publishing Ltd 2024

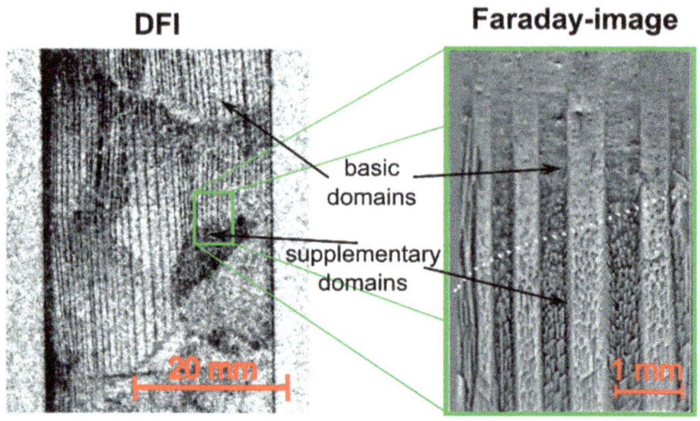

Figure 13.1. Comparison of neutron dark-field contrast image and Faraday image of magnetic domain structure of a Goss textured electric steel lamination (figure 4.10 in reference [14], reproduced with permission).

resolves the domain walls and domain wall density in the case of the supplementary domains, which are beyond direct spatial resolution of the method. The Faraday image displays the stray fields of a small region on the surface, while the response of the neutron image covers the full volume. It has to be noted that the typical isolating coating on electric steel laminates hinders the Faraday method, but not neutron dark-field imaging, and removal of the coating alters the stress state and thus the underlying domain structure significantly [15]. This chapter will outline examples of neutron imaging studies in the field of magnetism and thus also the potential of neutron imaging for the study of magnetic materials and devices.

13.1 Soft magnetic materials

Bulk ferromagnetic domains play a key role in technological relevant soft magnetic materials for electric machinery such as in grain oriented (GRO) and non-oriented (NOR) electric steels which constitute more than 95% of the produced soft magnetic materials with a yearly production of about 12 million tons [16]. The volume domains of these materials which are used, e.g. in transformers and electro-motors determine the most important magnetic properties of the materials including hysteresis, remanence, saturation and losses. Electric steels typically contain up to 3.2% Si, which increases the electric resistivity by about half an order of magnitude and thus lowers the core losses correspondingly through narrower hysteresis and reduced eddy currents. While NOR steels are magnetically isotropic, GRO steels display a strong directional dependence, i.e. an about 30% increased magnetic flux density in rolling direction is achieved by a well-controlled texture referred to as Goss texture. Goss texture implies the [001] direction of the cubic crystal lattice being parallel to the in-plane rolling direction and the (110) plane aligned with the surface of the rolled sheet material. The preferred orientation of Goss electric steels (GRO) is exploited in particular in transformer applications, while motors and generators, requiring changing magnetization directions, profit from non-grain oriented steels.

Industrial production usually provides large sheets with thicknesses of a few 100 μm which are typically coated, depending on the final application, which helps to improve the electrical resistance between laminations, to limit eddy currents, and to avoid corrosion. These sheets have to be cut to the required shape for specific applications for which they are stacked to the required size in the respective machine or device. However, the impact of cutting leads to deterioration of the magnetic characteristics in extended affected zones in the vicinity of the cut. The coatings often hinder straightforward application of conventional local 2D domain characterization. Neutron imaging on the other hand could be shown to provide valuable respective insights.

Initially, a detailed *in situ* dark-field contrast neutron imaging study could demonstrate the severe effect of removing the coating of laminates on the underlying domain structure and thus the bias induced by such approach in order to apply conventional techniques [15]. The measurement was performed on Goss oriented electric steel laminates which display magnetic domains of several 100 μm across, which can thus be directly resolved by neutron imaging. In fact, the applied technique, dark-field contrast neutron imaging rather resolves the domain walls, the spacing of which reveal the domain size and morphology. The coatings of the laminates induce stresses which impact the domain structure. In order to investigate the influence of such stress states on the domain structure, a Goss steel sheet was measured first with coating. After chemical removal of the coating (glassy magnesium silicate) external tensile loading was applied until the initial domain structure of the coated state was recovered. The steel sheets are installed with their domains, i.e. [001], along the lines of the gratings of the grating interferometer used for dark-field contrast imaging (section 19.7). The tensile force is aligned in the same direction.

Figure 13.2 displays dark-field contrast images, the main results including measurements performed at 0, 1, 2.5, 7.5, 20 and 20 MPa load and the unloaded

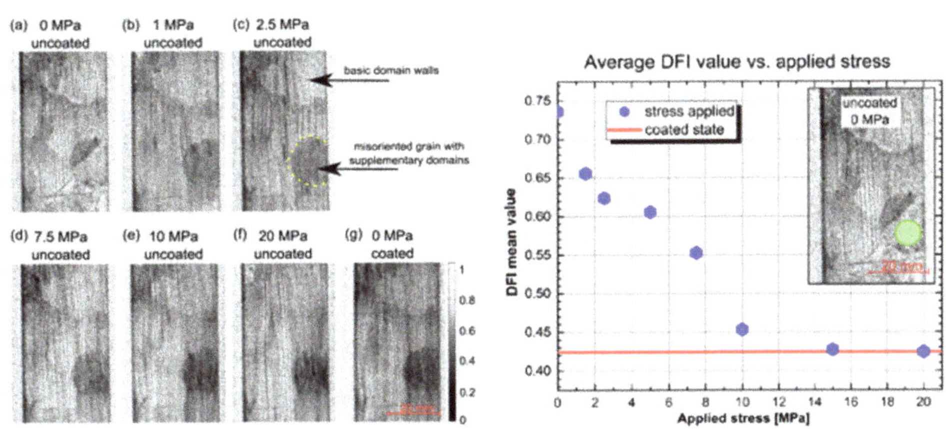

Figure 13.2. Dark-field contrast images [15], reprinted with permission of AIP Publishing, of uncoated Goss steel sheet under different tensile loads (vertical direction [001]) (a)–(f) juxtaposed to unloaded coated state of the same sheet (g); right-hand side: extracted mean dark-field contrast value of images in indicated position (inset).

coated state for comparison [15]. Based on the average dark-field contrast values in a region of significant supplementary domains it is concluded that the coating equals an effect of about 15–20 MPa loading and distorts the domain structure considerably. Thus, removal of the coating for investigations such as Kerr microscopy leads to an obvious significant bias.

The deterioration of magnetic properties of electric steel sheets during manufacturing, in particular when cutting the desired shape required in specific machines, is detrimental to the efficiency of electric machines and devices. Thus, it is of outmost importance to gain insights that enable optimizations of the machining. Overall magnetization measurements provide no information on the local effects responsible for deterioration. A combination of magnetization measurements with local microstructure investigations point to either thermally or mechanically induced stresses for laser and mechanical cutting, respectively, to cause magnetic property deterioration, somewhat irrespective of the different influence of both methods on grain morphology changes [17]. However, such studies are unable to provide a local picture of magnetic property changes in the vicinity of cut edges and clear correlations of changed permeability as a function of distance from the actual cut edge depending on cutting process and parameters. Such information is only accessible today by neutron imaging, as has been demonstrated in various studies, where the dark-field contrast measured can be related, depending indeed on numerous other material parameters, to the state of magnetization [18–21].

A seminal neutron imaging study [18] demonstrated this potential investigating the effect of cutting and comparing several cutting techniques including guillotine, punching, solid-state laser and CO_2 laser cutting on a non-oriented electric steel grade M330-35 A. Figure 13.3 displays some of the results of strips cut on both sides (cut 1 left (0 mm), cut 2 right (10 mm)). Line profiles of the dark-field contrast at different applied magnetic fields are displayed.

The unique spatially resolved results provide insights with respect to the local deterioration of magnetic structures and properties. While annealed sheets displayed a uniform dark-field contrast value of 0.3 at the maximum applied field

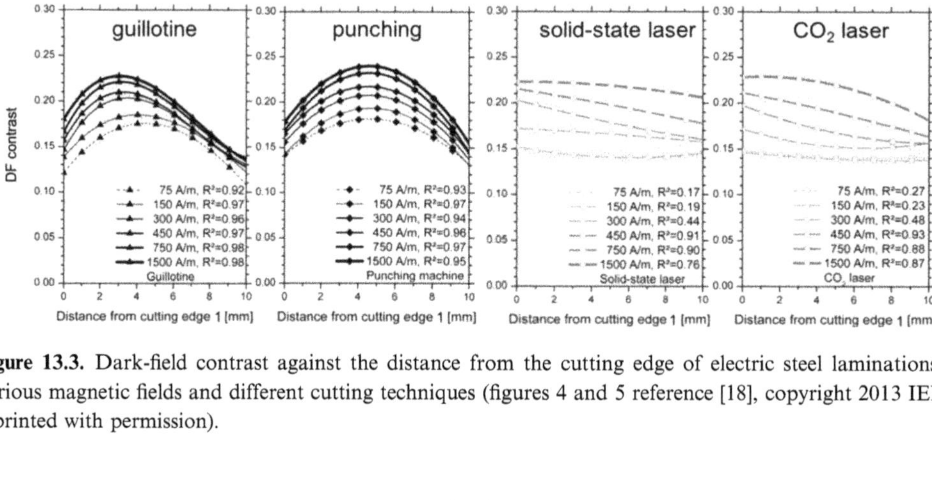

Figure 13.3. Dark-field contrast against the distance from the cutting edge of electric steel laminations at various magnetic fields and different cutting techniques (figures 4 and 5 reference [18], copyright 2013 IEEE, reprinted with permission).

corresponding to 1500 A m^{-1}, the cut sheets display distinct profiles at no point reaching the value of 0.3. This implies relatively long-ranging effects of the cutting. In addition, some profiles display asymmetries between the two cut edges, which for the guillotine case can be explained by mechanical asymmetries in the cutting process, while for the laser cutting the thermal history might play a role, a question which, however, remains unanswered yet. A more recent study utilizing improved instrumentation focusses on punching and investigates a series of different electric steel grades, sheet thicknesses and grain sizes [21]. Through carefully correlating dark-field contrast and polarization independently for each material, the study succeeds in mapping the local polarization dependent on the applied field for a number of different cutting parameters. The study finds that more severe deterioration is found for the thicker sheets, due to more stress induced by mechanical punching and that the effect reaches about a value of three times the thickness into the material when measured from the cutting edge.

In electric motors, field guidance is required in the electric steel sheets forming the cores of such engines in order to concentrate the magnetic field at specific locations, thereby maximizing the electromagnetic forces driving the motor. This field guidance is typically achieved by introducing cut-outs in the material. However, such cut-outs significantly weaken the mechanical stability of rotating parts of the motor, thereby ultimately limiting the maximum rotation speeds based on the arising centrifugal forces thus limiting the efficiency of the motor. Hence, it is of great interest to develop magnetic flux barriers that do not compromise the structural integrity of the electrical steel sheets. A novel approach to guide the magnetic field by deliberately introducing residual stress through embossing is currently being examined by combining neutron grating interferometry (NGI) experiments with magnetization measurements and finite element simulations [22]. Residual stress in the electrical steel sheets reduces the mobility of the magnetic domain walls due to the magneto-elastic effect, resulting in a reduced response of magnetic domains to external magnetic fields [23], thereby effectively decreasing the local magnetic permeability. By performing neutron grating interferometry on embossed non-oriented electric steel samples a direct connection between the residual stress states from embossing and the magnetic properties of the electric steel was shown. Moreover, the influence of different shapes [24], orientations [25] and arrangements of embossing [26] on the efficiency of the flux barriers was investigated. Other works deal, e.g. with the optimization of the domain structures in electric steels in order to optimize performance [27].

While these examples have shown that the effect of magnetic deterioration can be studied dependent on an applied external field, for Goss textured steel laminations neutron imaging enables to observe, *in situ*, the domain morphology variations as a function of alternating applied fields. Initial measurements of this kind were performed with varying static fields applied [28], soon the potential for observations in alternating fields was recognized and utilized to investigate, e.g. domain mobility dependencies on applied frequencies and field strengths. Such attempts resulted first in the visualization and characterization of frequency-induced freezing phenomena of bulk magnetic domain walls [29] (figure 13.4 left-hand side panel) and ultimately in direct time-resolved observations of domain wall dynamics [30] (figure 13.4 right side top panel).

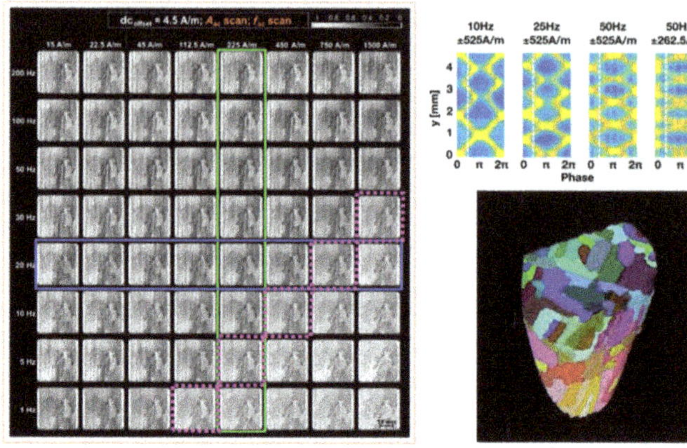

Figure 13.4. Panel showing time-averaged dark-field contrast imaging observations of a GOS steel at different frequencies and applied currents, indicating frequency/current combinations under which the domains cannot follow and thus remain static and can be resolved in a time-averaged image (left side), reprinted with permission from [29], copyright (2016) by the American Physical Society; the domain movements resolved by time-resolved dark-field imaging in the same material (right top). reproduced from [30] CC BY 4.0; a three dimensional magnetic domain reconstruction from dark-field contrast imaging of a bulk FeSi specimen with 8 mm diameter (right bottom), reprinted from [33], copyright (2010), with permission from Springer Nature.

While in the former case static, time-averaged, measurements were performed, in which the appearance of well-resolved domain walls indicated a freezing of domain movement [29], in the latter experimental work a detector with sufficient time resolution was utilized to capture domain movements with sub-millisecond resolution through a quasi-stroboscopic data recording throughout many cycles of the repetitive process (figure 13.4 right, top) [30]. That means that the recorded images were sorted after the experiment according to their respective phase of the alternating field and of the phase stepping required for Talbot–Lau grating interferometry. This eventually allowed observation and analysis of the movements of individual domains and the influence of the frequency and current value on the domain mobility.

Given the fact, that the so far discussed dark-field contrast modality of neutron imaging (compare section 2.2) provides contrast for the domain walls only, thus, does not allow quantifying domain orientations and field strengths, attempts exist to utilize neutron polarization measurements for such purpose. Principle feasibility has been demonstrated particularly for the large domains of Goss oriented steels [31, 32] but so far no corresponding applications have been reported in the context of soft magnetic domain investigations.

On the other hand, it was shown that visualizing magnetic domains through dark-field contrast neutron imaging is also possible in three dimensions (3D). In a first work it proved possible to depict the individual volume domains of a bulk FeSi crystal (figure 13.4 right, bottom). The results [33] are the first reported observation of the domain network in a centimetre-sized object, i.e. beyond the micron-scale of the top surface of an opaque magnetic material, which other methods achieve. Given the relevance of the bulk magnetic domain structure for the magnetic properties of materials this constitutes a seminal progress for the characterization toolbox of magnetic materials.

13.2 Ferromagnetic phase transitions

The depolarization of a neutron beam after transmission of a sample (section 2.2.6) has already been identified as a valuable tool to study ferromagnetic properties many decades ago [34]. The neutron depolarization allows accessing information about the presence of magnetic domains, the average domain size or a preferred orientation of the domains at different length and time scales paired with the spatial resolution of neutron imaging [35]. Unlike most other available techniques, neutrons are highly sensitive for probing the magnetic properties inside the bulk of a sample. Moreover, the neutron depolarization technique can be applied to samples which are placed inside complex sample environment devices such as cryostats, furnaces, magnets, pressure cells and stress rigs [36], where classical bulk measurement techniques such as PPMS (Physical Property Measurement System) frequently fail due to the weak signals. The combination of the neutron depolarization technique with neutron imaging denoted as neutron depolarization imaging (NDI) provides the unique possibility to access such information with spatial resolution across the sample in 2D and 3D. A prerequisite for the technique is the implementation of a spin polarizer, analyser and a spin flipper into a collimated beam for neutron imaging (compare section 19.6). As a consequence of the fact that a bulky analyser needs to be placed between the sample and the detector, the achievable spatial resolution with the NDI technique is typically lower than in standard neutron imaging (compare equation 2.1).

A classical model for the description of ferromagnets in which the magnetic moment is carried by the conduction electrons has for long time been the Landau–Fermi-liquid (FL) theory. Despite its good description of many ferromagnetic materials, recently deviations from this classic FL model have been found in several materials. Particular focus was put on materials with a very low Curie temperature where the ferromagnetic phase transition is driven by quantum instead of thermal fluctuations, thereby exposing the underlying mechanisms of the phase transition, which is denoted as quantum phase transition (QPT). In such materials a variety of novel phases like unconventional superconductivity or partial order have been found to arise, giving these materials a boost in interest. In this context, the Curie temperature of ferromagnetic materials can be tuned by several non-thermal control parameters such as pressure or chemical composition. In a series of studies, NDI was employed to investigate the ferromagnetic state of a variety of materials in the vicinity of a QPT by means of adjusting such control parameters. Compositional doping was used in the case of the materials $Pd_{1-x}Ni_x$, $CePd_{1-x}Rh_x$ and $Nb_{1-y}Fe_{2-y}$ to identify spatial phase separation of ferromagnetism using NDI [37]. These materials show a strong dependence of the magnetic ordering temperature T_C on even small changes of the composition. Here, NDI is a unique tool for direct verification of the sample quality and the magnetic properties with sub-mm spatial resolution. It was shown that the preparation of perfectly homogeneous samples is of utmost importance for any measurement techniques inherently averaging over the entire sample volume such as neutron scattering or bulk property measurements.

An example of NDI measurements on two polycrystalline $Pd_{1-x}Ni_x$ samples produced at University of Augsburg, Germany with nominal Ni concentrations of $x = 5\%$ and 2.5%, respectively, is shown in figure 13.5. A clear neutron depolarization signature with a certain degree of spatial inhomogeneity is observed in the sample with higher Ni concentration. The Curie temperature of this particular sample was found to be approximately 70 K. The second sample with a Ni concentration of 2.5% did not show any depolarization down to the lowest achievable temperature in the measurement of 4 K. This is in good agreement with the phase diagram of the material where ferromagnetism is proposed to set in at a critical concentration of $x = 2.6\%$ [38].

Another investigation demonstrates the potential of the NDI technique, a tomographic scan of the depolarization signature at $T = 8$ K was acquired for another polycrystalline $Pd_{1-x}Ni_x$ sample grown by the Czochralsky technique after observing strong spatial inhomogeneities of the magnetic properties in the sample [39]. The combination of the reconstructed tomography volumes of the absorption data and the NDI data allow for a clear identification of coexisting ferromagnetic and paramagnetic regions within the sample volume. This information can

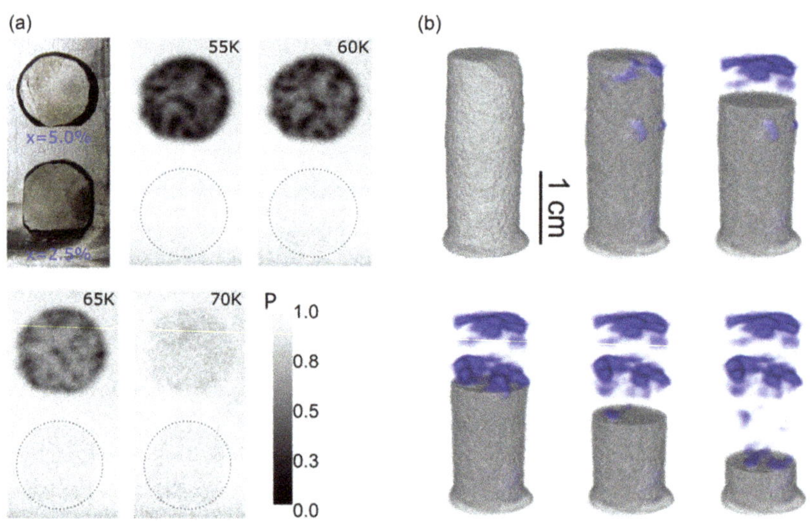

Figure 13.5. Photo and NDI measurements of two coin shaped $Pd_{1-x}Ni_x$ samples with nominal Ni concentrations of $x = 5\%$ and 2.5%. Spatial inhomogeneity is observed in the depolarization of the sample with higher Ni content, while no depolarization was found for the sample with lower concentration down to the lowest accessible temperature of 4 K (a), reprinted from [37], copyright (2010), with permission from Springer Nature (left). Tomographic reconstruction of the NDI data acquired for a cylindrical sample at 8 K (right). Significant inhomogeneity is observed in the distribution of ferromagnetic regions (blue). The attenuation contrast tomography is rendered in grey (b). Reproduced from [39], copyright IOP Publishing Ltd. All rights reserved.

consequently be used to extract smaller samples with homogeneous magnetic properties from the large crystal for further investigations.

Generally, a high degree of inhomogeneity was observed for samples where compositional changes were used to tune the Curie temperature. Consequently, the application of hydrostatic pressure came into focus as a more reliable and easier to tune control parameter to adjust the ordering temperature. However, relatively high pressures are required to significantly suppress the Curie temperature of materials such as Ni_3Al or $Hg_2Cr_2Se_4$. As a trade-off for increasing pressure the maximum sample size becomes smaller and smaller. In a first approach, clamp type pressure cells which can achieve pressures of up to 12 kbar while still providing reasonable neutron transmission were used with a Ni_3Al sample [40]. Such pressure cells still provide enough sample volume for cubic centimetre-sized samples, which is a reasonable size considering the sub-mm spatial resolution provided by NDI. Even higher pressures are needed to explore the magnetic phase diagram of other materials. Pressures of 100 kbar and more are typically achieved in diamond anvil pressure cells with a reduced sample volume of the order of a few cubic millimetres. Here, novel approaches are required to provide sufficient spatial resolution across the sample volume. In a pioneering study on $Hg_2Cr_2Se_4$ inside a clamp type pressure cell at 15 kbar, a magnifying Wolter-type neutron optic (compare sections 2.1 and 19.1) was used to magnify the image of the sample by a factor of four onto the neutron detector while at the same time decoupling the achievable spatial resolution from the distance between the sample and the detector [41]. Using this approach, the spatial resolution could be improved by a factor 5 to ~100 μm while at the same time reducing the required acquisition time, enabling resolution of the morphological features and evolution of the ferromagnetic phase transition.

Ferromagnetic phase transitions are also studied in engineering, where for example the martensitic phase transition in transformation induced plasticity (TRIP) steels plays a central role in work hardening. The martensitic phase is ferromagnetic and even small amounts are efficiently detected by NDI. Good time resolution enables *in situ* load and deformation studies triggering the phase transition [42]. The phase transition also plays a role in some shape memory alloys. These cases relating to engineering materials are, however, presented in chapter 5.

13.3 Superconductors and Skyrmion systems

A hallmark of superconductivity is the Meissner effect, which manifests by a complete expulsion of magnetic flux from the bulk of a superconductor (SC) and ideal diamagnetic behaviour in weak applied magnetic fields. In most elemental superconductors, superconductivity breaks down when the applied magnetic field exceeds a critical value B_C. Such materials are denoted as type I superconductors. In contrast, type II superconductors show a different behaviour where particle-like vortex lines, which each carry one flux quantum may enter the superconductor, thus permitting the magnetic field to penetrate the superconductor while remaining in the SC state. Generally, these vortex lines arrange in the form of a vortex lattice. In both types of SC phase coexistence and domain formation has been observed. In a type I SC the energy of the system may be minimized by splitting up the volume of the SC into normal conducting and superconducting domains denoted as the intermediate state (IS). Similarly, domain formation of vortex lattice domains and field-free Meissner regions has been observed in type II SCs in the intermediate mixed state (IMS). Moreover, superconductivity strongly depends on the purity of the employed materials and when cooled below critical temperature for superconductivity T_C in an applied magnetic field, flux trapping may be observed after removing the SC from the external field.

Imaging with polarized neutrons [35] (compare sections 2.2 and 19.6) has been employed to visualize the manifestation of the Meissner effect and of flux trapping based on the precession of the neutron spin in magnetic fields. This approach of neutron imaging was introduced in a pioneering work where a permanent magnet was levitating above a YBaCuO SC. The dipole field around the permanent magnet was visualized with polarized neutrons and compared with simulations of the field integral experienced by the neutrons while traversing the sample [43]. Moreover, in this study flux trapping in a superconducting Pb sample after field cooling was visualized in a tomographic experiment demonstrating the possibility to measure the 3D distribution of magnetic fields oriented perpendicular to the neutron polarization.

Further studies using this approach were performed to identify Meissner effect or partial suppression of the Meissner effect in samples of Nb, Pb, $La_{2-x}Sr_xCuO_4$ (x = 0.09) [44–48]. In all these works reported, simulations of the magnetic field strength and geometry are required as a basis for the data analysis. Based on these simulations the acquired field integral along the transmission path through and around the sample and consequently the spin rotation can be calculated and compared with the measured data. Adapting the parameters in the calculation to fit the measured images in principle enables quantitative analyses of the magnetic fields. However, while this is straightforward to achieve for *a priori* known principle field geometries of well characterized samples, it is not always possible to find a unique field constellation for the interpretation of the measured data, or the uniqueness of the achieved and presented solution cannot be fully established. This limitation could be overcome in the first application of polarimetric neutron tomography for the study of magnetic vector fields [49] which was focused on the visualization of the trapped field in a superconducting lead sample measured at $T = 4.3$ K [50]. Different representations of the trapped magnetic field that have been reconstructed from the tomographic scans for each polarization matrix element (section 2.2) can be seen in figure 13.6. It was found that the trapped field strongly varies locally in intensity and field orientation displaying a filament-like network structure with numerous maxima throughout the sample. It was found that the sample consisting of a variation of grain sizes displays more flux trapped in regions with finer grains and thus more grain boundaries which channel the flux.

Another recent study combining polarized neutron imaging and SANS measurement was able to visualize the compensating currents of a type-II/1 SC in high field cooling [51] and thus provided new insights into the thermodynamics of low purity SCs. The Abrikosov vortex lattice was characterized throughout the thermodynamic path through the respective phase transition and the occurring compensating currents and flux line closure was characterized in detail.

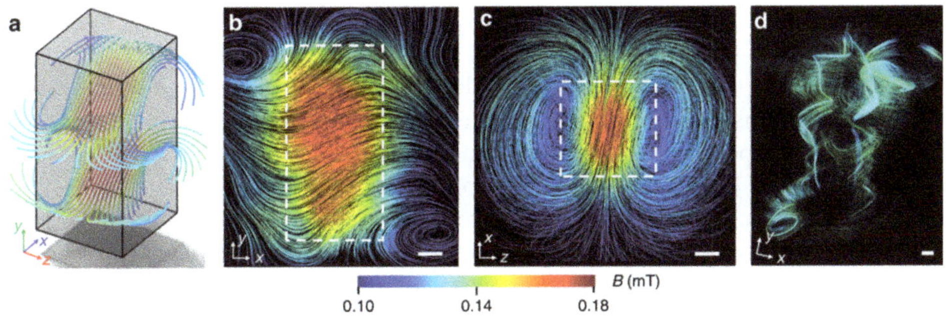

Figure 13.6. Magnetic vector field reconstruction from polarimetric neutron tomography of the field trapped in a superconducting lead sample. Reproduced from [50] CC BY 4.0.

In addition to imaging with polarized neutrons, the dark-field contrast image (DFI) (section 2.2) obtained from NGI (section 19.7) measurements has proven to be an excellent modality to study domain formation in type I and II SCs with spatial resolution. Here, the sensitivity of DFI with conventional Talbot–Lau interferometers to ultra-small angle neutron scattering (USANS) of neutrons from the domain structure has proven to be suited to probe respective structural correlations of the spatial distribution of micrometre-sized domains in the sample. While not providing equally detailed information about the domain size and shape as conventional USANS measurements in double crystal diffractometer instruments [52], the DFI image provides the unique possibility to visualize the domain distribution across the sample cross-section as a function of external parameters such as applied magnetic field or temperature. In contrast to all other available spatially resolved techniques, the DFI probes the domain distribution in the bulk volume of the sample and not only the domains on the surface. In particular, a combination of NGI with SANS and USANS measurements was found to provide a complete picture of the SC domain morphology from the nm to the cm length scale. Most notably, NGI was able to identify unconventional volume filling in a high purity sample with IMS domains based on its geometric shape, which was verified with local SANS and VSANS measurements [53]. Such studies of domain morphology in SCs have analogy to other investigations related to the general phenomenon of domain formation in two-phase systems, which are not easily accessible in many materials [54, 55]. Here, the ease of tuning the boundary conditions for domain formation of the IMS in Niobium through the applied magnetic field and temperature facilitates such studies of domain formation. A number of additional studies on domain formation in SC and their spatial distribution have been performed, in particular taking into account the sample purity and the resulting pinning [56, 57] as well as the formation of IS domains in different materials such as Pb [58].

Another prominent example of the formation of unconventional magnetic domain structures was the discovery of magnetic skyrmions in the chiral magnet MnSi using neutron scattering [59]. With their particle-like properties and ease of manipulation through small electric currents, skyrmions are considered promising systems for application in future data storage or spintronics devices. For example, in a small region of the phase diagram of MnSi, denoted as the A-phase, arising skyrmions form a hexagonal skyrmion lattice (SkL) (see figure 13.7(b)). While magnetic skyrmions are found in an increasing number of materials and meanwhile even at room temperature, diffractive neutron imaging (compare section 2.2) was used to resolve the debate on the nucleation and decay of the skyrmion lattice [60]. Below the magnetic transition temperature the skyrmion phase is surrounded by the conical phase. However, measurements of specific heat and magnetization indicate that the expected first order phase transition is indeed a broad crossover region. This crossover region had been intensely debated, often attributing the observations to exotic physics also due to the fact that existing spatially resolved measurement techniques were only sensitive to the sample surface and could not unambiguously disentangle effects introduced by the sample geometry from fundamentally new physical behaviour. Here, diffractive neutron imaging was found to be an ideal tool for a spatially resolved investigation of the distribution of the SkL over a disc shaped MnSi sample of 9.3 mm diameter and 2.1 mm thickness when crossing the boundary regions of the A-phase in the phase diagram. The setup used in this study is shown in figure 13.7(a). Using a cryomagnet that allows to reach the A-phase pocket of MnSi, the sample was placed in front of an imaging detector. A crucial aspect of this study was the application of a micro-channel plate collimator (MCP) [61] in order to separate the small angle scattering image generated by the SkL domains under typical scattering angles of 1.2° from the transmission image of the sample (compare section 19.8). By rotating the MCP collimator to the scattering angle of the SkL, spatially resolved information about the distribution of the SkL over the sample volume was obtained at a constant temperature of 28 K as a function of applied magnetic field including the controversially discussed boundary regions between the

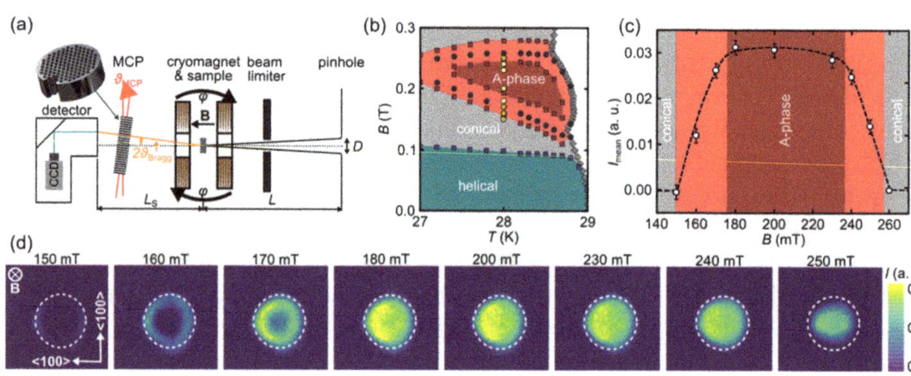

Figure 13.7. Setup used for diffractive imaging of the SkL in MnSi at ANTARES. The setup employs an MCP collimator to separate small angle neutron scattering off the SkL from the direct transmission image (a). Phase diagram of MnSi showing the pocket of the A-phase (dark red) and border regions (light red). Measurements were performed as a function of magnetic field at a temperature of 28 K as indicated by the round points. (b). The mean scattered intensity off the SkL is proportional to the fraction of the sample in the A-phase (c). Images of the scattering off the SkL showing evidence of phase coexistence at magnetic fields at the border of the phase pocket (d) [60].

conical and the A-phase. The mean scattering intensity measured for these magnetic fields, which is proportional to the fraction of the sample in the A-phase is shown in figure 13.7(c). A plateau indicating that the sample is fully in the SkL phase is only observed in the centre of the phase pocket. A macroscopic phase separation of the SkL and the conical phase is observed in the scattering images of the SkL presented in figure 13.7(d) for magnetic fields at the border of the phase pocket. The SkL nucleation begins at the edges of the sample and only covers the full sample volume in the central part of the phase pocket. This data confirms the spatial coexistence of conical phase and SkL due to demagnetization effects arising in macroscopic samples, which is in full agreement with the phenomena observed at the border regions of the phase pocket. This study underlines the importance of taking demagnetization and geometric effects into account when it comes to the conclusive interpretation of bulk measurements in the vicinity of the border regions of the A-phase pocket.

13.4 Macroscopic magnetic fields and devices

Remote measurements of magnetic fields, in particular inside of bulk objects or devices still pose a significant challenge. While large efforts have been made to simulate complex magnetic fields, very often it is still challenging to verify such simulations in real world conditions. Neutrons indeed are a unique tool for remote measurements given their penetration characteristics, and in the case of magnetic fields their inherent magnetic moment. While simple magnetic fields of electric coils were used to develop and demonstrate polarised neutron imaging [43, 62, 63] and polarimetric imaging [49, 64] (sections 2.2 and 19.6), imaging with polarized neutrons was applied to sense e.g. magnetic fields in electric machines and electric currents in conductors.

The magnetic field distribution and evolution in an operating electromagnetic engine is of utmost importance for the efficiency of the machine. However, generally such fields are not accessible but only explored and evaluated through simulation. In order to verify the predicted field distribution in an electric motor polarized and in particular polarimetric imaging has been applied [65]. Due to the repetitive nature of the process of the revolution of the rotor a stroboscopic approach allows for sufficient time resolution, despite the relatively long exposure required per image for polarisation contrast. Images of the same phase of the process can be accumulated till sufficient signal to noise ratios are reached [66]. While especially a single projection does not allow to extract the full vector field information, it could be shown that the method allowed to assess the quality of the simulated field distribution, respectively to highlight deviations of the real field distributions from the assumptions of the simulation.

For many electric components and devices including but not limited to high current applications, superconducting components, inductors, transformers, electric shielding, low-loss connectors and motors assessing electrical current distributions within them is of key importance for optimisations and energy efficiency. It is well understood, that electric current distributions in conductors depend foremost on the material but also shape and contacting of the conducting structure, but equally on the current function in terms of shape, amplitude and frequency. For example, the induced eddy currents in large bulk conducting parts lead to preferential current paths in the vicinity of the surface of the material and a current depletion in the central regions for alternating currents (AC). This so-called skin effect limits the efficiency of conduction as it limits the utilized cross-section. The thickness of the layer efficiently transporting the current, i.e. the skin depth, depends on the material parameters of conductivity and absolute magnetic permeability as well as on the AC frequency. An increase of the two latter parameters reduces the effective cross-section, while higher conductivity improves the situation. A polarized neutron imaging experiment was conducted to investigate the influence of the skin effect for a current through a thick cylindrical rod when contacted symmetrically and asymmetrically [67]. The skin effect and its frequency dependence were clearly observed and analysed for the investigated system and geometries. The results further reveal that contacting, which was extending over about 1 cm^2 was not uniform, which lead to asymmetric current distributions and it was concluded correspondingly that the availability of such method probing actual distributions in bulk components is indispensable, where actual current distributions are critical to performance and efficiency.

This potential of neutron imaging of magnetic phenomena is particularly important in fields of research and development relevant to today's societal challenges such as for electrochemical energy conversion devices, namely batteries and fuel cells. However, the often very small local currents are still challenging to be resolved and observed. Approaches to assess low signals with regards to neutron polarisation include assessing the spin phase, i.e. small changes of the polarization vector, through respective scans [68] or through a smart perpendicular spin analyses, where the signal variation in a sinusoidal is the strongest [69].

Also, strong fields, which can lead to a beam depolarization due to limited monochromatisation or strong gradients are a challenge to polarized neutron imaging. A spin-echo approach has been proposed and tested to overcome corresponding limitation [70], but in particular differential phase contrast based on the magnetic contribution to the refractive index (section 2.2) showed potential [62, 71] for such cases.

Finally, dark-field contrast imaging with polarized neutrons and polarization analyses [72] was presented recently to have the potential to investigate in more detail the arrangement of magnetic microstructures, because it enables to probe independently spin–flip and non-spin–flip scattering in its sensitivity remit of very-small and ultra-small angle neutron scattering [73]. This implies that dark-field contrast imaging becomes sensitive to the magnetic interaction vector [74] enabling quantitative studies of magnetic small angle scattering [75].

References

[1] Marshall W and Lowde R D 1968 *Rep. Prog. Phys.* **31** 705

[2] Price D and Fernandez-Alonso F (ed) 2015 *Neutron Scattering—Magnetic and Quantum Phenomena* 1st edn eBook

[3] Schlenker M and Shull C G 1973 Polarized neutron techniques for the observation of ferromagnetic domains *J. Appl. Phys.* **44** 4181
[4] Baruchel J, Schlenker M and Roth W L 1977 Observation of antiferromagnetic domains in nickel oxide by neutron diffraction topography *J. Appl. Phys.* **48** 5
[5] Schlenker M, Bauspiess W, Graeff W, Bonse U and Rauch H 1980 Imaging of ferromagnetic domains by neutron interferometry *J. Magn. Magn. Mater.* **15–8** 1507–9
[6] Podurets K M, Somenkov V A and Shilstein S S 1989 Refractioncontrast radiography *Zh. Tekh. Fiz.* **59** 115–21
[7] Podurets K M, Somenkov V A, Chistyakov R R and Shilstein S S 1989 Visualization of internal domain structure of silicon iron crystals by using neutron radiography with refraction contrast *Physica* B **156 and 157** 694–7
[8] Podurets K M, Petrenko A V, Somenkov V A and Shil'shtein S S 1994 Neutron radiography with depolarized contrast *Zh. Tekh. Fiz.* **67** 134–6
[9] Hochhold M, Leeb H and Badurek G 1996 Tensorial neutron tomography: a first approach *J. Magn. Magn. Mater.* **157–158** 575–6
[10] Leeb H, Hochhold M, Badurek G, Buchelt R J and Schricker A 1998 Neutron magnetic tomography: a feasibility study *Aust. J. Phys.* **51** 401–13
[11] Hubert A and Schäfer R 1998 *Magnetic Domains* (Berlin: Springer) ed Y Zhu *Modern Techniques for Characterizing Magnetic Materials* (Berlin: Springer)
[12] Kronmüller H and Parkin S S P 2007 *Handbook of Magnetism and Advanced, Magnetic Materials* **vol 3** (New York: Wiley)
[13] Grünzweig C, David C, Bunk O, Dierolf M, Frei G, Kühne G, Schäfer R, Pofahl S, Rønnow H and Pfeiffer F 2008 Bulk magnetic 221 domain structures visualized by neutron dark-field imaging *Appl. Phys. Lett.* **93** 112504
[14] Betz B 2016 Visualisation of magnetic domain structures and magnetisation processes in Goss-oriented, high permeability steels using neutron grating interferometry *PhD Thesis* (EPFL, Lausanne)
[15] Betz B, Rauscher P, Harti R P, Schäfer R, Van Swygenhoven H, Kaestner A, Hovind J, Lehmann E and Grünzweig C 2016 In-situ visualization of stress-dependent bulk magnetic domain formation by neutron grating interferometry *Appl. Phys. Lett.* **108** 1
[16] Varga L K and Davies H A 2008 Challenges in optimizing the magnetic properties of bulk soft magnetic materials *J. Magn. Magn. Mater.* **320** 2411–22 *Proc. of the 18th Int. Symp. on Soft Magnetic Materials*
[17] Araujo E G, Schneider J, Verbeken K, Pasquarella G and Houbaert Y 2010 Dimensional effects on magnetic properties of Fe–Si steels due to laser and mechanical cutting *IEEE Trans. Magn.* **46** 213–6
[18] Siebert R, Wetzig A, Beyer E, Betz B, Grunzweig C and Lehmann E 2013 Localized investigation of magnetic bulk property deterioration of electrical steel: analysing magnetic property drop thorough mechanical and laser cutting of electrical steel laminations using neutron grating interferometry *2013 3rd Int. Electric Drives Production Conf. (EDPC)* pp 1–5
[19] Siebert R, Schneider J and Beyer E 2014 Laser cutting and mechanical cutting of electrical steels and its effect on the magnetic properties *IEEE Trans. Magn.* **50** 1–4
[20] Betz B, Rauscher P, Siebert R, Schaefer R, Kaestner A, van Swygenhoven H, Lehmann E and Grünzweig C 2015 Visualization of bulk magnetic properties by neutron grating interferometry *Phys. Proc.* **69** 399–403

[21] Weiss H A, Steentjes S, Tröber P, Leuning N, Neuwirth T, Schulz M, Hameyer K, Golle R and Volk W 2019 Neutron grating interferometry investigation of punching-related local magnetic property deteriorations in electrical steels *J. Magn. Magn. Mater.* **474** 643–53
[22] Vogt S *et al* 2019 Extent of embossing-related residual stress on the magnetic properties evaluated using neutron grating interferometry and single sheet test *Prod. Eng.* **13** 211–7
[23] Moses A *et al* 2000 Magnetostriction in non-oriented electric steels: general trends *J. Magn. Magn. Mater.* **215** 669–72
[24] Neuwirth T, Backs A and Gustschin A *et al* 2020 A high visibility Talbot-Lau neutron grating interferometer to investigate stress-induced magnetic degradation in electrical steel *Sci. Rep.* **10** 1764
[25] Schauerte B, Leuning N, Vogt S, Moll I, Weiss H, Neuwirth T, Schulz M, Volk W and Hameyer K 2020 The influence of residual stress on flux-barriers of non-oriented electrical steel *J. Magn. Magn. Mater.* **504** 166659
[26] Gilch I *et al* 2021 Impact of residual stress evoked by pyramidal embossing on the magnetic properties of non-oriented electrical steel *Arch. Appl. Mech.* **91** 3513–26
[27] Rauscher P *et al* 2016 The influence of laser scribing on magnetic domain formation in grain oriented electrical steel visualized by directional neutron dark-field imaging *Sci. Rep.* **6** 38307
[28] Betz B *et al* 2016 Magnetization response of the bulk and supplementary magnetic domain structure in high-permeability steel laminations visualized *in situ* by neutron dark-field imaging *Phys. Rev. Appl.* **6** 024023
[29] Betz B, Rauscher P, Harti R P, Schäfer R, Van Swygenhoven H, Kaestner A, Hovind J, Lehmann E and Grünzweig C 2016 Frequency-induced bulk magnetic domain-wall freezing visualized by neutron dark-field imaging *Phys. Rev. Appl.* **6** 024024
[30] Harti P *et al* 2018 Dynamic volume magnetic domain wall imaging in grain oriented electrical steel at power frequencies with accumulative high-frame rate neutron dark-field imaging *Sci. Rep.* **8** 15754
[31] Hiroi K, Shinohara T, Hayashida H, Parker J D, Su Y H, Oikawa K, Kai T and Kiyanagi Y 2018 Study of the magnetization distribution in a grain-oriented magnetic steel using pulsed polarized neutron imaging *Physica* B **551** 146–51
[32] Dhiman I, Ziesche R, Riik L, Manke I, Hilger A, Radhakrishnan B, Burress T, Treimer W and Kardjilov N 2020 Visualization of magnetic domain structure in FeSi based high permeability steel plates by neutron imaging *Mater. Lett.* **259** 126816
[33] Manke I *et al* 2010 Three-dimensional imaging of magnetic domains *Nat. Commun.* **1** 125
[34] Halpern O and Holstein T 1941 On the passage of neutrons through ferromagnets *Phys. Rev.* **59** 960–81
[35] Strobl M, Heimonen H, Schmidt S, Sales M, Kardjilov N, Hilger A, Manke I, Shinohara T and Valsecchi J 2019 Topical review: polarisation measurements in neutron imaging *J. Phys.* D **52** 12
[36] Schulz M, Schmakat P, Franz C, Neubauer A, Calzada E, Schillinger B, Böni P and Pfleiderer C 2011 Neutron depolarisation imaging: stress measurements by magnetostriction effects in Ni foils *Phys.* B **406** 2412–4
[37] Pfleiderer C *et al* 2010 Search for electronic phase separation at quantum phase transitions *J. Low Temp. Phys.* **161** 167–81
[38] Nicklas M, Brando M, Knebel G, Mayr F, Trinkl W and Loidl A 1999 Non-Fermi-liquid behavior at a ferromagnetic quantum critical point in Ni_xPd_{1-x} *Phys. Rev. Lett.* **82** 21

[39] Schulz M, Neubauer A, Masalovich S, Mühlbauer M, Calzada E, Schillinger B, Pfleiderer C and Böni P 2010 Towards a tomographic reconstruction of neutron depolarization data *J. Phys.: Conf. Ser.* **211** 012025

[40] Schulz M, Neubauer A, Böni P and Pfleiderer C 2016 Neutron depolarization imaging of the hydrostatic pressure dependence of inhomogeneous ferromagnets *Appl. Phys. Lett.* **108** 202402

[41] Jorba P *et al* 2019 High-resolution neutron depolarization microscopy of the ferromagnetic transitions in Ni_3Al and $HgCr_2Se_4$ under pressure *J. Magn. Magn. Mater.* **475** 176–83

[42] Busi M, Polatidis E, Sofras C, Boillat P, Ruffo A, Leinenbach C and Strobl M 2022 Polarization contrast neutron imaging of magnetic crystallographic phases *Mater. Today Adv.* **16** 100302

[43] Kardjilov N, Manke I, Strobl M, Hilger A, Treimer W, Meissner M, Krist T and Banhart J 2008 Three-dimensional imaging of magnetic fields with polarized neutrons *Nat. Phys.* **4** 399–403

[44] Treimer W, Ebrahimi O and Karakas N Oct. 2012 Observation of partial Meissner effect and flux pinning in superconducting lead containing non-superconducting parts *Appl. Phys. Lett.* **101** 162603

[45] Treimer W, Ebrahimi O, Karakas N and Prozorov R May 2012 Polarized neutron imaging and three-dimensional calculation of magnetic flux trapping in bulk of superconductors *Phys. Rev.* B **85** 184522

[46] Aull S, Ebrahimi O, Karakas N, Knobloch J, Kugeler O and Treimer W Feb. 2012 Suppressed Meissner-effect in niobium: visualized with polarized neutron radiography *J. Phys. Conf. Ser.* **340** 012001

[47] Dhiman I, Ebrahimi O, Karakas N, Höppner H, Ziesche R and Treimer W Dec. 2015 Role of temperature on flux trap behavior in <100> Pb cylindrical sample: polarized neutron radiography investigation *Phys. Proc.* **69** 420–6

[48] Dhiman I *et al* Sep. 2017 Thermodynamics of Meissner effect and flux pinning behavior in the bulk of single-crystal $La_{2-x}Sr_xCuO_4$ ($x = 0.09$) *Phys. Rev.* B **96** 104517

[49] Sales M, Strobl M, Shinohara T, Tremsin A, Theil Kuhn L, Lionheart W, Desai N, Bjorholm Dahl A and Schmidt S 2018 Three dimensional polarimetric neutron tomography of magnetic fields *Sci. Rep.* **8** 2214

[50] Hilger A, Manke I and Kardjilov N *et al* 2018 Tensorial neutron tomography of three-dimensional magnetic vector fields in bulk materials *Nat. Commun.* **9** 4023

[51] Valsecchi J *et al* 2020 Visualization of compensating currents in type-II/1 superconductor via high field cooling *Appl. Phys. Lett.* **116** 192602

[52] Strobl M, Harti R P, Grünzweig C, Woracek R and Plomp J 2018 Small angle scattering in neutron imaging—a review *J. Imaging* **3** 64

[53] Reimann T *et al* 2015 Visualizing the morphology of vortex lattice domains in a bulk type-II superconductor *Nat. Commun.* **6** 1–8

[54] Bacak M *et al* 2020 Neutron dark-field imaging applied to porosity and deformation-induced phase transitions in additively manufactured steels *Mater. Des.* **195** 109009

[55] Kim Y, Valsecchi J, Oh O, Kim J, Lee S W, Boue F, Lutton E, Busi M, Garvey C and Strobl M 2022 Quantitative neutron dark-field imaging of milk: a feasibility study *Appl. Sci.* **12** 833

[56] Reimann T *et al* 2017 Domain formation in the type-II/1 superconductor niobium: interplay of pinning, geometry, and attractive vortex-vortex interaction *Phys. Rev.* B **96** 144506

[57] Backs A, Schulz M, Pipich V, Kleinhans M, Böni P and Mühlbauer S 2019 Universal behavior of the intermediate mixed state domain formation in superconducting niobium *Phys. Rev.* B **100** 064503

[58] Reimann T *et al* 2016 Neutron dark-field imaging of the domain distribution in the intermediate state of lead *J. Low Temp. Phys.* **182** 107–16

[59] Mühlbauer S, Binz B, Jonietz F, Pfleiderer C, Rosch A, Neubauer A, Georgii R and Böni P 2009 Skyrmion lattice in a chiral magnet *Science* **323** 915

[60] Reimann T, Bauer A, Pfleiderer C, Böni P, Trtik P, Tremsin A, Schulz M and Mühlbauer S 2018 Neutron diffractive imaging of the skyrmion lattice nucleation in MnSi *Phys. Rev.* B **97** 020406(R)

[61] Tremsin A S and Feller W B 2006 The theory of compact and efficient circular-pore MCP neutron collimators *Nucl. Instrum. Methods Phys. Res.* A **556** 556–64

[62] Strobl M, Treimer W, Keil S, Walter P and Manke I 2007 Magnetic field induced differential neutron phase contrast imaging *Appl. Phys. Lett.* **91** 254104

[63] Wada N, Shinohara T, Sato H, Hasemi H, Kamiyama T and Kiyanagi Y 2015 Evaluation of magnetic vector field by polarization analysis using pulsed neutrons at HUNS for magnetic field imaging *Phys. Proc.* **69** 427

[64] Strobl M, Kardjilov N, Hilger A, Jericha E, Badurek G and Manke I 2009 Imaging with polarized neutrons *Physica* B **404** 2611–4

[65] Hiroi K, Shinohara T, Hayashida H, Parker J D, Oikawa K, Harada M, Su Y and Kai T 2017 Magnetic field imaging of a model electric motor using polarized pulsed neutrons at JPARC/MLF *J. Phys.: Conf. Ser.* **862** 012008

[66] Tremsin A, Kardjilov N and Strobl M *et al* 2015 Imaging of dynamic magnetic fields with spin-polarized neutron beams *New J. Phys.* **17** 043047

[67] Manke I, Kardjilov N, Strobl M, Hilger A and Banhart J 2008 Investigation of the skin effect in the bulk of electrical conductors with, spin-polarized neutron radiography *J. Appl. Phys.* **104** 1

[68] Piegsa F M, van den Brandt B, Hautle P, Kohlbrecher J and Konter J A Quantitative radiography of magnetic fields using neutron spin phase imaging *Phys. Rev. Lett.* **102** 145501

[69] Jau Y -Y, Chen W C, Gentile T R and Hussey D S July 2020 Sensitive neutron transverse polarization analysis using a ^{3}He spin filter *Rev. Sci. Instrum.* **91** 073303

[70] Strobl M, Pappas C, Hilger A, Wellert S, Kardjilov N, Seidel SO and Mank I 2011 Polarized neutron imaging—a spinecho approach *Physica* B **406** 2415–8

[71] Valsecchi J, Harti R P and Raventós M *et al* 2019 Visualization and quantification of inhomogeneous and anisotropic magnetic fields by polarized neutron grating interferometry *Nat. Commun.* **10** 3788

[72] Valsecchi J, Makowska M G, Gruenzweig C, Piegsa F M, Kim Y, Lee S W, Thijs M A, Plomp J and Strobl M 2021 Decomposing magnetic dark-field contrast in spin analyzed Talbot-Lau interferometry—a Stern–Gerlach experiment without spatial beam splitting *Phys. Rev. Lett.* **126** 070401

[73] Valsecchi J, Kim Y, Lee S W, Saito K, Gruenzweig C and Strobl M 2021 Towards spatially resolved magnetic small-angle scattering studies by polarized and polarization-analyzed neutron dark-field contrast imaging *Sci. Rep.* **11** 8023

[74] Halpern O and Johnson M H 1939 On the magnetic scattering of neutrons, *Phys. Rev.* **55** 898

[75] Sebastian M *et al* 2019 Magnetic small-angle neutron scattering *Rev. Mod. Phys.* **91** 015004

Chapter 14

Soft matter

Markus Strobl

Typical soft condensed matter investigations with neutrons deal with nanometer-sized structures which are the domain of small-angle neutron scattering (SANS) and neutron reflectometry. Soft matter systems are in most cases studied in hydrogenous solutions, which are not favourable for transmission imaging, though deuteration is an appropriate tool used not only in scattering, but also in neutron imaging. Also, some of the presented application cases in other chapters are certainly related to soft matter, like, in particular, biomaterials. Here we shall present cases, which do not fit the other chapters and are potentially somewhat related to more fundamental soft matter science. These cases are subsumed in sections and albeit not without overlap, present studies on foams, gels, food and rheology.

14.1 Foams

14.1.1 Polymere foaming

The physical properties and cellular structure of polymeric foams depend strongly on the foaming process to which numerous mechanisms are contributing. However, the study of phenomena such as nucleation, growth, drainage, coalescence, coarsening and solidification, which are most relevant to understand and control foaming processes and thus foam properties [1–3] is limited by conventional non-destructive *in situ* methods like infrared and Raman spectroscopy, optical testing, infrared thermography, ultrasonic imaging, x-ray imaging, x-ray diffraction. In particular, x-ray imaging has been applied to follow in time the foaming dynamics and internal cellular structure evolution with respect to relative density, cell size and cell density on both thermoplastic and thermosets. However, x-ray imaging approaches are limited and do not allow for observing foaming under real technologically most relevant conditions such as in typical gas-tight metallic mould enclosures enabling pressures up to 4 MPa, due to the high x-ray attenuation by such structures hindering imaging of the latent and evolving subtle foam structures inside. Polymeric foams are typically produced by roto-moulding, reactive injection

doi:10.1088/978-0-7503-3495-2ch14

© IOP Publishing Ltd 2024

moulding, injection moulding, improved compression moulding, all of which require elevated pressure inside containments typically made of steel or aluminium alloys. Apart from the moulds providing the pressure containment and ensuring resulting foam parts with reduced densities and specific shapes, the mould impacts the foaming process in manyfold ways. On the one hand, the foam is contained and macroscopic foam collapse is prevented, and on the other hand, the foam liquid fraction is constant independent of composition and temperature. Other important aspects of the influence of the moulds on the foaming process such as constrained growth and anisotropic cellular structures or integral foam structures through local densification due to mould wall temperature and surface effects are considered [4–6] but could not be probed without neutron imaging. A neutron imaging study performed in 2015 was intended to shed light on technological as well as scientifically interesting phenomena under realistic foaming conditions [7]. The study investigated the behaviour of four different polymers (100, 50, 25, 0 wt% long-chain branched and high melt strength (HMS) PP (Borealis (Daploy WB 135 HMS)) blended into linear PP (Total Petrochemicals PPH 4070)), different densities (0.2%, 0.25%, 0.3%) and blowing agent (azodi-carbonamide (ADC) Porofor M-C1, Lanxess) contents (1%, 2%, 7.5%).

The foaming process (figure 14.1) in terms of cell size evolution and coalescence events were studied with neutron imaging at elevated temperatures comparing the different polymeric blends. Differences in the foam stability could be clearly

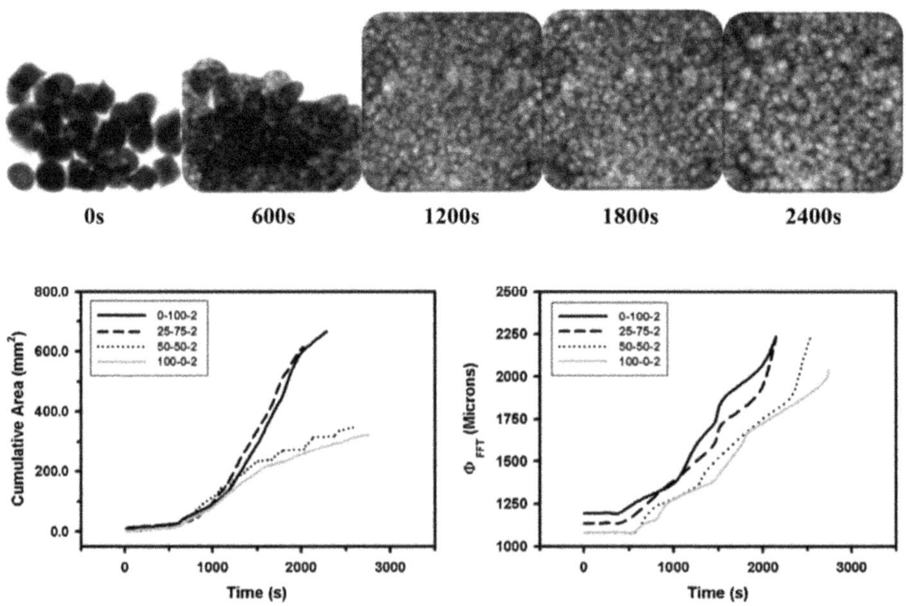

Figure 14.1. Top: neutron imaging time series foam expansion and ageing for polymere 100% PP plus 2% ADC sample with a relative density of 0.2. Bottom: effect of polymer type on the cell coalescence and cell size evolution. (a) Changes in the cumulative area of coalescence events with time. (b) Changes in the cell size with time (2 pph ADC, relative density 0.25). Reprinted from [7], copyright (2015), with permission from Elsevier.

identified in the mould cavity heated up to 250 °C. The coalescence rate and cell growth rate could also be related to rheological measurements of the strain hardening coefficient of the different mixtures of polymers with a clear decrease of both with increasing hardening coefficient.

The microscale evolution evaluation not only demonstrated the unique potential of neutron imaging in this specific field, but also shed light on the higher influence of the relative density than the strain hardening coefficient on the foam stability. Excessive addition of blowing agent demonstrated inducing significant cell degeneration rates.

14.1.2 Aqueous foams

A liquid foam or froth consists in principle of gas bubbles in a liquid medium, where the liquid is confined in liquid films and interstitial channels, also referred to as Plateau borders, between the gas bubbles [8, 9]. This heterogenous structure defines particular mechanical and rheological properties and evolves by the liquid distribution and transport, which is determined by capillary forces and gravity but also viscosity. However, the liquid flow through a froth is a highly non-trivial phenomenon influenced, amongst others, by the bubble size distribution, liquid content, surface mobility, foaming solution and chemical composition, but also external stimuli such as mechanical stress [10, 11]. Drainage through froth is not only interesting from a scientific point of view, but also technologically, as it is applied, e.g. in mineral processing, paper deinking and waste water treatment [12–14]. Thus, it is important to understand drainage flow and related to it the transport of particles through a froth. A number of neutron imaging studies have tackled this problem and demonstrated once more the value of neutron imaging for the investigation of complex structures and related processes.

A recent neutron imaging study focused on drainage flow in froths under shear strain and observed not only the changing alignment of the structural elements of the foam but also the impact on the liquid distribution under shear stress when liquid is supplied to the system from the top. The work is testing predictions based on numerical simulations [15] and considers open discussions on the liquid transport through the liquid film as well as reported anisotropies in drainage [16–18]. A deflection of the distribution of drainage liquid is observed experimentally and strain in the system even without applying external stress is found through producing texture tensor maps from the imaging data. These strains are measureably influencing the inner drainage distribution as well as the bubble size. The study of Heitkam *et al* on pseudo 2D liquid foam implies significant potential gains in understanding of drainage flow by applying on-the-fly time-resolved neutron tomography [19] (compare e.g. chapters 7 and 19).

Other studies [20, 21] have focused on particle motion in aqueous foams, which is related to the drainage flow and is particularly interesting for applications such as mineral processing as floatation is a very important separation technique [12]. Neutron imaging can track particles and their motion through the froth as well as resolve the local froth structure and quantify the liquid fraction. In mineral

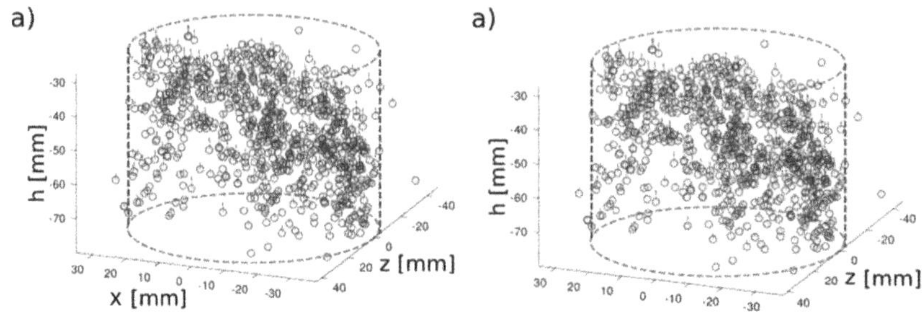

Figure 14.2. Tracking results of Gd particle tracking in froth showing detected particle positions and vertical velocity in 3D and a 2D projection [21] John Wiley & Sons. Copyright 2019 WILEY-VCH Verlag GmbH & Co. KGaA, Weinheim.

processing the floatation of particles is used to separate valuable minerals from valueless gangue. Hydrophobized gadolinium particles in the range of 100 mm size have been used in such studies to trace particle movement in froths of different quality and under varying conditions with controlled liquid content. This enables mapping particle velocities in the volume as represented in figure 14.2. It could be shown, for example, that in vertical direction the particle motion decreases, while particle diffusion in horizontal direction increases with decreasing foam stability leading to bubble ruptures. With regards to hydrophobicity of the particles and drainage flow, different rates for wash out of the particles could be measured and an avalanche-like flow of clusters of particles could be observed.

14.2 Gels

Gel is a colloidal dispersion consisting of a solid and a continuous liquid phase and which has become viscous enough to behave similarly to a solid. Gels are of interest for a wide range of applications and, e.g. hydrogels [22, 23] are applied as bioadhesives, chemical sensors and responsive coatings, in drug delivery and food products but also as potential scaffolds for artificial organs and tissue. In this context the diffusion properties of hydrogels for particles and molecules is highly relevant and thus intensely studied [24].

Neutron imaging has been shown to be able to play a role in understanding solvent diffusion and its correlation with solute motion. To this end a study by Wagner *et al* [25] concurrently combined neutron imaging with fluorescence and optical brightfield transmission imaging [26] to elucidate the swelling and deswelling as well as diffusion of solvent, fluorescent dyes, and macromolecules into crosslinked polyacrylamide hydrogels with varying solute size and network properties (figure 14.3).

The study found, for example, that solvent diffusion was slightly accelerated and supported solvent uptake in a partially swollen state of the hydrogel. In contrast, solute diffusion was found impeded by partial drying of the hydrogel and by increasing the crosslinking density in the hydrogel as both affect the pore size distribution. It is also concluded that the approach involving neutron imaging is a

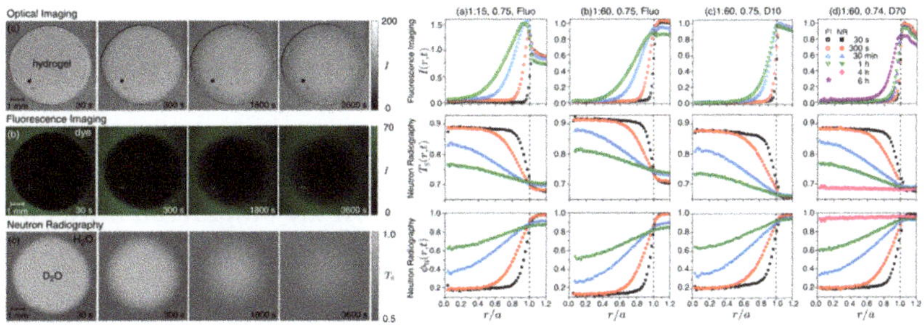

Figure 14.3. Left: optical, fluorescence and neutron images of D_2O-hydrogels with different crosslinking ratios and degree of swelling when contacted with H_2O containing dye. Right: corresponding intensity profiles of fluorescence imaging, neutron imagging and H_2O volume fraction profiles, ϕ_H, derived from neutron images as a function of normalized radius and time. Reprinted from [25] with the permission of AIP Publishing.

valuable tool well suited to investigate quantitatively the diffusion and evolution of the distribution of different phases present in the gel simultaneously.

A similar study in 2016 investigated the liquid ingress into a casein film [27] on the technologically relevant macroscopic scale of a thick film by means of neutron imaging [28]. It concluded that the water ingress has not only a dominating imbibition, but also a diffusional component and a diffusion coefficient could be defined based on a diffusion-like equation which can be used for diffusion and imbibition [29].

These exemplary studies underline the potential of neutron imaging in investigating imbibition and diffusion, not only of water but also of specific molecular species in gels and thus constitutes a powerful tool to characterize and subsequently understand and tune diffusion characteristics of gels, although such applications are not yet numerous.

A completely different way to look at gels and gelation processes with neutron imaging is established by advanced modalities, in particular neutron dark-field contrast imaging (compare section 2.2). The sensitivity to small-angle scattering enables insights into the microscopic network of gels and can be used to investigate and quantify spatial heterogeneities of microstructures on the macroscopic scale of direct spatial resolution. The ability of neutron dark-field contrast imaging to provide insights into the gel network was recently demonstrated in a feasibility study to characterize milk gel networks induced by acid and rennet gelation, respectively [30]. However, as milk gels relate to dairy products, this study shall be discussed rather in the next section on neutron imaging in food science.

14.3 Food

Milk products and milk itself do not only play an important role in global food industry and nutrition, but their structure has also been studied intensively by neutron scattering techniques in past decades. Studies of milk components using neutron scattering techniques are well documented in the literature [31–33] and might contribute the largest part of neutron studies on food. Small-angle (SANS) and ultra-small-angle neutron scattering (USANS) are applied to resolve the structural network of casein micelles in the range of 10–100 nm and fat globules of around 1000 nm in size. Dark-field contrast neutron imaging was applied in attempting to resolve these structural features, with the aim to subsequently observe heterogeneous structural evolutions in dairy samples exposed to external and internal stimuli [30]. It was found that on the one hand the sensitivity was not sufficient to resolve the small casein micelles, but on the other hand it was well suited to studying the fat globule size distribution and its dependence on the fat content of milk (figure 14.4).

However, the situation changes when the milk is gelled through addition of acid or enzymes and the casein micelles form extended networks. These networks appeared detectable and quantifiable through dark-field contrast neutron imaging even in the presence of stronger scattering fat content. These results hold significant promise for the application of neutron imaging to dairy gel structure and structural evolution and similar studies.

Beyond this advanced characterization of dairy systems neutron imaging has been applied to a variety of food-related investigations. We exclude here studies of plants which are covered in chapter 7 but also food-related studies on food packaging [34]. Most neutron imaging studies related to food are referring to processing of food such as drying [35–37], freeze-drying [38–41] or even cooking [42, 43].

Several neutron imaging studies engaged in the study of drying fruit tissue [35, 36] in order to inform and verify corresponding models through radiographic and tomographic time series and utilizing Black Body background correction, which is particularly important for water quantification. The neutron imaging results were in good agreement with total water loss measurements in a climate chamber and verified corresponding dehydration models (figure 14.5) and hygro-stress modelling of shrinkage but implied also the need for coupled mechanical-water transport models.

Another drying process relevant to food, but also to pharmaceutical applications and others is freeze-drying [44–46]. Several neutron imaging studies providing insights with regards to the sublimation front *in situ* and in the volume have been reported [38–40]. Knowledge about the sublimation front in lyophilisation or freeze-drying is crucial as it not only determines the material properties after drying but also relates to the overall energy consumption, drying velocity and ultimately the

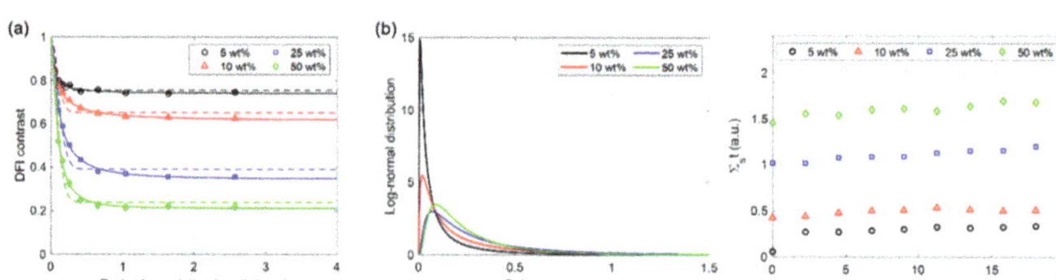

Figure 14.4. Dark-field curves of milk samples with different fat content. (Left) analyses yielding log-normal fat globule size distribution (mid) and variation of fat scattering over the height of the samples, implying a slightly increasing fat content towards the top of the sample for high fat content. Reproduced from [30] CC BY 4.0.

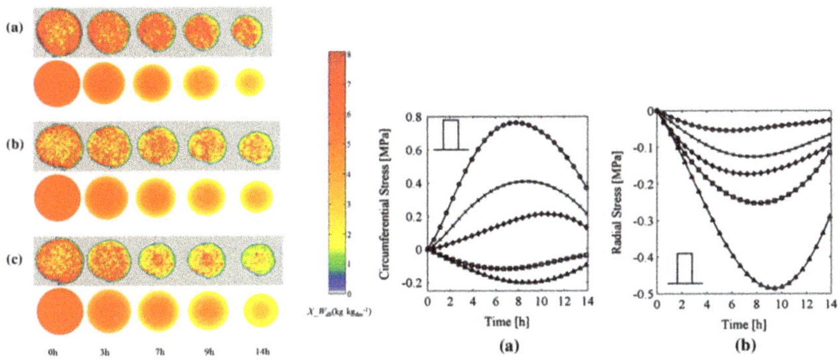

Figure 14.5. Left: slices at different height through drying apple tissue and corresponding modelling as well as, right, corresponding calculations of circumferential and radial stresses induced by dehydration and shrinkage. Reprinted from [35], copyright (2013), with permission from Elsevier.

drying time. Thus, product quality, throughput, and energy expense depend on defining optimum conditions. Only limited theoretical studies describe freeze-drying behaviour [47, 48] and experimental validation of applied models is challenging with conventional means. X-rays are less sensitive to water distribution and might struggle with the bulk environment for freeze-drying. The neutron experiments were successful in observing complex sublimation fronts in 3D and in identifying highly heterogeneous sublimation fronts with distinct dry fingers covering the sample, in addition to fractal peripheral sublimation fronts deviating substantially from ideally assumed flat fronts with drying parameters at the limit of material collapse in a maltodextrin solution.

Another example is neutron imaging measurements on meat during frying that have been undertaken [42, 43]. Those aimed to study the migration of water (meat juice) and related morphology changes based on the contraction of meat fibres caused by cooking.

14.4 Physical chemistry and rheology

Rheology deals with a wide range of responses of liquid to soft solid materials in the presence of forces in terms of stresses applied to these materials. While a number of applications described earlier in this chapter but also elsewhere, and in particular the work on foams in section 14.1, are strongly related to rheology, here we collect especially a number of applications which refer to rheology, but do not fit well in the above, more specific, topics. Furthermore, we include here soft matter work in the remit of physical chemistry, in particular that on liquids, soft solids and diffusion.

One of these topics concerns the behaviour of liquids and gas diffusion into those under high pressure. Neutron imaging has been reported to be a powerful method for time-resolved capturing of surface phenomena, liquid swelling and concentration profiles, e.g. in deuterated liquid ethanol (C_2D_6O) and n-decane (n-$C_{10}D_{22}$) during absorption of supercritical methane (CH_4) [49]. Measurements under industrially relevant conditions revealed anomalously slow diffusion in the two liquids in the early stages of the absorption experiments while contact angle and liquid level could be monitored continuously. Another, related investigation concerned the concurrent observation of solubility, surface tension, partial molar volume and diffusivity for normal and supercooled liquid solutions of methane in perdeuterated p-xylene (p-C_8D_{10}) (figure 14.6), and, for comparison, o-xylene (o-C_8D_{10})[50]. A significant increase of diffusivity, swelling and partial molar volume of methane were found in the supercooled p-xylene under pressures of interest for the production of liquefied natural gas (7.0 °C–30.0 °C, 1.0–101.1 bar) and the related freeze-out of p-xylene.

Another example demonstrating how high-resolution neutron imaging can reveal the change of the distribution of a specific phase in a nonequilibrium system of multiple components is an observation of the evaporation and condensation of a sessile droplet of a liquid mixture, i.e. H_2O and D_2O [51]. While the study investigated the phenomenon of water evaporation and simultaneous condensation and mixing (figure 14.7) the use of a deuterated component can also here be extended to a large range of binary or multi-component mixtures of scientific and technological relevance. Insights into the evolution of phase distributions in space

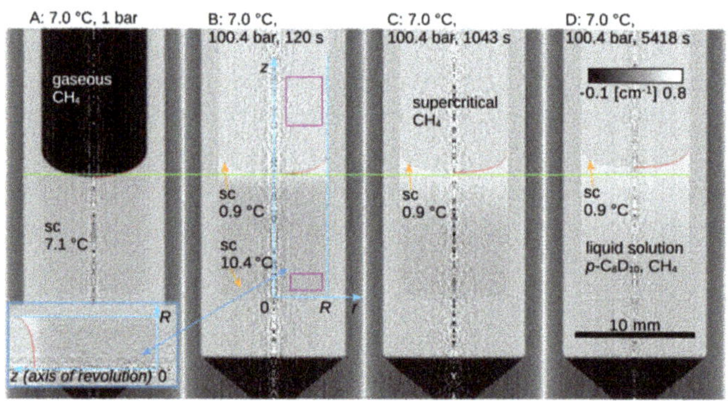

Figure 14.6. Slices of tomographic reconstructions of pressure cell with supercooled perdeuterated p-xylene (p-C_8D_{10}) exposed to methane at 1.0 bar (A) and at 100.4±0.2 bar at 7.0±0.2 °C at different times (C)–(D). Reproduced from [50] CC BY 4.0.

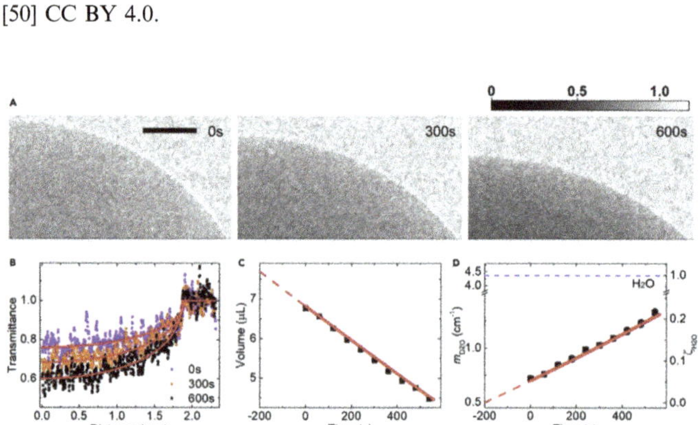

Figure 14.7. Neutron images of an evaporating sessile D_2O droplet (A) and extracted attenuation profiles (B), volume change (C) and attenuation coefficient evolution (D). Reprinted from [51], copyright (2021), with permission from Elsevier.

and time through neutron imaging can thus be considered to be also applied to prominent processes such as skin formation at liquid–gas interfaces, a phase separation occurring under certain conditions of heterogeneous solute distribution.

Melting polymers that were in the focus of a study finding 'kinks in experimental diffusion profiles of a dissolving semi-crystalline polymer' [52] by time-resolved neutron imaging shall be mentioned. When taking into account the measured time evolution of the total polymer concentration during dissolution of polyethylene oxide (PEO) tablets, where the crystalline polymere is melting when reaching a certain water concentration, it was found that the process could be well described by Fickian diffusion equations when applying concentration-dependent diffusion coefficients. The relevance is in poly(ethylene oxide) (PEO) being a model substance for tablet dissolution and drug delivery but it is also of interest for rather fundamental studies [53].

Another exemplary study illustrating the use of neutron imaging in rheology-related research was dedicated to the study of vortex driven thin films in a vortex fluidic device, which is a multi-purpose thin-film processing device able to instigate chemical reactivity, probe the structure of self-organized systems and process materials. Applications range from refolding of proteins [54], fabricating different nanocarbon materials [55], incorporating drugs into bilayers, accelerating enzymatic and other chemical reactions [56] etc. The respective study was aiming at enabling a potential upscaling of organic synthesis through improved mathematical modelling of the process. Neutron imaging was successful in informing the model through measurements of the film thickness in the rapidly rotating quartz tube of this thin-film microfluidic platform [57].

Finally, taking a look at the microscopic scale, the application of dark-field contrast (compare section 2.2) neutron imaging has paved the way to a wide range of topics. Sedimentation as a fundamental problem in studying how a suspension moves under gravity has become an evident phenomenon to study with quantitative dark-field contrast neutron imaging already in its first demonstration [58], when a change in particle density over time was observed. Consequently, subsequent experiments were designed to study sedimentation of a colloidal suspension not only to contribute to the understanding of the formation of quasi-crystalline sediments of colloids but as an ideal model system mimicking crystal nucleation on an easier accessible length and time scale [59]. To this end, spherical polystyrene particles with a few micrometer diameter were dispersed in a mixture of water and heavy water with a density slightly lower than that of the colloids in order to trigger slow sedimentation.

The measurements revealed the height-dependent concentration and eventual ordering in a quasi-crystalline state at the bottom. A map of the measured correlation functions (figure 14.8) clearly reveals a distinct depletion zone (denoted with 'C') where the particle concentration slightly decreases below the onset of increasing concentration and just above a rapidly increasing density towards an ordered densely packed state. While these data were largely interpreted qualitatively, in more recent studies modelling the scattering curves including form and structure factor has been enabled through adaptation of small-angle scattering models

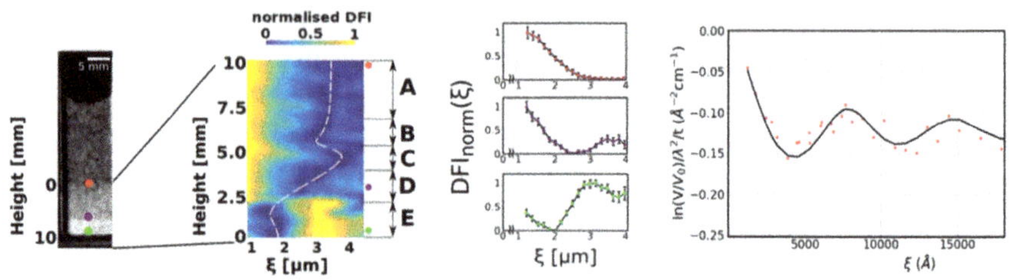

Figure 14.8. Neutron dark-field contrast image of a colloidal suspension sedimenting at the bottom of a quartz-cuvette and a map over the sample height of measured correlation functions such as three examples of these representing on top an isolated sphere state of dilute particles, a liquid like distribution with the development of a structure factor contribution and a quasi-crystalline state in the bottom, reproduced from [59] CC BY 4.0. On the right hand side a local correlation function measured at the cold neutron imaging instrument ICON at PSI by dark-field contrast imaging of densely packed particles in a suspension fitted by a model containing form and structure factor.

according to dark-field contrast theory [60]. Such a fit relating to densely packed spherical SiO particles after centrifugation in the context of studies of the packing behaviour depending on surface roughness [61] is shown in figure 14.8 on the right hand side.

A last example in the field of rheology shall be given with the structural evolution of a fractal cohesive powder under compression as seen by neutron dark-field contrast imaging [62]. The fractal structure of the cohesive powder defines its behaviour when a force is applied. The force is transferred through the sample following an initial compaction and breakdown of the specific microstructure. This breakdown, thus, does not happen simultaneously throughout the material, but starts where the force is applied, forming a compaction front moving through the sample. Therefore, the observation of this process requires macroscopic spatial resolution to observe the moving front, and microscopic resolution to understand the nature of the structural breakdown in space and time. This could be achieved, again, by dark-field contrast neutron imaging being sensitive to both length scales.

Through dark-field contrast but also high-resolution neutron imaging, many related soft matter systems and stimulated processes and phenomena involving micro- and mesoscale structures and domains with variations on a macroscopic scale including flocculation and coagulation, but also, e.g. the shear behaviour of suspensions etc, are moving into the focus of advanced neutron imaging.

References

[1] Khemani K C 1997 *Polymeric Foams an Overview* ed K C Khemani (Washington, DC: Polymeric Foams: Science and TechnologyACS Symposium Series) pp 1–7

[2] Eaves D 2004 *Handbook of Polymeric Foams* (Shrewsbury: Rapra Technology)

[3] Klempner D and Sendijarevic V 2004 *Handbook of Polymeric Foams and Foam Technology* 2nd edn (Munich: Hanser Publishers)

[4] Patterson B M, Henderson K, Gilbertson R D, Tornga S, Cordes N L, Cahvez M E and Smith Z 2014 Morphological and performance measures of polyurethane foams using x-ray CT and mechanical testing *Microsc. Microanal.* **20** 1284–93

[5] Hamilton A R, Thomsen O T, Madaleno L A O, Jensen L R, Rauhe J C M and Pyrz R 2013 Evaluation of the anisotropic mechanical properties of reinforced polyurethane foams *Compos. Sci. Technol.* **87** 210–7

[6] Solorzano E and Rodriguez Perez M A 2013 Cellular materials *Structural Materials and Processes in Transportation* ed D Lehmhus, M Busse, A S Herrmann and K Kayvantash (Weinheim: Wiley-VCH Verlag GmbH & Co. KGaA)

[7] Solórzano E, Laguna-Gutierrez E, Perez-Tamarit S, Kaestner A and Rodriguez-Perez M A 2015 Polymer foam evolution characterized by time-resolved neutron radiography *Colloids Surf. A: Physicochem. Eng. Aspects* **473** 46–54

[8] Dollet B and Raufaste C 2014 Rheology of aqueous foams *C. R. Phys.* **15** 731–47

[9] Hoehler R and Cohen-Addad S 2005 Rheology of liquid foam *J. Phys.: Condens. Matter* **17** R1041–69

[10] Saint-Jalmes A 2006 Physical chemistry in foam drainage and coarsening *Soft Matter* **2** 836 Köhler S A, Hilgenfeldt S and Stone H A 2000 A generalized view of foam drainage: experiment and theory *Langmuir* **16** 6327–41

[11] Safouane M, Durand M, Saint Jalmes A, Langevin D and Bergeron V 2001 Aqueous foam drainage. role of the rheology of the foaming fluid *J. Phys. IV* **11** Pr6-275–80

[12] Wills B A and Finch J A 2015 *Wills' Mineral Processing Technology: An Introduction to the Practical Aspects of Ore Treatment and Mineral Recovery* (Oxford: Butterworth-Heinemann)

[13] Rubio J, Souza M L and Smith R W 2002 Overview of flotation as a wastewater treatment technique *Miner. Eng.* **15** 139–55

[14] Kemper M 1999 State-of-the-art and new technologies in flotation deinking *Int. J. Miner. Process* **56** 317–33

[15] Neethling S J 2006 Effect of simple shear on liquid drainage within foams *Phys. Rev.* E **73** 061408

[16] Carrier V, Destouesse S and Colin A 2002 Foam drainage: a film contribution? *Phys. Rev.* E **65** 061404

[17] Köhler S A, Hilgenfeldt S and Stone H A 2004 Foam drainage on the microscale *J. Colloid Interface Sci.* **276** 420–38

[18] Carrier V and Colin A 2002 Anisotropy of draining foams *Langmuir* **18** 7564–70

[19] Zarebanadkouki M, Carminati A, Kaestner A, Mannes D, Morgano M, Peetermans S and Trtik P 2015 On-the-fly neutron tomography of water transport into lupine roots; E H Lehmann, A P Kaestner and D Mannes *Physics Procedia: Proc. of the 10th World Conf. on Neutron Radiography (WCNR-10) (Grindelwald, Switzerland, October 5–10, 2014) vol 69 pp 292–8*

[20] Heitkam S, Rudolph M, Lappan T, Sarma M, Eckert S, Trtik P, Lehmann E, Vontobel P and Eckert K 2018 Neutron imaging of froth structure and particle motion *Mineral. Eng.* **119** 126–9

[21] Heitkam S, Lappan T, Eckert S, Trtik P and Eckert K 2019 Tracking of particles in froth using neutron imaging *Chem. Ing. Tech.* **91** 1001–7

[22] Hydrogels, Wang W, Narain R and Zeng H 2020 Hydrogels *Polymer Science and Nanotechnology* ed R Narain (Amsterdam: Elsevier) ch 10, pp 203–44

[23] Lapidot S A and Kost J 2001 Hydrogels ed K H Jürgen Buschow, R W Cahn, M C Flemings, B Ilschner, E J Kramer, S Mahajan and P Veyssière *Encyclopedia of Materials: Science and Technology* (Amsterdam: Elsevier) pp 3878–82

[24] Sandrin D *et al* 2016 Diffusion of macromolecules in a polymer hydrogel: from microscopic to macroscopic scales *Phys. Chem. Chem. Phys.* **18** 12860–76

[25] Wagner D, Burbach J, Grünzweig C, Hartmann S, Lehmann E, Egelhaaf S U and Hermes H E 2016 Solvent and solute ingress into hydrogels resolved by a combination of imaging techniques *J. Chem. Phys.* **144** 204903

[26] Wagner D, Börgardts M, Grünzweig C, Lehmann E, Müller T J J, Egelhaaf S U and Hermes H E 2015 Neutron, fluorescence, and optical imaging: an *in situ* combination of complementary techniques *Rev. Sci. Instrum.* **86** 093706

[27] Holt C 1985 *Developments in Dairy Chemistry-3* (P. F. Fox London: Elsevier Applied Science Publishers)

[28] Metwalli E, Hermes H E, Calzada E, Kulozik U, Egelhaaf S U and Müller-Buschbaum P 2016 Water ingress into a casein film quantified using time-resolved neutron imaging *Phys. Chem. Chem. Phys.* **18** 6458–64

[29] Crank J 1975 *The Mathematics of Diffusion* (Oxford: Oxford University Press)
Sahimi M (ed) 2011 *Flow and Transport in Porous Media and Fractured Rock* (Oxford: Wiley-VCH)

[30] Kim Y, Valsecchi J, Oh O, Kim J, Lee S W, Boue F, Lutton E, Busi M, Garvey C and Strobl M 2022 Quantitative neutron dark-field imaging of milk: a feasibility study *Appl. Sci.* **12** 833

[31] Smith G N 2021 An alternative analysis of contrast-variation neutron scattering data of casein micelles in semi-deuterated milk *Eur. Phys. J.* E **44** 5

[32] Physics World 2017 Land of milk and neutrons *Phys. World* https://physicsworld.com/a/land-of-milk-and-neutrons/

[33] Smith G N 2021 Revisiting neutron scattering data from deuterated milk *Food Hydrocoll.* **113** 106511

[34] Lagorce-Tachon A *et al* 2015 The cork viewed from the inside *J. Food Eng.* **149** 214–21

[35] Aregawi W, Defraeye T, Saneinejad S, Vontobel P, Lehmann E, Carmeliet J, Derome D, Verboven P and Nicolai B 2013 Dehydration of apple tissue: intercomparison of neutron tomography with numerical modelling *Int. J. Heat Mass Transf.* **67** 173–82

[36] Defraeye T, Aregawi W and Saneinejad S *et al* 2013 Novel application of neutron radiography to forced convective drying of fruit tissue *Food Bioprocess. Technol.* **6** 3353–67

[37] Tanoi K, Hamada Y, Seyama S, Saito T, Iikura H and Nakanishi T M 2009 Dehydration process of fish analyzed by neutron beam imaging *Nucl. Instrum. Methods* A **605** 179–84

[38] Vorhauer-Huget N, Mannes D, Hilmer M, Gruber S, Strobl M, Tsotsas E and Foerst P 2020 Freeze-drying with structured sublimation fronts—visualization with neutron imaging *Processes* **8** 1091

[39] Gruber S, Vorhauer N, Schulz M, Hilmer M, Peters J, Tsotsas E and Foerst P 2020 Estimation of the local sublimation front velocities from neutron radiography and tomography of particulate matter *Chem. Eng. Sci.* **211** 115268

[40] Foerst P, Gruber S, Schulz M, Vorhauer N and Tsotsas E 2020 Characterization of lyophilization of frozen bulky solids *Chem. Eng. Technol.* **43** 789–96

[41] Thomik M, Gruber S, Kaestner A P, Foerst P, Tsotsas E and Vorhauer-Huget N 2022 Experimental study of the impact of pore structure on drying kinetics and sublimation front patterns *Pharmaceutics* **14** 1538

[42] Scussat S, Ott F, Hélary A, Desert S, Cayot P and Loupiac C 2016 Neutron imaging of meat during cooking *Food Biophys.* **11** 207–12

[43] Scussat S, Vaulot C, Ott F, Cayot P, Delmotte L and Loupiac C 2017 The impact of cooking on meat microstructure studied by low field NMR and neutron tomography *Food Struct.* **14** 36–45

[44] Hua T-C, Liu B-L and Zhang H 2010 *Freeze-Drying of Pharmaceutical and Food Products* (Burlington, VT: Elsevier Science)
[45] Kharaghani A, Tsotsas E, Wolf C, Beutler T, Guttzeit M and Oetjen G-W 2000 Freeze-drying *Ullmann's Encyclopedia of Industrial Chemistry* (Weinheim: Wiley-VCH Verlag GmbH & Co. KGaA) pp 1–47
[46] Haseley P and Oetjen G-W 2017 *Freeze-Drying* 3rd edn (Newark, NJ: Wiley)
[47] Vorhauer N, Först P, Schuchmann H and Tsotsas E 2018 Pore network model of primary freeze-drying *Proc. of the International Drying Symp. (Valencia)* pp 11–4
[48] Rasetto V, Marchisio D L, Fissore D and Barresi A A 2010 On the use of a dual-scale model to improve understanding of a pharmaceutical freeze-drying process *J. Pharm. Sci.* **99** 4337–50
[49] Vopička O, Čihal P, Klepić M, Crha J, Hynek V and Trtík K *et al* 2020 One-pot neutron imaging of surface phenomena, swelling and diffusion during methane absorption in ethanol and n-decane under high pressure *PLoS One* **15** e0238470
[50] Vopička O, Durdáková T M and Čihal P *et al* 2023 Absorption of pressurized methane in normal and supercooled p-xylene revealed via high-resolution neutron imaging *Sci. Rep.* **13** 136
[51] Im J K, Jeong L, Crha J, Trtik P and Jeong J 2021 High-resolution neutron imaging reveals kinetics of water vapor uptake into a sessile water droplet *Matter* **4** 2083–96
[52] Hermes H E, Sitta C E, Schillinger B, Löwen H and Egelhaaf S U 2015 Kinks in experimental diffusion profiles of a dissolving semi-crystalline polymer explained by a concentration-dependent diffusion coefficient *Phys. Chem. Chem. Phys.* **17** 15781–7
[53] Trotzig C, Abrahmsén-Alami S and Maurer F H J 2007 Structure and mobility in water plasticized poly(ethylene oxide) *Polymer* **48** 3294–305
[54] Yuan T Z *et al* 2015 Shear-stress-mediated refolding of proteins from aggregates and inclusion bodies *ChemBioChem* **16** 393–6
[55] Vimalanathan K, Chen X and Raston C L 2014 Shear induced fabrication of intertwined single walled carbon nanotube rings *Chem. Commun.* **50** 11295–8
[56] Britton J, Meneghini L M, Raston C L and Weiss G A 2016 Accelerating enzyme catalysis using vortex fluidics *Angew. Chem. Int. Ed.* **55** 11387–91
[57] Solheim T E, Salvemini F, Dalziel S B and Raston C L 2019 Neutron imaging and modelling inclined vortex driven thin films *Sci. Rep.* **9** 2817
[58] Strobl M, Betz B, Harti R P, Hilger A, Kardjilov N, Manke I and Gruenzweig C 2016 Wavelength dispersive dark-field contrast: micrometer structure resolution in neutron imaging with gratings *J. Appl. Cryst.* **49** 569–73
[59] Harti R P, Strobl M, Betz B, Jefimovs K, Kagias M and Gruenzweig C 2017 Sub-pixel correlation length neutron imaging: spatially resolved scattering information of micro-structures on a macroscopic scale *Sci. Rep.* **7** 44588
[60] Strobl M 2014 General solution for quantitative dark-field contrast imaging with grating interferometers *Sci. Rep.* **4** 7243
[61] Zanini M, Marschelke C, Anachkov S E, Marini E, Synytska A and Isa L 2017 Universal emulsion stabilization from the arrested adsorption of rough particles at liquid-liquid interfaces *Nat. Commun.* **8** 15701
[62] Harti R P, Valsecchi J, Trtik P, Mannes D, Carminati C, Strobl M, Plomp J, Duif C P and Grünzweig C 2018 Visualizing the heterogeneous breakdown of a fractal microstructure during compaction by neutron dark-field imaging *Sci. Rep.* **8** 17845

Part VI

Cultural and natural heritage

IOP Publishing

Neutron Imaging
From applied materials science to industry
Markus Strobl and Eberhard Lehmann

Chapter 15

Cultural heritage

David Mannes, Eberhard Lehmann, Francesco Grazzi and Francesco Cantini

The use of neutron imaging techniques as a method for non-invasive investigation of hidden properties and structures of museum and art objects from the human history has quite a long tradition [1–3]. In this way, the high penetration power of mainly thermal neutrons for heavy (metallic) elements was used in combination with the detectability of organic components inside the object under investigation. The resulting neutron images have therefore often a nearly inverse contrast scheme compared to the more common x-ray methods, however, depending always on the particular object. An example is given in figure 15.1, showing clearly the mentioned difference in the use of the two kinds of radiation.

Due to lack of alternatives, the very early investigations with neutrons were carried out as radiographies using film methods. They were limited in spatial resolution, valid dynamic range and contrast and were not useful for a precise quantification of the content of the material within the samples. The radiography provides an overlay as integration of all materials along the penetrating beam direction. It is therefore impossible to obtain any depth information of the features inside the objects.

Furthermore, these early studied objects were arbitrarily and randomly selected in order to demonstrate and showcase the performance and applicability of neutron radiography and were by far not the content of systematic studies by archaeologists and museums experts. Hence, these investigations were driven and motivated mainly by the operators of neutron imaging facilities and the actual scientific question was in many cases only of secondary importance.

With the introduction of digital neutron imaging at the end of the last century, new methods like tomography, time-resolved imaging and high-resolution imaging were enabled. In this course, also dedicated studies for cultural heritage objects could benefit. Much deeper insights into the materials in 3D or even 4D (3D + time dependency) were from then on possible and the data were stored, transferred, processed with dedicated tools and circulated on a digital level around the world.

doi:10.1088/978-0-7503-3495-2ch15

© IOP Publishing Ltd 2024

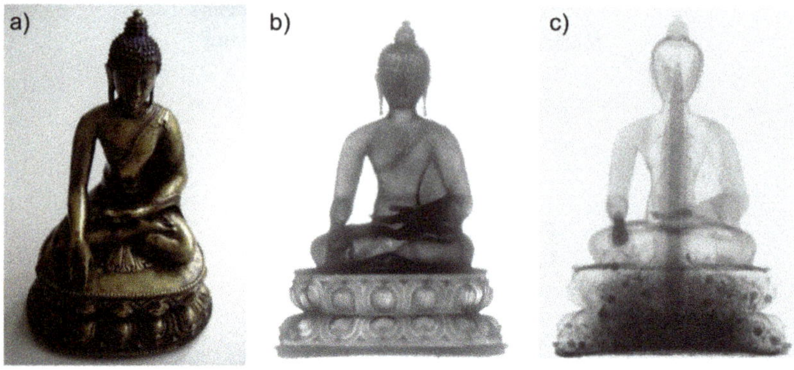

Figure 15.1. Different contrasts shown on the example of a Buddhist sculpture (Buddha Shakyamuni, Tibet, approximately 15th century (photo right), private collection M Speidel); (a) photo, (b) x-ray (120-kV high-voltage tube) transmission image: figure appears to be empty, (c) neutron transmission image (thermal neutrons) of the statue organic filling (e.g. wooden pillar, textile layers around, plants in the lotus throne base, etc) become visible as consequence of the complementarity of the two types of radiation (source: [16] modified, John Wiley & Sons. Copyright University of Oxford, 2009).

In this way, networks for systematic usage of neutron methods were established on a national, continental and global level [4, 5]. However, it was necessary to bring together two communities with completely different mindset, working procedures and focus—researchers for neutrons on one side and art and historical scientists from humanities on the other side. Not only the kind of thinking, but also the particular knowledge about historical objects, their value and importance have to be communicated and adapted in the best way. This 'moderation' process is still needed today in each newly joined project of investigations of cultural heritage objects. Museums experts must be convinced about the neutron imaging technique performance and its possibilities. The facility operators, who most often provide also the final data processing, need to understand the particular problems and the interpretation of the obtained results. A tight collaboration is therefore needed for successful studies.

Also in the recent time, facility operators [6–8] published results of cultural heritage objects studies and demonstrated in this way their performance and the established methods. It is to underline, that this can be only the first step into continuous collaborations with museums experts. The further steps should be more systematic investigations, devoted to the needs and questions from the side of the archaeologists and historians in solving their real problems. In the best way, neutron imaging methods have to be adapted and improved to the needs of the particular study. This has to be done by choosing the most suitable experimental set-up with respect to, e.g. neutron energies, the best field-of-view and spatial resolution, adequate climate conditions, the utilization of contrast agents or by comparing/combining neutron to x-ray data (data fusion).

This chapter will highlight some main aspects for studies with cultural heritage objects and show examples of investigations.

15.1 Historic artworks

What remains from previous cultures after many centuries of material decay and degradation? Only the most robust and durable materials—metals, stones, ceramics and in a few cases organic material (such as wood, bone or leather) have survived. In this first paragraph, the focus lies on studies of bronzes sculptures. For such objects, the 'survival chances' are best due to the limited corrosion rate and their high material values in the time of their production and usage.

Copper alloys, such as bronze have been in use by humanity for millennia (cf 'Bronze age') and have been ubiquitous in many parts of the world (e.g. Africa, Asia, Europe). Bronze (or other copper alloys), which was considered in general as a very precious material was used for a multitude of purposes and objects, such as weapons, jewellery, decorative objects, artefacts for cultural or religious activities, etc. As before-mentioned such objects are often still in very good condition even after thousands of years.

In Europe for example, many of such artefacts remain from the Roman period. They are now stored and exhibited in famous museums and collections around the world [9]. For a deeper analysis of structure, content and production techniques, neutron imaging methods are very useful. Bronze and brass are copper alloys which can contain besides Cu, portions of Zn, Sn and Pb. The particular composition can be determined, e.g. by x-ray fluorescence methods, even if these are only sensitive to layers close to the sample surface. Other neutron methods like neutron activation analysis (NAA) or prompt-gamma activation analysis can be used as support of the x-ray methods. Because neutrons can penetrate deeper into the materials, these methods are also able to analyze bulk properties and inner material distributions.

The advantage of neutron imaging compared to x-ray imaging can be demonstrated by the comparison of values of linear attenuation coefficients of the alloy's components, available in corresponding data libraries [10, 11]. The linear attenuation coefficient is a measure for the reduction of the radiation intensity by a layer of the exposed material. The much lower values for the neutron case are the indications for the much higher penetrability through even thick bronze layers (see figure 15.2).

15.1.1 Studies about bronze artefacts from the Roman period, found in parts of Switzerland

Initiated by the Swiss National Museum, Zurich, and in collaboration between the University Zurich, Department of Archaeology and the Paul Scherrer Institut, supported by the Swiss National Science Foundation, a representative number of small bronze sculptures from the Roman period, available from different museum's collections within Switzerland, was investigated by neutron imaging methods.

From the start of the project it became clear that standard x-ray methods would fail in most cases because of the limited transmission power for the materials used for the manufacturing of the objects, as a consequence of the material composition as explained already above. Because the Romans used also lead to lower the melting point of the bronze alloy, the x-ray transmission for Roman bronze objects is most often very challenging. The example in figure 15.3 demonstrates this fact and

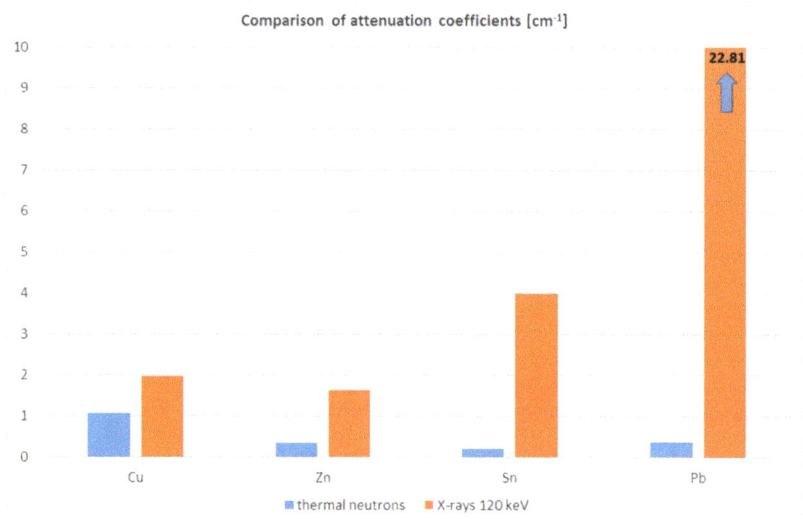

Figure 15.2. Comparison of linear attenuation coefficients of the bronze alloy constituents for thermal neutrons and x-rays, respectively.

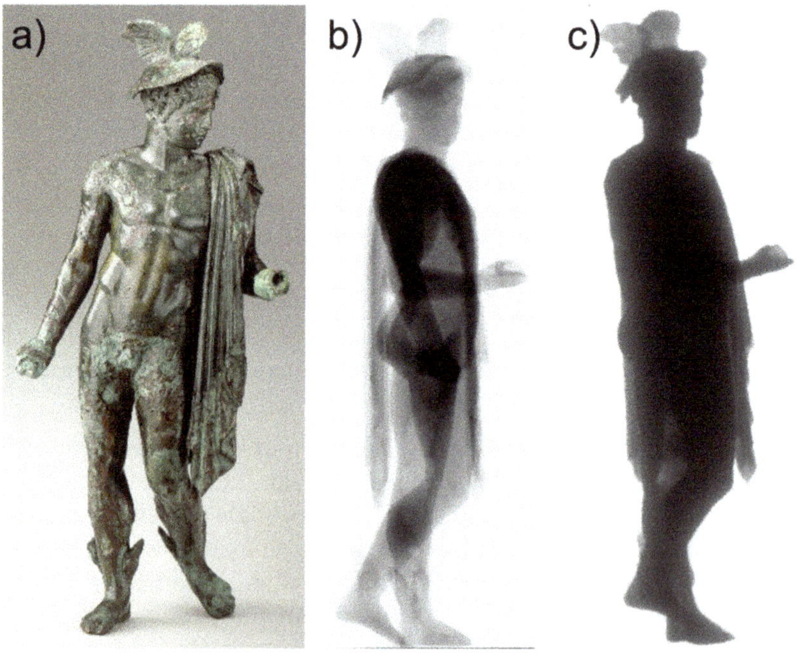

Figure 15.3. Bronze statuette depicting the Roman deity Mercury ((a) photo); comparison of neutron (b) and x-ray (c) images (info on the sculpture: Mercury from Thalwil, exhibit from the Swiss National Museum, Inv. A-3447, height: ca. 12 cm).

underlines the reason why in such a case the utilization of neutrons shows some advantages when compared to x-rays (as for other bronze artefacts—see below).

Studies with neutron imaging methods (radiography, tomography) allow an insight into the objects. For practical reasons (e.g. weight) and because the base material bronze was valuable, most of such sculptures as shown in the example, were cast as hollow structures and not as entire solids. The casting process was already quite sophisticated and continued to be used over centuries.

Whereas simple radiographies give first information about composition and inner structures, tomography data allow slices at arbitrary positions through the object, where precise values of the dimensions and the local compositions can be derived. In this way, also the casting quality, cavities in the metal and remains of casting cores can be determined, visualized and measured dimensionally. Because neutron tomography requires at least a few to several hundred views through the objects from different angular positions around the object, this method is much more demanding than simple radiography. In addition, the volume reconstruction and data visualization requires computation efforts and qualified manpower for the deeper analysis.

This is the reason why only about 50 out of 200 pieces from the different Swiss collections were analyzed by neutron tomography within this project. In the remaining cases, easy radiography images gave enough information because no inner properties and interesting features were expected.

Results of this study were presented during the 18th International Congress on Antique Bronzes [9] and should be published in the Catalogue of Roman Bronzes in Switzerland.

The investigated samples represent often mythological characters or deities (such as Mercury, Minerva, Venus, Mars, …), displays of important persons (emperors) or artefacts for handling spiritual activities. Often, these objects were in use in private environments. An example of a successful tomography analysis is shown in figure 15.4 by comparing the reality (photo) with the 'virtual reality' of the outer tomography view. Artificial slices are enabled in this way in all directions.

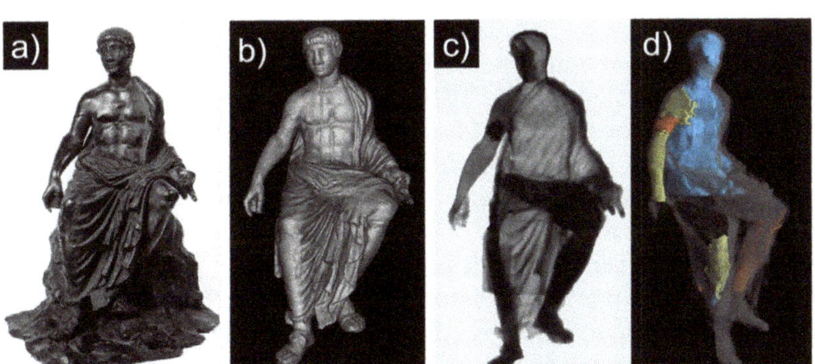

Figure 15.4. Example of the 'Mercury from Ottenhusen, a bronze object from the Roman period, comparison between photo (a), outer neutron tomographic view (b), neutron radiography (c) and 3D rendering view of the object with half-transparent surface and regions with different attenuations on the inside (blue, void; yellow, core material; orange, higher attenuating material) (modified from [12], copyright (2022), with permission from Springer Nature).

The neutron imaging data are very helpful for the understanding of origin, purpose and usage of the different found pieces.

15.1.2 Tomography studies of renaissance bronzes from the Rijksmuseum Amsterdam, the Netherlands

The famous Rijksmuseum in Amsterdam, the Netherlands, owns an important collection of bronze sculptures from the Renaissance era, dated between 15th and 17th century. The objects originate mostly from Italy and Northern Europe and exhibit outstanding quality [13].

It was a unique opportunity when the Rijksmuseum undertook a huge renovation campaign in the years from 2003 to 2013. During that period all collections were stored outside the exhibitions in depots, which allowed carrying out dedicated studies with neutron methods on parts of the collection. By the initiative of the chief restorer of the Rijksmuseum, R van Lang, neutron tomography was used for investigating some of these bronze sculptures for a better understanding of the manufacturing techniques, the inner structures and composition of the casting products and their conservation status. The objects were selected in order to fit the beam size of the used neutron imaging facility (NEUTRA at SINQ, PSI) with a maximum diameter of 40 cm. In some cases, it was necessary to scan the object at different height positions and to stitch the resulting image data in a later data processing step.

Similar to the example of the Roman bronze objects, x-ray imaging tests failed to a large extent for these larger samples, which is why neutron imaging was preferred for these investigations. Fourteen objects were successfully studied by means of neutron tomography. Before the actual tomography run (with total exposure time of several hours), overview radiography images were acquired in order to confirm a sufficient transmission in all observation directions. Only a few objects could not be taken into account for the investigation campaign, due to lacking neutron transparency; the main reason was in most cases remains of the bulky casting core, which hindered the penetration by thermal neutrons.

Exemplarily and representative for all other objects, we show here the case of the 'striding Nobleman', dated around 1610, inventory number BK-16083, originating from the Netherlands or Germany. It was produced using the method of 'indirect cast'; arms and lower legs were cast separately and added subsequently to the sculpture by brazing. For stabilization, some reinforcing rods were added in the range of shoulders and knees.

All these hidden, inner structures become visible with the neutron tomography using suitable visualization tools (figure 15.5). Because the brazing material has a different composition compared to the bulk alloy, the contrast is so deviating allowing to clearly distinguish between them. The data also enables the measurement of the wall thickness of the object at any position with precision of the spatial resolution of the method (in this example 0.2 mm). Furthermore, the quality of the casting can be evaluated by inspecting the untreated inner surface of the object (roughness, holes, supporting structures). Finally, the tomography data allow

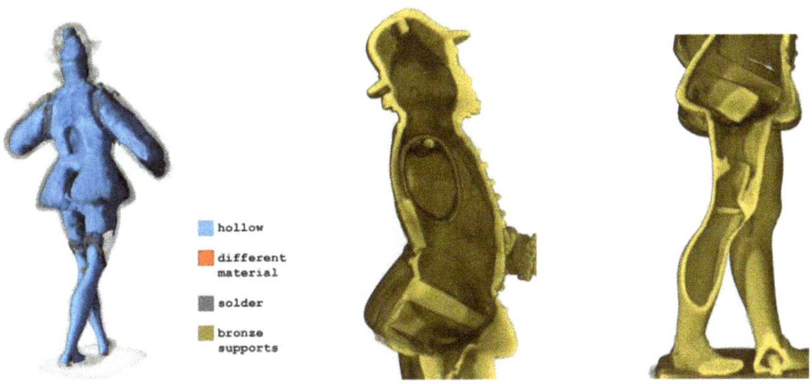

Figure 15.5. The 'Striding Nobleman' (inventory number BK-16083), manufactured around 1610, tomographic view of empty regions and solder connections (left), slices of the sculpture enables to see the inner reinforcement with bronze rods in the shoulders and knees (middle and right). Adapted from [14], Copyright (2017), with permission from Springer Nature.

deriving a 3D-model for CAD representation, which can be used for building copies, e.g. with laser printing, either as 1:1, minified or even magnified.

Further views and animations of the 14 objects were published [14] and links to 3D movies via YOUTUBE are given. The PhD thesis by R van Lang [15] also presents an example linking neutron imaging data to neutron diffraction results; this yields spatially resolved information on the crystalline structure and the casting process. This all gives insight to the manufacturing methods, common several hundred years ago.

15.1.3 Observation of the hidden content of Buddhist sacral figures of Tibetan origin

Buddhist religion and culture is accompanied by images of Buddha often in the format of sculptures made out of bronze alloys. Spread over all countries with Buddhist traditions (India, Tibet, Nepal, Burma, China, …), these sculptures have been produced for more than 500 years and have been stored in monasteries, temples and in rich human buildings. In addition to depictions of Buddha in his manifold appearances, there are also Stupas, pyramidal casted objects and displays of important priests or lamas—all made of bronze alloys.

Similar to the examples presented in the previous paragraphs, most of these objects have been cast hollow with a few millimeters wall thickness, in order to save the relative expensive alloy material—this is a demanding and sophisticated procedure, taking into account the technological level at the time of their creation.

However, the hollow structure has given the opportunity to fill these objects in a dedicated way with important material of spiritual importance. After the filling procedure within a particular ceremony, the samples were closed and sealed—for ever!?

Initiated by private, well-experienced collectors [16, 17] first investigations were performed with both x-rays and thermal neutrons [18]. It was proven in several cases that thermal neutrons enabled a high penetration of the metal cover while the

Figure 15.6. The Buddha sculpture BA133 from the Aschmann Foundation within the Rietberg Museum Zurich, representing the Great Lama Pandita of the Sa-skya school (1182–1251), left: outer tomographic view, middle: neutron transmission radiography, right: vertical tomographic slice in the middle of the object (reprinted from [19], copyright (2018), with permission from Elsevier).

content (filling of mainly organic materials) became clearly visible. On the other hand, x-rays have much lower chance to transmit the metal; when using higher x-ray energies, which allow penetration of the metal, the mostly organic filling material seems to be invisible because of only low attenuation w.r.t. the x-rays.

Based on this experience, connections to further collections and museums within Switzerland were established with the aim to learn more about the religious content of bronze statues, created, filled and consecrated many centuries ago.

An example is given in figure 15.6 as the result of a tomography study of the Great Lama Pandita of the Sa-skya school, dated 13th century from Tibet. With the help of software tools, the data can be used to navigate through the sample and to extract hidden parts of the filling. Because each filling has been made individually, many different objects have to be compared to find systematics and variations.

What can be derived from these data on Buddhist bronzes?

1. Material properties of the cover bronze (or brass) material: the homogeneity, the casting quality (inclusions, voids, bubbles, …), the wall thickness and repair work can be made visible and measurable.
2. Content of the consecrate filling: identifying of 'ingredients', their arrangement and amount. Because the materials have been stored for centuries, their status of conservation and performance can be checked non-invasively—and compared to modern methods and materials of fillings, because these procedures were common until today.
3. The question of whether the sculpture is a copy, a fake or true historical sample can be answered better based on the knowledge about the inner content and the material distribution, provided by the neutron imaging data. This is important because a mechanical opening of the object is not allowed for religious reasons. Such 'inspections' would desecrate the sculpture and reduce its (spiritual and monetary) value dramatically.

4. Because figures of this cultural area and age have a high impact and value on the art market they can reach selling prizes on the order of millions of US dollars. Non-destructive inspection results have been taken as arguments on whether the object satisfies expectations on its provenance.

Until today and just taking into account investigations carried out at the neutron imaging facilities of PSI, over 80 Buddhist objects were studied, most of them with neutron tomography, all at least with neutron radiography. The results of these studies will be made public within a database, to be publicly accessible via the internet.

15.1.4 Preliminary studies about the casting methods in Southern India Chola bronze statues

Some results related to one of the four bronze statues from the Asian Pavilion of Rijksmuseum investigated through a diagnostic study are reported in this short contribution. The study of these masterpieces was conducted by neutron imaging at Helmholtz-Zentrum Berlin (HZB) on CONRAD-2, thanks to an informal agreement between the Italian National Research Council (CNR) and the Rijksmuseum in Amsterdam. The data shown and discussed here as part of the chapter are the result of the work of the Master's Thesis in Conservation Science at the University of Florence by Francesco Cantini, and published in a dedicated paper [36].

The four bronze statues were analyzed with white beam neutron tomography and monochromatic neutron radiography; the results were discussed and interpreted in collaboration with the conservators of the Rijksmuseum Metal Conservation Department: Dr Sara Creange, Dr Joosje van Bennekom. In the following we present some results relating to a solid cast bronze statuette representing the Shiva Indian deity.

The goal of the diagnostic investigation was trying to answer some of the technological and conservation issues raised by conservation staff:

- identify the weakest areas of the metal and assess the conservation state;
- evaluate any phenomena of diffusion of cracking inside the statues;
- identify interventions of restoration, replacement or integration of parts of the statues;
- characterize the alloy microstructure in the different areas of the whole statues;
- obtain clues to be able to confirm or guess details on the casting process, especially the orientation of the mould.

Neutron imaging techniques were able to answer all the questions with different levels of completeness.

The Shiva with the moon in his hair (AK-MAK-1291) statuette (figures 15.7(A) and (B)) is a Chola-style bronze originally from Tamil Nadu supposed to be cast in the period 1000–1200 AD. According to the curators of the Rijksmuseum, the sculpture and the pedestal should be a single casting, while the flame cornicing

Figure 15.7. 'Shiva with the moon in his hair' (AK-MAK-1291). (A) Front, (B) side, pictures of the statuette. (C), *xz* neutron tomography slice; (D) *yz* neutron tomography slice. The statue is a solid cast bronze characterized by a ubiquitous porosity along the central axis of the body. It is also evident the presence of areas with different attenuation power, which can be correlated to different composition and/or dendritic micro-structure of the various portions. Reconstruction artefacts due to the halo shadowing effect are visible in (D).

structure surrounding Shiva (halo), was produced separately and mechanically fixed to the pedestal.

White beam tomography confirms that Shiva is a solid bronze statue characterized by a ubiquitous porosity along the central axis of the body, as can be seen from figures 15.7(C) and (D). There are no evident fractures inside the statue; the presence of corrosion patinas is limited to small interstitial areas, such as the attachment of the ear lobes (particularly attenuating area, white in the picture).

Concerning the morphology of the *halo*, the attenuation power of the metal is comparable with that of the body of the statue, but no sign of porosity can be detected from the imaging data.

Monochromatic radiography evidenced the presence of large size almost perfect single crystal (oligo- crystal) within the solid bronze matrix, of both the statue's body and the most part of the *halo*. They appear as dark spots and the phenomenon is related to a strong coherent scattering effect removing neutron from the primary beam. Oligo-crystals can certainly be attributed to the bronze microstructure since they are no longer visible in radiographs taken at wavelengths over the (111) family of copper lattice planes (FCC), where the scattering effect of such a phase no longer contributes to the attenuation.

Two radiographs showing this phenomenon are reported in figure 15.8(A).

The presence of such oligo-crystals, identified by the monochromatic neutron radiographs and of dendrites, visible from the tomographic sections, suggests that the statue has undergone a slow cooling suggesting the statue was cast underground [20].

The *halo* radiographs show the presence of oligo-crystals affects the entire development of the arch but their size abruptly decreases at the height of the right forearm (indicated by a red line in figure 15.8(B-I)). It is also evident that the decrease in the size of the crystals appears to follow a certain angle (figures 15.8(B-III) and (B-IV)).

The size of the grains is strictly related to the speed of the solidification process; this implies that part of the *halo* must have cooled down faster since the grains are

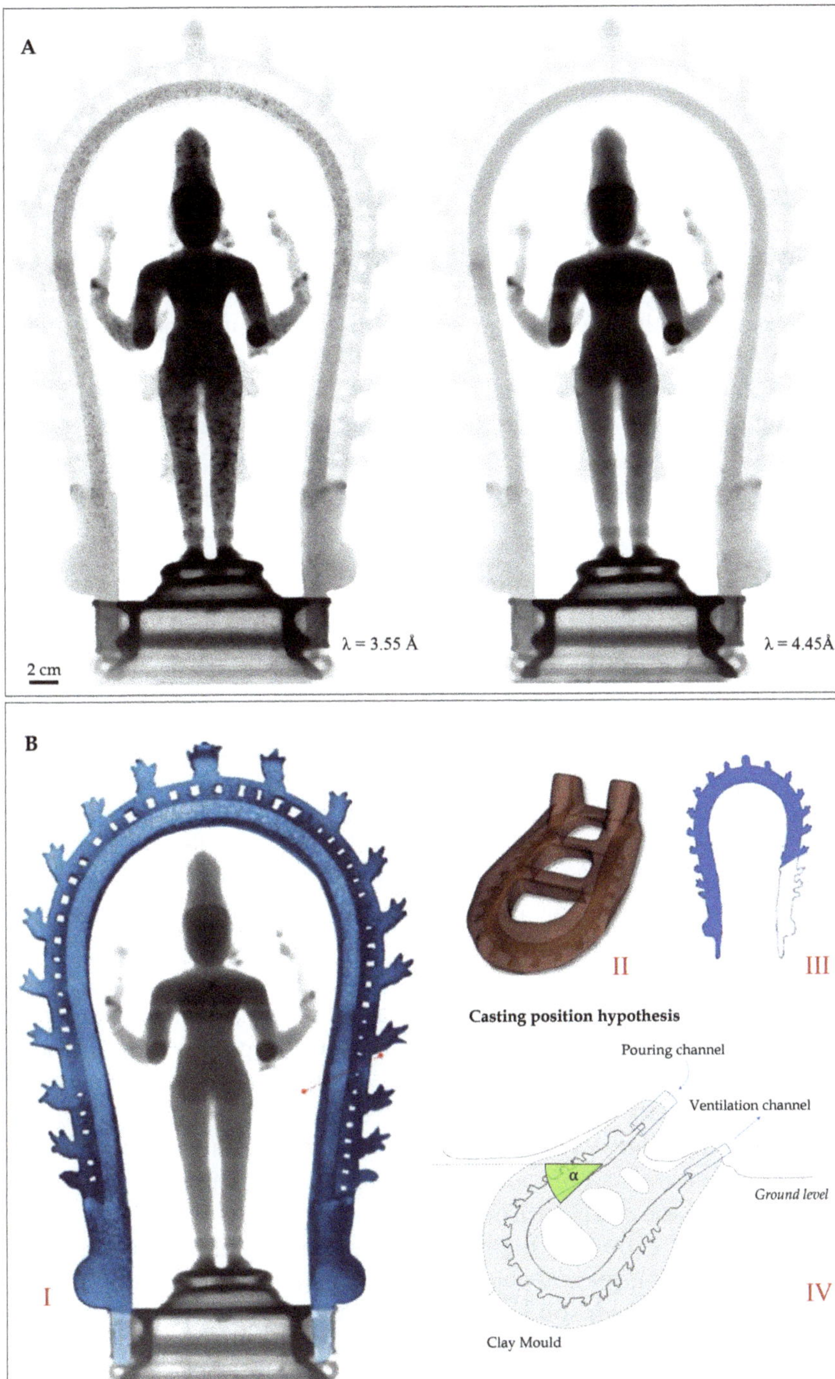

Figure 15.8. Shiva with the moon in his hair (AK-MAK-1291). (A) Monochromatic neutron radiographs acquired at $\lambda = 3.55$ Å and $\lambda = 4.45$ Å, before and after the last Bragg Edge of the copper (FCC structure)

lattice planes: [111] family. As is evident from the $\lambda = 3.55$ Å radiograph there is wide diffusion of bronze oligo-crystal grains appearing as dark spots (size for the body of Shiva ~1 mm; for the halo ~100 μm). (B) Hypothesis of positioning of the mould for casting the halo, based on the oligo-crystal spatial distribution: (I) enhancement of the halo structure emphasizing the visualization of oligo-crystal distribution. (II) Schematic diagram of the clay moulds for decorative elements (adapted from [21]). (III) Scheme showing in blue the halo area probably subjected to slower cooling: (IV) Casting position hypothesis diagram of a possible arrangement of the mould in the ground.

smaller. Based on the models described in the literature and on the techniques of the artisan tradition [21], a casting model for the *halo* is proposed (figure 15.8(B-IV)). It is hypothesized that the mould was placed in the ground, perpendicular to the ground level.

The *halo* and the Shiva figure were cast separately, possibly partially or completely underground; the different metal thickness brought to the formation of different size oligo-crystals due to the different cooling rate, induced by size effect. The only exception is constituted by one end part of the *halo*, which was likely cast overground.

15.2 Historic weapons

15.2.1 Neutron imaging capability to reveal morphological and microstructural features in iron and steel historical artefacts

Historical and archaeological weapons made of iron and steel represent a very interesting set of artefacts in the research of the evolution and development of metallurgical technologies in different historical periods and geographical areas. In fact, iron and steel represented, for several centuries, the metals exhibiting the best performances in terms of hardness, resilience and workability, combined with an abundant distribution of raw materials. The most intense efforts in optimizing such performances followed different paths and brought to the development of unique and peculiar methods of refinement, and of application of thermal and mechanical treatments.

Iron ore can be refined using a reducing furnace and the temperature range, the size, the ratio between iron ore and charcoal fuel, the purity level of the ore and several other minor parameters can strongly affect the quality of the produced metal in terms of homogeneity in carbon content, presence of solid non-metal inclusion generally classified as slags. The reduction in size, and number, and the neutralization of the detrimental effect of such slags within a metal artefact is among the most interesting topics in the study of iron and steel technology evolution.

Unless steel has an extremely high carbon content, it is almost impossible to melt iron and steel using furnace- and charcoal-based historical technology so all artefact production always relied on solid state treatments.

Since the iron production usually provided small size blocks, in order to produce large artefacts, the welding technology was developed, in which separate metal cakes were connected together trying to strengthen at the most the joining surface among them. These areas usually contain a lot of non-metal inclusions describing the shape of the welding surface.

Iron (and steel) follows a phase transition at about 900 °C (730 °C in steel) turning from the body-centred cubic structure called ferrite (plus a second phase called cementite in steel) to a face-centred cubic structure called austenite (still in a mixture with ferrite, depending on the carbon content). Austenite is easier to be worked through plastic deformation, so that the most part of iron and steel smithing has been historically performed at high temperature, over the ferrite-austenite transition, which can be visually identified by the weak reddish light emission of metal at such temperature. Some final mechanical adjustments in artefacts could have been performed at room temperature instead.

Because of such temperature-related phase transition, the cooling rate can affect the grain size and then the mechanical properties of the room temperature artefacts. For this reason there was the development of the technology of thermal treatments, able to modify the grain size and allow the formation of different low temperature phases, containing carbon. These processes were empirically related to the cooling medium (air, calm water, running water and other liquids) and to the starting high temperature of the artefact. A very hard and brittle phase called martensite (rich in carbon) was possible to be formed on the surface of fast cooling steel artefacts.

Neutron imaging has the capability of identifying in a univocal way the most part of the aforementioned features since the attenuation power of steel artefacts is influenced by the composition (metal, smelting slag inclusions, welding slags), the grain size (higher attenuation for smaller grains), the different phase arrangement (i.e. martensite exhibits a larger attenuation coefficient than any other steel related phase) and the application of cold work treatments (leaving structural stress visible thanks to coherent scattering attenuation effects). Then neutron imaging represents one of the most relevant (if not the best) investigation techniques able to provide important technological features in iron and steel artefacts in a completely non-invasive way. Moreover, the activation level reached by neutron irradiated carbon steel historical artefacts is almost always negligible and they can be recovered by users, a few hours after the end of the experiment.

Here we report a set of three groups of historical steel samples, analyzed using cold neutrons in three European neutron facilities: ICON at the Paul Scherrer Institut, CONRAD-II at the Helmholtz-Zentrum Berlin and ANTARES at Meier Leibnitz Zentrum.

The first example is the cold neutron white beam tomography of the blade of a Japanese halberd known as *naginata* and constituted by a single-edge steel blade, similar to that of a long sword, with a very long tang and mounted on a long wooden pole. Up to now, only very few metallographic analyses were performed on a small section of naginata blades [22] and literature is only available in Japanese. Neutron tomography was performed at the CONRAD-II beamline in 2017, on a blade provided by Mr David Edge, the former head of conservation of the Wallace Collection Museum in London. The blade is believed to be not older than the last decades of the 19th century so it is likely it is produced in a hybrid way with respect to traditional steel forging typical of Japanese swords, with some modern concept forging procedure. The white beam tomography was performed on the upper part of the blade, including the tip, and it allowed us to extract several slices aligned along the main geometrical directions of the blade itself, which revealed a very homogeneous composition, with almost total absence of defects and slag inclusions and definitely no visible welding lines. This demonstrates that the blade was either forged from a single piece of metal (which is not typical of traditional Japanese sword forging) or made by different parts welded together using modern clean welding techniques which is not compatible with a late 19th century artefact (figure 15.9).

The artefact exhibits large surfaces with high attenuation power, especially on the sides, testifying to the presence of areas rich in martensite which is, instead, almost absent on the cutting edge, where the highest amount of this hard phase is expected (figure 15.10). Figure 15.10 also shows that, stretching the contrast, the metal grain of the core steel is characterized by some attenuation inhomogeneities, which follow a very regular periodic pattern on the blade length, which could be attributed to the forging process: the regular path could be related to a modern early industrial production method in which a trip hammer and a mechanically repeated blade sliding process was applied.

Concerning two Japanese blades provided by a professional sword polisher from Genève (Switzerland) and dated 16th (long sword) and 18th Century (short sword), their microstructure appears completely different [23]. They were measured, through white beam cold neutron tomography, on the ICON beamline at the Paul Scherrer Institut. Selected tomography slices show the presence of several features typical of a traditionally hand forged artefact (figure 15.11).

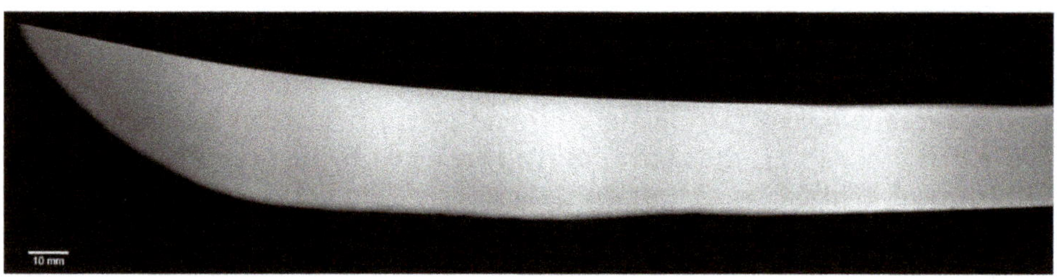

Figure 15.9. Side view neutron tomography section of the naginata blade.

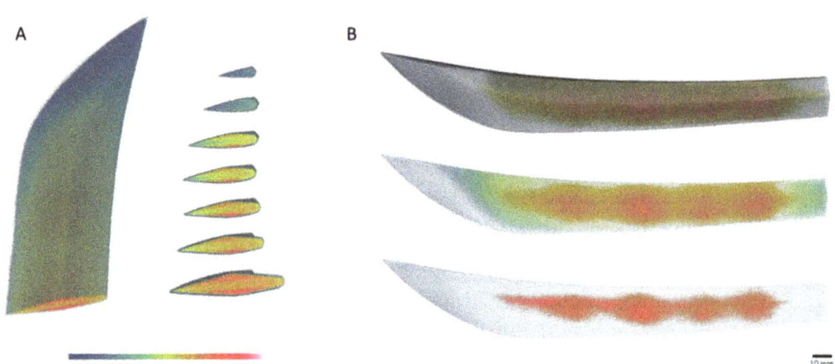

Figure 15.10. (A) 3D false coloured rendering of the tip of naginata blade. (B) Side view of the 3D false coloured naginata blade stretching the contrast within the steel.

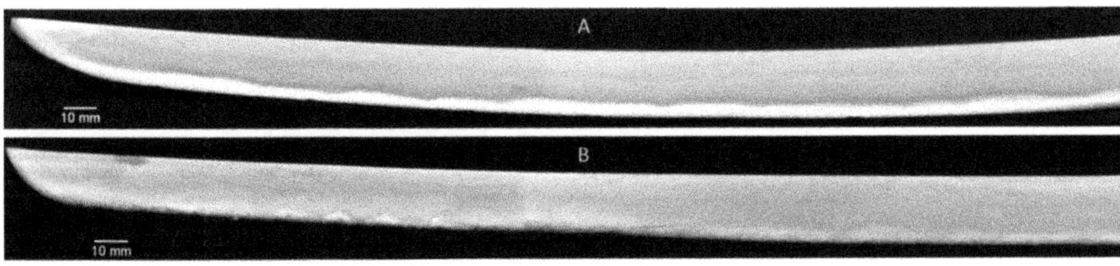

Figure 15.11. Neutron tomography slices of two Japanese blades: (A) 18th century short sword; (B) 16th century long sword.

There are several welding lines showing that the blades were both forged using different steel sheets probably employed for different functional behaviour (soft resilient core, surrounded by hard edge and sides). This process generated several cracks and flaws within the blades. The 16th century blade appear less homogeneous than the 18th century one. Both blades also show a very well pronounced martensitic edge but the relative concentration of martensite (level of brightness) appear different. The older long sword has a thin martensitic edge which is mechanically well connected to the blade core thus confirming the hypothesis this is a very solid and reliable blade employed for fighting. The short sword, in contrast, shows a very wide martensitic edge which has a dark area in the volume immediately behind it towards the blade core. The edge looks less connected to the core. This phenomenon can be interpreted as the exasperation of the quenching in a way to maximize phase spatial separation and enhance the beautiful chromatic effect of the temper line (*hamon*) visible thanks to the traditional polishing on the blade surface. This procedure could be applied only to blades which are not intended to be used in battle (as it was in 18th century when the blades only constituted a status symbol) since the mechanical stability of the blade was seriously compromised by the porous rich area behind the hard edge. This effect is due to the mechanical shrinking of the metal during a very fast and severe quenching in which the thermal contraction of martensite with respect to the rest of the blade is so different as to strip out some areas in the connecting part, thus generating micro-pores which optically enhance the *hamon* line after polishing. The short blade also shows the signs of cold working in the tip confirming that the quenching procedure brought to some negative effects such as the tip bending, which was then recovered by plastic deformation at room temperature. The hits remained impressed in the microstructure and revealed by neutron imaging phase contrast effects.

False colour sections and 3D views enhancing all these effects are shown in figure 15.12.

The section of the two traditional Japanese blades and the modern *naginata* are shown in figure 15.13 to ease the comprehension of the differences in metal arrangement and microstructure.

The last example devoted to steel arms and armour characterization shows the microstructure of authentic and modern replicas of wootz steel [24]. The experiment was performed on the ANTARES beamline at the Meier Leibnitz Zentrum in Garching by taking white beam tomography of a set of steel fragments extracted

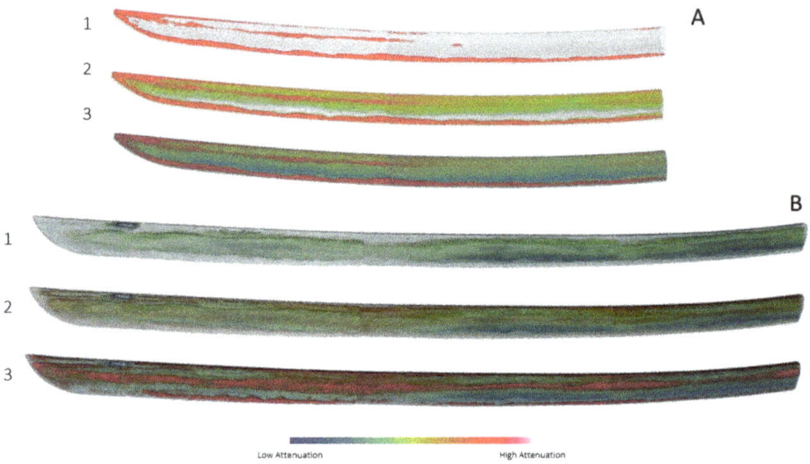

Figure 15.12. 3D false coloured rendering of the Japanese blades. (A) 18th century short sword: 1, 2, 3 martensite enhancement (red) and lower density area (blueish grey). (B) 16th century long sword: 1, 2, 3 martensite enhancement (red) and welding lines and defect rich area (ranging from blueish grey to green).

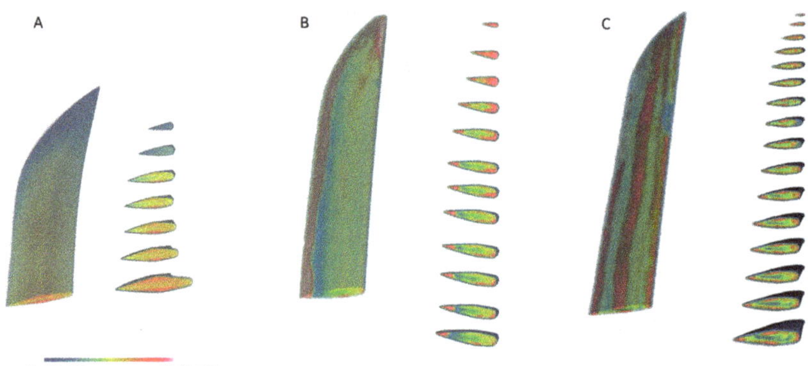

Figure 15.13. 3D false coloured rendering of the Japanese weapons. (A) 'Modern' naginata blade. (B) 18th century short sword. (C) 16th century long sword.

from a group of weapons provided to the Wallace Collection in London by the Nizam Historical Arsenal of Hyderabad (India). The samples are several cm in size and were extracted as the only good quality parts of the blades which were provided as sacrificial specimens to learn about Indian historical metallurgy and wootz steel production [25, 26]. The measurement revealed that white beam tomography on a high intensity source and maximizing the statistic is capable of distinguishing the cementite Fe_3C volume clusters within the steel (figure 15.14), that are geometrically arranged on a dimensional scale of a few hundred micron, typical of wootz, as also observed by metallographic analysis on the same samples. In order to learn about manufacturing procedure of wootz (a technology lost before 19th century) we asked a professional blacksmith from the US, which is promoting himself as able to replicate wootz and describing his procedure, to provide us some of his test

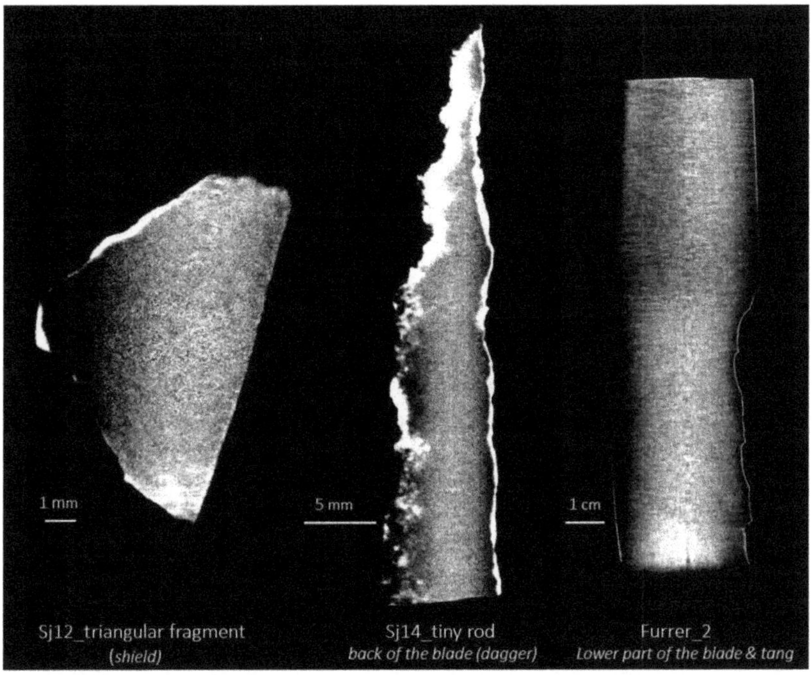

Figure 15.14. Neutron tomography slices of two historical wootz samples (Sj12, Sj14) and a wootz-imitating modern replica. Contrast is strongly enhanced to visualize the weak attenuation difference between cementite and ferrite. White border areas represent the corroded surface of the specimens. The size scale of the cementite structure is growing from left to right samples (reprinted from [26], copyright (2018), with permission from Elsevier).

experiments specimens to be analyzed. The tomography data show that wootz replicas possess a similar coherence in microstructure of cementite but they look less complex in arrangement and the dimensional scale of the plates is at least a factor 5 coarser than the original wootz.

There are several other examples showing the power of neutron imaging in revealing and characterizing traditionally produced steel artefacts spacing from armour and helmets [27], to European [28] and South East Asia weapons [29] thus attesting to the use of neutron imaging as one of the best methods for the characterization of steel artefacts and definitely the best one using a non-invasive approach.

15.2.2 The Swiss Degen—combining x-ray and neutron tomography data for the investigation of a mixed-material object

Besides the metallurgical characterization described in the previous section, many other aspects and questions can be addressed when studying ancient metal artefacts and weapons. Also, for many of these other subjects, neutron imaging techniques prove to be very well suited. Nevertheless, a single method or modality is often not able to deliver a satisfying answer, especially in the case of mixed-material objects.

In this chapter, we will demonstrate it on the example of a medieval sword found by divers at the bottom of lake Zug, Switzerland. The sword, which had been lying for centuries in the lake, was still in exceptionally good condition, when it was found. The highly ornated wooden sword hilt and the general assembly of the sword make it a very special weapon. It was identified as 'Swiss degen' a weapon type, from which only two other examples still exist. From these two specimen, one was visibly repaired and modified repeatedly over the centuries, while the other specimen had been found in a burnt layer of an archaeological excavation and no traces of the hilt had been preserved. The sword found in the lake, hence represented a unique chance to study the assembly of this very special type of weapon in its original state. The hilt itself was richly decorated with a multitude of small nails and metallic inlays covering the wooden hilt in spirally patterns. The small inlays consist of an amalgam, i.e. an alloy of mercury. This posed a problem, when the sword was first examined by means of x-ray at the museum. The multitude of small inlays overshadowed the signal of the wooden matrix in the resulting x-ray images. To have a chance to obtain information on the wooden components of the hilt, neutron tomography was carried out at the thermal neutron imaging facility NEUTRA (PSI) in addition to an x-ray tomography. The projections with the two modalities show already the expected differences (figure 15.15). While the x-ray image clearly shows all the metal parts, the wood can scarcely be seen. In the neutron image it can be seen that the wooden hilt attenuates the neutrons to a high extent. The comparison of the two modalities underlines the complementarity of the types of radiation for this type of material combination of metal and organic material. When comparing slices through the reconstructed tomography data sets, the complementarity becomes even more obvious (figure 15.16). In the reconstructed x-ray data only the metal parts can be seen, while there seem to be no signs from the wood. The reconstructed neutron tomography data on the other hand shows almost the opposite: the wood and even the wood structure (at least down to the annual rings and some knots) can be seen as well as the bigger metal parts. The small amalgam inlays appear to have attenuation coefficients in the same order of magnitude as the surrounding wood and could hence not, or only scarcely, be identified. Neither of the two modalities provides the full information on the investigated object. To obtain a satisfying result both methods have to be applied. For the presented case, the 3D volume data of both modalities was combined in a very simple way using a tomography evaluation and visualization software (VG Studio max) (figure 15.17). The material was segmented in each of the data sets and subsequently combined and virtually reassembled in the software. All the wood components and large metal parts were taken from the neutron tomography data set, while all the smaller metal pieces where derived from the x-ray data set. Only by combining the two data sets was it possible to obtain the complete information on the sword, which then allowed documenting the way such a type of weapon, notably the hilt, was assembled. In the course of the project the experimentally gained insights on the assembly were used to manufacture a replica of the sword using experimental archaeology methods.

More detailed information on the project can be found in [30–32].

Figure 15.15. (a) Photograph of a Swiss degen found in lake Zug (Switzerland). (b) X-ray transmission image. (c) Neutron transmission image (modified from [30], copyright (2015), with permission from Elsevier).

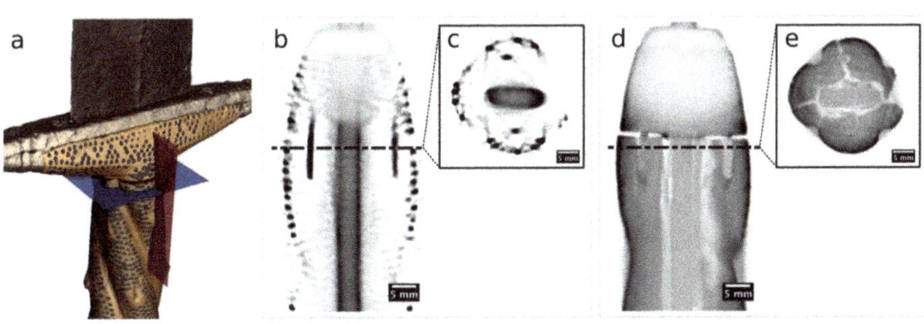

Figure 15.16. Panel (a) shows a rendered 3D view of the combined CT data sets indicating the position of the slices: (b) and (d) vertical plane, (c) and (e) horizontal plane. Panels (b) and (c) show slices with inverted grey scale through the x-ray, (d) and (e) neutron tomography data sets of the sword; the x-ray slices show all the metallic parts including the ornamental inlays on the surface; the neutron slices show the wood structure as well as the main metal parts (reprinted from [30], copyright (2015), with permission from Elsevier).

Figure 15.17. 3D-visualization of the combined x-ray and neutron CT data sets; information of the small metal pieces such as ornaments pins and nails originate from the x-ray CT; the wooden parts and larger metal parts originate from the neutron CT (reprinted from [30], copyright (2015), with permission from Elsevier).

15.3 Relics and human bones

For ethical reasons and in order to not disturb the peace of the dead, special care has to be taken when it comes to the study of relics or other human remains.

In a Christian, or more specifically a Catholic context, human remains from saints, i.e. holy persons (bones, tissue, and other parts of the body) are considered to be relics of 'first class'. The 'second class' of relics consists of things, which were in contact with holy persons during their lifetime.

The adoration, presentation and spiritual usage of such relics have changed dramatically over the centuries. While the more reformed churches of Christianity reject the celebrations around relics more or less, some Christian confessions such as the Catholic part still perform dedicated celebrations in a traditional way.

Relics are stored in specific covers (reliquary) of mostly metal or stone for their protection over long periods, often well decorated and exhibited. Although the trade with relics is forbidden from early times, there are collections with many of them available.

The scientific study of relics has different aspects. On the one hand, it might be important to know about the real existence inside the cover and the current substantial status. On the other hand, it is claimed to avoid the disturbance of the dead body.

Only in cases of common interests are such investigations allowed and they should be performed in a completely non-invasive way. This kind of investigation can be done either with x-rays or with neutron imaging methods—or in combination. Often information on whether inside a covering a structure is a human part is enough to handle the object accordingly.

Such investigations were done using both x-rays and thermal neutrons for an altar plate from the monastery of Freiburg, Switzerland [33]—see figure 15.18. It was assumed one would find in the centre of the plate a hollow space, covered by an insert stone plate, barely distinguishable from the base stone material. Some parts of bones were expected to be found in the best case inside the cavity. The performed x-ray and neutron radiography investigations enabled verifying this hypothesis in a non-invasive way. While the x-ray image does not show any organic material, the neutron image clearly visualized the bone parts. In the way of quantification of the content, it was derived that neither leather, wax nor teeth are hidden, but there are bones of about 4–5 mm, presumably from a wrist or ankle.

When it comes to the investigation of cultural heritage objects, the number of possible questions and hence applications seems almost innumerable. Neutron

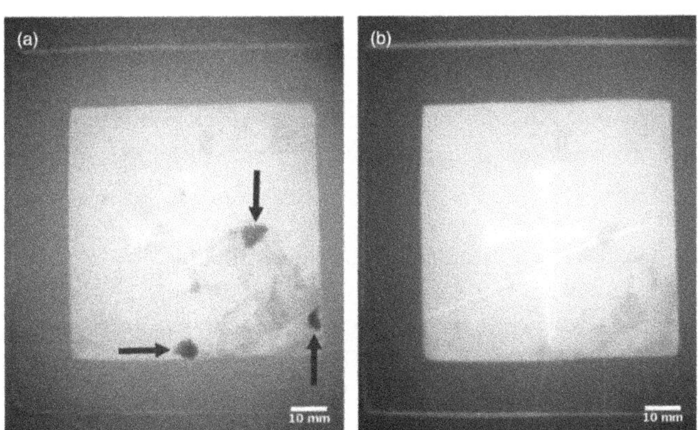

Figure 15.18. Transmission images (cropped) of the altar stone using (a) neutrons and (b) x-rays; in the neutron radiograph, three objects with high contrast are visible (arrows), while these do not appear in the x-ray radiography. Imaging of an altar stone from Fribourg, Switzerland (source: [33] John Wiley & Sons. Copyright 2013 University of Oxford).

imaging methods prove to be very suitable for a broad variety of such investigations and allow an alternative and sometimes unique insight on some of the studied objects and related questions. In this and the following two chapters (chapters 16 and 17) only a small fraction of the possible application of neutron imaging methods could exemplarily be highlighted. With the ever increasing number of new archaeological discoveries, the number of potential topics and questions that can be tackled with neutron imaging methods is rapidly increasing. Hence, ever new areas of application for neutron imaging are opened, such as the study of ancient Egyptian animal coffins [34] or the virtual unwrapping of medieval lead amulets [35]. These investigations shed new light on aspects of the common religious beliefs as well as the daily life of our ancestors.

References

[1] Rant J *et al* 2006 Neutron radiography examination of objects belonging to the cultural heritage *Appl. Radiat. Isot.* **64** 7–12

[2] Casali F 2006 X-ray and neutron digital radiography and computed tomography for cultural heritage *Physical Techniques in the Study of Art, Archaeology and Cultural Heritage* (Elsevier) vol 1 pp 41–123

[3] Stanojev Pereira M A *et al* 2013 The neutron tomography facility of IPEN-CNEN/SP and its potential to investigate ceramic objects from the Brazilian cultural heritage *Appl. Radiat. Isot.* **75** 6–10

[4] Denker A *et al* 2006 *COST Action G8: Non-destructive Testing and Analysis of Museum Objects* (Stuttgart: Fraunhofer IRB Verlag)

[5] 2010 Wood science for conservation of cultural heritage *Proceedings of the International Conference held by COST Action IE0601* ed L Uzielli (Firenze University Press)

[6] Di Martino D *et al* 2018 From tiny gold filigrees to majestic iron tie rods: neutron, facilities for the benefit of cultural heritage *Eur. Phys. J. Plus* **133** 371

[7] Kasztovszky Z and Rosta L 2012 How can neutrons contribute to cultural heritage research? *Neutron News* **23** 25–8

[8] Mongy T 2014 Application of neutrontomography in culture heritage research *Appl. Radiat. Isot.* **85** 54–9

[9] Lehmann E H, Wörle M and Deschler-Erb E 2014 Using neutron imaging methods for the non-destructive investigation of large ancient bronze artefacts *Proc. 18. Internationaler Kongress über antike Bronzen (Zurich)*

[10] https://ncnr.nist.gov/resources/n-lengths/

[11] https://physics.nist.gov/PhysRefData/Xcom/Text/intro.html

[12] Mannes D and Lehmann E H 2022 Neutron imaging of cultural heritage objects *Handbook of Cultural Heritage Analysis* (Cham: Springer International Publishing) pp 211–37

[13] Scholten F and Verber M *From Vulcan's Forge* (D. Katz Exhibition Catalogue)

[14] Lehmann E H, van Lang R, Estermann M, Hartmann S, LoCelso F, Kardjilov N and Tusa S 2017 Bronze sculptures and lead objects tell stories about their creators: investigation of renaissance sculptures and ancient ingots by means of neutron tomography *Neutron Methods for Archaeology and Cultural Heritage* ed N Kardjilov and G Festa (Berlin: Springer) pp 19–39

[15] van Lang R 2012 Technical studies of Renaissance bronzes *Dissertation* University of Delft

[16] Lehmann E H, Hartmann S and Speidel M 2009 Investigation of the content of ancient tibetan metallic Buddha Statues by means of neutron imaging methods *Archaeometry* **52** 416–28

[17] Henss M and Lehmann E 2016 The scanned Buddha *Orientation* **47** 77–81

[18] Lehmann E *et al* 2022 Ancient Buddhist metal statues using neutron tomography *Handbook of Cultural Heritage Analysis* (Berlin: Springer) ch 11

[19] Lehmann E H 2018 Using neutron imaging data for deeper understanding of cultural heritage objects experiences from 15+ years of collaborations *J. Archaeol. Sci.: Rep.* **19** 397–404

[20] Craddock P 2014 The metal casting traditions of South Asia: continuity and innovation *Ind. J. Hist. Sci.* **50** 55–82

[21] Levy T, Levy A and Sthapathy D R 2007 *Master of Fire—Hereditary Bronze Caster of South India* (Bochum: Deutsches Bergbau-Museum)

[22] Tawara K 1953 Nippon-to no Kogataki Kenkyu *Scientific Study of Japanese Swords* (Tokyo)

[23] Grazzi F, Salvemini F, Kaestner A, Civita F, Lehmann E and Zoppi M 2015 Characterization of two Japanese ancient swords through neutron imaging *Japan Neutron Sci. Soc. J.* **25** 206–13

[24] Williams A 2012 *The Sword and the Crucible: A History of the Metallurgy of European Swords up to the 16th Century* (Brill)

[25] Grazzi F, Civita F, Williams A, Scherillo A, Barzagli E, Bartoli L and Zoppi M 2011 Ancient and historic steel in Japan, India and Europe, a non-invasive comparative study using thermal neutron diffraction *Anal. Bioanal. Chem.* **400** 1493–500

[26] Grazzi F, Cantini F, Salvemini F, Scherillo A, Schillinger B, Kaestner A and Williams A 2018 The investigation of Indian and central Asian swords through neutron methods *J. Archaeol. Sci.: Rep.* **20** 834–42

[27] Salvemini F, Grazzi F, Fedrigo A, Williams A, Civita F, Scherillo A and Zoppi M 2013 Revealing the secrets of composite helmets of ancient Japanese tradition *The Eur. Phys. J. Plus* **128** 1–10

[28] Salvemini F, Grazzi F, Peetermans S, Gener M, Lehmann E H and Zoppi M 2014 Characterization of European sword blades through neutron imaging techniques *Eur. Phys. J. Plus* **129** 1–8

[29] Salvemini F, Grazzi F, Kardjilov N, Manke I, Scherillo A, Roselli M G and Zoppi M 2020 Non-invasive characterization of ancient Indonesian Kris through neutron methods *Eur. Phys. J. Plus* **135** 1–25

[30] Mannes D, Schmid F, Frey J, Schmidt-Ott K and Lehmann E 2015 Combined neutron and x-ray imaging for non-invasive investigations of cultural heritage objects *Phys. Proc.* **69** 653–60

[31] Bernasconi G L, Binggeli M and Sager F 2014 Kopie des Degens von Oberwil *Zeitschr. Schweizerische Archäol. Kunst.* **71** 141–8

[32] Schmidt-Ott K, Hunger K and Mannes D 2014 Die Konservierung des Degens von Oberwil unter Einbeziehung aktueller Analyseverfahren *Zeitschr. Schweizerische Archäol. und Kunst.* **71** 129–40

[33] Mannes D, Benoît C, Heinzelmann D and Lehmann E 2014 Beyond the visible: combined neutron and x-ray imaging of an altar stone from the former Augustinian Church in Fribourg, Switzerland *Archaeometry* **56** 717–27

[34] O'Flynn D, Fedrigo A, Perucchetti L and Masson-Berghoff A 2023 Neutron tomography of sealed copper alloy animal coffins from ancient Egypt *Sci. Rep.* **13** 4582

[35] Wilster-Hansen B, Mannes D C, Holmqvist K L, Ødeby K and Kutzke H 2022 Virtual unwrapping of the BISPEGATA amulet, a multiple folded medieval lead amulet, by using neutron tomography *Archaeometry* **64** 969–78

[36] Cantini F, Creange S, Li Y, van Eijck L, Kardjilov N, Kabra S and Grazzi F 2024 Morphological and microstructural characterization of an ancient Chola bronze statuette by neutron based non-invasive techniques *Archaeol. Anthropol. Sci.* **16** 45

IOP Publishing

Neutron Imaging
From applied materials science to industry
Markus Strobl and Eberhard Lehmann

Chapter 16

Natural heritage

Eberhard Lehmann, David Mannes and Florencia Malamud

The term ‘natural heritage’ includes all natural objects and samples protected from the past that are not human made. Their preservation for future generations often needs professional measures of treatment and dedicated storage.

The investigation of objects with natural heritage importance needs to be done non-invasively if they are unique and not available in large amounts. In this way, deeper understanding about structure, age, composition and function can be derived from the examinations.

Neutrons, and in particular the method of neutron imaging, can contribute to specific studies due to the provided contrast in the transmission data. In this way, involved materials can be distinguished and their content can be determined quantitatively.

However, neutron imaging is only one among many other investigation techniques and can be combined with these, if needed and possible.

This chapter gives an overview about selected investigations of fossils, wood, meteorites, gemstones and pearls. The given examples are not conclusive and present a data base about what information can be derived from such studies—and where the limitations are.

16.1 Fossils

According to a definition [1], fossils are objects of evidence about past life on earth, which have been preserved in its crust and which date back to past geologic ages, i.e. often millions of years ago. During this time, organic substances were often replaced by stones created in a mineralization process. Scientifically, palaeontology is dealing with such fossilized objects.

Most spectacular are find pieces from extinct dinosaurian due to their size and by the general popularity of the ‘jurassic time’.

But also mineralized human and hominid bones have a high importance because they can give information about the origin and development of humankind.

doi:10.1088/978-0-7503-3495-2ch16 © IOP Publishing Ltd 2024

Studies with x-rays and even neutrons were already done in the last century. Because the background by the stony matrix in which the fossils are embedded often limits the visibility, such radiographic approaches were not very helpful for deeper understanding of the materials and their composition.

A real breakthrough was the introduction of tomography methods with both x-rays and neutrons. Even with the limitations in the spatial resolution of the virgin neutron tomography by 0.25 mm pixel size, Schwarz *et al* [2] derived already interesting information in their comparative study of dinosaur bones as vertebrate remains. Besides the structure and density of the samples, the resin and wax from the consolidation process was possible to be distinguished.

In the meantime, neutron (and x-ray) tomography has been developed towards a routine method at several institutions with high flexibility regarding sample size and spatial resolution. Depending on the finally possible transmission, objects of up to about 40 cm in diameter can be investigated. On the other hand, the current inspection limit for very small samples with respect to the achievable spatial resolution is in the order of a few micro-meters, however, for samples in the millimeter size range.

Based on this methodical progress, several studies have been performed which can here only be sketched with reference to the original publications.

Bevitt *et al* [3] provided a very detailed recent investigation of an object with a dinosaur who is on the way to eating a crocodile. The two objects were studied with neutron tomography and the many individual samples later composed by means of image processing.

De Beer *et al* [4] give an overview about their own and other published data with the focus on examples of dinosaur eggs, embryos, sculls and vertebrae.

A very intensive and systematic study is provided by Laass and coworkers [5–9] with studies at several neutron imaging facilities about the development of the dinosaur acoustic organ (figure 16.1). Also here, different tools of image processing and visualization were applied successfully.

Because the teeth of animals have a high endurance, their remains can be found even after thousands of years. This makes them interesting for comparative studies of different species in different lifespans. However, a mechanical extraction is accompanied by the risk of partial or total destruction which is why non-invasive studies are more useful. Neutron imaging has been found very useful and applicable in studies by Schillinger *et al* [10] and Zanolli *et al* [11].

There are other attempts of investigation of fossils, done in Brazil [12], Korea [13], Hungary [14] or the Czech Republic [15], sometimes only for demonstration of their performance on a 'relevant' sample. The combination with other analytical methods like PGAA or optical microscopy are interesting approaches.

Besides the fossils of animals or parts of animals, also fossilized plant materials play an equivalent role in palaeontology. Also in this domain, neutron imaging can play an important role as the study of Kach *et al* [15] on conifer cones shows. Neutron data shows very high contrasts between different areas, while x-ray data of the identical objects appears relatively homogenous (figure 16.2).

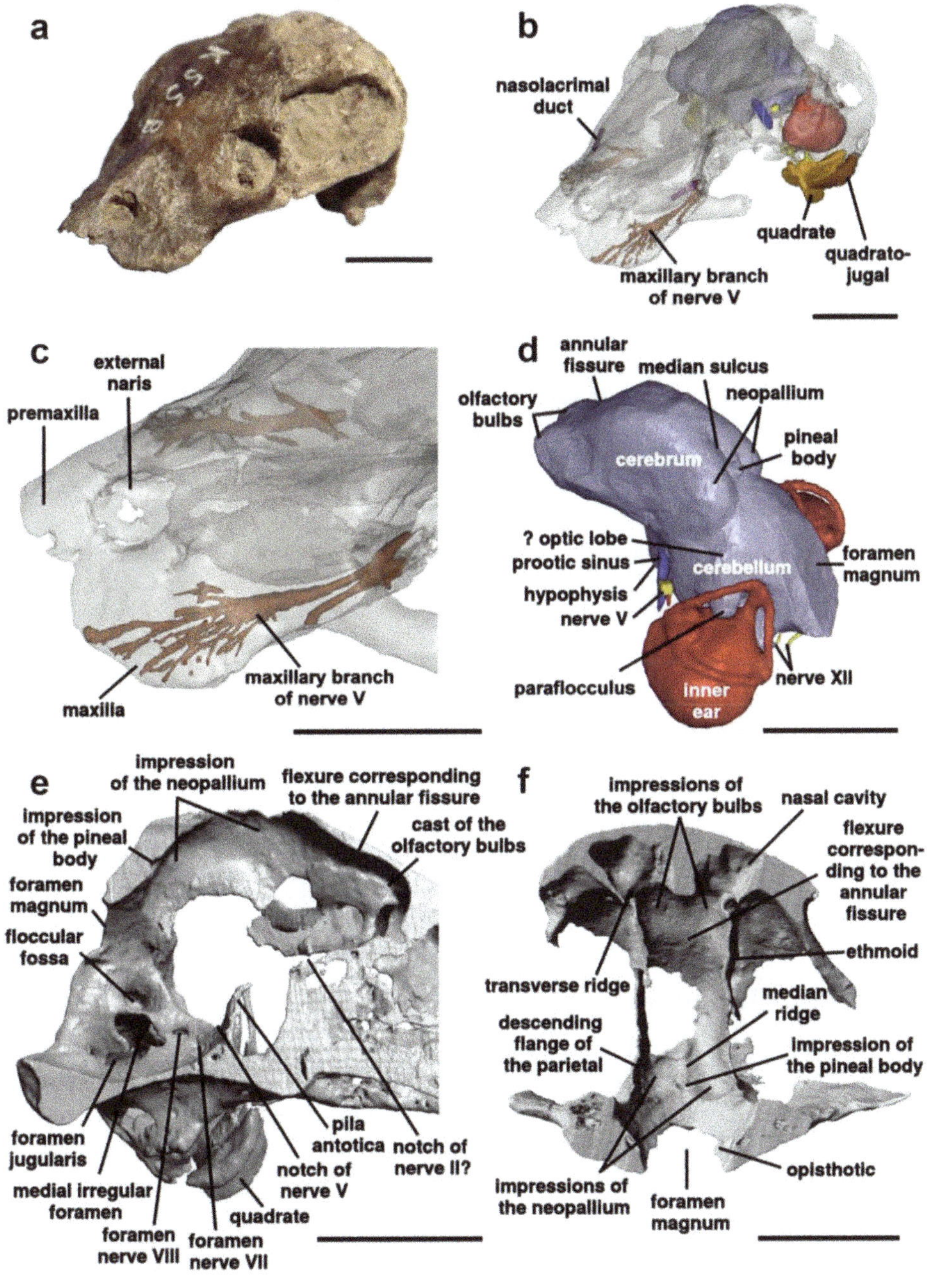

Figure 16.1. The skull of Kawingasaurus fossilis (GPIT/RE/9272). (a) Photograph. (b) Virtual 3D-model of the skull. (c) The maxillary branch of nerve V. (d) Virtual endocast of the brain and inner ears. (e) Virtual sagittal section through the skull showing the brain cavity. (f) Virtual coronal section through the skull showing the dorsal part of the brain cavity including the impressions of the neopallium. Scale bars: 5–10 mm (source: [8] John Wiley & Sons. Copyright 2017 Wiley Periodicals, Inc.).

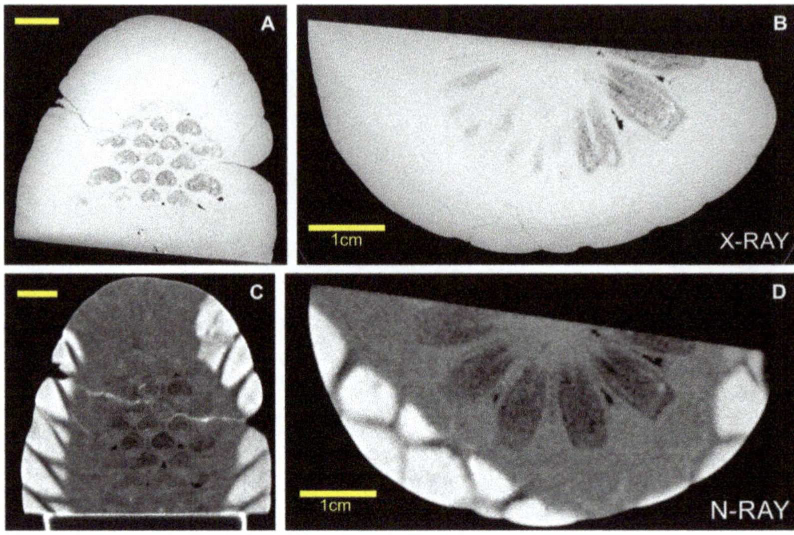

Figure 16.2. Dammarites albens, stem—x-ray micro-tomography of silicified cone-like stem (A,B). Presented sagittal (A) and transversal (B) slices show good contrast in seeds with high detail. Compared to that neutron CT—sagittal (C) and transversal (D) slices of silicified cone-like stem provide high contrast between petrified sediment and the cone itself. Such contrast is unachievable using standard x-ray CT. Voxel size of reconstructed micro-CT is 34.82 μm. Sample measured at Perkin Elmer flat panel detector at CET, measurement parameters: tube voltage 200 kV, tube current 400 μA, acq. time 1 s, 1200 projections, 0.3-degree angle step. Voxel size of the reconstructed neutron CT is 105.4 μm. Sample measured at ICON-M CCD detector at BOA beam in PSI, measurement parameters: acq. time 20 s, 751 projections, 0.48-degree angle step. Reproduced from [15], copyright 2017 IOP Publishing Ltd and Sissa Medialab. All rights reserved.

In any cases of deeper understanding of fossilized objects it is important and necessary to have a close collaboration of palaeontology experts with operators of imaging labs, who understand the content and conclusion of the image data.

16.2 Dendrochronology

Wood is a relative long-lasting material used for construction, tools and cultural objects over the ages. Because the trees where these wooden samples have been made from grow irregularly in 'annual rings', depending on the environmental conditions, the sequence of the ring thickness (and density) gives information about the tree's age.

This ring structure is characteristic for the specific kind of wood, the local climate of the region of growing and other impacts to the trees.

Well established tables of the sequences of annual rings have been created in the framework of 'dendrochronology', which enable determining the era of growth of a specific wooden sample. Depending on the conservation status of aged wood, dating back over centuries is quite common and possible.

In order to determine data of the ring structure, several detection methods are established. Next to photographic methods, some transmission techniques are common by using x-rays. Recently, also neutron imaging has been introduced as alternatively usable procedure.

Mannes [16] showed on the example of degraded archaeological wood samples from excavations in and near Hallstadt (Austria), that it was possible to use neutron radiography (figure 16.3) as well as neutron tomography (figure 16.4) to study annual ring sequences.

When compared to the more common x-ray techniques it could demonstrated that neutron imaging can provide similar results [17, 18]. An example, comparing x-ray and neutron measurements for an identical sample is shown in figure 16.5.

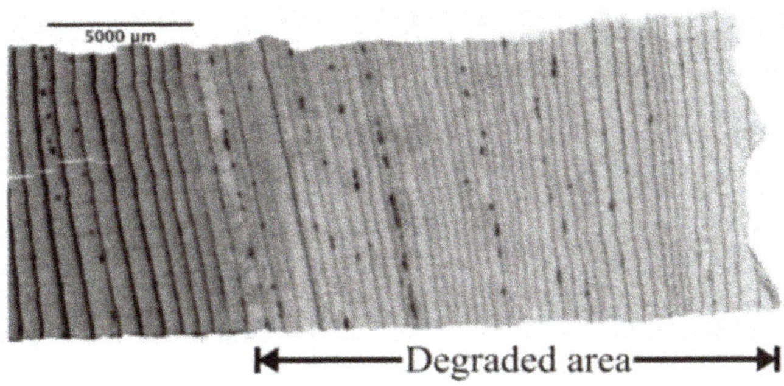

Figure 16.3. Section of a partially degraded soft-wood sample in a neutron transmission image. The 500-year-old sample originates from an excavation near Dachstein (Austria). Reproduced with permission from [16].

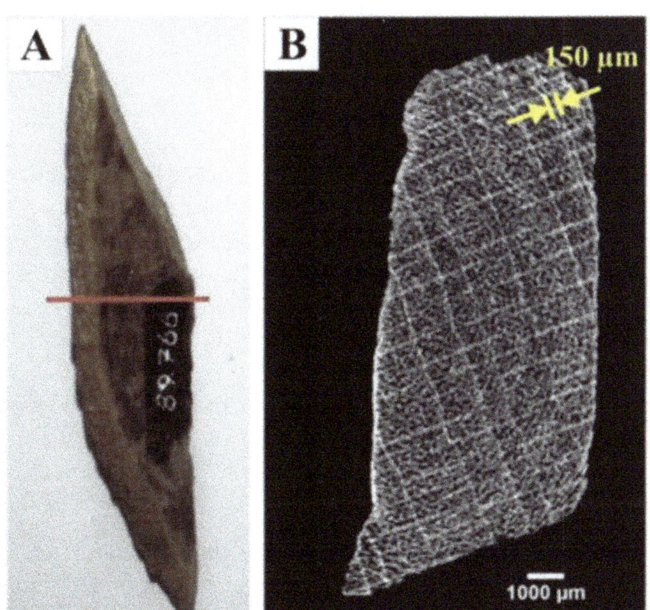

Figure 16.4. (A) Fragment from a wooden bowl found in a prehistoric salt mine near Hallstatt (Austria). (B) tomographic section through the fragment (red line); even very narrow growth rings, width of 150 µm (yellow arrows), can be resolved. Reproduced with permission from [16].

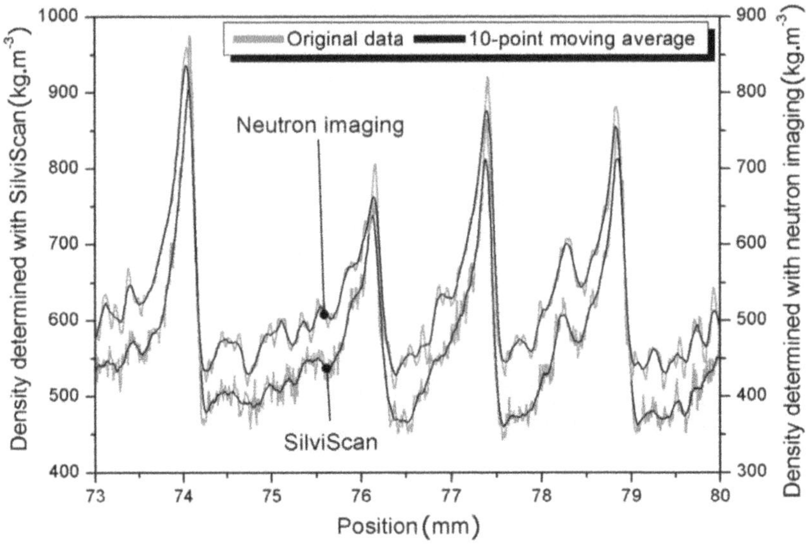

Figure 16.5. Radial density profiles of an identical specimen section, determined by x-ray (SilviScan) and by NI. Reproduced with permission from [18]. Published by Wood Research.

Besides these proof-of-principle experiments, neutron imaging was also used in dendrochronological studies aiming at a more precise dating of the Santorini eruption, a volcanic incident which had occurred in Antiquity in the Mediterranean sea [19, 20].

Even though the neutron imaging and standard x-ray techniques can deliver equivalent results, neutrons are only very seldomly used for such kinds of investigations. This is mainly due to the much higher availability and often better spatial resolution of the x-ray techniques. The only cases where neutron imaging has advantages with regards to dendrochronological studies would be the case of more complex archaeological objects, consisting, for example, of a combination of metal and wood. In such cases x-rays are hindered by the metal parts, hence shielding the information on the wood structure; contrary to that, neutrons can penetrate the metal parts relatively easy while still providing enough contrast to retrieve information on the wood structure [21, 22].

More information about neutron imaging applications in wood research is provided in chapter 8.

16.3 Meteorites

Meteorites are fragments of planetoids travelling through space that were captured by the Earth's gravity, cross our atmosphere as meteors, and fall to the Earth's surface. In particular, metallic meteorites are assumed to be from asteroid cores created early in our Solar System's history. By crossing the Earth's atmosphere, the iron meteors experience heating effects, which for large meteorites (more than 10 cm^3) only affect their very external parts, leaving the interior unaffected.

The study of the internal crystalline structure of metallic meteorites could therefore provide fundamental information to understand the geochemical and geophysical conditions present during the early formation of the Solar System.

Metallic meteorites contain a combination of several FeNi alloys, mainly kamacite (BCC crystal structure with up to 7.5% nickel in solid solution) and taenite (FCC structure with nickel fluctuating from 25% to 70%), together with a a few other compounds such as sulphides, phosphides, and oxides. Usually, iron meteorites are classified based on their structures, shape, and size of the kamacite lamellae, and on chemical classification based on Ni, Ge, Ga, and Ir content. In this sense, the study of the meteorite micro- and macrostructure can provide valuable information about the process by which the material formed, the composition of the parent asteroid, and the phenomena that happened after meteorite crystallization.

Over the past years, neutron imaging has been proved to be an excellent non-destructive technique to study meteorites. Peetermans and collaborators [23] employed neutron tomography by means of energy-selective neutron imaging to identify the morphology and location of mineral inclusions and oxidation crust of the Mont Dieu meteorite. Using neutron transmission radiographies and tomographies of the meteorite for different wavelengths, they distinguished two different types of morphology, convex crystallites of centimeter size and a lamellar Widmanstätten patterns of taenite and kamacite). Following these pioneer studies, Caporali and collaborators [24], employed neutron tomography to reconstruct the 3D distribution, size, and shape of the kamacite lamellae of different meteorites (Seymchan, Sikhote-Alin, Agoudal, Campo del Cielo, and Muonionalusta meteorites), allowing the classification of the meteorites without any damage to the sample. The internal structure of pallasites, a stony-iron meteorite consisting of olivine grains and FeNi metal alloy, such as the Marjalahti [25] or the Seymchan [26] meteorites, and chondrites, such as the Chelyabinsk meteorite [27] or the Allende meteorite [28], were also studied using neutron tomography. The difference in the neutron attenuation coefficients of iron and nickel in comparison with the olivine components allowed the refinement of petrological and morphological features, like the anisotropy of nickel distribution within the iron component of the Seymchan meteorite or the presence of large kamacite grains in the Chelyabinsk meteorite.

Neutron tomography, combined with x-ray tomography, was also employed to detect hydrogen-bearing materials in meteorites [29] and tektites, natural glasses formed during meteorite impacts [30], and to understand the complex build-up of meteoritic impacts [31, 32]. Fedrigo and collaborators investigated samples of impactites from the Wabar meteorite impact in the uninhabited Empty Quarter of Saudi Arabia [31], and the Monturaqui impactite, formed by shock metamorphism during the impact of an iron meteorite with the target rocks in the Monturaqui crater in Chile [32]. The applied bimodal methodology enabled the segmentation of a large number of different materials, their morphology as well as distribution in the specimen including the quantification of volume fractions and demonstrated the potential of combined neutron tomography and x-ray tomography for non-destructive characterization of complex multi-phase objects such as impactites.

On the other hand, wavelength-resolved neutron imaging was used to study the uniformity of the bulk microstructure within iron and stony meteorites. Santisteban and collaborators studied two meteorites from the Sikhote-Alin fall, displaying single crystal type and textured polycrystalline microstructures across the samples [33]. The lattice parameter, crystalline orientation and mosaicity of the single crystal sample were also characterized using wavelength-resolved neutron imaging [34]. Kockelmann and collaborators [35] demonstrated that it is possible to reconstruct maps of elemental composition (for a limited set of elements, which have large enough resonance absorption cross-section) through the analysis of neutron resonance absorption in the epithermal range of energies.

16.4 Gemmology

Gemmology comprises all methods used to investigate and assess all kinds of gemstones and pearls used in the jewellery industry. Most common are optical inspection methods using visible light up to UV light for transparent samples. This is not applicable for pearls, where other kinds of radiation are needed to look into internal structures and to visualize hidden properties.

Unlike x-ray imaging methods, which already represent a standard testing method in this field, neutron imaging has scarcely been used for investigations so far. First tests using neutron radiography for the investigation of pearls were carried out by Okamoto *et al* [36]. In the cited work, black and blue pearls were inspected using neutron radiography; these results were compared to radiographs based on x-ray and Gamma-ray transmission. However, although it stated that the neutron radiographic results were sharper and yielded more contrast, these remained the only neutron imaging investigations for a while.

In a more extensive work, Hanser [37], Mannes *et al* [38] and Hanser *et al* [39] investigated the possibilities and limitations of different imaging methods for the study and assessment of natural and cultivated pearls (figure 16.6). Besides the more common x-ray methods, neutron radiography, neutron tomography and neutron grating interferometry were applied to a variety of pearl samples. Here, neutron radiography and tomography were considered valuable additional non-destructive testing methods, delivering complementary information, when compared to x-ray methods. The results from neutron grating interferometry on the other hand were considered too noisy, at least with the used set-up, yielding no additional benefit compared to standard neutron radiography and computed tomography. The strength of standard neutron imaging methods is again in the high sensitivity for hydrogen and hence the possibility to discriminate between areas with organic and inorganic matter inside the pearls. These allow clearly stating if small cracks and fissures are filled with organic material or are empty [39, 40]. One step further in the examination of pearls was carried out by Vitucci *et al* [41], who performed energy-resolved neutron tomography on pearls. These studies were performed at the IMAT beamline fed by the pulsed spallation neutron source ISIS using microchannel plate

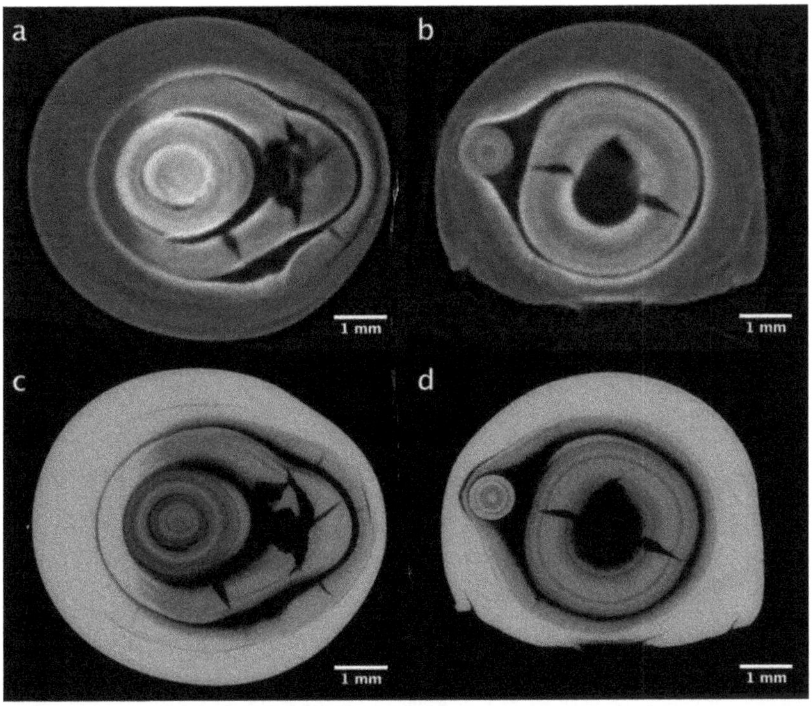

Figure 16.6. Virtual sections through a eadles saltwater cultured pearl by means of neutron tomography (a, b) and x-ray tomography (c, d) highlighting the complementarity of the two methods (i.e. high attenuation for neutrons in regions containing organic (hydrogeneous) material) (source: Hanser 2015 modified). Reprinted from [38], copyright (2017), with permission from Elsevier.

detector. This new approach allowed assessing and visualizing the crystallographic orientation of aragonite inside the sample.

Beside pearls, all other gems and coloured stones are also subject to the field of gemmology but unlike the former these have scarcely been studied using neutron imaging methods. Mannes *et al* [39] studied emeralds using neutron tomography and neutron grating interferometry and compared the results to x-ray tomography data of the same samples. It was stated that especially by combining neutron and x-ray tomography it was possible to better study '...specific growth patterns (such as 'Trapiche' growth patterns in emeralds), jadeite grain boundary textures and the permeation of filler substances along these grain boundaries ....' [39] (figure 16.7).

The applicability of neutron imaging in the field of gemmology has been studied using a variety of different techniques yielding results which were to a large extent promising. Nevertheless, neutron imaging techniques are still scarcely used in this field, with exception of the listed reference studies. Because the spatial resolution in neutron tomography has now be improved down to the micro-meter level, some new approaches are worth consideration.

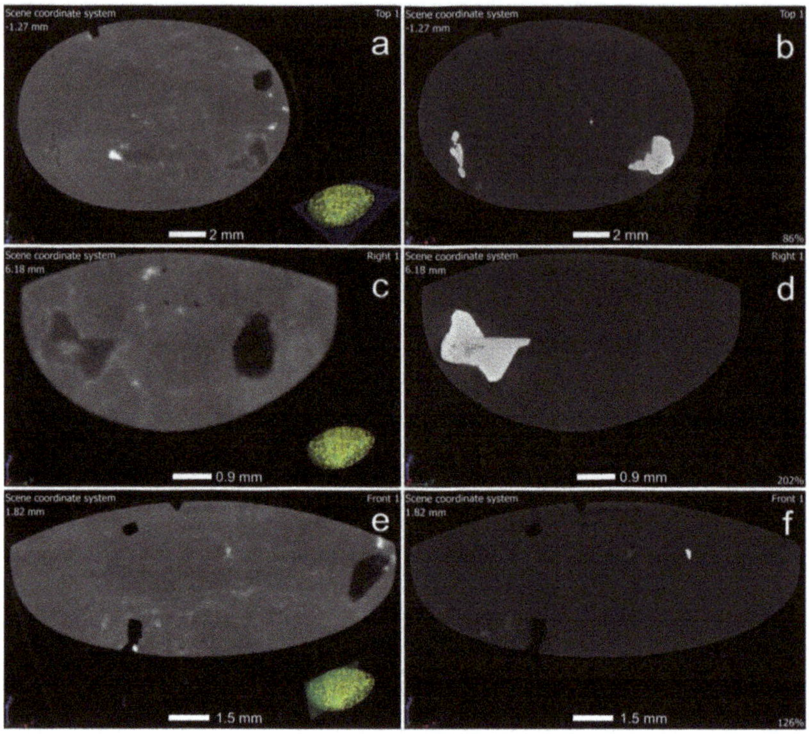

Figure 16.7. Comparison of neutron (left; (a), (c), (e)) and x-ray (right; (b), (d), (f)) CT sections in the three orthogonal planes; while metallic inclusions are highlighted in the x-ray images they only appear with a weak contrast in the neutron images. Here, organic inclusions, e.g. from resins or oils used during the emerald preparation, highlight microcracks below the actual spatial resolution. Repinted from [38], copyright (2017), with permission from Elsevier.

References

[1] Encyclopaedia Britannica 2022 'fossil' (https://britannica.com/science/fossil) (Accessed 22 March 2023)

[2] Schwarz D *et al* Neutron tomography of internal structures of vertebrate remains: a comparison with x-ray computed tomography *Palaeont. Electr.* **8** 30A:11 800KB http://palaeo-electronica.org/paleo/2005_2/icht/issue2_05.htm

[3] Bevitt J J 2018 Discovering dinosaurs with neutrons *Nat. Rev. Mater.* **3** 296–8

[4] De Beer F 2017 Paleontology: fossilized ancestors awaken by neutron radiography, in neutron methods for archaeology and cultural heritage *Neutron Methods for Archaeology and Cultural Heritage* ed N Kardjilov and G Festa (Cham: Springer) ch 7

[5] Laaß M 2015 Virtual reconstruction and description of the cranial endocast of P risterodon mackayi (T herapsida, A nomodontia) *J. Morphol.* **276** 1089–99

[6] Laaß M and Schillinger B 2015 Reconstructing the auditory apparatus of therapsids by means of neutron tomography *Phys. Proc.* **69** 628–35

[7] Laaß M, Schillinger B and Kaestner A 2017 What did the 'unossified zone' of the non-mammalian therapsid braincase house? *J. Morphol.* **278** 1020–32

[8] Laaß M and Kaestner A 2017 Evidence for convergent evolution of a neocortex-like structure in a late *Permian therapsid J. Morphol.* **278** 1033–57

[9] Laaß M, Schillinger B and Werneburg I 2017 Neutron tomography and x-ray tomography as tools for the morphological investigation of non-mammalian synapsids *Phys. Proc.* **88** 100–8

[10] Schillinger B *et al* 2018 Neutron imaging in Cultural Heritage Research at the FRM II Reactor of the Heinz Maier-Leibnitz Center *J. Imaging* **4** 22

[11] Zanolli C *et al* 2017 Exploring hominin and non-hominin primate dental fossil remains with neutron microtomography *Phys. Proc.* **88** 109–15

[12] Pugliesi R *et al* 2019 Study of the fish fossil Notelops brama from Araripe-Basin Brazil by neutron tomography *Nucl. Inst. Methods Phys. Res.,* A **919** 68–72

[13] Maroti B *et al* 2020 Joint application of structured-light optical scanning, neutron tomography and position-sensitive prompt gamma activation analysis for thenon-destructive structural and compositional characterization of fossil echinoids *NDT&E Int.* **115** 102295

[14] Grellet-Tinner G 2011 Description of the first lithostrotian titanosaur embryo in ovo with Neutron characterization and implications for lithostrotian Aptian migration and dispersion *Gondwana Res.* **20** 621–9

[15] Karch J, Dudák J, Žemlička J, Vavřík D, Kumpová I, Kvaček J and Trtík P 2017 . X-ray micro-CT and neutron CT as complementary imaging tools for non-destructive 3D imaging of rare silicified fossil plants *J. Instrum.* **12** C12004

[16] Mannes D C 2009 Non-destructive Testing of Wood by Means of Neutron Imaging in Comparison with Similar Methods *Doctoral Dissertation* ETH Zurich

[17] Mannes D, Lehmann E, Cherubini P and Niemz P 2007 Neutron imaging versus standard x-ray densitometry as method to measure-tree-ring wood density *Trees Struct. Funct.* **21** 605–12

[18] Keunecke D, Mannes D, Evans R, Lehmann E and Niemz P 2010 Silviscan vs. neutron imaging to generate radial softwood density profiles *Wood Res.* **55** 49–60

[19] Cherubini P, Humbel T, Beeckman H, Gärtner H, Mannes D, Pearson C, Schoch W, Tognetti R and Ley-Yadan S 2013 Olive tree-ring problematic dating: a comparative analysis on Santorini (Greece) *PLoS ONE* **8** e54730

[20] Cherubini P, Humbel T, Beeckman H, Gärtner H, Mannes D, Pearson C, Schoch W, Tognetti R and Lev-Yadun S 2014 The olive-branch dating of the Santorini eruption *Antiquity* **88** 267–73

[21] Mannes D, Schmid F, Frey J, Schmidt-Ott K and Lehmann E 2015 Combined neutron and x-ray imaging for non-invasive investigations of cultural heritage objects *Phys. Proc.* **69** 653–60

[22] Masalles A, Lehmann E and Mannes D 2015 Non-destructive investigation of 'The Violinist' a lead sculpture by Pablo Gargallo, using the neutron imaging facility NEUTRA in the Paul Scherrer Institute *Phys. Proc.* **69** 636–45

[23] Peetermans S, Grazzi F, Salvemini F, Lehmann E H, Caporali S and Pratesi G 2013 Energy-selective neutron imaging for morphological and phase analysis of iron–nickel meteorites *Analyst.* **138** 5303

[24] Caporali S, Grazzi F, Salvemini F, Garbe U, Peetermans S and Pratesi G 2016 Structural characterization of iron meteorites through neutron tomography *Minerals* **6** 14

[25] Kozlenko D P, Kichanov S E, Lukin E V, Rutkauskas A V, Bokuchava G D, Savenko B N, Pakhnevich A V and Rozanov A Y 2015 Neutron radiography facility at IBR-2 high flux pulsed reactor: first results *Phys. Proc.* **69** 87–91

[26] Kichanov S E, Kozlenko D P, Lukin E V, Rutkauskas A V, Krasavin E A, Rozanov A Y and Savenko B N 2018 A neutron tomography study of the Seymchan pallasite *Meteorit. Planet. Sci.* **53** 2155–64

[27] Kichanov S E, Kozlenko D P, Kirillov A K, Lukin E V, Abdurakhimov B, Belozerova N M, Rutkauskas A V, Ivankina T I and Savenko B N 2019 A structural insight into the Chelyabinsk meteorite: neutron diffraction, tomography and Raman spectroscopy study *SN Appl. Sci.* **1** 1563

[28] Canella L, Kudějová P, Schulze R, Türler A and Jolie J 2009 *Appl. Radiat. Isot.* **67** 2070–4

[29] Treiman A H, LaManna J M, Anovitz L M, Hussey D S and Jacobson D L 2018 Neutron computed tomography of meteorites: Detecting hydrogen-bearing materials *49th Annual Lunar and Planetary Science Conf. No. 2083* p 1993

[30] Hess K-U, Flaws A, Mühlbauer M J, Schillinger B, Franz A, Schulz M, Calzada E, Dingwell D B and Bente K 2011 Advances in high-resolution neutron computed tomography: Adapted to the earth sciences *Geosphere* **7** 1294–302

[31] Fedrigo A, Marstal K, Bender Koch C, Andersen Dahl V, Bjorholm Dahl A, Lyksborg M, Gundlach C, Ott F and Strobl M 2018 Investigation of a Monturaqui impactite by means of bi-modal X-ray and neutron tomography *J. Imaging* **4** 72

[32] Fedrigo A, Marstal K, Dahl A B, Lyksborg M, Gundlach C, Strobl M and Koch C B 2017 A tomography approach to investigate impactite structure *Proc. 3rd International Conference on Tomography of Materials and Structures (Lund, Sweden, 26-30 June 2017)* ICTMS2017-110

[33] Santisteban J R, Edwards L and Stelmukh V 2006 Characterization of textured materials by TOF transmission *Physica* B 385–6 Part 1 636–8

[34] Malamud F and Santisteban J R 2016 Full-pattern analysis of time-of-flight neutron transmission of mosaic crystals *J. Appl. Crystallogr.* **49** 348–65

[35] Kockelmann W *et al* 2018 Time-of-flight neutron imaging on IMAT@ISIS: a new user facility for materials science *J. Imaging* **4** 47

[36] Okamoto S, Hiraoka E, Tsujii Y and Furuta J 1983 Neutron radiography of pearls *J. Gemmol. Soc. Jpn.* **10** 59–65 (in Japanese with English abstract)

[37] Hanser C 2015 Comparison of imaging techniques for the analysis of internal structures of pearls *Master's Thesis* University of Freiburg, Germany

[38] Mannes D, Hanser C, Krzemnicki M, Harti R P, Jerjen I and Lehmann E 2017 Gemmological investigations on pearls and emeralds using neutron imaging *Phys. Proc.* **88** 134–9

[39] Hanser C S, Krzemnicki M S, Grünzweig C, Harti R P, Betz B and Mannes D 2018 Neutron radiography and tomography: a new approach to visualize the internal structures of pearls *J. Gemmol.* **36** 54

[40] Micieli D, Di Martino D, Musa M, Gori L, Kaestner A, Bravin A and Gorini G 2018 Characterizing pearls structures using x-ray phase-contrast and neutron imaging: a pilot study *Sci. Rep.* **8** 12118

[41] Vitucci G, Minniti T, Di Martino D, Musa M, Gori L, Micieli D and Gorini G 2018 Energy-resolved neutron tomography of an unconventional cultured pearl at a pulsed spallation source using a microchannel plate camera *Microchem. J.* **137** 473–9

IOP Publishing

Neutron Imaging
From applied materials science to industry
Markus Strobl and Eberhard Lehmann

Chapter 17

Conservation

David Mannes and Eberhard Lehmann

All materials in Nature undergo specific changes in the interaction with environmental damaging conditions (temperature, humidity, sunlight, animal attacks or mechanical actions). There are only a few materials with extreme high durability like precious metals, gemstones, or stones.

The aim of conservation is to extend the lifetime of important materials and objects of the human culture as long as possible to make them accessible over generations in best conditions.

There are different stages to handle culturally important objects with respect to their protection:

1. To describe the current status of the object regarding composition, degree of degradation, completeness, age and context;
2. Observation of previous attempts for restauration and conservation and their success;
3. Definition of the final status to achieve after the conservation procedure;
4. Definition of the best treatment procedure, including agents and application techniques;
5. If the object is at a stage where a complete rebuild of the initial status is impossible, the production of a suitable replica/copy can be considered, based on the investigations, e.g. with the help of tomography data.

Most of the conservation experts are linked to or employed by museums, where the most important objects of our cultural heritage are stored and exhibited. Their knowledge has grown over centuries and been applied in the best possible practice. However, there are also new conservation materials and application techniques under development, which have to be tested and compared to the traditional ones.

For all stages, mentioned above, analytical methods are needed to describe the original status, the degree of degradation, the impact of treatment procedures and the result of conservation work. All this inspection work should be done non-destructively

 © IOP Publishing Ltd 2024

or with minimal invasive methods in order to maintain the integrity of the object or at least to minimize the damage as much as possible.

Neutron imaging methods with their specific abilities to 'see' material distributions can play an important role for conservation work on the different levels of treatments. In this chapter, we will show with examples of investigations how such data can be used to improve the knowledge about objects of cultural importance and to extend the ability for their protection.

We will start with a typical example from excavation work, where the hidden content of an assembly—sealed by a gypsum layer—should be analyzed regarding the content, before the contained pieces are separated and treated.

With a Roman sword (gladius type) it should be demonstrated how a previously done treatment can be analyzed with the aim to perform an improved consolidation work under state-of-the-art conditions.

Even modern art is affected by corrosion. This will be demonstrated by the example of the 'violinist' from Gargallo. The investigation showed the degree and positions of lead corrosion around the wooden core and delivered the basis for the strategy of the later conservation procedures.

Although stone is considered as a relatively stable material, as important monuments from centuries ago demonstrate, degradation also happens when the environmental conditions are too harsh. The need for treatment and consolidation has a more methodical character, made on representative samples, in order to find out the best methods and agents for consolidation.

This is also valid for many wooden objects. Some of them were already stabilized with materials which were common at the time of first treatments. Over the years, some further degradation can happen, also by the chemical interaction between the wooden matrix and the consolident. New and advanced methods have to find out.

The study of corrosion of metals like iron, bronze or other common ones, used in the past for artifacts and daily life, needs deeper research regarding corrosion status, the time-dependent corrosion process and the development of counter measures. Some broader projects in this respect have been started recently.

17.1 Neutron imaging as a tool in conservation work

As explained in more detail in earlier chapters, e.g. chapters 1 and 2, the interaction of (thermal and cold) neutrons with hydrogen is quite high with the consequence of high contrast for hydrogenous materials even in small amounts. At the same time, metallic samples, in particular precious metals like Au, Pt or Ag, can be penetrated in certain layer thicknesses. This behaviour is quite the opposite for x-ray investigations. Figure 17.1 compares the attenuation behaviour in the format of half-value layer thicknesses for metals relevant for many cultural heritage objects for both kinds of radiation in comparison to hydrogen.

With respect to conservation, the contrast for agents like resin, wax, polymers or impregnating agents—all with a high hydrogen content—enables their visibility even through thick layers of the structural materials, which should be consolidated. By comparing a sample before and after the treatment a further increase in the

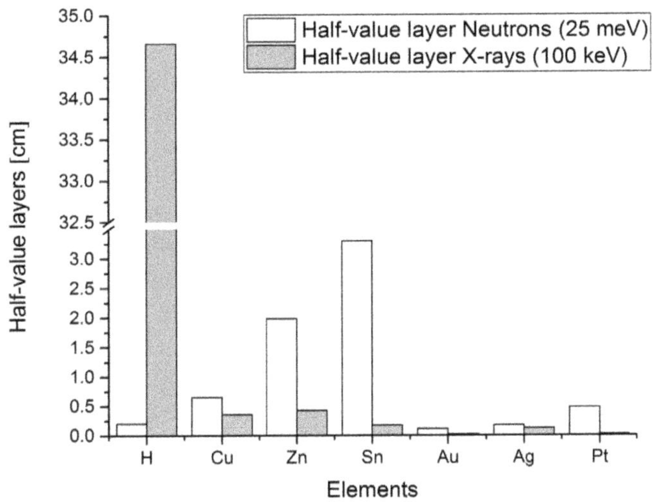

Figure 17.1. The penetration power through layers of hydrogen and (heavy) metals, expressed by the half-value thickness, for neutrons and x-rays; metal layers of mm to cm can be transmitted only with neutrons, while providing simultaneously high contrast for hydrogenous material (modified from [1], copyright (2022), with permission from Springer Nature).

information of the agent's distribution can be achieved. The division of images with a well registered pixel matrix helps to 'remove' the base material while the added material is enhanced. This is demonstrated by the injection of a protective solvent into a wooden sample (see the example in section 17.5).

Even the corrosion process itself can be investigated in detail due to the content of hydrogen in many corrosion products (hydroxides). Because of the lower density of the corrosion products compared to the bulk metal, x-ray techniques see in, contrary to neutrons, only the gaps of missing metallic structures, based on the limited contrast for hydrogen and oxygen.

Studies of consolidated museum objects with x-rays are often likely to fail also due to the limited contrasts for the hydrogenous agents. If wood, stone or even metals are studied, mainly the gaps will be visible, not the agent itself.

One disadvantage of neutron imaging compared to x-ray imaging is the immobility of the facilities. Whereas x-ray tubes are in use also in museums and are semi-mobile, neutron imaging facilities rely in general on large scale neutron sources such as research reactors or spallation neutron sources. In the consequence, the samples for inspection and observation have to be moved to the neutron source. If this happens on national level, the effort is reasonable. However, if samples have to be transported over national borders, assurance, customs and special regulations for cultural heritage objects have to be considered carefully.

17.2 Block excavations of find objects

Most ancient objects with cultural importance are still found in or near human settlements from the past (Roman, Celtic, Middle Ages,...). Often, such places have been in use repeatedly or continuously over many generations, resulting in many succeeding occupation layers. As a consequence, on new building sites in such historical settlement areas, possibly relevant archaeological finds are made. When a deeper investigation of samples on-site is not feasible due to the missing local laboratory condition or if there is no time for a thorough on-site examination, parts of the excavation are sealed and transported away for further study with best possible performance.

The strategy of further treatment, including the concept for conservation depends much on the specific material within the excavated ensemble. Metals have to be handled differently from organic material (bones, wood, leather, …). Often, the conservation work is critical in time after opening the hidden sample and hence exposing the objects to the ambient air.

Therefore, non-destructive inspections of the content of the block excavation can be a valuable tool for the determination of the inner content of the blocks, when delicate objects and material combinations are expected. Because x-rays and neutrons can deliver different contrast features, both kinds of radiation should be applied in the best case together. As modern digital imaging detectors have about the same performance regarding spatial resolution, the image data can directly be compared.

Figure 17.2 demonstrates these facts by comparing the neutron and the x-ray images of the block excavation sample shown on top. Impressively, the x-ray image

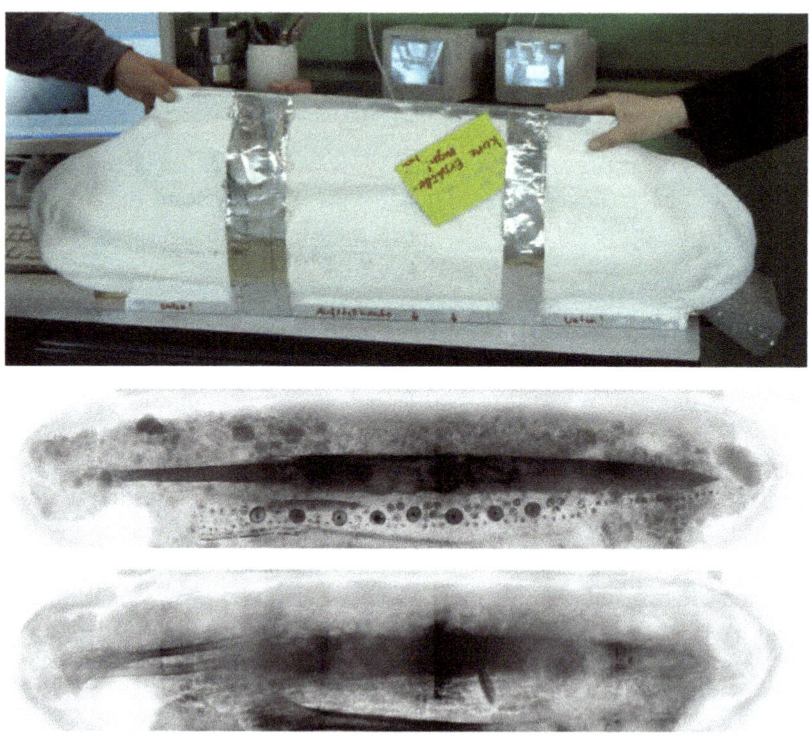

Figure 17.2. Results of the inspection of a block excavation sample by x-ray (above) and neutron (below) imaging; metallic parts (sword, decoration) can be identified in the x-ray image while the neutron image delivers bones and leather.

shows the metallic part from a sword and a knife and the metallic decorations of a belt. On the other side, the neutron image shows the leather of the belt and in addition some human bones. The complete picture of this sample is only possible with the help of the two kinds of radiation, complementing each other in the best way.

Based on this knowledge about the content of the block excavation sample, the experts for conservation can decide if and how the assembly can be treated and the different content separated and protected in the best possible way.

17.3 The gladius from Vindonissa

In the neighbourhood of the current town Brugg-Windisch in Switzerland, a Roman short sword of the type 'Gladius- type Mainz 1' was found during excavation work in 1990. This site represents the location of the former 'Vindonissa' a little south of the Rhine river.

The sword is relatively well preserved, considering its age of about 2000 years, lying underground. Only the wooden handle and parts of the metal structure are missing. The object's surface was cleaned from corrosion products, the object was then restored with some consolidating resin and sealed afterwards with lacquer. All details about the excavation and common archaeological analysis is given in [2]. The sword is part of the exhibition of the Vindonissa-museum in Brugg [3].

Non-destructive tests with x-rays and neutrons were performed some years later with suitable beamlines at the Paul Scherrer Institut, Switzerland, after their availability since 1998. The results are given in figure 17.3.

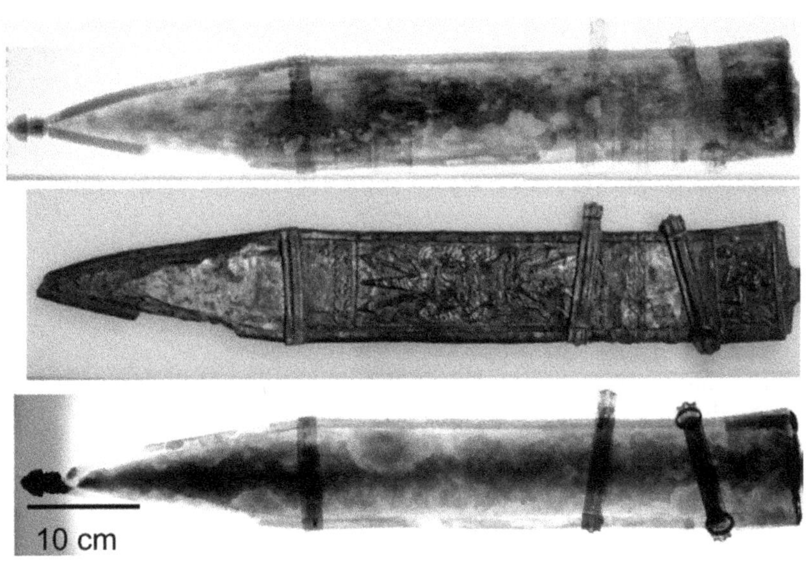

Figure 17.3. The gladius sword from 'Vinonissa' (photo-middle), inspected with neutrons (top) and 150 kV x-rays (bottom) (source: [2] John Wiley & Sons).

At first glance, both images looks similar and show the structure of the sword well. Looking in more detail, the x-ray image delivers the remaining steel of the damaged sample. In the neutron image, the steel is visible with lower contrast. In addition, we can see the distribution of the resin, which was used for conservation. Furthermore, a streaky structure is enhanced along the sword, which is interpreted as remaining wood material, used for the sword's scabbard.

This additional information about the gladius can now be used to complete the picture of the sample. The investigation using neutron imaging yields important information on the behaviour of consolidating and conserving agents on such an object. Nevertheless, such investigations should be carried out BEFORE any conservation treatment is applied. In this way, it might be possible to minimize the amount of conservation materials and to optimize the consolidation techniques necessary for a specific object. Hence, the long-term consolidation of fragile samples could benefit from an optimized, minimally invasive consolidation approach, still resulting in the best possible conservation [4].

17.4 The 'violinist' from P Gargallo

Another example where neutron imaging successfully helped with regards to a cultural heritage object, is the sculpture 'the violinist' by the artist Pablo Gargallo (ca. 1920). The sculpture, which is part of the collection at the Museu Nacional d'Art de Catalunya, Barcelona (Spain) consists of numerous lead sheets which are fixed on a wooden core. On this object, the responsible conservators of the museum found indications of corrosion and decay; the covering lead material had bulged out in some areas and white powder, which turned out to be mainly lead carbonate had gathered below the sculpture. In order to assess how far the corrosion process had already progressed on the inside of the sculpture as well as to obtain information on the build-up of the sculpture (i.e. fixation of lead sheets, shape of the inner core, etc) the object was studied using neutron tomography.

Investigations with neutron tomography seemed to be the only technique to solve this problem because the lead layer of the object on the outside (few millimeters) would hinder the penetration of x-rays completely. On the other hand, the massive wooden core might attenuate the neutron beam in such an amount that a tomography reconstruction is disabled.

In this respect, the investigation was a challenge at the limit of the neutron tomography performance. Fortunately, lead has such a low attenuation coefficient for thermal neutrons ($\sum_{tot} = 0.38\ cm^{-1}$) that layers of some centimeters can easily be penetrated.

The object, shown in figure 17.4, is higher than the field-of-view of the neutron beam and the detectors and was therefore investigated in two steps—the higher and the lower part. In order to have the full object in one dataset, the stacked volume data were combined into only one. More details of the investigation are summarized in the related publication [5].

Based on the non-invasive reconstruction of the object's status it is now possible to plan the treatment and reconstruction. All corroded regions could identified and allowed for a

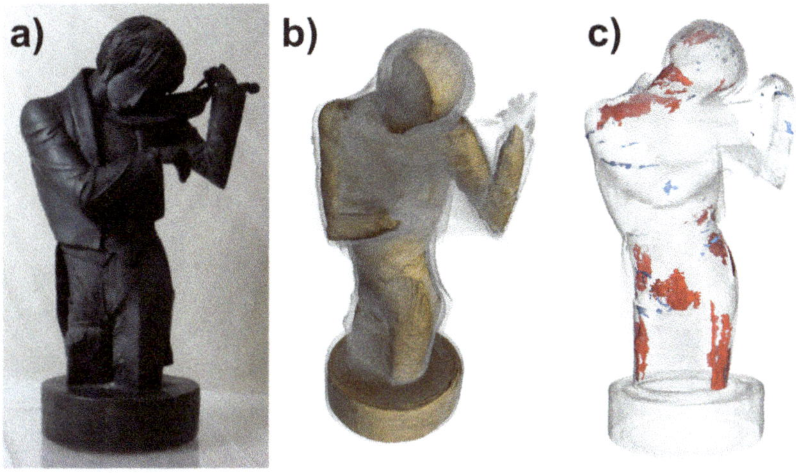

Figure 17.4. The 'violinist' by P Gargallo (photo (a)), a lead covered sculpture with a wooden core, made 1920 (55.3 × 31.8 × 21.6 cm and weighs 11.9 kg); with the help of neutron tomography the contours of the core (b)) and the positions with a high degree of lead corrosion (c)) were identified –marked in red (reproduced with permission from [6]).

better appraisal of the urgency for conservatory measures. The identification of the fixation points would furthermore allow disassembly of the individual lead sheets in a minimally invasive way. Because the chemical interaction with solvent portions of the wood is made responsible for the lead corrosion from inside, the strategy will be to replace wood by an inert (plastic) material. For this purpose, the volume data of the inner wood structure taken from the tomography run can be used best to manufacture the 'inert' replacement core.

17.5 Wood conservation

Wood as a grown organic material represents a more complex topic. It is and has been used in many different areas of application: as structural component of buildings, for tools and weapons as well as for art works. Compared to stones, wood structures are often less durable for different reasons:

1. As hygroscopic material, wood reacts under the influence of moisture much more by volume changes than stones and undergoes cycles of swelling and shrinking, which handicaps the mechanical stability.
2. It is affected by attacks from animals, fungi and other microorganisms.
3. Wood is flammable and many monuments and objects have been lost to fire.
4. Next to water influence, the influence from the light of the Sun should not be underestimated, which damages at least the surface properties.

Nevertheless, the protection measures of wooden objects and structures are often similar to those used for stone. Wood is more porous than stone and has with its fibre structure and anisotropic behaviour preferred directions regarding moisture (and protective agent) uptake.

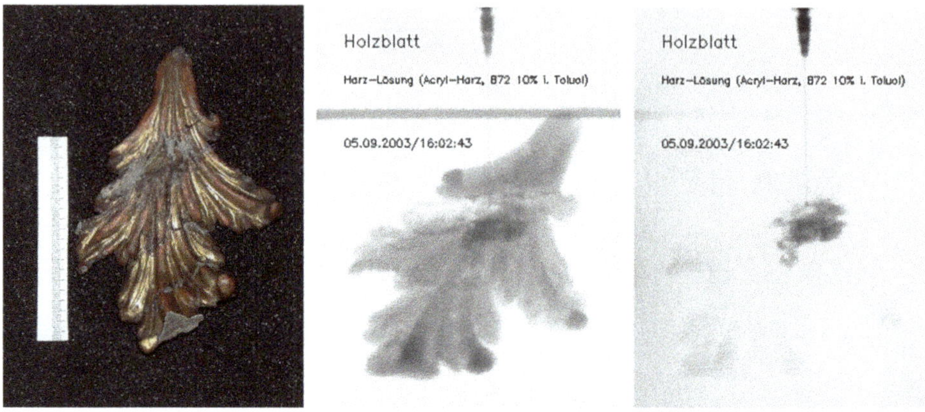

Figure 17.5. Into the wooden test piece (photo, left) a certain amount of the resin/solvent solution is injected; removing the wooden structure by image processing, the netto amount of the solution can be obtained (right). Reprinted from [12], copyright (2005), with permission from Elsevier.

The behaviour of wood and the studies with neutron imaging methods are given in chapter 8 in more detail. Here, we focus on the protection and treatment measure alone by some dedicated work in this direction. Some methodical and systematic work was done by Kucerova *et al* [7]. This team focused later also on the treatment with agents for the protection against fire [8, 9] with ammonium phosphate and sulfate-based fire retardants.

Compared to stone, the neutron attenuation for wood is higher, given by the hydrogen content of about 6%. This limits on the one hand the thickness of samples under investigation in most cases to a few centimeters. Furthermore, the visibility and detectability of protective solvents is also reduced as a consequence of the low contrast, between agent and wood matrix. Therefore, some attempts were done by using fast neutrons for the inspection of huge samples (about 20 cm in diameter = full trees). Because fast neutrons are much less sensitive for hydrogen than thermal neutrons, only the density gain by uptake of the protective agent can be observed. Some results of studies on the treatment of large wooden art objects are summarized in [10, 11], but are not really convincing compared to studies with slow neutrons. Furthermore, there are only very few facilities for imaging with fast neutrons available.

More convincing results are available in [12], where figure 17.5 is taken from. The use of x-ray techniques in addition to neutron imaging is described in [13]. However, it has to be considered that the wooden matrix changes its dimensions during the treatment with liquid agents more than stone and the image comparison becomes more difficult.

17.6 Stone conservation/consolidants

Although stones have very high durability, they undergo some degradation and even destruction by environmental impacts. Also, recent and future climate change and air pollution might advance stone damage in several regions. This problem is mainly of relevance for historical buildings standing unprotected with influence from moisture, temperature changes and chemical attacks. In the case of a wet climate combined with temperatures below 0 °C, the ice impact can destroy even stone materials easily.

The stone matrix is not as compact and stable as metallic structures, but has some porosity and even voids. In addition, stones are not as homogenous as cast metals and undergoes density variations. These two aspects are also the reason for influences from outside: moisture can migrate into stony structures; temperature changes can induce strain in the material due to the density differences.

For the protection, conservation and consolidation of stony objects from our cultural heritage we have different options available:

- To bring the samples into a controlled and stabilized environment. This is not possible for big monuments, but for pieces of high relevance like the Pergamon altar in Berlin or in the museum of the Acropolis in Athens.
- To fill the porosity of the stone matrix with agents, applied from outside into the sample or through the building surface.
- To apply a sealing and protective layer onto the surface of the stony structure. Thorough coating is required for a long-lasting solution.

The two last options are topic for further developments regarding the selection of the best agents (chemical composition, effectivity, durability, costs, …), the application techniques and the compatibility with the stony matrix—over a long period.

Neutron imaging can be well used in this development process because the transparency of stone is given by the low attenuation coefficients for many stone components such as Si, Ca, O, Al, Na, K,…. On the other hand, most of the protective agents have a relative high content of hydrogen, which delivers high contrast for neutron image applications.

The experimental work with stones with respect to cultural heritage protection measures can be divided into three classes:

1. To investigate characteristic stone samples regarding their water uptake and diffusion;
2. To study the ingress of consolidating materials into the stone matrix;
3. To check the effectivity of the applied agents (e.g. water repellence, …).

All these investigations can be done in time sequences and quantitatively. Because stones are not swelling like wooden structure (see section 17.5), a virtual removal of the initial stony matrix can easily be done with image processing tools. Figure 17.6 demonstrates the high sensitivity for the detection of the agent distribution within the stony/wooden matrix [14–18].

The first topic is covered in this book with other, more general considerations about geology (chapter 9) where the water uptake and distribution has also been studied, in some cases also in all three dimensions (tomography). In order to enhance the visibility of moisture in the samples, reference images with x-rays can be performed which do not have a high sensitivity for water (unlike neutrons) and reflect mainly the stony matrix—which can be subtracted from neutron imaging data.

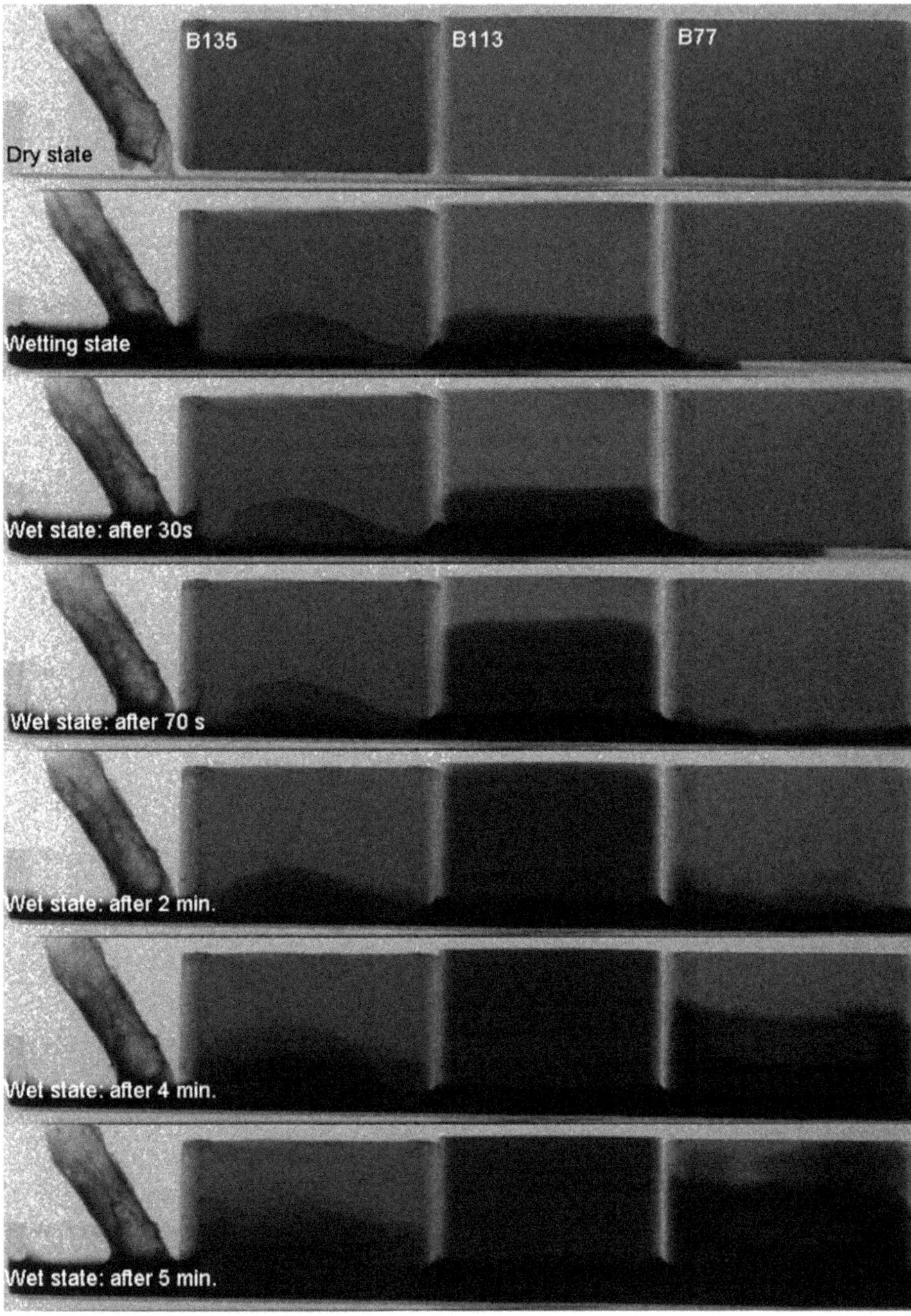

Figure 17.6. Sequence of radiographs taken for the Bray sandstone samples B135 (left column), B113 (middle column) and B77 (right column) in function of time, while absorbing water by capillarity. Reprinted from [18], copyright (2007), with permission from Elsevier.

17.7 Paintings—monitoring of moisture content changes in canvas

Changes in moisture content play an important role when it comes to the preservation of easel paintings, where, e.g. oil paint, was applied on canvas. Besides the fact of being a piece of art, such paintings also represent a complex system where many different materials are combined, such as canvas, ground paint, animal glues (e.g. hide glue, bone glue, …), the actual oil paint etc (not to mention the frame of the picture, which is in many cases made of wood and hence also susceptible for moisture changes). Moisture content changes influence a multitude of possible degradation occurring in paintings including chemical changes, pigment decomposition and mechanical damage to the canvas. A better understanding of the processes happening with changing ambient climatic conditions (especially changing relative humidity) can also benefit the preservation and conservation of such paintings.

Boon *et al* [19] used neutron radiography to carry out time-dependent studies on the water uptake in mock-up painting. The study proved that it was possible to quantitatively determine the moisture content changes in the different layers of the painting with sufficient precision. In another study, Hendrickx *et al* [20] investigated the changes in the moisture distribution in mock-up samples when they are exposed to a sorption/desorption (drying) cycle. They could demonstrate how the different materials showed different sorption/desorption behaviour, especially with respect to the rate of change for these values (figure 17.7). The results were furthermore complemented and

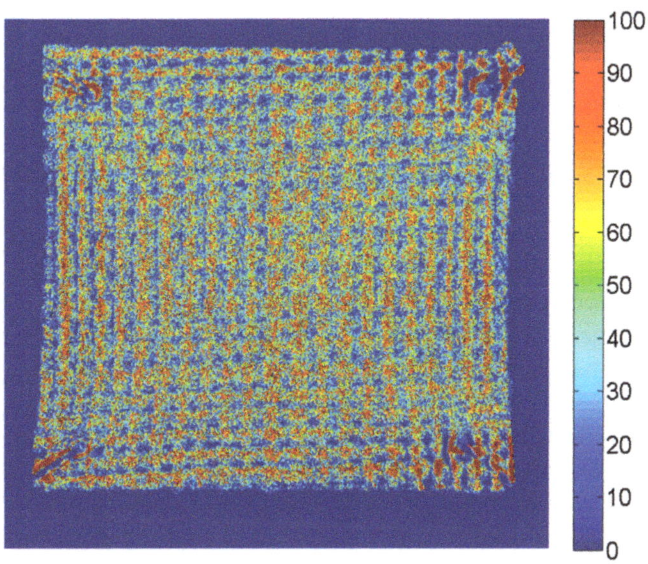

Figure 17.7. Neutron radiography image of the moisture distribution over a reconstructed painting (sample type 1) in frontal position, 10 expressed in μm of moisture. The thickness of the sample is ~500 μm. Thus, a value of 50 can correspond to a volumetric moisture 11 content of 10%. Reprinted from [20]. Copyright The International Institute for Conservation of Historic and Artistic Works, reprinted by permission of Informa UK Limited, trading as Taylor & Francis Group, http://www.tandfonline.com on behalf of The International Institute for Conservation of Historic and Artistic Works.

verified with NMR. The results of both methods show very high consistency. To compare different approaches to consolidate paintings on canvas, Bridarolli *et al* [21] used neutron imaging to assess how different nanocellulose treatments influence the hygroscopic behaviour of canvasses under cyclic climatic changes.

Despite the relatively low number of investigations in this field, it shows how neutron imaging can help in this field of conservation. Not only when it comes to a better basic understanding of the processes occurring in the material, but neutron imaging can also help to develop and assess new methods for consolidation and conservation.

17.8 Brass wind instruments—handling of historical objects

Neutron imaging can not only help with delivering important information when it comes to the documentation of the status of an object or to define the best way to preserve or restore an object, it can also help to determine how historical objects should be handled. Mannes and Lehmann [22] showed how different neutron imaging approaches can be used to monitor the impact of regular playing on historical brass wind instruments. The study was part of an interdisciplinary project with different Swiss Museums and Research institutions (Bern University of the Arts, ETH Zurich, Swiss national museum). It is a very controversially discussed topic whether a historical instrument can or should still be used to play music. Old instruments sound different and are different to play when compared to modern instruments. This means to really hear and understand what a composer wanted to express with their music, the piece has to be played on contemporary historical instruments (or proper replicas). Many brass wind instruments appear still to be in very good condition, which would still allow musicians to play music on them. From the side of the museums, to which these historical objects most often belong, this is a considerable risk. Playing of a brass wind instrument introduces moisture in the instrument and increases the risk of corrosion. In the before-mentioned project, different historical brass wind instruments were studied with respect to the changes occurring when being played on a regular basis over a period of almost a year. Neutron imaging was used to monitor possible changes (e.g. corrosion) along the course of the project, as a consequence of the regular play or the subsequent handling during that period (different ways of cleaning, drying, etc). Neutron tomography was performed on the tuning slides of several instruments at the start and the end of the project. In a first evaluation approach the inner surfaces of the tuning slides were compared and checked for material accumulation (e.g. as consequence of newly formed corroded areas) and displacement. This approach yielded only limited information as the spatial resolution was relatively low as a consequence of the relatively large size of the objects. In addition, the accumulation of newly corroded material was occurring mostly below the spatial resolution. As a third problem, some of the tuning slides showed plastic deformation; due to inserting, tuning, etc the slides had partially deformed. Even if this was ever so slightly, the deformation was larger than any potential newly appeared corrosion layer. In a different approach the neutron CT data was registered, virtually unrolled

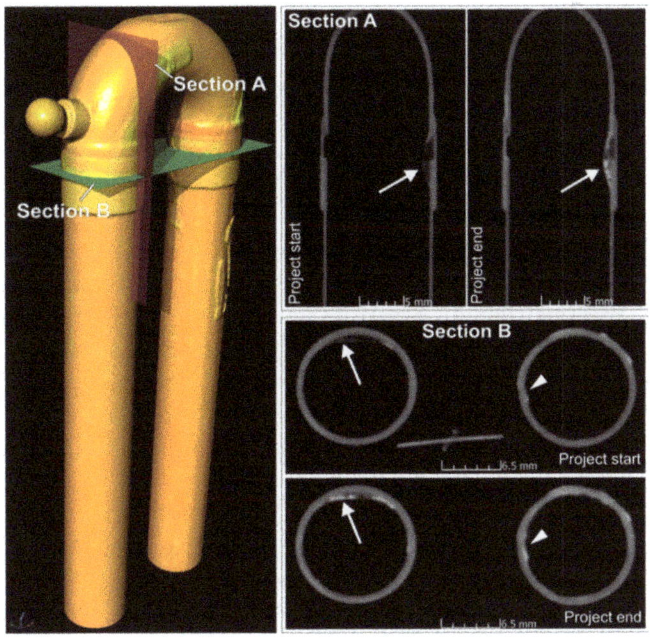

Figure 17.8. Reconstructed tomography data of a tuning slide from a historical brass wind instrument. CT data allow to section the reconstructed 3D data in arbitrary positions and directions; the sections show two viewing planes in the instrument at the beginning and the end of an experimental project on the playability of historical instruments; arrows and arrow heads show where higher attenuating materials (presumably corrosion products) have accumulated in the course of the project. Reproduced with permission from [22].

and compared piece by piece. This way regions could be identified which showed considerable change in their attenuation coefficient during the course of the project and were presumed to be corrosion (figure 17.8).

17.9 Appraisal of the effectivity of conservation treatment

Neutron imaging is not only helpful when it comes to assessing the impact and effectivity of conservation treatments. Such studies can also work in an opposite direction and help to develop and push the methods used for the investigation. In the following example, we show how only the combination of x-ray and neutron tomography was able to assess the effectivity of a conservation treatment.

Archaeological iron objects, which are found in soil are inevitably in contact with Cl ions, contained in the surrounding soil. When excavated the system is exposed to air and hence oxygen, which reacts with the chlorides, can start an accelerated corrosion. One approach to stop this process is a desalination process, which aims at removing Cl ions from the object. To assess whether this process is successfully finished, the Cl concentration in the desalination bath is determined on a regular basis. So far it was nevertheless not clear how deep the treatment was penetrating the treated object or if it was only working on the object's surface.

Jacot Guillarmod *et al* [23] proposed a combination of neutron and x-ray tomography to tackle this problem. By using both modalities, it is possible to take advantage of their complementarity and hence create synergetic effects. While x-rays in principle only show the density of the object (with the sound metal as highly attenuating region and corroded regions with lower attenuation) the sound metal is relatively easily penetrated by neutrons while the corroded regions are highlighted due to the content of hydrogen and chlorides (figure 17.9). For the project several samples (corroded nails from the Roman period in Switzerland), were studied with neutron and x-ray tomography before and after a conservation treatment. The reconstructed neutron and x-ray data was then evaluated. Combining both modalities allowed not only the comparison of slices side by side, but it also opened up new possibilities. By using a bivariate histogram as basis it was possible to obtain a more precise sophisticated segmentation of the 3D object (figure 17.10). This allowed better assessment and confirmation of the effectivity of the conservation treatment.

Neutron imaging can be a very powerful tool, when it comes to the field of conservation and preservation of cultural heritage artefacts. As we could show, it can help on every level of the topical area, starting from the excavation and the best way to extract and stabilize sensitive material from the soil. Neutron imaging can also help to assess the state of an object and the urgency and best way to start with the conservation and preservation treatment. Furthermore, it provides unique

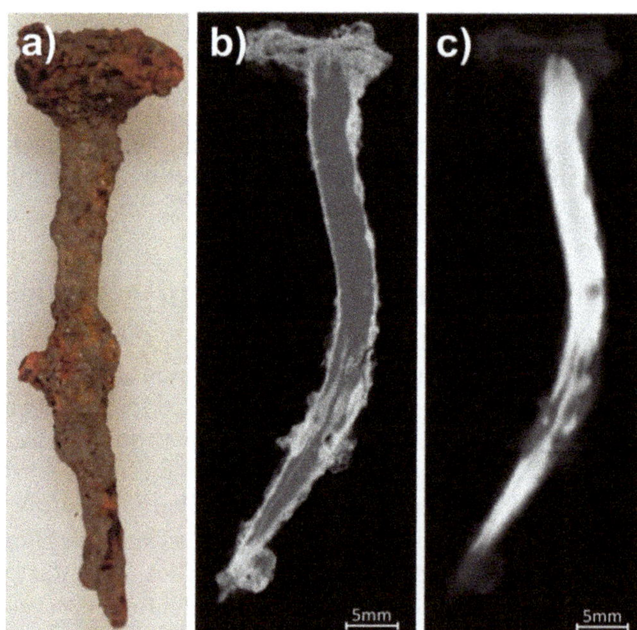

Figure 17.9. Corroded Roman nail, stored at the Swiss National Museum, Zurich; (a) photography, (b) and (c) central slices through the tomography data for neutrons (b) and x-rays (c) (source: [19] modified). (Reproduced from reference [23] CC BY 4.0.)

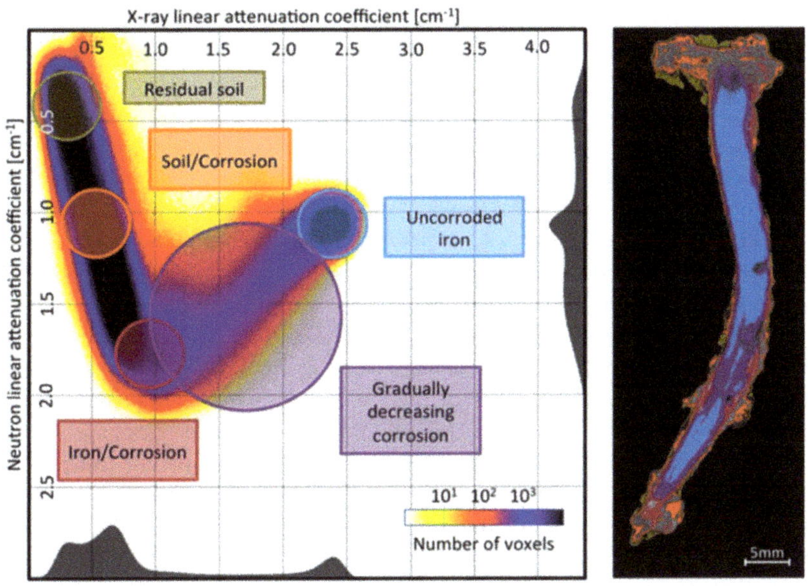

Figure 17.10. Bivariate histogram (left) combining the histograms of the neutron and x-ray CT data allows to assign certain materials/compositions to specific regions within the histogram; these regions can be transferred back to the actual volume data allowing a localization of the different materials in the object (right) (modified from [23] CC BY 4.0).

insight on many fundamental working principles of the underlying phenomena, which have to be understood to best preserve cultural heritage objects. This allows optimizing handling instructions and developing and validating new conservation treatments.

In an interdisciplinary project funded by the Swiss Science Foundation teams from PSI; ETH Zurich, EPF Lausanne, HE-Arc are taking advantage of the opportunities that the bimodal approach described in 17.19 hinted at. The goal of the project is to optimize a multimodal approach (with a combination of neutron and x-ray tomography at its centre) to study corrosion processes in opaque media. This includes the corrosion of rebars in reinforced concrete as well as iron objects (mainly nails) from the Roman period. The project should yield an insight and better understanding of the corrosion processes under conditions unexposed to ambient atmosphere but should then also show what processes happen, after and while the objects are eventually exposed. The aim is, to allow for a better guidance for the preservation of cultural heritage objects as well as recommendations for excavations and the preservation of engineering structures.

References

[1] Mannes D and Lehmann E H 2022 Neutron imaging of cultural heritage objects *Handbook of Cultural Heritage Analysis* (Cham: Springer International Publishing) pp 211–37

[2] Deschler-Erb E 1996 Vindonissa: ein Gladius mit reliefverzierter Scheide und Gürtelteilen aus dem Legionslager *Jahresbericht/Gesellschaft Pro Vindonissa*

[3] Vindonissa Museum Brugg (https://museumaargau.ch/vindonissa-museum)

[4] Deschler-Erb E, Lehmann E H, Pernet L, Vontobel P and Hartmann S 2004 The complementary use of neutrons and x-rays for the non-destructive investigation of archeological objects from Swiss collections *Archaeometry* **46** 647–61

[5] Masalles A, Lehmann E and Mannes D 2015 Non-destructive investigation of 'The Violinist' a lead sculpture by Pablo Gargallo, using the neutron imaging facility NEUTRA in the Paul Scherrer Institute *Phys. Proc.* **69** 636–45

[6] Mannes D, Lehmann E, Masalles A, Schmidt-Ott K, Schaeppi K, Schmid F and Hunger K 2014 The study of cultural heritage relevant objects by means of neutron imaging techniques *Insight, Non-Destr. Test. Cond. Monit.* **56** 137–41

[7] Kucerová I, Ohlídalová M, Novotná M and Michalcová A 2007 Examination of damaged wood by ammonium phosphate and sulphate-based fire retardants—the results of the Prague Castle *Florence: Proc. of the Int. Conf. Held by Cost Action Ie0601 (Florence)*

[8] Kucerová I, Michalcová A, Novotná M and Ohlídalová M 2007 Examination of damaged wood by ammonium phosphate and sulphate-based fire retardants: the results of the prague castle roof timber examination *Examination of Damaged Wood by Ammonium Phosphate and Sulphate-Based Fire Retardants* (Firenze University Press) pp 1000–5

[9] Kučerová I 2012 Methods to measure the penetration of consolidant solutions into 'dry' wood *J. Cult. Herit.* **13** S191–5

[10] Osterloh K R and Nusser A 2014 X-ray and neutron radiological methods to support the conservation of wooden artworks soaked with a polluting impregnant 'Carbolineum' *Proc. of the 11th European Conf. on Non-Destructive Testing (ECNDT 2014)*

[11] Osterloh K, Bellon C, Hohendorf S, Kolkoori S, Wrobel N, Nusser A and Tittelmeier K 2015 Computed tomography with x-rays and fast neutrons for restoration of wooden artwork *Phys. Proc.* **69** 472–7

[12] Lehmann E, Hartmann S and Wyer P 2005 Neutron radiography as visualization and quantification method for conservation measures of wood firmness enhancement *Nucl. Instrum. Methods Phys. Res.* A **542** 87–94

[13] Triolo R, Giambona G, Celso F L, Ruffo I, Kardjilov N, Hilger A and Paulke A 2010 Combined application of x-ray and neutron imaging techniques to wood materials *Conserv. Sci. Cult. Herit.* **10** 143–58

[14] Conti C, Colombo C, Festa G, Hovind J, Cippo E P, Possenti E and Realini M 2016 Investigation of ammonium oxalate diffusion in carbonatic substrates by neutron tomography *J. Cult. Heritage* **19** 463–6

[15] Kis Z, Sciarretta F and Szentmiklósi L 2017 Water uptake experiments of historic construction materials from Venice by neutron imaging and PGAI methods *Mater. Struct.* **50** 1–14

[16] Hameed F, Schillinger B, Rohatsch A, Zawisky M and Rauch H 2009 Investigations of stone consolidants by neutron imaging *Nucl. Instrum. Methods Phys. Res.* A **605** 150–3

[17] Realini M, Colombo C, Conti C, Grazzi F, Perelli Cippo E and Hovind J 2017 Development of neutron imaging quantitative data treatment to assess conservation products in cultural heritage *Anal. Bioanal. Chem.* **409** 6133–9

[18] Cnudde V, Dierick M, Vlassenbroeck J, Masschaele B, Lehmann E, Jacobs P and Van Hoorebeke L 2007 Determination of the impregnation depth of siloxanes and ethylsilicates in porous material by neutron radiography *J. Cult. Herit.* **8** 331–8

[19] Boon J J, Hendrickx R, Eijkel G, Cerjak I, Kaestner A and Ferreira E S B 2015 Neutron radiography for the study of water uptake in painting canvases and preparation layers *Appl. Phys.* A **121** 837–47

[20] Hendrickx R *et al* 2017 Distribution of moisture in reconstructed oil paintings on canvas during absorption and drying: a neutron radiography and NMR study *Stud. Conserv.* **62** 393–409

[21] Bridarolli A, Odlyha M, Burca G, Duncan J C, Akeroyd F A, Church A and Bozec L 2021 Controlled environment neutron radiography of moisture sorption/desorption in nanocellulose-treated cotton painting canvases *ACS Appl. Polym. Mater.* **3** 777–88

[22] Mannes D and Lehmann E 2023 Monitoring the condition of played historical brass instruments by means of neutron imaging *To Play or Not to Play. Corrosion of Historic Brass Instruments. Romantic Brass Symposium 4* ed A v Steiger, D Allenbach and Schliengen M S: Argus 2023 (Musikforschung der Hochschule der Künste Bern, vol 15), pp 83–91

[23] Jacot-Guillarmod M, Schmidt-Ott K, Mannes D, Kaestner A, Lehmann E and Gervais C 2019 Multi-modal tomography to assess dechlorination treatments of iron-based archaeological artifacts *Herit. Sci.* **7** 1–14

Part VII

Industry

IOP Publishing

Neutron Imaging
From applied materials science to industry
Markus Strobl and Eberhard Lehmann

Chapter 18

Industry

Eberhard Lehmann and David Mannes

18.1 Introduction

Neutrons can penetrate material layers of certain thickness, depending on the particular composition. This allows visualizing and quantifying material distributions non-invasively, even in a complex and hidden environment. Hence, neutron imaging is available as a tool for material testing, similarly and complementarily to the common *x*-ray methods.

The complementarity of neutrons compared to *x*-ray methods makes it a particularly interesting tool for the study of industrial and engineering applications. As described in more detail in previous chapters, the attenuation properties of neutrons and *x*-rays differ in a systematic and deviating manner, resulting in a partial complementarity of the two radiation types. The interaction of *x*-rays is strongly correlated with the atomic number of the elements; hence, materials consisting of elements with a low atomic number such as hydrogen (e.g. water, organic materials, oils, resins etc) are easily penetrated; materials with higher atomic number (e.g. metals) are attenuating *x*-rays to a much higher extent, making it more difficult to study e.g. metallic objects. The complementary behavior of neutrons provides a high sensitivity for some light elements such as hydrogen, lithium or boron; already a few millimeters of water will stop nearly all thermal neutrons in a beam. At the same time, neutrons can penetrate some elements with high atomic number; even lead, bismuth, tin or precious metals can be penetrated in certain layer thickness.

Early applications were carried out using film-based methods; these investigations comprised non-destructive studies (e.g. defect analyses) on a variety of objects such as turbine blades, pyrotechnical equipment or nuclear fuel. Some standard applications, which have been certified as routines for quality control, are still practiced today using films.

However, neutron imaging methods have considerably evolved since the first attempts. Based on digital detection systems, it is now possible to perform neutron

doi:10.1088/978-0-7503-3495-2ch18

© IOP Publishing Ltd 2024

tomography routinely. Due to the high dynamic range of digital detectors, adequate regions of the dynamic range can be chosen, which enables a much higher visibility of tiny contrast variations.

Time dependent investigations of processes in real time or as a stroboscopic study of cyclic processes (e.g. running engine) are enabled by modern systems.

Recent improvements, such as the utilization of variable neutron energies, polarized neutrons or the application of phase-contrast methods have extended the possibilities of neutron imaging. More details about the state-of-the-art can be found in chapter 2.

Industry with its varying and manifold demands can take direct advantage from the methodical progress. A direct and focussed dialogue has to be established in each particular case, if and which neutron technique can be applied for the problem to be solved. This chapter demonstrates, based on practical examples, how neutron imaging methods can be applied successfully. Not all problems can be solved and not all questions answered by neutron imaging methods. There are many cases with advantages for x-rays—or in best cases, a combination of both imaging techniques. Because of the higher costs of neutron imaging, other methods should be tested first, before neutrons are used as an ultimate technique.

When industry is involved, the business aspect must not be ignored. On the one hand, the solution of a technical problem, the development and improvement of a product, or simply quality, control deliver advantages for the industrial customer/partner in the market. Therefore, it is fair and reasonable to charge industry at least for the effort involved during the inspection. On the other hand, the money earned for the investigation can be used to further improve imaging techniques, the equipment and to train and preserve qualified manpower. In an ideal case, this can be seen as an iterative process with questions and demands from the industry, method development and new methodological approaches to suggest solutions on a higher level, available and useful for industrial as well as scientific partners.

There are many overview papers available of the leading teams in neutron imaging, where options for collaboration with industry, including relevant examples are given [1–9]. Some of them are reflected in this chapter—without specific preference or exclusions.

Neutron imaging with the whole variety of possible options is mainly available at large-scale neutron sources, organized as user facilities. The beam quality is optimized and the detector techniques and the methodical standards are well established. Unfortunately, there are only a very limited number of such neutron imaging facilities available worldwide, and the access is quite limited and prioritized. But for many problems and questions from the industry, smaller neutron sources with their related detection systems and available methods can still deliver pragmatic and satisfying solutions, for industrial applications on a less demanding level.

18.2 Automotive

Individual and public transportation on the roads has been organized in the last century and the first years in the 21st century by combustion engines of different sizes. Attempts for electrical motorized engines failed for different problems until recently.

Therefore, this chapter summarizes neutron imaging studies mainly for 'conventional' engines and car components. Regarding battery research using neutrons, we refer the reader to chapter 11.

18.2.1 Fuel injection

Combustion engines were investigated quite early in the time of neutron radiography, using static film methods. In this way, all inner components are visible due to the high transmission power of thermal (or even fast) neutrons. The aim of such studies was to find out manufacturing errors or malfunctions under realistic conditions before dismantling.

With the introduction of digital neutron imaging techniques, two essential options became available: time-dependent investigations and tomographic studies for the 3D-visualization of all inner components in quasi-functional positions [10].

Because the outer motor cases are often made of aluminum, (mainly to reduce weight) the engines are quite transparent for neutrons and steel or brass parts can be distinguished easily. Rubber or plastic sealing are especially clearly visible due to their high hydrogen content. In particularl, the filling of liquids (for oil or water-cooling) give a high contrast even in small amounts. Leaks or disarrangements can be found in this way quite easily.

Time dependent investigations for running engines with about 1000 rpm in a real-time mode are more or less impossible under continuous neutron 'illumination'. The frame rate of current detection systems is not high enough to enable such investigations, and the dose rate per image is too small to enable clear images.

However, because combustion engines operate in a repetitive regime, the solution to follow their internal process is the synchronization of the detector with the engine by an electronic trigger signal. Stacking many short-time frames from identical positions in the engine's operation cycle provides valid images for each position. In this way, vibrations of the setup can also be cancelled out. Pictures from many different positions of the engine's operation cycle can be composed into a full movie of the combustion process.

Because a real water cooled combustion engine, fired with gasoline or diesel oil, cannot be penetrated due to the 'shielding' by the water channels in the combustion head, investigations about the cooling oil distributions can be carried out using an externally driven setup actuated by an electromotor device.

Such investigations were performed at a preliminary setup at the ILL reactor in Grenoble [11] and at the NEUTRA facility at PSI [12] with success. One single frame of the movie can be seen in figure 18.1.

Because most of the two-stroke engines are only air-cooled, the neutron transmission in this case is not hindered by a cooling liquid. Therefore, real fired engines can be studied with the same method as described above using a triggered and gated intensified digital camera detector. The resulting data can help to obtain a better insight on the distribution of the inner oil film build from the fuel–oil-mixture. Such

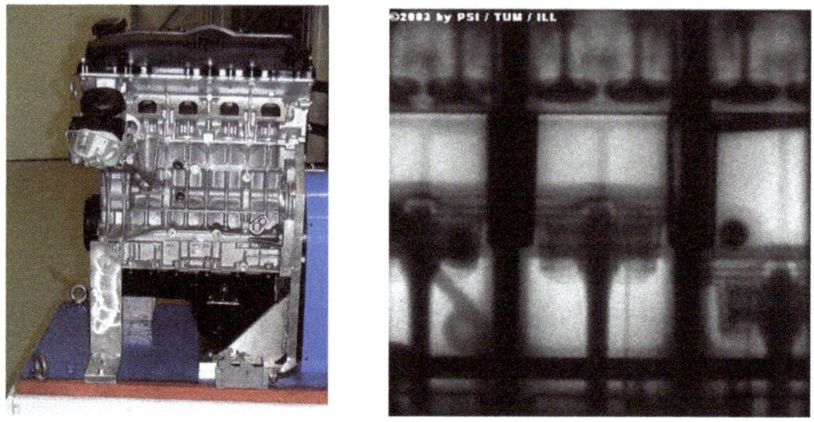

Figure 18.1. Photograph of a combustion engine of a motor car, driven externally (left); neutron imaging snapshot of the piston motion at several hundred of rpm—the oil jet from below is well visible. Reprinted from [11], copyright (2006), with permission from Elsevier.

Figure 18.2. Snapshot of a two-stroke engine under real operational conditions, fired, air-cooled. Reprinted from [14], copyright (2013), with permission from Elsevier.

an investigation was for example carried out at the cold neutron imaging beamline ICON at the spallation neutron source SINQ at PSI, Switzerland. One result was a short animated time series; the movie is shown in [13] and a single frame is given in figure 18.2.

Beside the knowledge about cooling oil distributions, the injection process of fuel into the combustion engines was a topic of investigations with neutron imaging methods. Even if the amount of gasoline or diesel oil per individual injection is quite low, attempts were made under quasi-real conditions [10]. In order to increase the contrast for the fuel, Gd-based tracer agent was added to the fuel. This emphasized the contrast and allowed a much higher visibility of the injected fuel during the experiments.

Also, the diesel injection nozzles themselves were a subject of neutron imaging studies. Because the steel components of the nozzle are relatively transparent for neutrons (better than for x-rays) and the small amounts of liquid fuel already give a good contrast, the nozzle was investigated by neutron tomography, in particular the head, where the fuel cloud is formed. However, the holes in the head have a diameter of only a few tenths of a millimeter. Fortunately, a high-resolution setup was available at the PSI neutron imaging facilities with pixel sizes of only 13.5 μm and the study of the injection holes was done in good performance [13].

18.2.2 Carburetors

Another way to generate the air–fuel mixture necessary for the combustion process is the utilization of carburetors. While most modern combustion engines rely on fuel injection, the carburetor is still widespread, for example, for two-stroke engines. Carburetors were the subject of neutron imaging studies quite early in simple radiography mode as well as in 3D using neutron tomography (see figure 18.3). These examples were limited to static observations of the objects. In a previously unpublished study, Grünzweig and Mannes (personal communication) examined how the flow distribution in carburetors dynamically evolves with different under-pressures and changing throttle positions. The study was carried out on different carburetors such as standard carburetors from chainsaws as well as on high

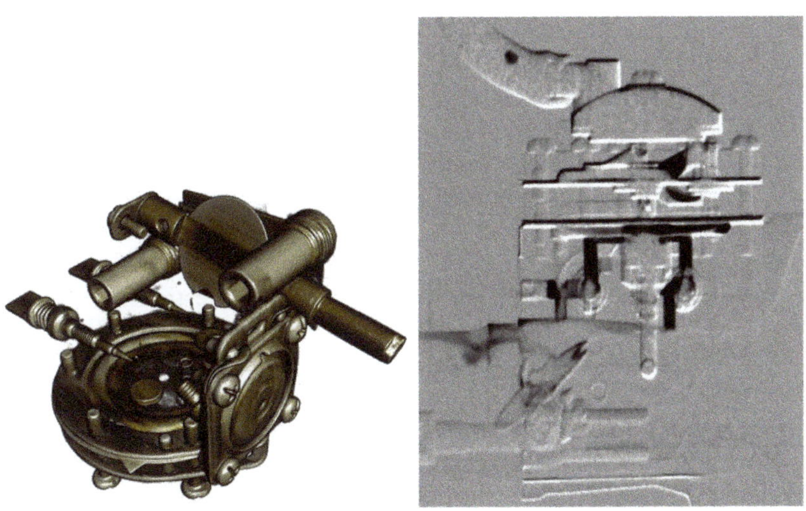

Figure 18.3. Neutron tomography image of a carburetor device (size about 10 cm, left) and the gasoline/air mixture flow inside, simulated by water. The image is taken in 'reference mode' in order to enhance the visibility of the flow. Reproduced from [1] CC BY 4.0.

performance carburetors as used in the engines of racing carts. For safety reasons, the fuel was substituted in the experiments by water. For the actual measurement the carburetors were fixed in front of a scintillator-digital camera detector and connected to a water supply replacing the fuel reservoir. The air flow necessary to produce the air–water mixture was generated by an industrial vacuum cleaner. The throttle position was modified using a small remotely controlled servomotor. Sequences of images with varying throttle positions were acquired with and without the water substituting the fuel. By normalizing/referencing the image sequence with water to the ‘dry’ sequence it was possible to emphasize and visualize the development of the flow distribution and compare the different types of carburetors tested (figure 18.3(right)).

18.2.3 Particulate filters

For automotive systems with combustion engines, the exhaust after treatment is a paramount topic. For diesel as well as for gasoline engines, particulate filter (PF) systems have become an often mandatory requirement. In order to be able to improve and optimize the particulate filter systems with respect to filtering efficiency, engine performance and longevity, it is of the utmost importance to assess the actual loading and regeneration of the PFs. Conventional methods prove to be rather limited in that regard. Destructive methods go along with massive manipulation during sample preparation (e.g. fixation of soot and ash with resins, cutting, etc) with the risk of introducing artifacts. A further disadvantage of destructive methods is that they allow only for a very restricted view on the surface of the samples prepared. Also, time-dependent studies such as the soot loading or regeneration processes are impossible.

Many non-destructive approaches such as weighing only yield very limited information; weighing allows investigation of the development of the overall weight over time, but does not allow localizing and discriminating between the different materials. Conventional *x*-ray tomography methods are already more promising. They allow for a general determination of ashes but are very limited for the assessment of soot, due to lacking contrast. This is particularly true, if the PF is still in its metal casing, which is a requisition to perform time-dependent tests under operating conditions.

The material composition of the filters together with the complementarity of neutrons make neutron imaging particularly suited to investigating PFs.

The filter substrate consists in general either of silicon carbide (SiC) or cordierite ($Mg_2Al_3[AlSi_5O_{18}]$), which are both composed of elements with low neutron attenuation coefficients. The particles which are filtered out of the exhaust gas stream and remain in the filter are mainly soot and mineral ashes. Both show relatively low attenuation coefficients due to their elemental composition and density. As soot contains hydrogen, neutrons allow one to distinguish it from the ash as well as from the filter substrate. As steel still provides a certain transparency for neutrons, it is even possible to carry out time-dependent experiments with the PFs in their metal canning (see figure 18.4).

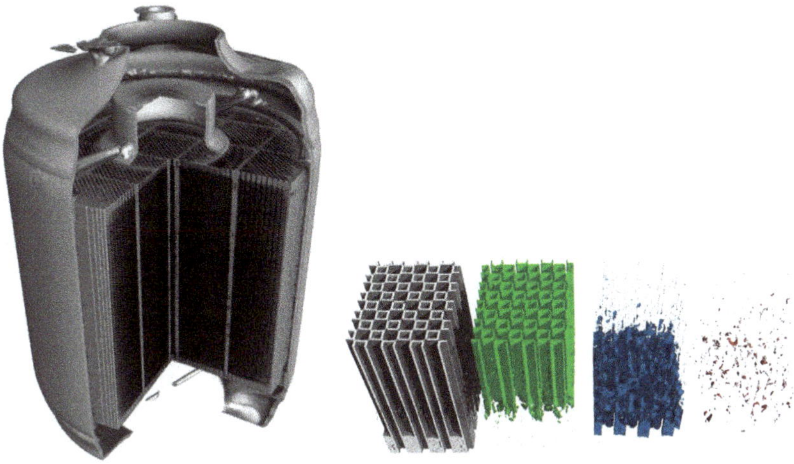

Figure 18.4. Neutron tomography view of the whole filter device (left) and of cut-outs with separated filter ceramics, soot, ash and metallic particles. Reprinted from [15], copyright (2012), with permission from Springer Nature.

While most experiments are conducted in tomography mode, there have been attempts to assess the layer thickness of soot deposits in diesel exhaust systems using neutron radiography. Ismael *et al* [16], for example, assessed thickness profiles of the soot that accumulated on the inside of metal pipes. To assess the soot distribution in PFs, radiography does not suffice. With neutron tomography it is nevertheless possible to assess the soot and ash distribution in the PF [17, 18]. While Strzelec *et al* [19] compared different diesel particulate filters, Manke *et al* [20] also compared different experimental setups and underlined the importance of spatial resolution for a proper identification of the soot layer as the contrast between the materials is relatively low. To enhance the contrast of the soot, Grünzweig *et al* [15] used Gd as tracing agent; a Gd-containing compound was added to diesel fuel, which was subsequently used to load diesel particulate filters. The tracer agent emphasizes the soot cake on the filter walls and makes it easily discernable. The downside of this approach is that the loading has to be carried out in a laboratory on an engine test bench and not in normal traffic.

In the same article, Grünzweig *et al* [15] extended the investigation to different length scales; they reported the investigation of a complete diesel particulate filter system inside the metal canning studied with a large experimental setup with large field of view and relatively coarse spatial resolution. Besides that, small sections of the filter system were studied using a small field of view and correspondingly small pixel size and accordingly much higher spatial resolution. While it was not possible to determine the exact layer thickness within the complete filter system, it was still possible to qualitatively assess the loading of the filter with soot; as fully functional canned PF systems can be studied this way, it would also be possible to investigate the loading and regeneration process with repetitive testing of the samples. The setup with high spatial resolution allows clearly discriminating between the different components, namely the filter substrate, the ash, the soot as well as some metallic particles. Toops *et al* [17] show that an iterative contrast enhancing reconstruction allows one to quantitatively analyse loading with particles in cordierite as well as SiC filters; furthermore, it also allows inspection of the coatings applied to the filter walls. Neutron tomography can also be used to study the loading and regeneration of PFs; in Toops *et al* [18], PFs are repetitively tested after loading and several steps of partial regeneration. Thus, it was possible to study and describe the thickness and morphology of the soot and ash over time. When soot is regenerated, ashes are produced and gather over time at the plugged ends of the filter channels. To extend the lifetime especially of the PFs of trucks and heavy duty vehicles, the PFs can be cleaned by removing the ashes using different methods. Grünzweig *et al* [21] compare the differences in the efficiency of such cleaning methods.

The above studies all describe investigations on diesel PF systems, but the utilization of neutron tomography was later extended to gasoline PFs [21, 22]. Moses-DeBusk [22] compared the production and reactivity of particulate matter from gasoline and several blends of gasoline with different alcohols.

Another problem of the exhaust after treatment for diesel engines is NOx; to reduce the NOx content in the exhaust stream an aqueous solution of urea can be injected. As urea is also hydrogen-containing, neutron tomography can be used to study its distribution on a filter substrate after it had been injected into the exhaust stream [23, 24].

Neutron tomography shows itself to be particularly useful to study a multitude of different questions concerned with particulate filters and exhaust after treatment. This is again mainly due the high sensitivity for hydrogen and the concurrent ability to penetrate metal.

18.2.4 Electric motors

Neutron imaging also shows high potential when it comes to investigations of electric motors and electro motor parts. This is again due to the particular material composition of electric motors, consisting of metals (most often copper is involved) and often contain some plastic hydrogenous material (e.g. resin to encapsulate the components). This material composition makes it hard for *x*-ray methods, as they have only limited sensitivity for resin between and behind the layers of metal and wire, while the complementarity of neutrons still allows for investigations.

In recent years, the demand after as well as the expectation from electric motors has increased. To fulfill all these demands for electric motors with high performance and high durability even in the harshest environments, special attention has to be given to the quality of the encapsulation of the components with potting resin. An even distribution of the resin filling all small cavities is paramount for good performance on behalf of high thermal conductivity, durability and longevity. As pointed out before, its high sensitivity for hydrogen allows neutron imaging for the investigation of the resin distribution in real-life components (cf figure 18.5). The discrimination of the areas with the different components (metal, resin, voids, etc) allows for a better understanding of the potting process. Neutron imaging can, in

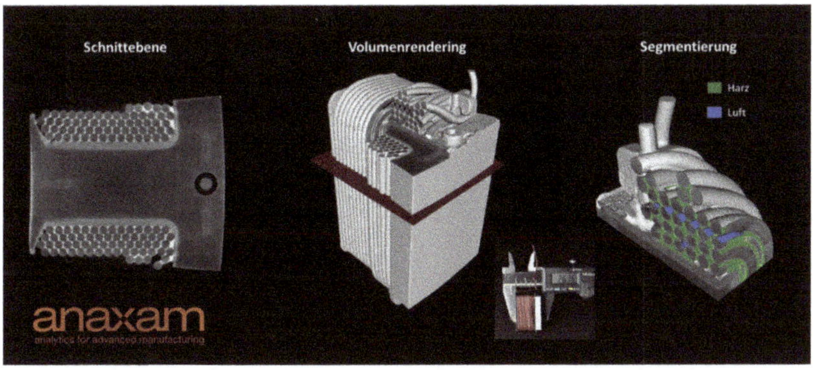

Figure 18.5. Results from a neutron imaging investigation of electric motor components; left: slice through the centre of the sample (position visible in the 3D rendered volume data set (middle)); right: 3D section with segmented resin (green) and trapped air voids (blue). [source: ANAXAM; www.anaxam.ch)

this context, be used to find defects in malfunctioning components but also to confirm the quality of potting processes or to optimize a whole process chain, hence allowing for more efficient and more competitive products.

18.3 Lubrication

Lubrication is needed in cases of motion of metallic parts between each other. In addition to reducing friction and overheating, such lubricant are also in use to reduce corrosion and to protect the surface image. The amount of lubricants must not be high and a thin film layer in the micrometer range of thickness is sufficient to enable its performance. Absence or loss of this protection layer, however, can result in performance reduction and even damage of the involved (metallic) parts.

Because lubricant layers are thin and their composites are of organic origin, their visualization by *x*-rays or other inspection techniques is difficult and often nearly impossible. In contrast, neutrons enable the detection of such thin organic layers due to the strong interaction with the involved hydrogen. In addition, the metallic structures of the operated device (engines, pumps, switches, metallic processors, …) are more or less transparent, which further enables the visibility of lubricants.

A very simple, but instructive example of the lubricant distribution is given in figure 18.6, a Swiss army knife, which was studied by means of neutron tomography [25]. Using special processing tools, the individual metallic and plastics parts were separated—and the small amounts of oil in between too. In total, less than 1 g of oil was used as initial lubrication and protection agent of the knife (figure 18.6).

Sakai *et al* [26] studied the fluidity of different types of greases in ball bearings using neutron radiography and tomography. In this study, two different Li-containing greases were tested and compared. The Li content additionally increases the high contrast of a hydrogenous lubricant even further. This allowed one to visualize and evaluate very thin lubricant layers and to describe the varying behavior of the two types of grease studied.

Figure 18.6. Distribution of the oil (yellow) inside a Swiss army knife. Reproduced from [21], copyright SAE International.

More demanding, and also important, is the observation of the lubricant distribution in operational engines, e.g. combustion-based, under real conditions. Next to the high speed of the moving parts, the motion of the whole assembly has to be considered and handled. It is impossible to follow the process in direct 'real time' with high-speed detection systems. The limitation is given here not by the detector, but by the 'illumination' from the neutron source. The problem can been solved by means of stroboscopic imaging methods, where the detection process is coupled to the motion of the observed engine. A stack of many frames at the identical position of the moving process delivers useful image quality. Further such superimposed stack-images at different positions in the motion process are used to generate a full movie of the repetitive cycle, while the distribution of the lubricants under the different speed levels of the process are possible to be detected.

Grünzweig *et al* [2] used this stroboscopic approach to study the oil distribution in a wet-running multi-disc motorcycle clutch (figure 18.7). In a specially designed neutron transparent test rig, the lubricant distribution was visualized in radial and axial direction. The clutch was operated in the rig at different speeds with up to 4000 rpm. By referencing on the dry state it was possible to visualize the oil distribution and hence to examine its dependence on different parameters such as groove geometry, rotation speed etc.

18.4 Pharmaceutical industry

For the pharmaceutical industry, neutron imaging is in general not so much used in the discovery and development of actual medications. It has nevertheless been successfully applied as a tool for better understanding freeze-drying processes which are an important process in the production of many pharmaceuticals. Gruber *et al* [3] used neutron radiography and tomography to study the sublimation front development in beds of maltodextrin particles with varying solid concentrations. It was hence possible to describe direction and speed of the sublimation front. Vorhauer *et al* [4] used neutron tomography to investigate the structure of the

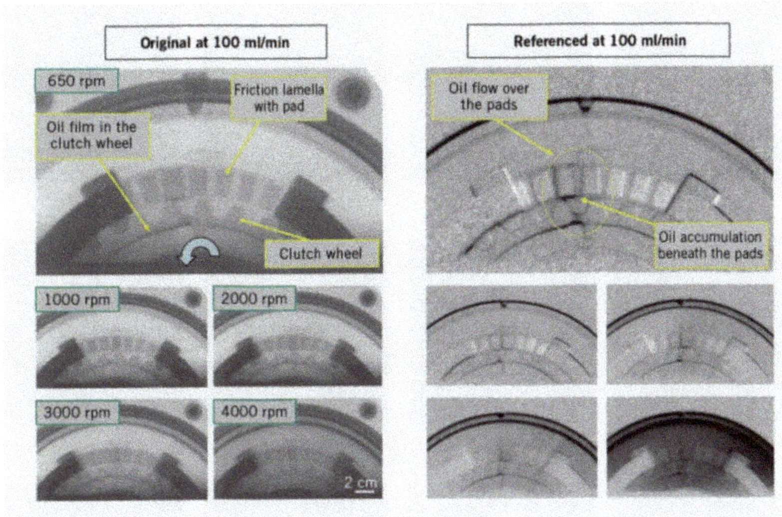

Figure 18.7. Results of the axial dynamic radiography investigations at a flow rate of 100 ml min^{-1}; the left side shows the original images and the right side shows the referenced ones; the oil film is detectable in the clutch wheel, the exit of the oil and its accumulation beneath the pad. Reprinted from [27], copyright (2013), with permission from Springer Nature.

sublimation front. The freeze-drying process was carried out *in situ* in a fully equipped environment in front of the neutron detector. The actual tomography was carried out 11.5 h after the start of the drying process. The results showed that in the studied case the sublimation front was not flat but showed on the contrary strong fractal structures (figure 18.8).

Besides this type of basic research touching steps in the pharmaceutical production process, neutron imaging has successfully been applied to more practical problems when it comes to the application of the pharmaceutical drugs. Here, neutron imaging was e.g. used to investigate the clogging of prefilled syringes. In Kaestner *et al* [5] the blocking of drug injection devices (i.e. autoinjectors) was studied. With standard neutron radiography it was possible to visualize and compare the behavior of the active drug product in comparison with a placebo within the syringe glass barrels and inside the needle at different temperatures. For these experiments the syringes were removed from the autoinjector. While these experiments merely served as proof of principle, the main investigation was carried out on the actual injector devices. Here, the injection process was studied using real-time imaging, reducing the exposure time to 200 ms. This allowed them to observe and describe the motion of fluid and piston inside the built-in syringe barrel with high temporal resolution and compare functioning with malfunctioning devices (figures 18.9–18.11).

Clogging is not just a problem occurring in autoinjectors, staked-in needle prefilled syringes (SIN-PFS) also show this phenomenon. One prerequisite for clogging of the needle is the presence of the liquid product inside the needle.

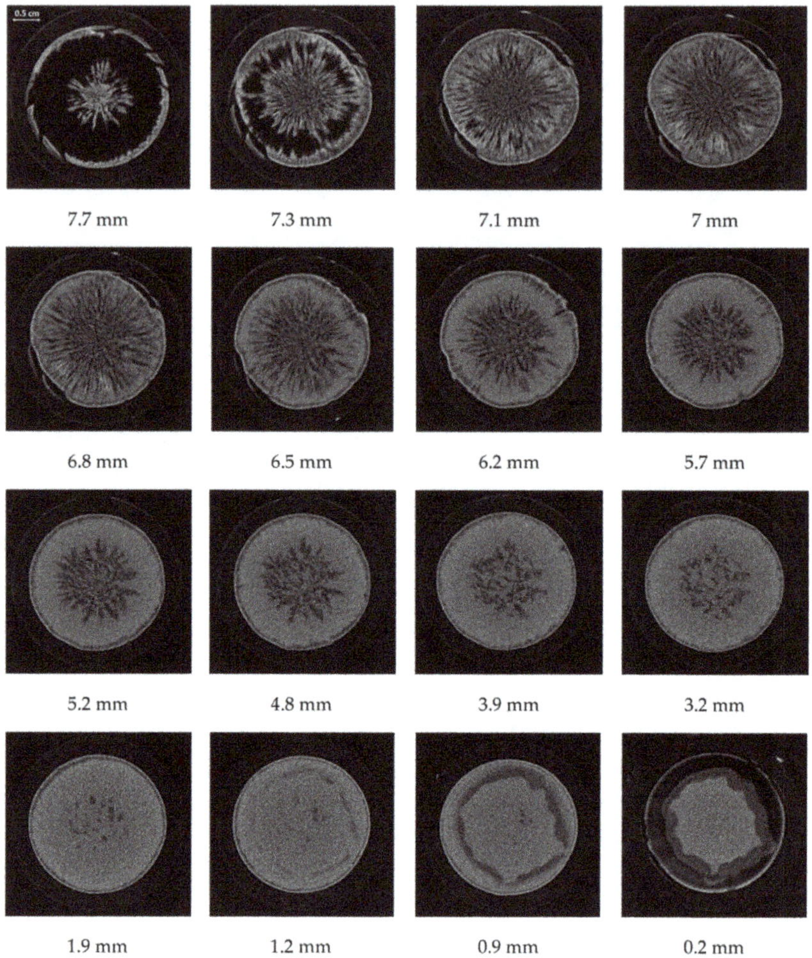

Figure 18.8. Series of horizontal cross-sectional slices of the tomographic volume reconstruction at different heights. The given positions are measured from the footprint of the sample (source: [3]).

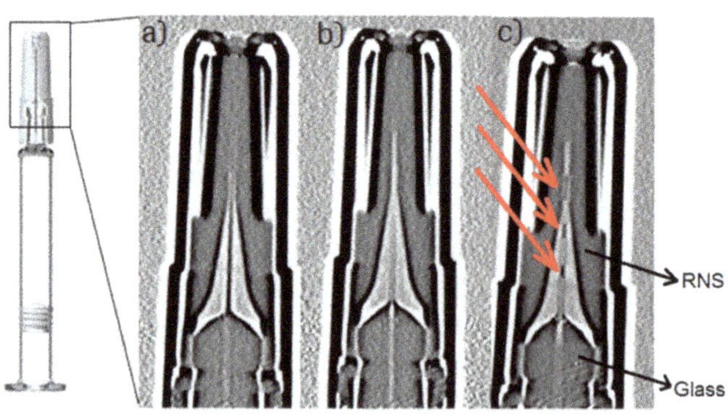

Figure 18.9. 2D reference samples obtained by neutron radiography: (a) needle filled with liquid, (b) empty needle and (c) partially full needle. Arrows indicate the segments of liquid in between air segments. Reprinted from [5], copyright (2016), with permission from Elsevier.

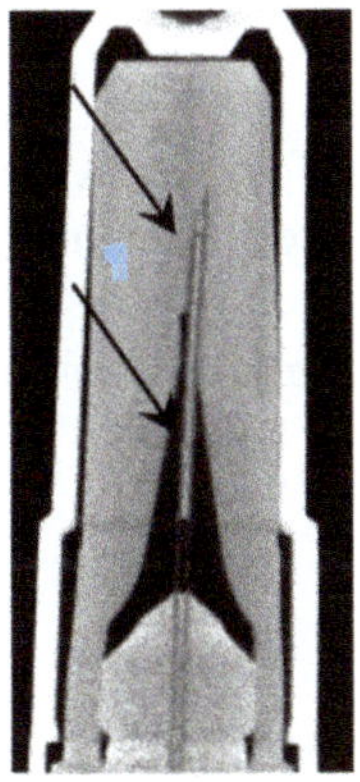

Figure 18.10. Vertical section through the neutron tomography data set of a syringe, which was stored at 40 ° C, 25% rH for 3 months. The arrows point to the position of two cracks, indicating that the material inside the needle is solidified drug product rather than liquid. Adapted from [5], copyright (2016), with permission from Elsevier.

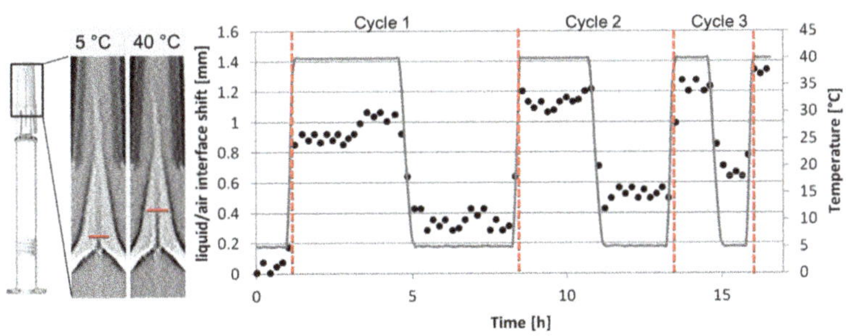

Figure 18.11. Left figure: neutron images showing the length of the liquid column in the same syringe stored at 5 °C and at 40 °C. Right figure: position shift of the liquid/air interface in the needle over time during temperature cycling (solid line represents temperature as secondary axis). Adapted from [5], copyright (2016), with permission from Elsevier.

Neutron imaging allows investigating this topic as it allows penetration of the steel needle and offers the high sensitivity for hydrogen and hence the detection of small amounts of pharmaceutical product inside the needle at the same time.

In a neutron radiography study, De Bardi *et al* [27] examined sets of SIN-PFSs after filling and after storage under different conditions. This allowed determining under which storing conditions liquid is more prone to enter the needles of the SIN-PFSs. De Bardi *et al* [28] also performed a neutron tomography of SIN-PFSs. In general, it is hard to decide whether the product in the needle is liquid or solid as the attenuation seems not to vary a lot. In the reported case, several cracks were visible in the product filling part of the needle, indicating that it had solidified at some point. In De Bardi *et al* [28], the impact of temperature changes on the presence of product inside the needles was tested. In a dynamic *in situ* experiment, several sets of SIN-PFSs were subjected to changing temperature conditions and the resulting changes in the filling level inside the needles was observed by continuous neutron radiography measurements.

Similar experiments were also performed by Scheler *et al* [29], who examined how different parameters influence the presence and movement of product in the needle of prefilled syringes.

Other attempts to use neutron imaging in biomedical applications are mentioned in [30].

18.5 Aerospace

18.5.1 Aircraft structures

One of the very first areas where neutron imaging methods were used to a large extent and literally on a bigger scale is for aerospace industry. Even dedicated facilities such as the ones at the McLellan Air Force Bases Nuclear Radiation Center (MNRC) (now part of UC Davis) in Sacramento (USA) were put into operation [31]. When it comes to aerospace applications, the main topics which can be studied using neutron imaging methods are: inspection of pyrotechnic equipment, turbine blades and search for hidden moisture or corrosion in the honeycomb structures in different parts of the aircraft [32–34]. While the most utilized imaging technique in this area is neutron radiography, the way the neutrons are produced can differ considerably and is not limited to large-scale facilities. Kiyanagi [35] reports on the utilization of compact accelerator driven neutron sources for the inspection of pyrotechnic equipment. Also, small californium-252 neutron sources can be used as was the case for the MNRC; here the small isotopic neutron sources were mounted at the ends of robotic arms, which could be moved around an aircraft to inspect the desired position [36, 37]. For applications where a higher neutron flux and higher collimation is needed, e.g. for higher spatial resolution the best choice would still be large-scale facilities, such as research reactors or spallation sources (figure 18.12).

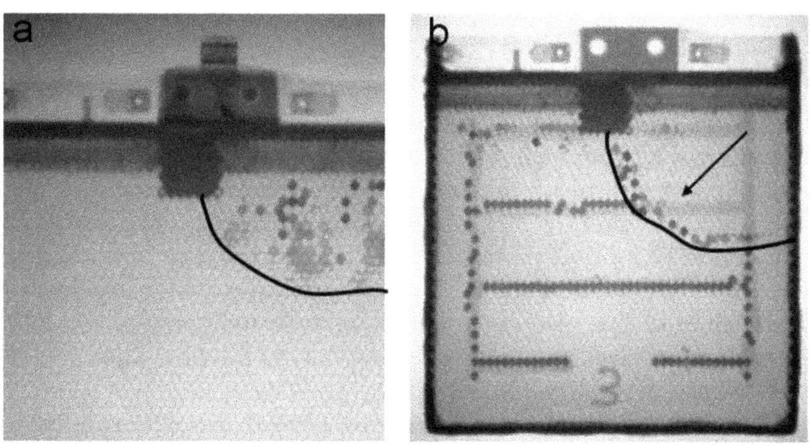

Figure 18.12. Evidence of bond degradation, where (a) is water ingress in U22–0180 prior to sectioning. Image (b) shows water movement in sectioned specimen. Water movement occurred under ambient conditions. Reprinted from [32], copyright (2009), with permission from Elsevier.

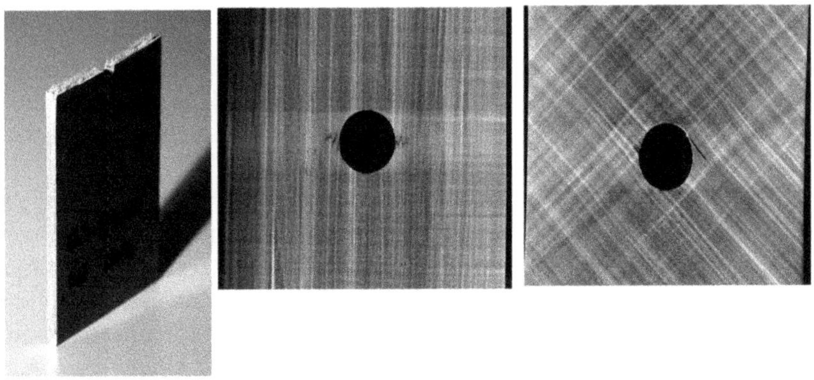

Figure 18.13. GLARE samples (left) can be inspected with neutrons concerning the fiber structure and the resin distribution between the fibers. (source: Paul Scherrer Institut, Applied materials group; www.psi.ch)

Neutron imaging can also be used, when it comes to the inspection of compound materials used in the aerospace industry. One such example is an investigation of glass laminated reinforced epoxy compound (GLARE), which is used as material for the production of aircraft fuselage. Neutron imaging allows penetration of the aluminum layers and inspection of the intermediate epoxy resin and glass fibre layers for defects and flaws (figure 18.13).

The inspection of pyrotechnical equipment is also an important topic when it comes to the utilization of neutron imaging in the aerospace industry [38, 39]. Here again the concurrent high contrasts for one component (i.e. hydrogen in pyrotechnically active material) together with the good penetrability for the other main component (i.e. the metal housing/containment) make the method so suitable for this kind of application. The method was so successfully applied, that it became a standard quality ensuring method for the European space industry. Here it is applied, e.g. as routine inspection for the pyrotechnical equipment used in ARIANE 5 heavy launcher system [38]. These inspections are still relying on analogue neutron radiography using *x*-ray film together with a Gd-foil or Gd-coated glass as converter material. Also, digital detectors have been used to demonstrate the possibilities that for example neutron tomography would allow. Baechler *et al* [40] demonstrated this on the example of a pyrotechnic cutter. The method showed to be a very powerful technique enabling investigation and segmentation of the different components, especially the hydrogen-containing parts such as o-rings and the actual explosive. Nevertheless, the utilizations of digital detectors and the inspection of this type of equipment using neutron tomography is still only restricted to a few examples. The standard approach remains the utilization of *x*-ray film as it allows for high spatial resolution for very large fields of view. Tomography is also not a widely used method, as it is too time-consuming for the routine testing of every single piece of equipment used, e.g. for the ARIANE 5 launcher.

18.5.2 Turbine blade inspection

Both military and civil jets operate with turbine engines where gasoline or other fuel is burned under very high temperatures. The involved turbine blades have to withstand these high temperatures and the involved stress over their full lifetime. Otherwise the risk of failure would initiate air crashes.

This is the reason to inspect all turbine blades carefully by non-invasive methods and by state-of-the-art devices. Neutron imaging has been found to be very useful because of high transmission probabilities for the metal structure of the blades and the feasibility to find residual ceramics from the casting process. In some applications, Gd solutions are applied as tracer agents in order to increase the contrast [40].

Because turbine blades were found to be more stable when mono-crystalline structures are applied, it is worth investigating the crystallinity. Whereas a broad neutron spectrum would not be able to 'see' such structural behavior, mono-energetic neutrons, where the diffraction at the oriented crystals delivers reflections into a determined direction can solve the problem.

In figure 18.14 we show a 'white beam' image, the neutron tomography and the reflected image with the distribution of the oriented crystals.

18.5.3 Inspection of initiators for aerospace mission

For manipulations of satellite components in space it is still very common to use explosives as initiators for cutting wires or other determined motions—very remote from the earth. These explosive components are well sealed in metallic covers in order to withstand the harsh conditions during takeoff and flight of the shuttles in space.

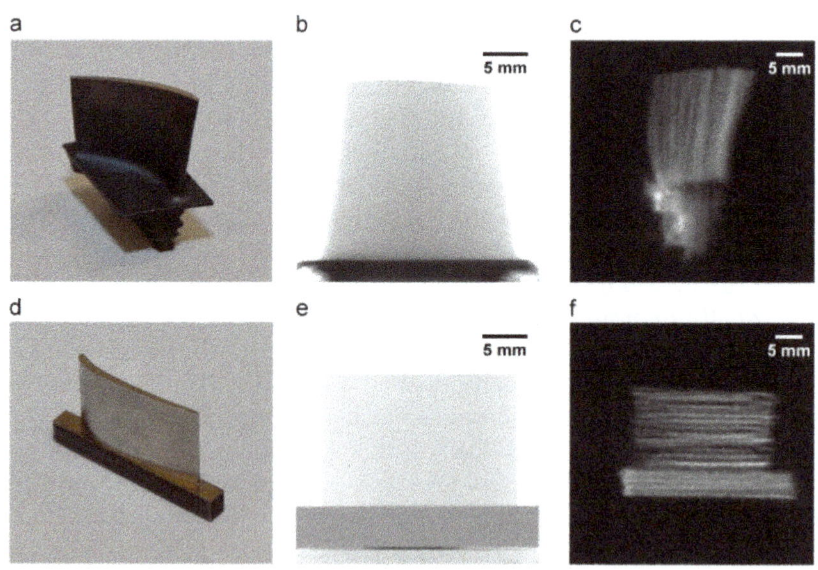

Figure 18.14. Inspection results of two different turbine blades (top row) and (bottom row): photograph (a,d), neutron transmission projection (b,e) and diffraction projection on the side detector (c,f). Reprinted from [8], Copyright (2016), with permission from Elsevier.

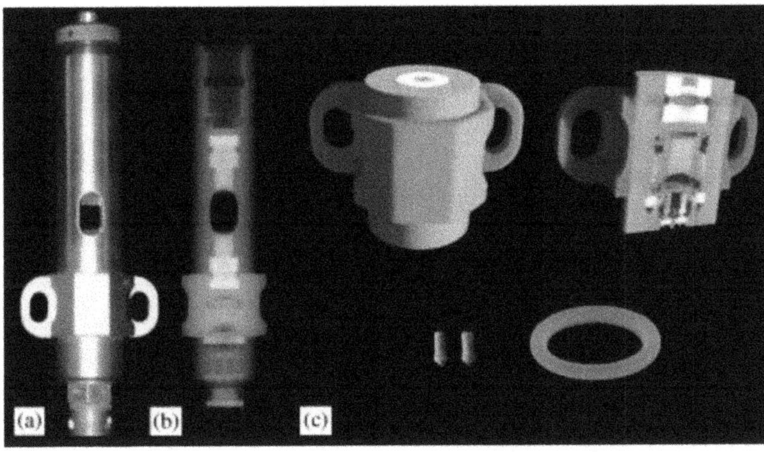

Figure 18.15. Quality control of a 12-cm-height pyrotechnic cutter (Dassault-Aviation): (a) picture; (b) neutron radiograph; and (c) partial reconstructed volume and segmented components. Reprinted from [40], copyright (2002), with permission from Elsevier.

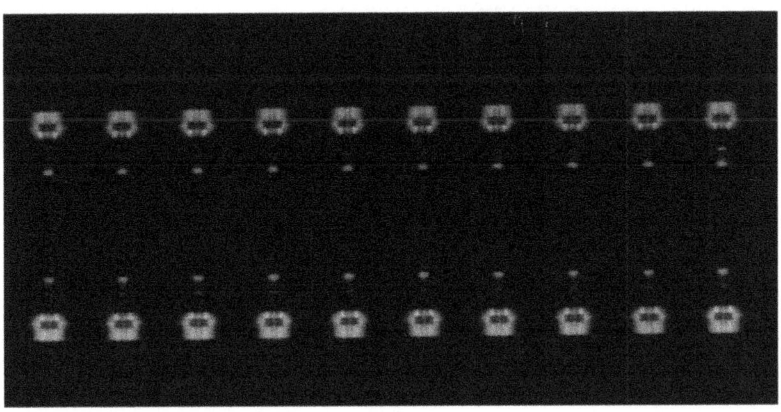

Figure 18.16. Neutron radiography of ARIANE 5 detonating cord ends (Dassault-Aviation). Reprinted from [38], copyright (1999), with permission from Elsevier.

Explosives are chemicals with a high amount of hydrogen and are therefore well visible with neutron imaging methods. On the other hand, the metallic cover of the explosives (steel, copper or even lead) are relatively transparent for thermal or cold neutrons. Therefore, neutron imaging methods (mostly only radiography) have been used in order to perform the investigations.

Because these studies require a high level of reliability and precision, a certified procedure with (traditional) film methods is still very common.

Examples of inspections of cosmic initiators are given in figure 18.15, also some tomography attempts (what is not included in the routine inspection work) (figure 18.16).

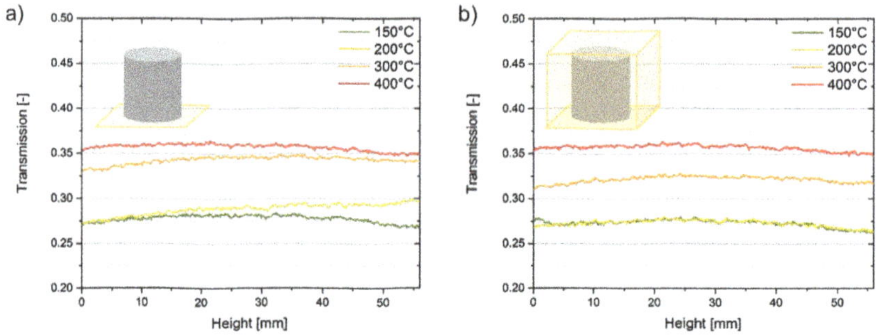

Figure 18.17. Height profile of the transmission values for different burnout temperatures. (a) Sample placed in the furnace ($0.18 m^3$ of air). (b) Samples placed in a sagger ($0.01 m^3$) to simulate a reduced amount of oxygen. Reprinted from [41], copyright (2018), with permission from Elsevier.

18.6 Ceramics

Ceramics, in particular, the production of ceramics is another topic where neutron imaging methods can be applied beneficially. Nevertheless not much is yet reported in literature.

Donzel *et al* [41] studied the binder burnout process in dry pressed ZnO ceramics, which are used for the production of varistors. Again, the high sensitivity for hydrogen that neutron imaging methods provide plays an essential role. In general, a binder is used for the dry-pressed green bodies to maintain form stability. This binder, which is usually hydrogen-containing, must be removed before the sintering. In general, this binder removal is obtained by thermal treatment. This step is crucial for the quality of the resulting product as all of the binder should be removed without inducing any fails or cracks caused by the treatment. In the presented study, neutron radiography and tomography were used to assess the distribution of the binder burnout under different conditions. Here, the conditions were varied with respect to the temperature regime, availability of oxygen and supporting structures. Neutron imaging allowed visualizing and quantitatively assessing and comparing the binder burnout. The results show the production parameters influence the binder burnout in axial and radial direction within the cylindrical samples (cf figure 18.17).

18.7 Bonding

In order to join and to compose components of industrial assemblies, connections are common and needed. Some of them can be reopened again like screw connections, other make a permanent link. It is practically impossible to disassemble such connections, e.g. for quality assurance purposes, as this involves the risk of major damage of the connection or even destruction of the whole assembly in the process of decomposition.

The more elegant way to characterize joins is to use non-destructive imaging methods. The very common and well-established way in industry is the application of *x*-ray techniques in the radiography and even tomography mode. While the results are satisfactory in many cases, there are limitations when thick layers of materials, and in particular heavy materials like metals, have to be penetrated.

Here, the advantage of neutron imaging comes into the game: better penetration through metal layers, higher contrast for the connection agents. We discuss and demonstrate here the following options for metal joining:

- Adhesives;
- brazing, soldering;
- welding.

18.7.1 Adhesive connections

Because of its straightforwardness in industrial applications, gluing with different suitable adhesive materials is very useful: it can be applied at moderate temperatures, no impact is given to the joined components, the joining process can be automatized. On the other hand, the distribution of the adhesive material is completely unknown after application and the curing process. It is quite interesting to learn that an ideal flat layer of the adhesive material is randomly changed after the combination with the other components. This has been tested and verified by test samples and neutron tomography studies afterwards (figure 18.18).

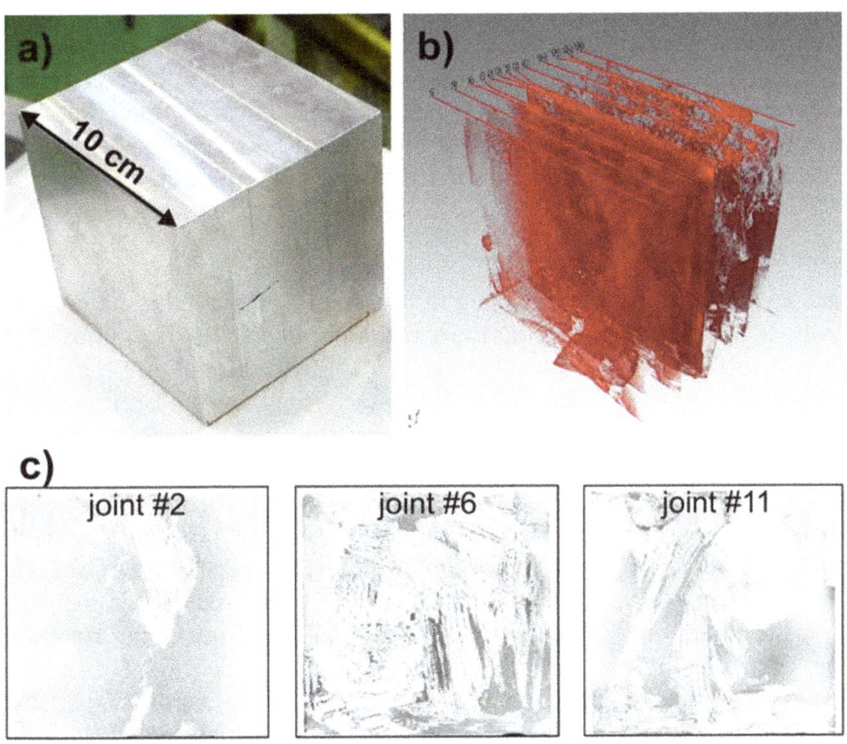

Figure 18.18. Distribution of the adhesives in a Al cube (a) after application: although the glue was distributed homogenously initially, the final distribution is not homogeny at all after solidification—a risk for the stability of the connection (b) tomography of all layers (c) separated individual layers. Reprinted from [14], copyright (2013), with permission from Elsevier.

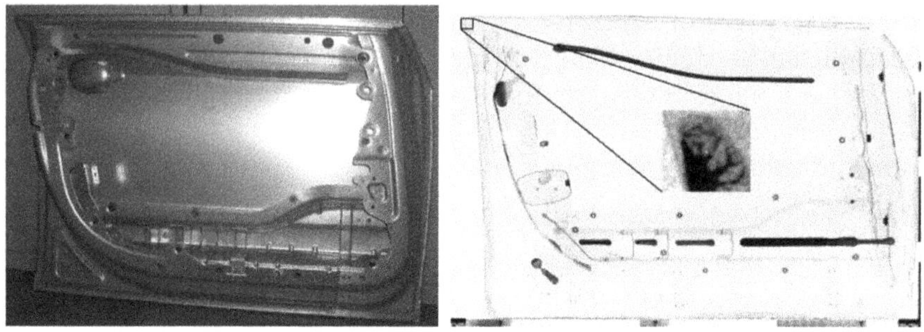

Figure 18.19. Inspection of a car door regarding the quality of the glue connection at the outer edges; cut-out for visualization of the inhomogenity. Reprinted from [42], Copyright (2024), with permission from Elsevier.

This example also demonstrates the power of neutron imaging for adhesive distribution detection: the bulk material (Al, steel, wood, Cu, …) can be penetrated quite easily and the hydrogenous compound of the adhesive can be observed even in very thin layers on the order of 0.1 mm.

The inhomogeneous glue distribution after application and solidification might give reasons for failures and attacks from moisture, corrosion and other environmental impacts [9]. One examples of practical relevance is shown in figure 18.19, where two parts of a motor car door are fixed together with adhesives. The scan of the bonds along the fitting line gave surprising results of imperfection. Based on these image data, the remaining adhesive layer thickness can easily be derived and it can be decided if the glue connection is still acceptable—or not.

18.7.2 Brazing, soldering

There are several techniques of brazing in use. In all cases, a filler material as liquid metal with a lower melting point than the components to be joined is applied between two surfaces. In the case of brazing, the filler material and the flux melting agent are applied together at and near the connection line. In a heating process, the whole assembly is set to temperatures where the agent becomes liquid and is distributed towards the 'right place' of the connection. This distribution cannot be treated and manipulated during the heating process. By means of destructive methods (slicing at relevant positions), the resulting agent position can be checked.

With the help of neutron imaging techniques it becomes possible to 'see' the agent distribution between metal sheets (see figure 18.20). The reason for the given high contrast is the content of boron in the filler material. Since boron is a strong neutron absorber, even very thin layers can be verified and inspected. This approach is completely non-destructive and enables studying the filler material distribution at arbitrary positions in the compound.

In the case of soldering, the compounding agent is liquefied alone and placed between metal parts. Also, in this case, the neutron imaging contrast between soldering and base material is high enough to distinguish them and to see defects and missing materials.

18.7.3 Welding

Welding connections are very stable and consists of the same or similar materials as the base material (most commonly: steel, aluminum). Because in the non-destructive inspection methods (x-rays, thermal neutrons) a visible contrast is not easily possible to achieve, only defects like missing material or inclusions, cracks or bubbles can be identified.

In the case of melting and solidifying metal, the resulting crystalline structure of the weld is important for the resulting stability. In the best case, the base components will have similar metallic properties like grain size, their orientation and texture.

In order to check or verify (non-destructively) such structural properties, energy-selective neutron imaging studies have to be performed [7] (see chapter 6). This approach is based on the fact of the neutron diffraction according to Bragg's law. If regions in the weld sample have different orientations in the crystalline structure, the neutrons are reflected accordingly while the image contrast in a quasi-homogenous sample will vary significantly. Depending on the used neutron energy in a mono-chromatized beam, the different Bragg peaks (see figure 18.20) become relevant. By scanning with different neutron energies, the regions with similar orientation will be enhanced.

In the case of a multiple energy 'white' beam, where all exited states are averaged out, only a homogenous neutron distribution within the weld can be obtained.

Energy-selective neutron imaging cannot only be applied for the weld region, but also to characterize the base material. Often it is pretreated by rolling, quenching or hammering, which can influence the crystalline structure and

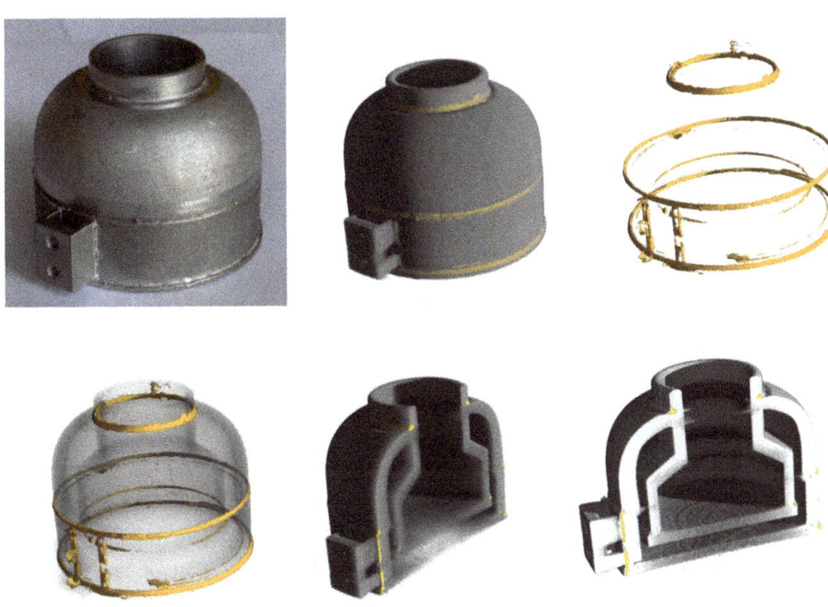

Figure 18.20. With neutron tomography means it is possible to separate the steel parts of the assembly from the brazing agents, showing their distribution clearly. Reproduced from [23].

orientation. Such material changes are not visible with other methods in the bulk (only with electron

Backscatter diffraction at polished surfaces), but with energy-selective neutron imaging methods (see figure 18.21).

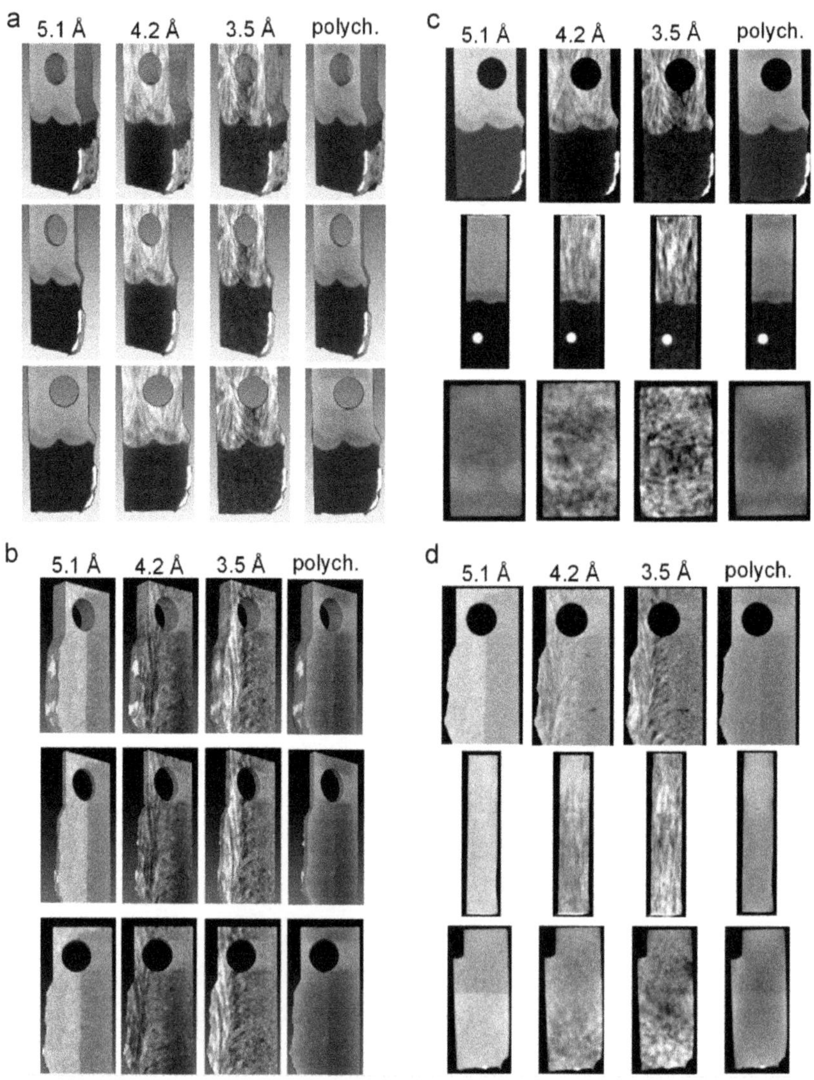

Figure 18.21. Energy selective neutron tomography of two welded stainless steel samples. (a–b) Reconstructed volumes at 5.1, 4.2 and 3.5 Å and in the polychromatic beam. Top: the tilted view of the whole reconstructed volumes. Middle: the tilted view of the thin extracted reconstructed volumes. Bottom: the frontal view of the whole reconstructed volumes. (c–d) Reconstructed slices from the bulk of the samples in three perpendicular planes at 5.1, 4.2, 3.5 Å and in the polychromatic beam. Reprinted from [43], copyright (2011), with permission from Elsevier.

18.8 Additional topics with relevance for industry

Without going into as much detail as in the above sections, we want to relate further options and features with industrial relevance using neutron imaging techniques:

- Neutron imaging investigations play a paramount role, when it comes to the investigation of energy and energy storage related questions such as batteries and fuel cells. Please consult the dedicated chapter about energy research (chapters in Part 4).
- Resin distributions in composites, dynamically: because organic resin for reinforced fiber structures delivers a strong contrast for neutrons, their distribution during and after the injection process can be followed in real time. In this way, the manufacturing process can be controlled, mistakes identified and models for such processes be validated.
- Defense: protective wests under moisture influence: it is well-known, that Kevlar based protection structures become less effective in a wet state. By means of test measurements with dry and moistured panels having different stepping lines the moisture distribution was precisely checked before and after the bullet impact.
- Single crystalline turbine blades, verification: high-performance turbine blades, e.g. in helicopter engines are produced as single crystals in order to provide best stability. By means of a scan of the reflections from neutrons in different angular directions, the resulting image data can be used to characterize the blades w.r.t. homogeneity of the phases and their orientations—see chapter 6.
- Borated steel/Al for nuclear fuel storage—criticality: In order to protect nuclear fuel assemblies from criticality and to make the storage devices as compact as possible, neutron absorbing plates are positioned between the fuel elements. To check the absorber (mostly B-10) efficiency and its distribution neutron imaging is a very useful tool with high spatial resolution even for extended panels [30].
- Casting porosity Al, steel: due to the high penetration power regarding thick metal layers, neutrons can be used more efficient than x-ray to inspect larger assemblies after the casting process for holes, cracks and inclusions.
- Corrosion of metals: metal hydroxides are formed during the corrosion attacks of metals. Even small traces can be detected in their distribution around the bulk material. The effectiveness of protective measures against corrosion can be validated in the same way (see chapters 15 and 17).
- Nuclear fuel: integrity, enrichment, burnup, cladding failure: nuclear fuel is quite transparent for neutrons (compared to failing x-ray methods). Because the two contained uranium isotopes U-235 and U-238 deviate strongly in their attenuation behavior, the fuel enrichment (U-235 content) can be obtained non-invasively. Furthermore, material changes from the burnup process and the fuel–cladding interaction behavior can be studied non-destructively. However, facilities to handle the highly activated materials have to be available [24] —see chapter 4.

18.9 Summary and conclusions

Industry can take profit from neutron imaging studies in many cases, as shown above. In particular, the higher penetration through metals and the visibility of low amounts of organic substances can be used with success. However, since neutron beam lines are stationary, samples and processes have to be moved to the imaging facilities. 'On-site' studies in the industrial environment are nearly impossible with quality, as shown within this chapter.

References are the same as in the pdf proof, except for the highlighted [44] below, so do not need to be typeset again if you want to use the existing reference list from the proof.

References

[1] Lehmann E H, Boillat P, Kaestner A, Vontobel P and Mannes D 2015 Neutron imaging methods for the investigation of energy related materials-Fuel cells, battery, hydrogen storage and nuclear fuel *EPJ Web of Conf.* **vol 104** (EDP Sciences) p 01007

[2] Grünzweig C, Wagner M, Ruf J and Helmer D 2013 Visualisation of the Oil Distribution in a Wet-Running Multi-Disc Clutch *ATZ worldwide* **115** 52–8

[3] Gruber S, Vorhauer N, Schulz M, Hilmer M, Peters J, Tsotsas E and Foerst P 2020 Estimation of the local sublimation front velocities from neutron radiography and tomography of particulate matter *Chem. Eng. Sci.* **211** 115268

[4] Vorhauer-Huget N, Mannes D, Hilmer M, Gruber S, Strobl M, Tsotsas E and Foerst P 2020 Freeze-Drying with Structured Sublimation Fronts—Visualization with Neutron Imaging *Processes* **8** 1091

[5] Kaestner A, Roth J and Gruenzweig C 2016 Real-time neutron imaging to detect origin of blocking in drug injection devices *PDA J. Pharm. Sci. Technol.* **70** 353–60

[6] Kornmeier J R, Hofmann M and Schmidt S 2007 Non-destructive testing of satellite nozzles made of carbon fibre ceramic matrix composite, C/SiC *Mater. Charact.* **58** 922–7

[7] Michaloudaki M, Lehmann E and Kosteas D 2005 Neutron imaging as a tool for the non-destructive evaluation of adhesive joints in aluminium *Int. J. Adhes. Adhes.* **25** 257–67

[8] Peetermans S and Lehmann E H 2016 Simultaneous neutron transmission and diffraction imaging investigations of single crystal nickel-based superalloy turbine blades *Ndt & E International* **79** 109–13

[9] Schillinger B, Brunner J and Calzada E 2006a A study of oil lubrication in a rotating engine using stroboscopic neutron imaging *Physica* B **385** 921–3

[10] Lehmann E H 2008 Recent improvements in the methodology of neutron imaging *PRAMANA J. Phys.* **71** 653–61

[11] Schillinger B, Brunner J and Calzada E 2006 A study of oil lubrication in a rotating engine using stroboscopic neutron imaging *Physica* B **385–386** Part 2 921

[12] Lehmann E H, Vontobel P and Wiezel L 2001 Properties of the radiography facility NEUTRA at SINQ and its potential for use as European reference facility *Nondestruct. Test. Eval.* **16** 191–202

[13] Lehmann E, Frei G, Kühne G and Boillat P 2007 The micro-setup for neutron imaging: A major step forward to improve the spatial resolution *Nucl. Instrum. Methods Phys. Res., Sect.* A **576** 389–96

[14] Grünzweig C *et al* 2013 Progress in industrial applications using modern neutron imaging techniques *Phys. Proc.* **43** 231–42

[15] Grünzweig C, Mannes D, Kaestner A and Vogt M 2012 Visualisierung der Ruß-*und* Ascheverteilung in Dieselpartikelfiltern mittels Neutronen-Imaging *MTZ-Motortechnische Zeitschrift* **73** 326–31

[16] Ismail B, Ewing D, Chang J S and Cotton J S 2004 Development of a non-destructive neutron radiography technique to measure the three-dimensional soot deposition profiles in diesel engine exhaust systems *J. Aerosol Sci.* **35** 1275–88

[17] Toops T J, Bilheux H Z, Voisin S, Gregor J, Walker L, Strzelec A and Pihl J A 2013 Neutron tomography of particulate filters: a non-destructive investigation tool for applied and industrial research *Nucl. Instrum. Methods Phys. Res., Sect.* A **729** 581–8

[18] Toops T J, Pihl J A, Finney C E, Gregor J and Bilheux H 2015 Progression of soot cake layer properties during the systematic regeneration of diesel particulate filters measured with neutron tomography *Emission Control Science and Technology* **1** 24–31

[19] Strzelec A, Bilheux H Z, Finney C E, Daw C S, Foster D E, Rutland C J and Schulz M 2009 Neutron imaging of diesel particulate filters (No. 2009-01-2735) *SAE Technical Paper*

[20] Manke I, Strobl M, Kardjilov N, Hilger A, Treimer W, Dawson M and Banhart J 2009 Investigation of soot sediments in particulate filters and engine components *Nucl. Instrum. Methods Phys. Res., Sect.* A **610** 622–6

[21] Gruenzweig C, Mannes D, Schmid F and Rule R 2016 Neutron Imaging: A Non-Destructive Testing Method to Investigate Canned Exhaust After-Treatment System Components for the Three Dimensional Soot, Ash, Urea and Coating Distributions (No. 2016-01-0985) *SAE Technical Paper*

[22] Moses-DeBusk M, Storey J M, Eibl M A, Thomas J F, Toops T J, Finney C E and Gregor J 2020 Nonuniform Oxidation Behavior of Loaded Gasoline Particulate Filters *Emission Control Science and Technology* **6** 301–14

[23] Lehmann E, Hartmann S and Haller M 1/2008 Zerstörungsfreie Prüfung von Lötstellen mittels Neutronenstrahlen *Schweisstechnik* 6–9

[24] Lehmann E, Thomsen K, Strobl M, Trtik P, Bertsch J and Dai Y 2021 NEURAP—A Dedicated Neutron-Imaging Facility for Highly Radioactive Samples *J. Imaging* **7** 57

[25] Sinha A 1999 Digital imaging of neutrons and its applications *BARC NewsLetter* 183

[26] Sakai K, Ayame Y, Iwanami Y, Kimura N and Matsumoto Y 2021 Observation of Grease Fluidity in a Ball Bearing Using Neutron Imaging Technology *Tribology Online* **16** 146–50

[27] De Bardi M, Müller R, Grünzweig C, Mannes D, Rigollet M, Bamberg F and Yang K 2018 Clogging in staked-in needle pre-filled syringes (SIN-PFS): Influence of water vapor transmission through the needle shield *Eur. J. Pharm. Biopharm.* **127** 104–11

[28] De Bardi M, Müller R, Grünzweig C, Mannes D, Boillat P, Rigollet M and Yang K 2018 On the needle clogging of staked-in-needle pre-filled syringes: Mechanism of liquid entering the needle and solidification process *Eur. J. Pharm. Biopharm.* **128** 272–81

[29] Scheler S, Knappke S, Schulz M and Zuern A 2022 Needle clogging of protein solutions in prefilled syringes: A two-stage process with various determinants *Eur. J. Pharm. Biopharm.* **176** 188–98

[30] Bastürk M *et al* 2005 Analysis of neutron attenuation in boron-alloyed stainless steel with neutron radiography and JEN-3 gauge *J. Nucl. Mater.* **341** 189–200

[31] Froom D A 1992 Nondestructive inspection perspectives *The 1991 Int. Conf. on Aging Aircraft and Structural Airworthiness* vol 3160 p 275

[32] Hungler P C, Bennett L G I, Lewis W J, Brenizer J S and Heller A K 2009 The use of neutron imaging for the study of honeycomb structures in aircraft *Nucl. Instrum. Methods Phys. Res., Sect.* A **605** 134–7

[33] Hungler P C, Bennett L G I, Lewis W J, Schulz M and Schillinger B 2011 Neutron imaging inspections of composite honeycomb adhesive bonds *Nucl. Instrum. Methods Phys. Res., Sect.* A **651** 250–2

[34] Shields K C and Richards W J 1995 Aircraft inspection using neutron radioscopic techniques *Nondestructive Evaluation of Aging Aircraft, Airports, Aerospace Hardware, and Materials***vol 2455** pp 133–44 (SPIE)

[35] Kiyanagi Y 2018 Neutron imaging at compact accelerator-driven neutron sources in Japan *Journal of Imaging* **4** 55

[36] Garbe U, Ahuja Y, Ibrahim R, Li H, Aldridge L, Salvemini F and Paradowska A Z 2017 Industrial application experiments on the neutron imaging instrument DINGO *Phys. Proc.* **88** 13–8

[37] Harvel G D, Chang J S, Tung A, Fanson P and Watanabe M 2011 Three-dimension deposited soot distribution measurement in silicon carbide diesel particulate filters by dynamic neutron radiography (No. 2011-01-0599) *SAE Technical Paper*

[38] Bayon G 1999 Present applications of neutron radiography in France *Nuclear Instruments and Methods in Physics Research Section A: Accelerators, Spectrometers, Detectors and Associated Equipment* **424** 92–7

[39] Viswanathan K 1999 Neutron radiography in Indian space programme *Nucl. Instrum. Methods Phys. Res., Sect.* A **424** 113–5

[40] Baechler S, Kardjilov N, Dierick M, Jolie J, Kühne G, Lehmann E and Materna T 2002 New features in cold neutron radiography and tomography: Part I: Thinner scintillators and a neutron velocity selector to improve the spatial resolution *Nucl. Instrum. Methods Phys. Res., Sect.* A **491** 481–91

[41] Donzel L, Mannes D, Hagemeister M, Lehmann E, Hovind J, Kardjilov N and Grünzweig C 2018 Space-resolved study of binder burnout process in dry pressed ZnO ceramics by neutron imaging *J. Eur. Ceram. Soc.* **38** 5448–53

[42] Schulz M, Lehmann E and Losko A 2024 Chapter 9—Neutron imaging *Non-Destructive Material Characterization Methods* ed A Otsuki, S Jose, M Mohan and S Thomas (Elsevier) pp 205–47 ISBN 9780323911504

[43] Josic L, Lehmann E and Kaestner A 2011 Energy selective neutron imaging in solid state materials science *Nucl. Instrum. Methods Phys. Res., Sect.* A **651** 166–70

[44] Lehmann E H, Boillat P, Kaestner A, Vontobel P and Mannes D 2015 Neutron imaging methods for the investigation of energy related materials-Fuel cells, battery, hydrogen storage and nuclear fuel *EPJ Web of Conferences* **vol 104** (EDP Sciences) p 01007

[45] Van Overberghe A 2005 *High Flux Neutron Imaging for highly dynamic and time resolved non-destructive testing* (Doctoral dissertation TU München)

[46] Schillinger B, Calzada E and Lorenz K 2006 Modern neutron imaging: Radiography, tomography, dynamic and phase contrast imaging with neutrons *Solid State Phenomena* **vol 112** (Trans Tech Publications Ltd.) pp 61–72

[47] Jones J D, Lindsay J T, Kauffman C W, Vulpetti A and Peters B D 1985 Real Time Neutron Imaging Applied to Internal Combustion Engine Behavior (No. 850560) *SAE Technical Paper*

[48] Lehmann E H, Frei G, Vontobel P, Josic L, Kardjilov N, Hilger A and Steuwer A 2009a The energy-selective option in neutron imaging *Nucl. Instrum. Methods Phys. Res., Sect.* A **603** 429–38

[49] Lehmann E H, Josic L and Frei G 2009b Material research with neutron imaging methods at SINQ *Neutron News* **20** 20–3

[50] Nguyen T, Vavrik D, Lehmann E H and Jeon I 2011 Neutron analysis for microvoids in an adhesive layer between high *x*-ray attenuation materials *Appl. Phys. Express* **4** 066401

[51] Vontobel P, Lehmann E and Frei G 2004 Neutrons for the study of adhesive connections *Annual Report NUM* 2003 (Switzerland: Paul Scherrer Institute, Villigen)

[52] Watkin K L, Bilheux H Z and Ankner J F 2009 Probing the potential of neutron imaging for biomedical and biological applications *Neutron imaging and applications* (Boston, MA: Springer) pp 253–64

[53] Wissink M L, Toops T J, Finney C, Nafziger E J, Splitter D A and Bilheux H Z 2018 Neutron Imaging of Advanced Transportation Technologies (No. ORNL/SPR-2018/1058) Oak Ridge National Lab.(ORNL), Oak Ridge, TN (United States)

Part VIII

Advanced methodology

Chapter 19

Instruments and methods

Markus Strobl

Neutron imaging is performed at a large variety of neutron sources including mobile and stationary, laboratory and large-scale sources. Neutron sources [1] convey isotope sources [2], photo-neutron sources [3], neutron generators [4], compact accelerator driven neutron sources (CANS) [5], nuclear reactors [6] and spallation sources [6] utilizing processes such as (α,n) [7], (γ,n) reactions, fusion, fission and spallation to produce neutrons. Most small and mobile sources are limited, however, to specific, often commercial, tasks, while advanced material and condensed matter science applications as well as R&D work is generally reserved for large-scale neutron sources and state-of-the-art neutron imaging instruments profiting in particular from thermal and cold neutron spectra with high brightness for enabling unrivalled image quality (table 19.1). The following sections will focus on such instrumentation.

19.1 Instruments at continuous and pulsed sources

Neutron imaging today is still generally based on the concept of a pinhole camera, constituting the state-of-the-art imaging geometry. Neutron lenses are not available yet, but some new concepts, in particular those based on Wolter optics [8–10], adapted from x-ray space telescopes, have been discussed intensely in recent years. However, in the typical pinhole geometry, spatial resolution is generated through collimation by a pinhole with diameter D at a distance of L upstream of the detector position, providing a collimation ratio of L/D. The sample is placed at a minimum distance of l from the detector to keep the geometric image blur $d = lD/L$ (chapter 2, equation (2.1)) as small as possible. This very basic relationship has to be considered very carefully for each imaging experiment and ideally the collimation is matched to the detector resolution and in particular the resolution required for a specific investigation. An excessive L/D ratio set for an experiment reduces the efficiency of the measurement proportional to $(L/D)^2$. A too small L/D on the other hand reduces the spatial resolution capability. Therefore, state-of-the-art neutron imaging

doi:10.1088/978-0-7503-3495-2ch19

© IOP Publishing Ltd 2024

Table 19.1. Neutron Imaging instruments with regular user program. S = steady, P = pulsed, E = epithermal (in some cases also fast), T = thermal, C = cold, VS = velocity selector, DCM = double-crystal monochromator, ToF = time-of-flight capability, NGI = neutron grating intereferometry, pol = polarized neutron imaging capability, x-ray = x-ray add-on; an overview of further facilities at smaller sources and without user program can be found, e.g. in reference [6]; table adapted from ISNR Newsletter 2023 [courtesy Kockelmann W, Tremsin AS, Strobl M].

Institute	Instrument	Neutron source	Neutron spectrum	Resolution (μm)	Wavelength resolution	NGI	Pol	X-ray
ANSTO	DINGO	S	T	50				
BNC	RAD	S	T	70				Yes
	NORMA	S	C					
CNEA	ASTOR	S	C	25	DCM	Yes		Yes
CSNS	ERNI	P	C	50	ToF	Yes		Yes
ESS	ODIN	P	C	10	ToF	Yes	Yes	Yes
FRM2/MLZ	ANTARES	S	C	20 100	VS/DCM	Yes	Yes	
	NECTAR	S	E, T					
ILL	NEXT	S	T	5	DCM	Yes	Yes	Yes
ISIS	IMAT	P	C	60	ToF			
JAEA/JRR-3	TNRF	S	T	140				
	CNRF	S	C					
JPARC MLF JRR-3 ?	RADEN	P	E, T, C	15	ToF	Yes	Yes	
LANSCE/LUJAN	ERNI/FP5	P	E, T		ToF	Yes		
NIST/NCNR	BT-2	S	T	10	DCM	Yes	Yes	Yes
	CNII	S	C					
ORNL	CG1D	S	C		ToF			
	SNAP	P	T, C					
	VENUS	P	E, T, C					
PSI	NEUTRA	S	T	5	VS/DCM/ToF	Yes	Yes	Yes
	ICON		C					
	BOA		C					
	POLDI		T					

instruments enable tuning the L/D ratio through both a selection of L, i.e. measurement position downstream of the pinhole, and through adjustment of the pinhole size D.

The most powerful neutron imaging instruments, in terms of flux, can be found and have been developed at continuous medium and high flux sources, in particular research reactors (table 19.1). They typically view a large-area thermal or cold moderator surface and are capable of utilizing a high divergence in favour of providing large fields of view (FoV), up to 0.5×0.5 m^2, to provide the capability to investigate large samples, profiting from the high penetration ability of neutrons. Many of these instruments have a direct view on the moderator (e.g. NEUTRA [11], ICON [12], ANTARES [13],BT-2 [14], DINGO [15]) resulting in highly homogeneous beam cross-sections, however, also a significant contribution of higher energy neutrons and gammas, which are largely suppressed in instruments situated at the end of a curved neutron guide(e.g. BOA [16], NEXT [17], CNII [18], CG1D [19]) further away from the source with the drawback, however, of limitations and inhomogeneities in the transported divergence and, thus, beam cross-section. While in conventional cold and thermal instruments the contribution of fast and epithermal neutrons is unwanted, a few instruments have specialized in the utilization of fast and epithermal neutrons (NECTAR [20], ERNI/FP5 [21]). At pulsed sources, ideally, these contributions can be separated, due to different arrival times of different energies at the detector or even at a T_0 chopper, a massive chopper closing the beam for the duration of the burst of the pulse to suppress the instantaneous pulse of high energy neutrons and gammas. The corresponding ability of energy and, thus, wavelength resolution is the stronghold of instruments at pulsed sources (e.g. LANL, SNS, JPARC, ISIS, IBR2), typically pulsed spallation sources.

Pulsed sources profit from high brightness as compared to high time averaged flux, which to date was the stronghold of continuous sources. While the newest generation of pulsed sources challenges this paradigm, the high brightness of pulsed sources can outperform high flux reactors for methods requiring significant wavelength resolution, i.e. for most scattering techniques. The time structure of a pulsed source enables, in contrast to continuous sources, nearly loss-free wavelength resolution in a white beam through the time-of-flight (ToF) approach. The efficiency depends in detail on numerous factors including the specific time structure, resolution and bandwidth requirements of specific techniques and experiments [22, 23]. Conventional attenuation contrast neutron imaging does not utilize wavelength resolution, and, thus, profits best from a high time-averaged flux typically offered by a powerful continuous source (SINQ, FRM2, HFIR, ILL HFR, OPAL, JRR-3, NCNR). However, numerous advanced contrast methods (compare section 2.2), spearheaded by diffraction contrast neutron imaging [24], initially developed at a pulsed source [25], profit from wavelength resolution and are therefore benefitting from the high brightness of e.g. pulsed sources [23]. This triggered the introduction of imaging and full-scale imaging beamlines at pulsed spallation sources (RADEN [26], IMAT [27],ODIN [28], VENUS [29], ERNI-CSNS [30]). The basic imaging principles remain the same, and therefore very similar instrumentation is found from pinhole to detector at pulsed and continuous sources. However, the choice of distance of the sample detector arrangement from the source is at pulsed sources defined by the ToF wavelength resolution envisaged. As the ToF resolution at pulsed sources depends first on the source pulse and the instrument length, also here instruments with [27, 28] and without guide [26, 29, 30] feeding the pinhole can be found. Additional challenges and choices for state-of-the-art imaging instruments at pulsed sources are the prompt pulse from the spallation process, often tackled with a massive so-called T_0-chopper, the definition of the wavelength band to avoid spectral overlaps requiring chopper systems and the choice of the moderator, offering a trade-off between pulse length, ToF resolution, and flux, and finally potential pulse shaping, in particular at a long pulse source such as the European Spallation Source [23]. Furthermore, pulse suppression choppers can offer a trade between bandwidth and flux in a specific bandwidth. On the other hand, short-pulse source instruments can provide access to resonances in the epithermal regime with sufficient wavelength, i.e. energy resolution (compare section 2.2).

The choice in wavelength resolution with pulsed beams deserves special attention and is in many respects quite analogue in considerations to spatial resolution, transferred, however, into the time domain. In general, adapting resolution to the needs of an experiment and aligning the contributions to the resolution is important for efficient measurements. While in the spatial domain the pixel size should match at least half the intended spatial resolution, in ToF the time pins should match at least half the intended time resolution. Comparable to D and L defining the beam collimation for spatial resolution the burst time τ and the length L_{ToF} from the pulse source to the detector define the time and, thus, the wavelength resolution

$$\Delta\lambda/\lambda = \tau/t_{\mathrm{ToF}} \tag{19.1}$$

where

$$t_{\mathrm{ToF}} = L_{\mathrm{ToF}} m\lambda/h \tag{19.2}$$

as $\lambda = h/mv$ according to de Broglie [31]. Thus, improving the wavelength resolution requires shortening the burst time or increasing the instrument length and, thus, shortening the bandwidth for a given repetition rate if not decreasing the repetition rate. All the latter actions decrease the flux linearly (which for the bandwidth is only true if the flux over wavelength could be considered constant) with increasing the resolution. This is analogue to trading flux for spatial resolution, where the relation is squared, however, due to the two spatial dimensions affected. Therefore, optimizing bandwidth and resolution is of utmost importance for an efficient experiment and in particular for time resolved studies, where short exposure times are key. Ironically it is best possible at continuous sources to tune all these parameters, but due to the high implied losses by chopping a continuous beam, pulsed sources are often still more efficient, even when tailoring these parameters is largely prevented in an existing instrument. Long pulse sources such as the European Spallation Source are an efficient alternative providing high brightness and high time-averaged flux with pulse widths (τ) that allow further tuning and tailoring to the respective requirements of a measurement and study [23, 28, 32, 33].

It has to be noted, that neutron imaging experiments are possible at numerous instruments including diffractometers [24, 25, 34–36], double-crystal diffractometers (USANS) [37], small-angle scattering instruments [38, 39], neutron reflectometers [40], spin-echo spectrometers [41, 42] etc, although these are not optimized for neutron imaging, which concerns often mainly the size of the FoV that can be achieved, as well as limited capabilities to tune L/D in some cases. Imaging detectors are generally relatively easy to implement. Such instruments are used in particular for testing and applications where capabilities, e.g. corresponding wavelength resolution means are or have been missing in dedicated imaging instruments.

Advanced imaging methods exploiting contrast mechanisms beyond conventional attenuation contrast require additional instrumentation and devices optimized for imaging purposes. In state-of-the art neutron imaging instruments these are often available in the form of optional add-on components. More details about such specific components and instrumental methods will be provided in the subsequent sections, including also advanced standard components such as the key device for imaging, namely imaging detectors. First, however, state-of-the-art instrumentation of different categories will be sketched and discussed.

19.2 Instruments with cold/thermal/fast neutrons

19.2.1 Thermal neutron imaging beamlines

Initially, neutron imaging was mainly performed at thermal neutron beamlines, with thermal neutrons providing better penetration characteristics. Most instrumentation described here for thermal beamlines is, however, equivalent for typical cold neutron imaging beamlines. Specific additional requirements for cold neutron imaging beamlines are described in the subsequent chapter. Spatial resolutions achieved with first generation digital detectors were limited to a few 100 μm and often large samples, profiting from the high penetration, have been studied. Standard thermal neutron imaging beamlines [11, 14, 15, 43, 44], like NEUTRA at the Paul Scherrer Institute (PSI) in Switzerland [11] are optimized for large homogeneous FoV and optimum flux conditions. A pinhole, or a pinhole exchanger is installed relatively close to the source, often already in the biological shielding. A direct view on a large part of the moderator surface is provided, which is projected through the pinhole onto the sample and detector. When L_{SP} is the source-to-pinhole distance, D_S is the source size, i.e. the viewed area of the thermal moderator, and L_{PD} is the pinhole to detector distance, then the FoV $\approx L_{PD}D_S/L_{SP}$ and the full utilized source divergence div $\sim D_S/L_{SP} = \text{FoV}/L_{PD}$, while the divergence of the beam in each point of the sample, responsible for image blur, thus defining the imaging capability of the instrument geometry is L/D, as defined previously. Indeed, such configuration creates a penumbra, i.e. the edges of the FoV are not sharp, which can be estimated by $P_{FoV} = L_{PD}D/L_{SP}$ and which typically is only used down to 50% or even only 80% of the plateau (peak) flux value, while the FoV is typically given to the 50% level. The direct view on the moderator implies also a significant contribution of unmoderated fast and epithermal neutrons as well as gamma radiation in the beam. The contribution depends amongst others on the viewing direction with respect to the primary neutron source, where a tangential view is favourable, the source-to-pinhole distance, and the length of the instrument. A number of filters (e.g. Pb, sapphire, Bi) are often used to optimize the corresponding beam spectrum and background, while carefully considering the impact on the thermal neutron flux. While the sensitivity of the detectors used is generally limited for fast and epithermal neutrons, gamma radiation can create white spots in the images, which need to be filtered digitally (compare chapter 20).

In many cases, several measurement positions are utilized downstream of the pinhole D, in order to provide additional flexibility in tuning flux versus resolution, given the beam size is sufficient also at upstream positions. The beam path from pinhole to measurement position is normally covered by mostly evacuated or He filled flight tubes including some beam limiters in order to avoid beam losses due to air scattering and background due to neutrons scattered at the edges of the beam. A slit system enables limiting the beam to the useful FoV for a specific experiment and detector system used. A sample stage allows for aligning and scanning samples, with a rotation stage for tomographic scans being the most basic installation, while translation across the beam, sometimes also along the beam and tilt stages are often helpful for alignment. A tilt is also required for neutron laminography, applied to relatively large but flat objects in order to avoid artifacts based on the sample geometry. All stages and detectors are remotely controlled.

For rapid tomography, i.e. with high time resolution, the method of on-the-fly tomographic measurements was pioneered at BOA [16] in a measurement on plant roots [45]. On-the-fly tomography allows rotating the sample continuously while projection images are recorded. This speeds up the imaging process and enables 4D spatio-temporal neutron imaging. This approach allowed overcoming limitations of fast measurements and enabled speeding up rapid neutron tomography at high flux sources by an order of magnitude when comparing approaches before [46] and after the implementation [47] of the on-the-fly technique, reaching about one tomogram per second. For enabling the continuous rotation when a sample has to be supplied, e.g. by current or liquids, a rotation stage with slip ring is recommended.

To apply up-to-date background correction through the black-body method [48] (compare chapter 20), the installation of a black-body grid between sample and detector and corresponding manipulation mechanics and potentially motorization is required. This is key for quantitative measurements with respect to attenuation coefficient values [49].

In addition, neutron imaging instruments with a wide range of applications require significant infrastructure, including e.g. electrical, gas and water supplies. A large sample stage for bulk and heavy samples of the order of $0.5{-}1t$ is often foreseen, while access is often required also through the roof with a crane. Otherwise, access is typically through a labyrinth entrance avoiding radiation streaming with a direct line of sight, or heavy sliding doors. Because samples are sometimes activated by neutron irradiation, corresponding sample handling and storage has to be considered. This requires often to carefully consider the measurement of valuable objects also with regards to their activation and decay. Space and supplies for large bulky and complex sample environments is typically also foreseen.

Finally, an imaging beamline requires in general heavy shielding, because the direct view or at least high flux of large beams implies also high primary and secondary radiation backgrounds around the measurement position. For thermal neutron imaging beamlines in close proximity to the source, thick concrete shielding of 500 mm and more is required. Alternative shielding materials are sometimes used, but the volume remains significant. Correspondingly access and supplies have to be planned carefully.

19.2.2 Cold neutron imaging beamlines

Cold neutron beamlines are dominating nowadays at state-of-the-art continuous and pulsed neutron sources [12, 13, 16–19, 26–30, 50, 51]. On the one hand, cold neutrons provide higher sensitivity paired with still excellent penetration capabilities, in particular, in view of significantly improved spatial resolution and thus decreasing typical sample sizes. On the other hand, cold neutrons enable a range of novel methods and contrast modalities (section 2.2), have advantages with regards to neutron optical devices, detection efficiency and last but not least host the most prominent Bragg edges (section 2.2) of many highly relevant materials for neutron imaging [24]. Also, in particular the latter potential of cold neutrons render cold neutron imaging beamlines especially attractive at pulsed neutron sources. All of this implies that cold neutron imaging beamlines often feature a number of additional and optional components as compared to thermal neutron imaging beamlines, while the basic instrumentation and requirements as described above remain largely the same.

At continuous sources cold neutron imaging beamlines might appear very similar to thermal neutron beamlines, in particular in their basic set-up and geometry. Also, here the capability for large FoVs is often preserved [12, 13, 26]. However, cold neutrons are also efficiently transported through conventional and supermirror coated neutron guides, which allow transporting sufficient beam divergence to neutron imaging stations at a farther distance from the source than in the basic configuration with a pinhole close to, within a few meters from, the moderator. The use of a neutron guide and the implied distance from the source has advantages and disadvantages for a neutron imaging instrument and the facility [16–18, 27, 28]. The additional distance creates more space and is suited to reducing the background of epithermal/fast neutrons and gamma radiation. The latter are further suppressed by a bent guide which loses line of sight to the moderator [16–18]. On the other hand, a neutron guide is associated with certain transport losses and wavelength dependent limitations associated with transport losses. This implies typically smaller FoVs of rather around 150 × 150 mm^2 than around 400 × 400 mm^2 as is the case in conventional set-ups. Additionally, there is a penalty on the homogeneity of the FoV [51], as dependent on the chosen guide geometry there are gaps in the transported divergence and dependent on curvature inhomogeneities in the spectral distribution across the beam. Flux variations across the FoV typically take the form of horizontal and vertical lines of reduced flux, which get sharper and deeper the more the pinhole at the exit of the guide is closed down. As a countermeasure, diffusers in the form of scattering powder like graphite powder are sometimes used in the vicinity of the pinhole. However, these attenuate the beam and might display significant Bragg edges in relevant areas of the spectrum, which in turn affects wavelength-resolved and in particular Bragg edge measurements (section 2.2). On the other hand a guide generally means that background is lower and only thermal/cold neutrons arrive in the endstation of the instrument. This eases the shielding needs in contrast to end stations in proximity of the source. Often a combination of led and boron carbide is sufficient to reduce radiation levels outside the shielding to low levels of no concern. Thus, pros and cons have to be weighted carefully.

The additional opportunities for imaging with cold neutrons also require the necessary preconditions to be met for installation and optional utilization of auxiliary devices. Sometimes a separated beam conformation room is available in the shielding in close proximity to the pinhole exchanger, which is the state of the art at cold neutron imaging beamlines. Auxiliary equipment installed in this region potentially includes, depending on the beamline:

- velocity selector;
- double-crystal monochromator;
- G_0 grating;
- Be filter;
- polarizer;
- chopper system.

Details and use of these devices will be discussed in more detail in subsequent sections.

The remaining part of the endstation in general matches that of the thermal counterparts, despite the fact that often auxiliary equipment has to be installed also in the vicinity of the sample position. Often a very flexible set-up with an optical bench is chosen, which allows quick and easy instalment and de-installation of equipment. Thus, flight tubes are often removable and measurement positions can be realized in different places along the optical bench and at its end for large samples and largest FoV. Figure 19.1 shows the prototypical installation of ICON at PSI [12].

Today, often also an x-ray imaging system can be installed across the neutron beam in the area of the sample position in order to enable concurrent *in situ* x-ray imaging, in particular tomography, as will be discussed in section 19.8.

Also, instruments at pulsed sources are built with [27, 28] or without a guide system [26, 29, 52], but here there is another consideration driving the instrument length and thus the potential need for a guide system, which is the required wavelength resolution, as outlined earlier. In addition, a chopper system is generally indispensable at a pulsed source, in particular, if it is operating at a high frequency. In such a case frame overlap might otherwise severely hinder correct wavelength resolution in ToF mode, the main advantage of such a source also in neutron imaging. In addition, often so-called T_0-choppers are employed in order to suppress the significant prompt pulse of gamma and high energy neutrons, which is particularly important if direct line of sight is kept and the instrument is short. Otherwise and apart from corresponding shielding needs, instruments are

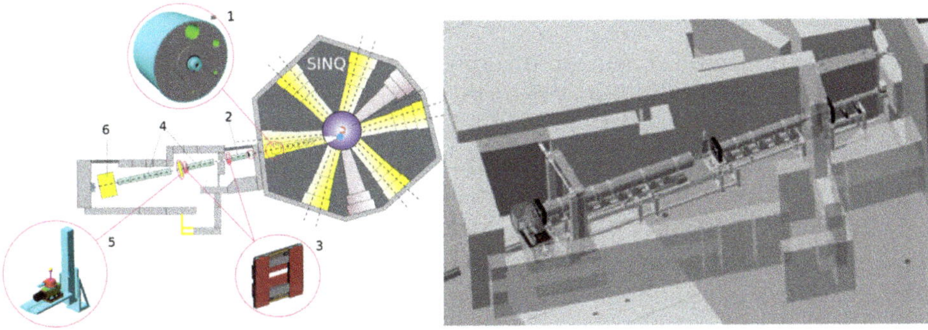

Figure 19.1. Left-hand side: sketch of the ICON [12] instrument relative to the SINQ spallation neutron source at PSI and highlighting the components: (1) pinhole exchanger wheel, (2) velocity selector position (optional) and first flight tube, (3) beam limiting slits, (4) flight tubes, (5) sample position and sample stage at mid position, (6) large sample stage in end position; right-hand side: 3D drawing of the instrument basic components with shielding partially invisible (upper front part and roof). Reprinted from [12], copyright (2011), with permission from Elsevier.

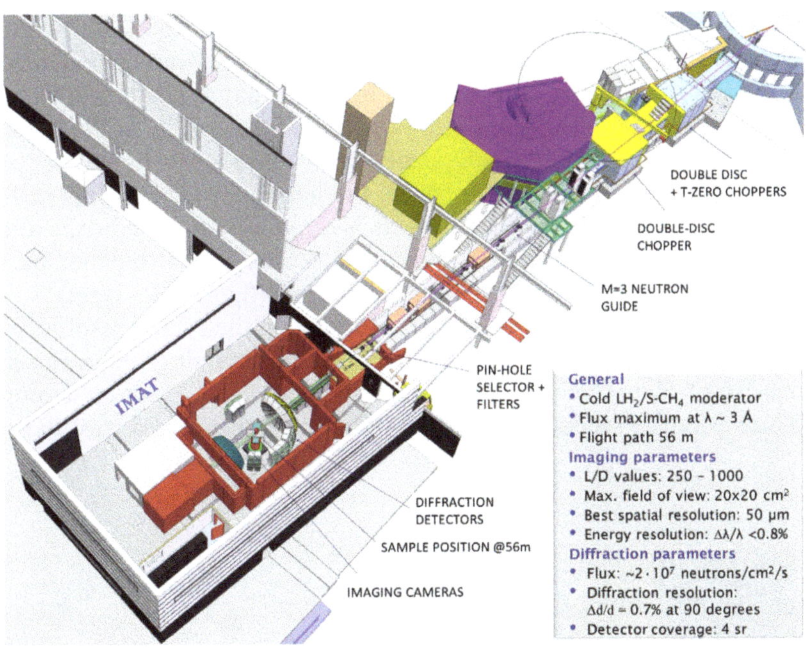

Figure 19.2. Shows the unique neutron imaging instrument IMAT, which is combined with a diffractometer at the ISIS pulsed neutron source target station 2 at the Rutherford Appleton Laboratory (RAL) in the UK [27]. The instrument features a neutron guide system as well as a frame overlap and T_0 chopper system. Image source: https://www.isis.stfc.ac.uk/Gallery/p1_IMAT_design_outline.jpg.

constructed similar to such at continuous sources. A potential extra at short pulse sources is access to ToF imaging with epithermal neutrons, providing access to resonance absorption contrast (section 2.2) (figure 19.2).

19.3 Detector technology

A key basic component for neutron imaging is the neutron imaging detector system. Spatial and temporal resolution in neutron imaging critically depend on this device and associated technologies. Earliest solutions were based on film [53], later, for time-resolved imaging also video cameras [54]. Also, imaging plates were in use sometimes [55]. However, today, digital detectors are state of the art and scintillators screens in conjunction with CCD and CMOS cameras still dominate the field. Other than, e.g. amorphous silicon flat panel detectors [56], which are rarely in use, such combined systems of scintillator and camera are highly versatile and allow adapting to different FoV and resolution requirements (figure 19.3) as well as to used spectra, flux and even radiation, in particular also when it comes to *n*/*x* bi-modal imaging.

The basic design of such a detector system (figure 19.3, inset) consists of a light tight detector box, which is covered with a scintillator screen at the side of the incoming beam. The scintillation light is collected and deflected by a mirror into an optical lens system mounted on the digital camera. An adjustable distance, exchange of lenses, respectively, refocusing allows adapting to different FoVs. The selected FoV together with the number of pixels results in an effective pixel size, i.e. pixel resolution, which is often mistaken as spatial resolution of the system. In general, and without special data reduction, the spatial resolution cannot be better than twice the effective pixel size. However, the actual intrinsic spatial resolution capability of the system also depends on the scintillator screen and the quality of the optics. While the latter, if well selected for the given purpose nowadays rarely limits the resolution, except when the system is not well focused, the light transfer capability of the lens system is essential. Often beamlines possess a range of detector boxes and digital cameras, which can be exchanged between the boxes, which together particularly cover the full range of FOVs and spatial resolutions (figure 19.3 right).

The used scintillator screen has a decisive impact on the spatial resolution and the image quality relative to a certain exposure condition. Standard neutron scintillator screens consist of a mix of converter material with high neutron capture cross-section, which produce secondary charged radiation that induces scintillation in the

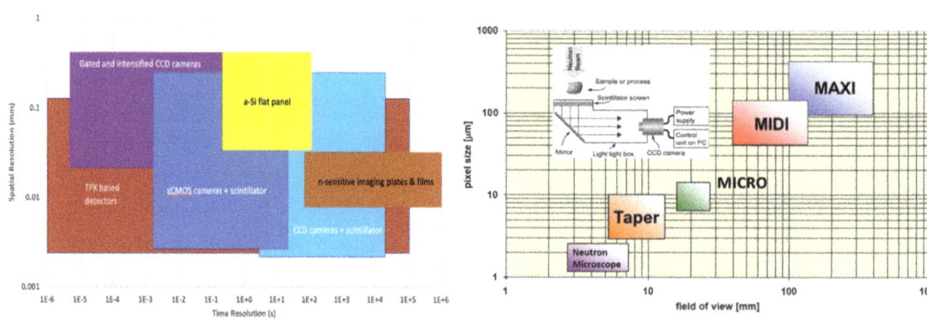

Figure 19.3. Spatial and temporal resolution of different detectors used in neutron imaging (left); suite of scintillator/camera imaging detector systems at PSI for different resolutions and FoV (right); schematic of a standard scintillator/camera detector (inset in right-hand side picture).

actual scintillator material. These are normally mixed with a binder material and coated on, e.g. a thin aluminium sheet, which is basically transparent to neutrons. For cold and thermal neutron scintillator screens typically ^{6}Li, $^{155/157}Gd$ and ^{10}B are used for capturing and converting neutrons and often ZnS is the actual scintillating compound [57, 58]. Gadox (gadolinium oxysulfide (Gd_2O_2S)) scintillators also used for x-ray imaging have been introduced to neutron imaging in the quest for highest resolution [59, 60]. Here the element with sufficiently high capture cross-section, Gd, is already integrated in the fluorescent compound. Additional elements are used to, e.g. shift the emitted photon wavelength to the maximum of the sensitivity of the utilized camera system. A traditionally widely used screen is LiF/ZnS where the ZnS is co-doped with Ag or Cu [57].

The efficiency of a scintillator screen is often assessed by its neutron capture efficiency, sometimes mistaken as the detection efficiency of the screen, and by the light output of the screen [61–63]. Both are important, indeed, but the combination of these characteristics, in the ideal case together with those of the full detector system, has to be accounted for [57]. This is best done through the concept of the detective quantum efficiency (DQE) for the cascade of processes that need to be considered from capture to image formation [64]. In many cases, in particular for high resolution and, thus, the small FoV Gadox is outperforming Li-based screens in such a comparison. However, Gd has a higher gamma sensitivity, which might stand against the use of such screens in cases of high gamma background. Currently, there is also increased attention on ^{10}B based scintillators which appear to have potential to overcome corresponding downsides of both Gd- and Li-based screens, if optimization of such screens is successful towards exploiting the full theoretical potential [58]. In addition, it has to be noted that novel approaches to neutron imaging detectors, e.g. through light sensitive timepix® (TPX) technology-based systems [65], have the potential to change the requirements to scintillators due to their ability to analyse the signature of a neutron event based on time and spatially resolved photon detection [66].

Also, the utilized light optics system focussing the light image onto the light sensitive chip of the detector is of significant importance for the spatial resolution capability of such a detector system. While in many cases standard photographic optics are sufficient, for highest resolution applications either dedicated magnifying high numerical aperture lenses [67, 68] or specific infinity corrected optics are utilized [69]. Also, diverging fibre optics tapers in direct contact with the scintillator screen have been applied to increase magnification and, thus, resolution [70].

The spatial resolution capability of an imaging detector system is ideally assessed by the concept of the modulation transfer function (MTF). The modulation transfer function represents the relative loss of contrast for a regular pattern of absorbing lines with 50% duty cycle due to decreasing period, i.e. increasing spatial frequency expressed in-line pairs per millimeter. With increasing frequency towards the resolution limit of an image detection system the modulation amplitude of the resulting sinusoidal image will decrease. Generally, resolution is expressed in neutron imaging as the spatial frequency where the MTF reaches 10%, though other thresholds are used sometimes. As this is not a unique resolution definition, the MTF value at which the spatial frequency is given should be provided. For simplicity often the edge spread function is measured instead of a modulation transfer. While the MTF can be calculated as the magnitude of the Fourier transform of the line spread function, the line spread function is the gradient of the edge spread function. Alternatively, a test structure or pattern, such as a Siemens star [71], is used to demonstrate specific resolution capabilities. In any case, it is important to have the test pattern in contact with the scintillator screen in order to measure the intrinsic detector resolution, while, when placing the test pattern in the sample position, the resolution of the whole system, beamline and detector is measured.

The highest spatial resolution to date is enabled by isotopically enriched ^{157}Gd Gadox screens [60], which have been developed specifically for this purpose. ^{157}Gd has a superior capture efficiency when compared to ^{nat}Gd. Such screens, with thicknesses of a few micrometers enabled scintillator/camera detectors in routine user operation to achieve spatial resolutions around 4 µm [67, 68], which currently is the state of the art for conventional applications. Further improvements with such detector systems can potentially be achieved by digital processing in the readout of the digital camera systems but also new sensor technologies contribute to such opportunities [66]. Higher spatial resolutions have been reported, but these are typically not fit for standard imaging applications, which can relate to efficiency/exposure times, no direct digital readout or other shortcomings with regards to general practical applications.

While time resolution was initially only relevant for time-resolved studies, the advent of the ToF approach in neutron imaging created the requirement of microsecond range time resolutions already short after the beginning of the Millennium [25]. This need gave rise to new imaging detector technologies being used, and in particular the use of event mode-based readouts, instead of full image readouts like those found in most camera systems. In particular, the TPX technology [65] provides promising opportunities and enables, e.g. in combination with ^{10}B doped microchannel plates (MCPs), a detector technology enabling around 100 µm spatial resolution at still relatively high detection efficiency [72], in contrast to many other approaches. This technology does not feature readout noise, in contrast to conventional camera systems. Accordingly, this technology was involved in a wide range of methodical developments in ToF neutron imaging and applications of such. The limiting factor was a relatively small FoV, first limited to the chip size of about 14×14 mm^2 later to four compound chips resulting in 28×28 mm^2. The temporal resolution in the sub-microsecond region enables even resonance absorption contrast imaging (section 2.2) at short pulse sources [72]. In addition, the event mode detection allows for signal processing such as centroiding, which enables sub-pixel resolution [73]. The TPX technology is under constant development, which enables increasing event rate capabilities and chip sizes. In addition, a light sensitive version entered the marked around 2020, which enables using TPX technology in conjunction with the conventional highly flexible scintillator camera-box designs. Furthermore, sophisticated software enables analysing and extracting selected information of single-neutron capture events on the basis of individual photon hits, which promises further improvements for image quality [66], but also new requirements for scintillator screens, as mentioned earlier.

However, frame rates of commercial CMOS cameras are also increasing, and first applications for low resolution ToF imaging have also been demonstrated with such systems [74].

19.4 Monochromatization

As outlined in section 2.2, many novel imaging techniques, i.e. contrast modalities, like dark-field contrast and diffraction contrast, also referred to as Bragg edge imaging, require wavelength resolution as can be achieved through monochromatization in particular of a continuous beam. As very well known from scattering techniques, either velocity selectors [75] or crystal monochromators can be used for this task. In some cases, even the utilization of a Be filter, which supresses the spectrum below a wavelength of 4 Å is sufficient, like for certain dark-field contrast studies with neutron grating interferometers (see section 19.7). Velocity selectors, in contrast can provide wavelength resolutions of 5%–20% depending on the wavelength and other factors such as the transmitted beam size and device specification. Crystal monochromators, in general pyrolithic graphite crystals (PGCs) of a specific mosaicity, are utilized for somewhat better wavelength resolution in the range of 3%–5%, again depending on the wavelength.

19.4.1 Velocity selectors

Due to the fact that neutrons are massive particles, their energy, and thus their wavelength, is associated with a certain velocity in the range of a few 1000 m s^{-1} in the cold/thermal wavelength range. Thus, it is possible to select certain velocities with fast rotating turbine devices with well-defined bent transmission channels. Such channels are realized either by exactly bent carbon fibre-based blades containing neutron absorber material, such as ^{10}B or Gd, forming numerous channels, which appear straight only to neutrons of a certain velocity at a selected operation frequency, or through numerous absorber coated discs with a large number of slits aligned serially in beam direction to only allow a specific neutron velocity and thus wavelength at a certain rotation condition along an axis parallel to the beam direction. Selecting the rotation frequency allows selecting a specific main transmission wavelength (figure 19.4). Typically, a few 100 Hz are possible, and beams of

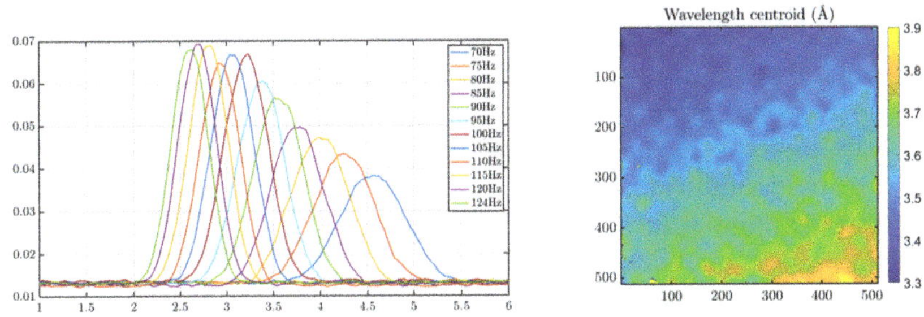

Figure 19.4. ToF measurement of transmitted wavelength spectra of a velocity selector (ICON) at PSI [12] at different rotation frequencies (left); distribution of peak wavelength on a field-of-view of 2.8 × 2.8 cm^2 from the same measurement (right).

some 10 cm^2 can be transmitted. High transmission values, in the 80% range, can be achieved. Performance is generally better for longer wavelengths and tilting the axis with respect to the beam direction allows additional tuning of resolution and range. The devices are relatively compact with tens of cm length and diameter.

However, in contrast to scattering instruments, neutron beams for neutron imaging carry a high divergence and the signal at the detector is measured with direct spatial resolution. Thus, also the transmission of the velocity selector, which is typically placed close to the pinhole of the imaging geometry, where the beam is still relatively compact, is mapped. As the transmission characteristics depend on the trajectories of neutrons travelling through the device, the main wavelength and wavelength resolution vary, often significantly, across the FoV of an imaging detector [76]. Figure 19.4 shows an example, where the wavelength transmitted by a selector for different frequencies (left), but also the peak wavelength distribution over the FoV of a ToF imaging detector, is measured through an additional ToF configuration with a chopper and an MCP/TPX ToF imaging detector [73]. Just across the relatively small FoV with a height of about 30 mm the maximum transmitted wavelength seems to vary by 0.7 Å, which equals about 20%. The direction of the variation over the height indicates that the window for the beam is placed at one side of the rotating turbine selector. For sensitive measurements this gradient in conditions has to be taken into account for the analyses. It has to be noted, however, as the measurement is based on a ToF measurement with a rotating disk chopper, corrections for the spatial ToF distribution would need to be applied additionally, depending on the position of the chopper set-up [77].

19.4.2 Crystal monochromators

Crystal monochromators are widely used in neutron scattering for selecting a specific wavelength for an experiment or for wavelength analyses of scattered neutrons. In the former case, the monochromatic beam is typically diffracted out of the incoming white beam towards a sample position according to the Bragg condition for the chosen wavelength in the monochromator crystal. The mosaicity is chosen corresponding to the resolution requirement of the instrument. In order to allow optional monochromatization in an imaging instrument which is designed to also support measurements with a white beam, typically a double-crystal monochromator set-up as developed by Treimer *et al* [78] is chosen. It allows deflection of a monochromatic beam out of the incoming beam direction and subsequently, with a second corresponding monochromator crystal back into the initial forward direction. Limiting the beam size accordingly in a position close to the pinhole allows one to only slightly off-set the beam to the initial trajectory which is also sufficient to block the transmitted spectrum of the initial incoming beam. For beamlines not using a bent neutron guide also the background of fast and epithermal neutrons has to be considered carefully, when blocking the direct beam. The described double-crystal monochromator geometry allows using all downstream installations, potentially only with a slight shift reflecting the off-set of the beam. The beam downstream of the monochromator has an altered divergence, which in particular in the direction perpendicular to the diffraction plane strongly depends on the mosaicity. In the direction of the beam cross-section in the diffraction plane the resulting divergence is reduced through the non-dispersive set-up, but is wavelength dependent resulting in a wavelength gradient at the detector in that direction in a similar way as already described for the velocity selector [79]. However, the gradient is typically significantly smaller for the double-crystal monochromator. The wavelength resolution is improving with lower mosaicity and ranges for a PGC at around 4 Å from about 1.5% at 1° mosaicity to about 4.5% at 3° mosaicity. However, while the distribution of wavelengths is narrower for smaller mosaicity, the flux density remains about the same or is even higher on a smaller beam area at the detector position.

Changing the angle of both crystals correspondingly and translating one crystal along the beam direction enables selecting a specific wavelength from a certain wavelength range, typically between about 2.5 and 6.5 Å for a pyrolythic graphite double-crystal monochromator [78]. This range is particularly relevant for, e.g. Bragg edge measurements, as it contains the most prominent edges of most materials suited for such investigations, like e.g. steels, Ni and Cu.

Care has to be taken for potential contamination of second order Bragg reflections of ½ of the desired wavelength, which becomes especially severe for longer wavelength choices, where the second order might be closer to the peak of the spectrum. Such contamination can effectively be avoided either by the use of a velocity selector, or a third crystal monochromator, which is aligned so as to deflect the undesired second order wavelength out of the beam. Both methods might imply a certain flux penalty due to transmission limitations.

Today such double-crystal monochromator devices are available in numerous leading state-of-the-art neutron imaging beamlines [16, 18, 79, 80].

A different implementation of a crystal monochromator for imaging can be considered with a single monochromator crystal in a beamline dedicated to monochromatic, wavelength selective measurements, as had been realized in the former instrument PONTO at the meanwhile shut down BERII reactor source in Berlin, Germany [81]. Here a monochromator deflects neutrons out of a white beam, which might continue to serve other instruments. In order to enable scanning wavelengths, the instrument has to be built up on an optical bench, which can be rotated around the monochromator in order to enable different diffraction angles and, thus, wavelengths to be utilized. This limits the instrument to a somewhat smaller scale and limited application field than the double-crystal set-up in a full blown state-of-the-art imaging instrument, although it might be able to coexist with such at a cold neutron guide providing additional capacity.

19.5 The time-of-flight technique

Wavelength resolved measurements without monochromatization can be performed when utilizing a ToF approach based on the de Broglie relationship, as outlined in section 19.1. When neutrons are emitted only at a certain point in time $t = t_0$ then neutrons of different wavelength, i.e. different energy and, thus, velocity, will arrive at a detector at a certain distance L_{ToF} at different times $t_{\text{ToF}}(\lambda) = t - t_0$. Principle contributions of the wavelength resolution are the burst time τ of the pulse, limiting the accuracy of t_0 for a detected neutron, as well as the length of the flight path L_{ToF}, defining the increment of neutron wavelengths relative to a burst time τ able to reach the detector at the same time t. This does not take into account yet the limited time resolution of a detector, and not the fact that source pulses are typically provided at a certain frequency. This repetition rate of neutron pulses, in particular such containing a wide spectral range, implies the possibility of frame overlaps at the detector, which increase with increasing length L_{ToF}, and therefore a potential bias in wavelength retrieval. This is typically avoided by the use of choppers referred to as bandwidth or fame overlap choppers. A neutron chopper is a rotating device, mostly a disk coated with neutron absorber rotating around an axis aligned parallel to the beam. The disk has one or several windows of a certain angular width, to allow neutrons to pass in certain time intervals. While choppers can be used to create a pulsed beam, when used as bandwidth choppers they allow defining a certain wavelength range not overlapping with previous and subsequent source pulses. It is obvious that such wavelength bandwidth decreases with increasing source frequency and increasing L_{ToF}. All these basic considerations are well depicted in ToF diagrams and are of outmost importance for efficient ToF instrumentation and measurements (figure 19.5).

While a ToF approach is an obvious choice at a pulsed source, also implementations at continuous sources have been demonstrated and are highly relevant [36, 38, 40, 41, 82, 83]. The dependence of the wavelength resolution on the burst time τ of

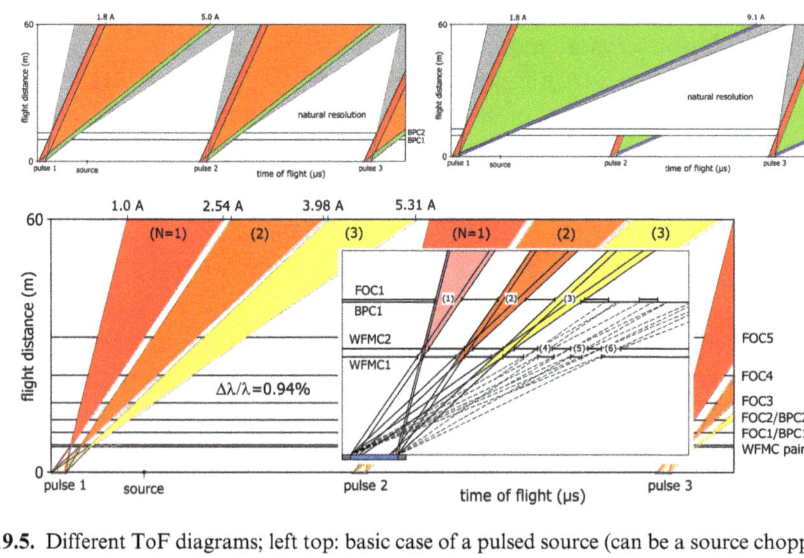

Figure 19.5. Different ToF diagrams; left top: basic case of a pulsed source (can be a source chopper) with two bandwidth (or band pass) choppers showing the useful wavelength range at a fixed instrument length and the penumbra leading to pulse overlap somewhat limiting the useful bandwidth; right top: example of pulse suppression at a pulsed source (same parameters than top left diagram) in order to gain bandwidth; bottom: ToF diagram of a highly complex chopper system at a long pulse source with wavelength frame multiplication utilizing a pair of optically blind pulse shaping choppers; examples taken from reference [33], copyright (2020), with permission from Elsevier.

the pulse and the flight path length L_{ToF} have been outlined in section 19.1 in the context of instrumentation at a pulsed source. Here we should consider potentials, limitations and specific ToF configurations at pulsed and continuous sources in more detail.

In general, it has to be assumed that a ToF measurement is efficient, when the wavelength resolution and the covered bandwidth meet the needs of a specific experiment. This implies that a monochromatic measurement at a pulsed source is as efficient as at a continuous source, and is often better carried out at a high flux continuous source, typically featuring a higher time averaged and thus also monochromatic flux. For low wavelength resolutions of around 10% a break even in efficiency between leading short pulse sources and high flux reactors is reached, due to the excessive resolution of the short pulse source at reasonable source-to-detector distance L_{ToF}. This can be formalized by the figure-of-merit definition introduced by Mezei [22, 84], where the peak brightness of the source is scaled with the ratio of the desired and delivered wavelength resolution.

$$\mathrm{FOM} = \Phi \cdot \min\left(1, \left(\frac{\partial\lambda}{\lambda}\right)_{\mathrm{delivered}} \Big/ \left(\frac{\partial\lambda}{\lambda}\right)_{\mathrm{required}}\right)$$

This implies that the FOM scales with the source peak brightness and not the time averaged flux as long as wavelength resolution is required, i.e. $\delta\lambda/\lambda < 1$. What this equation does not take into account yet, but is only stated in reference [84] where a simplified version is used, is a severe limitation of this FOM definition, as it only applies if the covered wavelength range equals the required wavelength range of an experiment. This can, however, be accounted for by simply defining $(\delta\lambda/\lambda)_{\mathrm{delivered}}$ more strictly by the resolution delivered for an instrument optimized for the bandwidth. At a pulsed source resolution, and in particular bandwidth, cannot be adapted as flexibly as at a continuous source, especially for individual measurements. Figure 19.6 depicts the definition of the FOM as given in references [22, 84] in terms of peak source brilliance. Most notably, the green and blue areas illustrate the efficiency loss for methods where the delivered resolution of the pulsed sources is excessive and the worst case considered is small-angle neutron scattering (SANS)

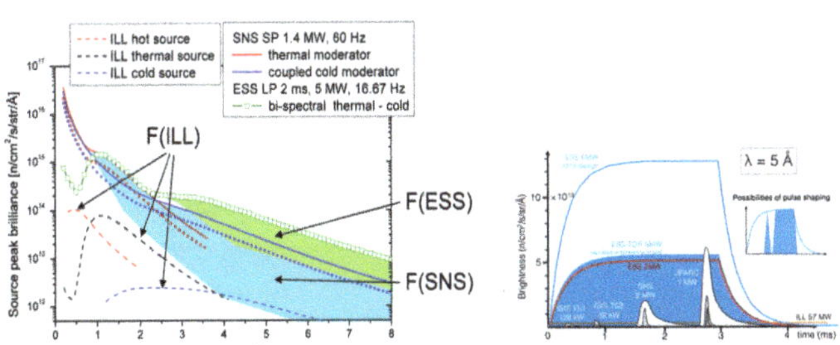

Figure 19.6. Source brightness and FOM reductions based on considerations of optimization to the required wavelength resolution (left), reprinted from [22], copyright (2007), with permission from Elsevier; time structure of leading neutron sources ILL, SNS, JPARC and the upcoming ESS (right), credit: ESS.

with a 15% resolution requirement, but full bandwidth acceptance. In such a case the most powerful short pulse source loses an order of magnitude in efficiency, putting it on bar with the most powerful continuous source (ILL).

In contrast to a pulsed source, at a continuous source bandwidth and resolution can be largely optimized independently and can even be tailored from experiment to experiment to match the specific requirements and measure with best possible efficiency [40]. The latter also entails the ability to keep the wavelength resolution constant throughout the utilized wavelength bandwidth through an optically blind double chopper producing the source pulse [85].

A key additional consideration has to be that ToF imaging requires an imaging detector that is capable of providing the respective required time resolution in the microsecond range. Currently available leading technologies, in particular those based on TPX chips are described in section 19.3.

19.5.1 ToF at short pulse sources

At a short pulse source the burst time τ is given by the source time structure, and depends in particular on the moderator. The moderation process extends the effective neutron pulse burst duration with respect to the typically very short proton pulses. In general, the burst time has a specific wavelength dependence and can be influenced by the moderator choice. At short pulse sources the burst time is typically a few 100 μs. The left and middle panels of figure 19.7 display the energy dependent brightness and burst times for different moderators available at the MW spallation neutron source of JPARC [86]. It becomes obvious that the choice of moderator provides a trade-off with respect to pulse length versus intensity by choosing to view a coupled, decoupled or even decoupled poisoned moderator, where the first provides the highest flux but longest burst times and the latter the shortest pulses but also lowest brightness.

The right-hand side panels of figure 19.7 show the specific situation for the combined neutron imaging and diffraction instrument IMAT [27] at the second

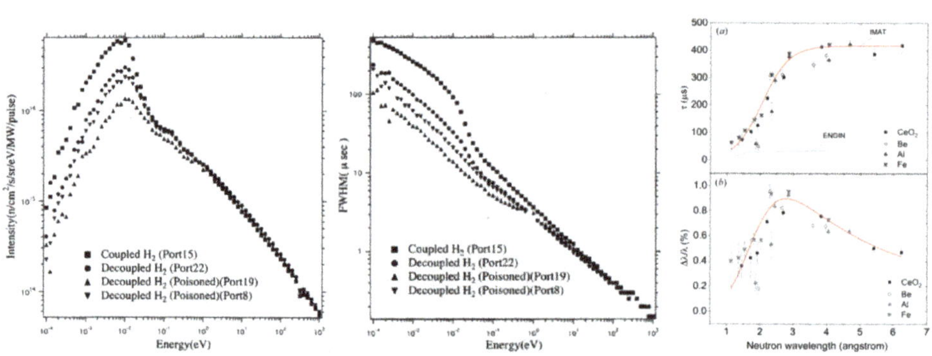

Figure 19.7. The left and mid panels display the energy spectra and the FWHM of the burst time t of the different moderators available at the Japanese Spallation Neutron Source (JSNS) at JPARC, respectively [86], copyright (2008), with permission from Springer Nature; the right hand panels display the wavelength dispersive burst time t and the resulting wavelength resolution for the imaging and diffraction instrument IMAT in England, reproduced from [87] with permission from the International Union of Crystallography.

target station (TS2) at ISIS, RAL [87]. It illustrates a rise in burst times for wavelengths between 1 and 3 Å, which then stabilizes at around 400 µs. For the wavelength dependent wavelength resolution of the about 55 m long instrument this implies that the wavelength resolution is best at shortest wavelengths with about $\delta\lambda/\lambda \approx 0.2\%$ and deteriorates to 0.8% at around 3 Å, before it improves again at about constant burst time but increasing t_{ToF} of longer wavelengths reaching a value of about 0.5% at 6 Å.

When the choice for a moderator is made, the only possibility to reach a target wavelength resolution is the instrument length L_{ToF}. Given the fixed repetition rate $1/P$ of the source, where P is the period between two pulses, L_{ToF} also defines the available wavelength bandwidth, according to equation (19.2)

$$\Delta\lambda = \mathrm{Ph}/L_{\mathrm{ToF}}m \tag{19.3}$$

Typical repetition rates of short pulse sources are 60 Hz at SNS, 25 Hz at JPARC, 50 Hz and 10 Hz at ISIS TS1 and TS2, respectively. Thus, the choice of moderator and instrument length have to be optimized together to provide the best combination of bandwidth and resolution for the instrument. While the instrumental wavelength resolution is completely fixed with this choice, the bandwidth can still be extended by shifting the band in subsequent measurements or by suppressing every other source pulse by a chopper. An excessive bandwidth on the other hand impacts the efficiency.

These limitations in flexibility bring around some limitations when considering the wide range of requirements of wavelength-resolved neutron imaging. While diffraction contrast imaging (section 2.2) for strain investigations (chapters 5 and 6) requires the highest wavelength resolution in the sub-percent range, typically around $\delta\lambda/\lambda \approx 0.5\%$, in most applications only a single Bragg edge is analysed, which limits the required wavelength range to a few tenths of an Ångstrom, for identifying and mapping crystalline phases in many cases resolutions of a few percent suffice, while a bandwidth of several Ångstrom to one Ångstrom might suffice depending on the specific use case. If texture is in the focus or other microstructural features such as grain size again a medium wavelength resolution is sufficient, and depending on the details of the study the required range can vary significantly, although for advanced analyses a broad band of several Ångstrom is preferable [88, 89]. For other wavelength-resolved methods (section 2.2), e.g. quantitative dark-field contrast imaging relaxed wavelength resolutions $\geqslant 10\%$ are adequate and in wavelength-resolved polarized neutron imaging resolutions of a few percent are used generally, depending, however, also on the specific technique and investigation. In contrast, inelastic scattering contrast can work even with some 30% wavelength resolution and beyond. Resonance absorption contrast imaging operates in a different energy range, where the pulse typically is significantly shorter than for moderated thermal neutrons and needs to be in the nano- to microsecond range for such application.

19.5.2 ToF at long pulse sources

At a long pulse source on the other hand, flexibility to tailor instrument resolution to the needs of specific measurements improves. However, the only real long pulse source, with a pulse length of a few milliseconds, i.e. about an order of magnitude longer than at common short pulse sources, is currently only under construction. The European Spallation Source (ESS) in Lund, Sweden, will feature a 3 ms burst time at a low frequency of 14 Hz. This implies on the one hand, that instruments requiring high wavelength resolution cannot work with the full source burst time, and that short instruments can on the other hand efficiently serve techniques with relaxed wavelength resolution requirements and assess quite a wide wavelength range, as typically advantageous for large-scale structure instruments like small-angle neutron scattering and neutron reflectometry.

In order to serve instruments with high wavelength resolution requirements, the long pulse source enables restructuring the source time structure through choppers (figures 19.5 and 19.6) in a similar fashion to how choppers are used to pulse a continuous beam. However, the initial time structure of the source has to be factored in, which results, e.g. in wavelength frame multiplication (WFM) and repetition rate multiplication (RRM) in order to utilize relatively broad wavelength bandwidths [22]. For neutron imaging in particular WFM is of interest [23, 32, 33].

The basic concept of WFM is to create a number of shorter pulses from a long pulse, taking into account a distance of several meters from the initial long pulse source, where a chopper can be placed, and thus covering a broad wavelength range by a sequence of frames of limited smaller wavelength bandwidth (figure 19.5). The shortest wavelength of each frame originates from the end of the long pulse and the longest from the beginning. These cross at an individual opening of a WFM chopper. Ideally the longest wavelength of a frame coincides with the shortest wavelength of the subsequent frame, but is separated in ToF by about the width of the long pulse. This in principle allows covering a wide wavelength range by n individual frames seamlessly in wavelength, but separated in ToF. A challenge of a real system is to keep the frames separated and avoid frame overlap, despite the blur in time induced by finite rise and fall times of the initial as well as the chopped pulse time windows. While the former is a source characteristic the latter is influenced by the finite beam cross-section and chopper speeds.

Utilizing a set of two optically blind WFM choppers enables achieving a constant wavelength resolution, because in such geometry the effective burst time is wavelength dependent and increases linearly with wavelength, thus compensating the longer ToF of longer wavelength neutrons resulting in $\tau(\lambda)/t_{\mathrm{ToF}}(\lambda) = \mathrm{const.}$ [85]. Thus, the wavelength resolution in such a case is only depending on the distance between the two WFM choppers. This in turn implies that when the distance between these choppers can be varied, the resolution can be adapted to the needs of a specific measurement, and, thus, the efficiency can be optimized. While the applicability of optically blind WFM to ToF neutron imaging has been postulated [23] and demonstrated in a test set-up [83], the first dedicated instrument is still under construction currently at the ESS in Sweden [28].

However, it has to be considered, that the full flexibility of the system can only be achieved within the limits of the initial time structure of the pulsed source and concerning the most relaxed resolution and the bandwidth as discussed before. For example, a coarser intrinsic wavelength resolution than given by the source burst time and the instrument length is indeed not achievable. The instrument ODIN (Optical and Diffraction Imaging with Neutrons) at the ESS and its chopper system are designed such that the wavelength resolution given by the source pulse and the instrument length is around 10% to efficiently enable techniques like neutron grating interferometry and some phase studies with diffraction contrast imaging, while the WFM system can tune this resolution to a few percent in order to enable better resolved diffraction contrast for phase and texture studies as well as polarization contrast imaging and further to sub-percent resolutions to enable also Bragg edge strain mapping. In addition, the chopper system allows utilizing for all resolution settings either a bandwidth of about 4 or 8 Å, where for the latter case every other source pulse is suppressed. Thus, the instrument achieves a very high flexibility to efficiently serve many modalities and applications profiting from the extremely high source brightness of the ESS long pulse source [23, 28, 33]. More flexibility can only be achieved at a continuous source, however, at typically lower source brightness.

19.5.3 ToF at continuous sources

At continuous sources all limitations in tailoring the time structure are absent. The continuous beam can be chopped by means of disk choppers at any position convenient and the repetition rate as well as the pulse burst time τ can be chosen in principle freely. While this is exploited extensively for scattering instruments, in particular SANS, reflectometry and quasi/inelastic ToF spectrometers at continuous sources, there is no dedicated installation for neutron imaging. However, the approach has been tested at scattering instruments and at a test beamline for the future ESS source [36, 38, 40, 82, 83]. This implies that the set-ups were not optimized for imaging. The general downside of such an approach at a continuous source is the high cost in flux, in particular for high wavelength resolution applications, so that monochromatization options were generally preferred so far with wavelength resolutions ranging from a few tens of percent (velocity selectors) to a few percent (crystal monochromators). Applications were accordingly chosen and tuned such that one or a few monochromatic measurements were sufficient, like e.g. in crystallographic phase mapping (compare chapter 5). ToF excels when a significant range of wavelengths or wavelength bins is required and/or the sample system evolves over time, so that scanning wavelengths successively is not practical. When a single wavelength is sufficient, the time average flux counts and continuous sources can compete despite lower peak brightness. When applying ToF, the peak brightness counts, in particular when the resolution and wavelength range can be matched to the need of a measurement. This is where continuous sources can regain some of the efficiency. Examples are the monochromatic case [90] and the so-called high duty cycle ToF approach [91] where the required wavelength resolution approaches around 30% and 2–3 wavelength bins can suffice. Another example could be strain mapping where the wavelength range could be limited to about 0.5 Å because only a single edge is typically analysed. That implies that the repetition rate at a continuous source could be set very high, i.e. a few 100 Hz, thus regaining about an order of magnitude in efficiency as compared to a typical pulsed source.

However, such an approach is difficult to realize even at a continuous source when considering the pulse overlap situation at such frequency and the issue to keep frames separated. For this purpose a pulse overlap approach was developed [36] at PSI, where such an approach was previously used in diffraction [92]. The method for imaging resembles earlier approaches for coded source/aperture imaging [93] in real space and transfers it to the time domain through a chopper with a pseudo-random arrangement of slits. Here care has to be taken, that the retrieved wavelength range is sufficiently broad to cover the whole range of the spectrum with significant flux through a suited choice of and sufficiently extended pseudo-random time structure. This method allows high wavelength resolution ToF imaging to be competitive with pulsed sources, e.g. in the domain of strain mapping. However, to date no dedicated system for the cold wavelength range, where Bragg edge contrast is best suited for strain mapping, is available, while a system is under construction at PSI in Switzerland.

The simplest approach, which was mainly used to measure the spectrum of instruments is to place a single disk at a certain distance of about 10 m from the detector and run at relatively low frequency (few 10 Hz) allowing acquisition of a full spectrum without relevant overlap. When aiming at focussing at a shorter wavelength range of interest, the distance from the chopper or the chopper frequency can be increased. To conserve resolution in both cases the chopper opening window has to be increased. This then requires a bandwidth chopper to be placed between the pulse chopper and the detector in order to prevent pulse overlaps and, thus, contamination with wavelengths beyond the chosen range. Different chopper windows, frequencies and distances allow tuning the resolution and bandwidth as needed. Chopper openings are typically tunable by using double choppers and phasing them to narrow or open up the opening time window. The wavelength resolution in all these configurations will depend linearly on the wavelength, with better resolutions at longer wavelengths. Such an approach has been exploited for imaging the first time at a ToF SANS instrument at HZB in Berlin [38]. A smart way to decouple bandwidth, wavelength and resolution is again the optically blind double chopper configuration [40, 85]. It not only allows for achieving a tunable but constant wavelength resolution, but also tuning the bandwidth through the chopper frequency, corresponding to the one of the bandwidth chopper, without impacting the resolution. Thus, it is the maybe most flexible and efficient realization for ToF imaging, for which it has been demonstrated at the ToF neutron reflectometer BioRef at HZB in Berlin [40]. The method is, however, limited in focussing on a very small wavelength bandwidth by limited chopper speeds and time collimation for avoiding frame overlap, when frame widths become very narrow. For a very small bandwidth a velocity selector used together with a chopper can potentially solve this issue.

It has to be noted, that the choppers rotating with limited frequencies, implying spatial correlations of opening and closing times, impact the wavelength resolution and required ToF calibration across the beam for neutron imaging. In neutron scattering methods such contributions are hardly considered due to spatial integration of measurements over the beam cross-section where spatial variations do not matter. However, the spatially resolved nature of neutron imaging makes a difference. The variation of ToF across the chopper cutting direction in ToF neutron

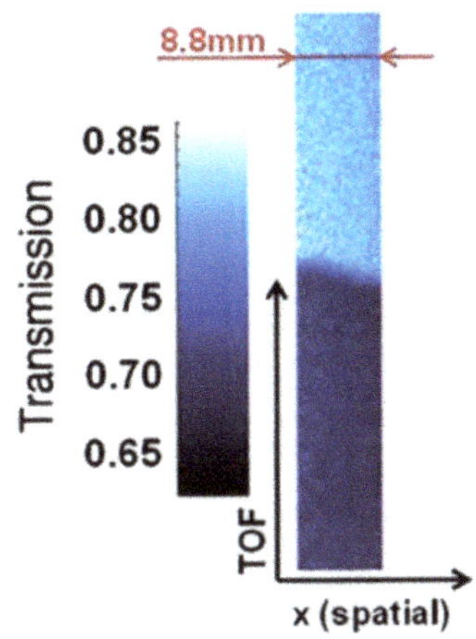

Figure 19.8. Variation of Bragg edge wavelength across the FoVw in a chopper-based ToF imaging measurement at a continuous source. Reprinted from [40], copyright (2012), with permission from Elsevier.

imaging is illustrated in reference [40] and displayed in figure 19.8. Although the shift is smaller than the nominal wavelength resolution of the set-up, it would strongly affect a strain-mapping application, if no correction based on the displayed reference were to be applied. Thus, the situation is again similar to that found with the velocity selector and crystal monochromators. This does not, however, equally apply at a short pulse source, but is created by a chopper system, while at the short pulse source an inclined view on the moderator surface can have similar effects [94]. Short pulse sources on the other hand typically also provide a pulse with a significant tail, which has to be accounted for in analyses as well [95].

19.6 Polarization

With the advent of polarization contrast neutron imaging (section 2.2), the need for suitable instrumentation to achieve, manipulate and control, conserve and analyse neutron spin polarization in neutron imaging arose. The large cross-sections, overall divergence as well as the wide spectrum assessed in neutron imaging applications imply difficult requirements to polarization instrumentation. The basic components comprise the polarizer and analyser, which are basically neutron spin filters (NSFs), which enable passage only of one spin state, typically the one aligned with their magnetic field orientation. NSFs are based either on the different refractive index of the parallel and anti-parallel spin which leads to deflection and transmission, respectively, at the interface of a magnetized neutron mirror after which one component typically meets an absorbing layer, or on the spin dependent absorption of neutrons in polarized ^{3}He. The difficulty with the latter is to maintain sufficient ^{3}He polarization, which implies a significant effort and infrastructure, while neutron optical NSFs are stable and independent components. While ^{3}He has also some advantages, e.g. for higher divergence, apart from a few exceptions [96–98], typically mirror-based NSFs are used in polarized neutron imaging [97–103]. A downside of the need of a polarization analyser between the sample and the detector is the increased sample-to-detector distance l inducing substantial additional image blur (compare section 19.1). The strong magnetic field of the NSF worsens the situation due to the need for a distance to avoid interference of the NSF stray field with the sample magnetic field. For the generally lensless nature of neutron imaging this implies a significant penalty in spatial resolution capability which can only be countered by additional collimation leading to significantly reduced flux. The latter is especially detrimental for polarized neutron imaging methods which require wavelength resolution. The polarization of the initial beam from the source comes at a cost of 50% of flux already. Solid state polarizing benders [104] have been utilized extensively for polarized neutron imaging because they come with a very compact design and good performance in the mainly exploited cold neutron range. To maintain the polarization throughout the set-up requires guide fields aligned with the polarization direction and homogeneous throughout the beam size. Guide fields span the flight path between components such as spin polarizer, spin flippers, sample and analyser. Typically, values of around 10 mT are sufficient, though this depends on the magnetic environment and potential stray fields in the area of the instrument (figure 19.9). An overview of polarized neutron imaging set-ups and corresponding literature is provided in reference [105].

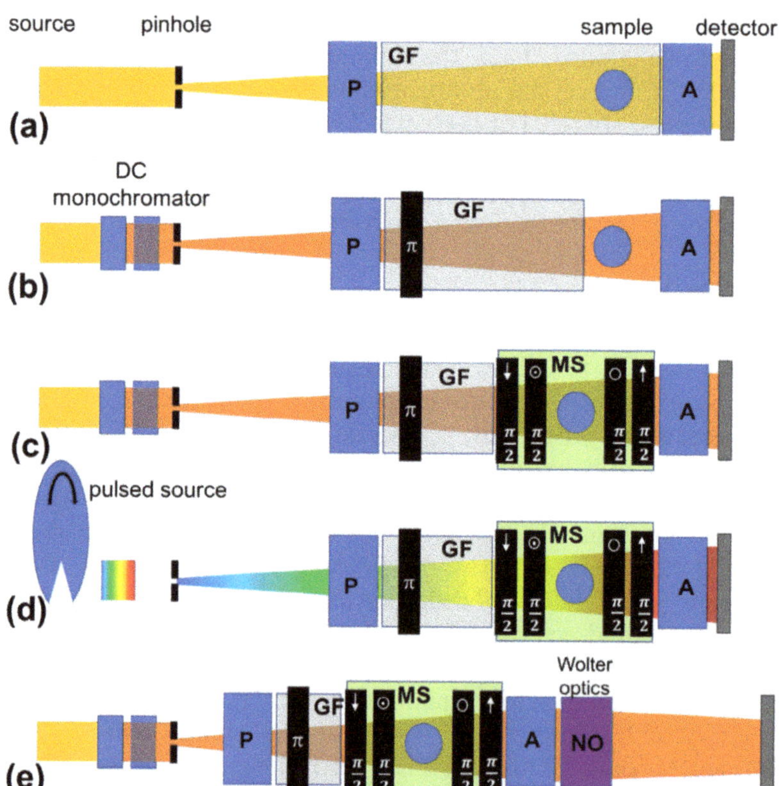

Figure 19.9. Schematics of different realizations and complexity of polarized neutron imaging instrumentation according to reference [105]. Copyright IOP Publishing Ltd. All rights reserved

The simplest set-ups to utilize polarization contrast neutron imaging add to the standard neutron imaging set-up merely a polarizer, a guide field and an analyser (figure 19.9(a)). Such implementation enables visualizing local depolarization and thus mapping, e.g. the Curie temperature distribution or in general the distribution of ferromagnetic phase in a sample (chapter 13). A more detailed analysis of domains is potentially possible when utilizing a monochromatic beam or ToF together with the polarized neutron imaging set-up, which particularly also allows mapping local spin precession [99, 103, 105]. For the latter it is an advantage if not a precondition to add a spin flipper to the set-up in order to measure local polarization variations quantitatively by correlating the impact on an incident spin-up and spin-down polarized beam [105]. Such a π flipper rotates or flips the polarization direction by 180°. A simple flat coil or current sheet can do this job for a monochromatic beam, but e.g. an adiabatic spin flipper, like an adiabatic fast passage (AFP) flipper [106], or a resonance spin flipper [107] is best suited for a white beam, like e.g. in ToF implementations. The most complex polarization contrast neutron imaging set-up is the one enabling polarimetric neutron imaging [101, 108, 109] (figures 19.9(c)–(e)) probing the full depolarization matrix P_{ij} where i and j denote the three perpendicular incident and analysed polarization directions, respectively. This requires two spin turners upstream and two spin turners downstream the sample position enabling one to either pass the original polarization direction, flip it by 90° in one direction with one turner switched on or 90° in two perpendicular directions with both spin turners in operation. For ToF applications such spin turners, typically flat coils transmitted by the neutrons, have to be synchronized with ToF in order to always create the right magnetic field for the specific wavelength passing to flip its spin by π. A sophisticated implementation of such a system can be found at the imaging instrument RADEN at the high power pulsed spallation neutron source of JPARC in Japan [109]. The influence of the complex stray field of this set-up is countered by the use of mu-metal shielding around the spin turners and sample position and the polarized beam is coupled in and out by coupling coils in order to prevent disturbance of the polarization in stray fields.

19.7 Beam modulation techniques for differential phase and dark-field imaging

The most powerful techniques for dark-field and phase contrast imaging with neutrons are based on transversal beam modulation in order to resolve (ultra) small angles of diversion of neutrons due to small-angle scattering and refraction. Various methods and variations of such for transversal beam modulation have been introduced since the pioneering works concerning differential phase and dark-field contrast neutron imaging with a Talbot–Lau interferometer [110, 111].

19.7.1 Talbot–Lau interferometer

Neutron Talbot–Lau interferometers typically comprise three gratings [110]. A source grating G_0, typically an absorbing line grating, which creates multiple line sources providing sufficient partial beam coherence at the second grating, the phase grating G_1, to induce strong interference effects manifesting at fractional Talbot distances downstream of G_1 corresponding to the Talbot–Lau effect. Typically, G_1 induces π phase shifts with the periodicity p_1 of the grating which leads to a beam modulation with a period of $p = p_1/2$ at fractional Talbot distances $d_m = mp_1{}^2/8\lambda$, where m is an odd integer denoting the order of the Talbot distance. The geometry is chosen such that all interference patterns of the partial waves created at G_0 are superimposed constructively at the chosen fractional Talbot distance which requires the distances relative to G_1 and periods, p_0 and p_2, to be proportional. While G_0 is required due to the limited neutron beam coherence, a third grating, again an absorption grating G_2 with period p is required to measure the beam modulation at the chosen fractional Talbot distance, because the typical periods of the order of μm cannot be resolved efficiently by the utilized neutron imaging detectors. This implies that in the case of sufficient beam coherence and spatial resolution capability only one grating, G_1, would be required. In the given three-grating set-up, which requires careful alignment to achieve good modulation visibility, the modulation is probed by a phase-stepping procedure applied to one grating which results in a sinusoidal intensity modulation, relative to the transversal grating stepping, in each pixel. It has to be noted that the grating set-ups do not impact the spatial resolution capability of the imaging geometry, which is still, independently defined by the L/D collimation ratio. Later also single-shot measurements and their analyses were introduced, based on the analyses of moiré fringes [112], which are visible for imperfect grating alignment. Those might, however, negatively impact the achievable spatial resolution, because the moiré fringes have periods significantly larger than the image resolution. However, they are highly desirable, e.g. for time-resolved studies. High time resolution studies have been demonstrated to be possible even with phase stepping, but only for repetitive processes, utilizing post-processing sorting sample phase and phase stepping in order to allow conventional analyses relative to each specific process phase [113].

The absorption gratings are mostly based on Gd patterns on silicon or quartz wafers, while the phase gratings are structured Si wafers. However, different designs and processing routes have been realized with time, often in order to optimize modulation visibility [114–117]. Also, varying the duty cycle for different gratings has been attempted to this end. High resolution remote controlled stages are required for alignment and phase stepping. The pioneering set-ups featured millimeter scale G_0 periods and micrometer scale modulation periods, with G_0 being meters upstream of G_1 and the utilized first fractal Talbot distance d_1 a few centimeters [110, 111, 114]. Such a grating interferometer is wavelength dependent in a twofold way. The correct phase shift at G_1 and the fractional Talbot distance are wavelength dependent. Thus, the set-up is optimized to a design wavelength and the performance is wavelength dependent, displaying repeating visibility maxima at longer wavelengths corresponding to the phase wrapping [117]. Utilizing a wide wavelength band [118] or white beam [119], the different weight of contributions of different wavelengths, thus, has to be considered. Samples were initially installed upstream of G_0 [110].

This approach changed somewhat with the advent of quantitative dark-field contrast imaging [120], probing microstructural features [121]. Placing the sample between $G1$ and G_2 enables scanning the probed correlation length ξ according to $\xi = \lambda L_s/p$ by scanning the sample-to-detector distance L_s, which requires a sample stage along the beam axis. Another approach to scan the probed correlation length is to vary the wavelength λ, which is, however, limited by the wavelength dependence of the set-up, or scanning both wavelength and distance, which has been done in the first reported quantitative dark-field imaging experiment [121]. However, with the initial set-up both wavelength range and sample distance are severely limited; at the first fractional Talbot distance in the pioneering set-up L_s could be varied on a range < 2 cm, and thus the total correlation length range probed did not exceed a few micrometers.

As a consequence, a number of different variations and approaches were demonstrated and introduced. The first of these simply utilized longer Talbot distances such as the 3rd or 5th fractional Talbot distance [122, 123]. This enabled three times or five times longer maximum L_s and thus correspondingly larger ranges of probed correlation lengths. However, the range was still limited to less than an order of magnitude, limited also by set-up and sample sizes, and to the structure sizes in the micrometer range, overlapping with best real space resolution of direct imaging.

In order to enter also the nanometer size range with Talbot–Lau interferometers for imaging, symmetric Talbot–Lau interferometers were introduced [115, 124]. Here the distance between G_0 and G_1 equals the distance G_1 to G_2. In such an arrangement the periods of all gratings are the same, including the period of the phase grating G_1 and the interference pattern p, due to a magnification of 2, which needs to be taken into account. The period p is significantly larger in such arrangements and ranging at some ten micrometers, implying the ability to probe nanometer range correlation lengths, while the significant Talbot distances realized, a few ten centimeters to a few meters, also enable addressing the micrometer range. Thus, these set-ups enable probing about two orders of magnitude in length scales, which is significant progress. The coherence enabling larger periods of phase gratings is established through the significantly smaller period of the G_0 gratings, when compared to the initial Talbot–Lau interferometers for neutron imaging.

A mixed approach of higher order fractional Talbot distances and moving in the direction of a symmetric set-up with somewhat larger modulation periods has also been presented [117], with advantages similar to the symmetric version. This work also introduced phase gratings with 3π phase shift, which allows for good modulation visibilities for a range of neutron wavelengths and, thus, appears to be an interesting Talbot–Lau option for ToF instruments. The ToF method can be used as additional means to vary the probed correlation length [125, 126]. Such an approach enables concurrently probing also diffraction contrast, i.e. Bragg edges, by the ToF wavelength-resolved attenuation contrast signal. This means that three different length scales can be probed simultaneously from Angstroms, via the nano and microscale to millimeters and centimeters.

Finally, Talbot–Lau interferometers have been combined with polarized neutrons and even polarization analyses [118, 127]. This requires the additional installation of a guide field throughout the set-up and an analyser on a remote controlled alignment

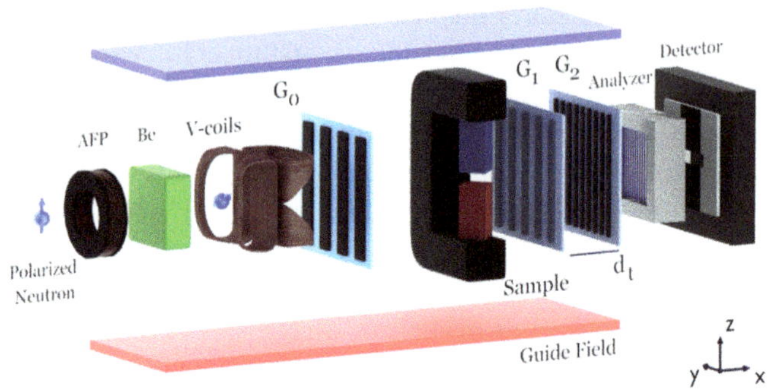

Figure 19.10. Representation of a neutron Talbot–Lau interferometer equipped for working with polarized neutrons and polarization analyses; this particular version includes V-coils to adiabatically turn the polarization of the incoming beam into the horizontal plane [127].

stage (figure 19.10). As polarization is analysed downstream the grating analyser it does not affect the working of the interferometer but allows distinguishing spin-up and spin-down dark-field signals.

19.7.2 Far-field interferometer

Another approach to dark-field contrast imaging has been taken with so-called far-field neutron interferometers [128]. Here two phase gratings positioned close to each other but relatively far from the detector create moiré fringes of relatively large period to be directly resolved with an imaging detector of medium resolution. The far-field moiré beam modulation period

$$p = Lp_g/D_G \tag{19.4}$$

decreases reciprocally with the distance D_G between the two typically $\pi/2$ phase gratings, which in general have the same period p_g. For the probed correlation length in such a far-field grating interferometer one can thus write

$$\xi = \lambda L_S D_G/\left(P_g L\right) = \lambda L_S/P \tag{19.5}$$

where L is the length from the collimation slit, required to provide the required coherence, to the detector. The equation simplifies to the conventional definition on the right-hand side when substituting the grating period p_g with p according to equation (19.5). Like in neutron Talbot–Lau interferometers, the slit can be replaced by a grating, if measurements are performed at constant D_G and thus constant period. In this case also an analyser grating can be utilized, rendering the set-up basically equivalent to a Talbot–Lau interferometer, in particular, of the symmetric or close to the symmetric kind discussed above. When utilizing large periods, directly resolved by the detector and typically of the order of >100 μm, the probed correlation lengths typically are rather small and in the range of tens of nm, except when the sample distance L_s becomes large, i.e. tens of centimeters. This in turn aversely affects the spatial resolution, except when the pinhole is closed down significantly at the cost of intensity. The technique is often referred to as achromatic, however, the visibility, like the phase shift in the gratings, depends significantly on the neutron wavelength.

19.7.3 Single grating

A different and much more simplified solution to beam modulation is found in the use of simple single absorption gratings [129]. No interference or alignment is required in this case, as the absorption pattern is simply projected on the detector. This is a truly achromatic approach, however, the distance between grating and detector is limited by the geometrical resolution, and, thus, the resolvable correlation length range is limited this way. Increasing the range requires closing down the pinhole, which, however, also restores the corresponding spatial resolution, equivalent to the far-field interferometer case. However, it has been found that, even beyond the geometrical spatial resolution limit, a modulation pattern is restored with maxima at regular distances, while the pattern is oscillating between a negative and positive modulation pattern with respect to the absorption grating [130]. Typically, the single absorption grating approach enables probing correlation lengths between a few tens and some hundred nanometers and thus well complements the range of the Talbot–Lau interferometers.

A particular strength of the single grating method, beyond the simplicity, is the potential to use 2D absorption patterns and thus to tackle anisotropic small-angle scattering and hence anisotropic structures [130, 131]. It is superior to an initial reported approach with a source and diffraction grating [132] in the way that it is much simpler to implement, while providing the same spatial and reciprocal space resolutions.

The single grating method nicely demonstrates that no interferometry is required to measure the corresponding signals and effects, but only transversal beam modulation is of interest for this kind of multi-modal imaging.

19.7.4 Spin-echo modulation

Consequently, spin-echo modulation constitutes yet another alternative. As proposed by Gaehler [133] a spatial beam modulation can be induced by spin echo, when utilizing inclined magnetic field faces in a polarized neutron beam such that a first field splits the spin-up and spin-down states transversally and a second field refocuses them on the detector. This way a modulation of the resulting spin states at the focal plane is achieved when $L_1B_1 = L_2B_2$, where L_1 and L_2 are the distances of the two magnetic field regions B_1 and B_2 from the detector. A polarization analyser anywhere between the second field region and the detector implies a transversal intensity modulation at the detector. A first basic experimental demonstration of such modulation [134] was followed by demonstrations of the capability to utilize such intensity modulation for SANS measurements, i.e. either monochromatic [39] or ToF spin-echo modulated small-angle scattering (SEMSANS) [41]. As these measurements are fully equivalent to dark-field contrast neutron imaging, the sample can be placed downstream of the polarization analyser close to the detector and the method is sufficiently robust to beam divergences typical for neutron imaging, it could be demonstrated shorty after, that quantitative spin-echo modulated dark-field imaging (SEM-DFI) is possible [135].

Spin-echo beam modulation is somewhat more sophisticated and technically challenging than gratings. Components developed for spin-echo small-angle scattering (SESANS) [136] can be utilized for constant fields B_1 and B_2 in the precession regions inclined by an angle of θ_0 against the incoming beam. Regarding these parameters and the constant $c = 4.632 \times 10^{14}\ T^{-1}\ m^{-2}$ the modulation period p will scale like

$$p = \pi \tan \theta_0 / (c\lambda(B_2 - B_1)) \tag{19.6}$$

over the wavelength λ and, thus, for a variation of wavelength, like especially in ToF applications [41], the probed correlation length

$$\xi = \lambda^2 L_S / P = c\lambda^2 L_S / (\pi \tan \theta_0) \tag{19.7}$$

scales with wavelength square. As the visibility is not affected by the wavelength this enables covering a broad range of correlation lengths through varying the wavelength. A limitation is indeed that the modulation period has to be resolvable by the imaging detector, which somewhat limits the potential range, even if the magnetic fields can be tuned sufficiently. The latter becomes possible with advanced SESANS technology [137]. An analyser grating can be used for smaller periods in monochromatic measurements [39] or when keeping the period constant through varying the fields B_1 and B_2 accordingly for changing wavelengths. A corresponding development has been demonstrated even for a pulsed beam ToF approach [138]. Overall, the method appears particularly well suited for ToF imaging at pulsed sources.

The instrumentation consists of polarization and polarization analyses, spin flippers, depending on the specific set-up, typically to couple into the precession field of the guide field and to flip in front of the second shaped field region, which often is aligned parallel to the first. An additional controllable precession field coil is typically helpful to set good echo conditions. For the shaped precession field regions a number of different solutions is available such as current sheets [41], triangular coils, bi-refringent Wollaston prisms [139], even superconducting Wollaston prisms [137] or radio frequency flippers in parallelogram shaped electromagnetic field regions [140]. Some methods are better suited for ToF and some for monochromatic measurements. Magnetic fields and flippers need to be controllable remotely. ToF measurements might require synchronized field controls and adiabatic flippers, V-coils etc. The technical requirements are significant, especially when considering a set-up as an add-on to a general purpose neutron imaging instrument.

19.8 Complementary x-ray installation

Another much used method requiring dedicated instrumentation is bi-modal neutron/x-ray imaging. The first approach combining neutron and x-ray imaging on the same sample and set-up in a single instrument consisted in an x-ray source installed meters upstream the sample position and could be moved into the neutron beam path for alternating use of neutrons and x-rays in the same imaging geometry. Such installation was pioneered at the NEUTRA instrument at PSI [141]. This solution profits from a remote controlled scintillator exchanger, which allows switching between an x-ray and neutron scintillator for best performance for both modalities, if not, a scintillator is used that works equally for both forms of radiation. The downside of this solution was that neutrons and x-rays could not be used simultaneously, e.g. on samples that evolved over time, like in the uptake of water in a porous medium like soil. This can be solved by an *in situ* set-up of an x-ray imaging device with a beam crossing the neutron beam and was pioneered at the cold neutron imaging instrument ICON at PSI [142, 143]. There are different ways to install such a crossed beam x-ray imaging system, which might be perpendicular or at an inclined angle. Care has to be taken, that the different types of radiation do not create additional background in the detector of the other. So far, the in-line version available at NEUTRA operates with a source of about 320 keV, while the microfocus sources of the *in situ* versions typically work rather at 120 keV, although larger values are currently also considered. Meanwhile the *in situ* bi-modal neutron/x-ray instrumentation can be found at several leading neutron imaging instruments [144, 145]

19.9 Diffraction detectors

Diffractive imaging is a very specific case as it is not necessarily performed at a dedicated neutron imaging instrument, because it requires far-field detectors detecting rather the diffracted than the transmitted intensity. However, there are also mixed cases where diffraction and transmission data is used in a correlated manner. In particular, the advent of ToF neutron imaging has spurred interest in combining transmission imaging with diffraction like that realized in the IMAT [27] instrument at RAL in the UK and considered for ODIN [28] at the ESS in Sweden. However, pure diffraction imaging was first explored at continuous sources, both with monochromatic neutrons and different spectral widths up to utilizing the full white beam spectrum.

The likely first approach to diffractive neutron imaging was neutron topography in analogy to the earlier established x-ray method. A monochromatic beam is generally used and the diffracted beam is imaged for a specific Bragg condition, i.e. at a specific diffraction angle, in order to observe crystal lattice distortions or, in particular with a polarized monochromatic beam, magnetic domains. Pioneering works were done by Japanese scientists in the 1970s [146, 147] while the historical development of this technique is thoroughly outlined in reference [148]. However, the method was not further exploited in modern neutron imaging with digital detectors. An individual white beam approach utilizing crossed collimators at a specific diffraction direction to avoid cross talk from multiple diffraction signals, enabled by the wide incident spectrum, was reported in 2005 [149]. While the work did not generate resonance in the community at that time, the approach of collimating the beam downstream of the interaction with the sample either in-line with the incident beam [150, 151] or inclined to a specific scatter direction [152] has generated interest again recently. The availability of compact microchannel plate collimators [150] might play a role in this development.

A novel development was 3D neutron diffraction, where, in contrast to topography, not one specific diffraction direction is chosen, but a large, uncollimated fraction of solid angle is probed in order to resolve the grain morphology and orientation distribution of polycrystalline samples. This approach typically utilizes a continuous broad incident neutron spectrum and large-area neutron imaging detectors positioned around the sample recording diffraction spots at multiple sample projections, i.e. in a tomographic manner [153]. Typical FoVs range around a few tens of centimeters at distances of a few tens of centimeters from the sample. A relatively short sample-to-detector distance enables one to also still ensure sufficient spatial resolution for grain morphology reconstructions with resolutions of a few 100 μm [35, 153, 154]. Some of these measurements were initially performed at dedicated Laue diffractometer set-ups. These can, however, also be installed as add-ons on other white beam instruments such as neutron imaging beamlines [153]. Latest implementation, in particular of the demanding forward modelling data analyses, enable indexing several hundred grains [155]. In ToF, large-grained crystals can also be indexed and reconstructed utilizing mainly a transmission detector, i.e. without additional diffraction detector installation [34]. However, this comes at the cost of significantly longer exposures. Most likely, a combination with far-field diffraction detectors can aid indexing while the near-field imaging detector can aid higher resolution morphology reconstruction.

This is one motivation for the installation of diffraction detectors on imaging beamlines, though the development is not yet conclusive on the best setting for this and whether one setting of distance, angular coverage and resolution can satisfy the requirements of all potential applications. The use of imaging detectors, however, also here enables ample flexibility.

Yet another application of detectors in diffraction geometry is correlated diffraction and imaging measurements, where diffraction is more efficiently sensitive to crystalline phase evolution, while imaging provides the corresponding high spatial resolution [156].

Thus, current neutron imaging instrument concepts, such as ODIN at the ESS [28] include the potential of adding on diffraction detectors.

19.10 Infrastructure

A state-of-the-art neutron imaging instrument has substantial requirements to infrastructure and media supply when considering the needs of different methods and set-ups as well as the wide range of applications and potential sample environments. Also, safety installations have to be considered carefully.

19.10.1 Instrument

The basic instrument requires all standard supplies for operating its equipment and remotely controlling it, including electricity, motion control electronics, which should generally be ready to control additional axis or other devices even beyond the add-ons of gratings, for polarized imaging, like for user supplied and ad-hoc devices and environments. Connections for computer driven and controlled devices and data transfer from mobile and user supplied devices and metadata have to be considered. Additionally, instrument components might require vacuum, He or Ar atmosphere, water cooling, chillers etc. Standard supplies in a state-of-the-art instrument include:

- cooling water;
- compressed air;
- nitrogen;
- argon.

which might be available from a grid in a typical neutron instrumental hall.

Other gasses like H_2, CO_2, O_2 etc are typically supplied from gas flask cabinets outside the experimental area and require a corresponding infrastructure and piping. Handling gases and liquids also require corresponding exhaust and return piping. The position of supplies in the instrumental area has to be chosen carefully with respect to different measurement positions.

The use of add-ons and dedicated sample environments, but also activation or invaluable sample material implies the need for corresponding storage and transport conditions. Heavy equipment requires crane access to the experimental area and/or in the experimental area.

Remote instrument control requires an operator and user space. Camera surveillance of the experiment positions and key equipment is recommended where possible in such areas with high radiation rates. Laser systems for alignment are a convenient extra for every complex experimental set-up.

19.10.2 Sample environment

Sample environments for neutron imaging are numerous and often unique pieces of equipment. This is due to the wide range of possible science cases and applications, where sample environment is often dedicated for one specific (kind of) measurement or application. It is thus impossible to provide a comprehensive list of sample environments, but particular consideration should be made on cases such as for electro-chemistry involving fuel cell and battery test benches including the required hydrogen and electric lines as well as sensors related to the systems. In engineering load rigs, as such enabling tomography have to be considered as well as connected heating systems, digital image correlation systems and strain gauges etc. Studies involving water penetration and uptake require dedicated systems as well as potentially slip-ring systems for sample rotation. Often, specific humidity and temperature require a climate chamber and in general furnaces, cooling or even cryo-equipment as well as different magnetic environments like aligned and regulated fields or yokes and their specific remote controlled supplies are of particular use.

19.10.3 Safety

Handling of gases and specifically such involving hazards requires the corresponding sensors and exhaust systems, as well as safe handling and storage capabilities.

Remotely controlled moving parts in many facilities are subject to laser aided safety installations and camera surveillance. Also, like any other instruments in a neutron lab, access to the instrumental area has to be controlled and safe, involving infrastructure and specific procedures and shutter interlocks, search buttons etc for safe operation. This includes also procedures and detectors for sample handling after exposure, as samples might get activated and require specific cool-down times and handling.

Some experiments might also require access to a lab, like a chemical or mechanical lab, which are often provided in the vicinity of the instruments and have their specific safety procedures established.

Training is typically required to work in these environments.

References

[1] Faw R E and Kenneth Shultis J 2003 Radiation sources *Encyclopedia of Physical Science and Technology* ed R A Meyers (New York: Academic) 3rd edn pp 613–31

[2] Joyce M J, Agar S, Aspinall M D, Beaumont J S, Colley E, Colling M, Dykes J, Kardasopoulos P and Mitton K 2016 Fast neutron tomography with real-time pulse-shape discrimination in organic scintillation detectors *NIM* A **834** 36–45

[3] Sari A, Carrel F and Lainé F 2018 Characterization and optimization of the photoneutron flux emitted by a 6- or 9-MeV electron accelerator for neutron interrogation measurements *IEEE Trans. Nucl. Sci.* **65** 2018

[4] Williams D L, Brown C M, Tong D, Sulyman A and Gary C K 2020 A fast neutron radiography system using a high yield portable DT neutron source *J. Imaging* **6** 128

[5] Kiyanagi Y 2018 Neutron imaging at compact accelerator-driven neutron sources in Japan *J. Imaging* **4** 55

[6] Lehmann E H 2017 Neutron imaging facilities in a global context *J. Imaging* **3** 52

[7] Carrel F, Makil H, Lainé F, Coulon R, Sari A *et al* 2021 Metrological characterization of intense (α, n) neutron sources by coupling of non-destructive measurements *Nucl. Instrum. Methods* A **987** 164818

[8] Mildner D F R and Gubarev M V 2011 Wolter optics for neutron focusing *Nucl. Instrum. Methods* A **634** S7–11

[9] Khaykovich B, Gubarev M, Bagdasarova Y, Ramsey B and Moncton D 2011 From x-ray telescopes to neutron scattering: using axisymmetric mirrors to focus a neutron beam *Nucl. Instrum. Methods Phys. Res.* A **631** 98–104

[10] Liu D, Hussey D, Gubarev M, Ramsey B, Jacobson D, Arif M, Moncton D and Khaykovich B 2013 Demonstration of achromatic cold-neutron microscope utilizing axisymmetric focusing mirrors *Appl. Phys. Lett.* **102** 183508

[11] Lehmann E H, Vontobel P and Wiezel L 2001 Properties of the radiography facility NEUTRA at SINQ and its potential for use as European reference facility *Nondestruct. Test. Eval.* **16** 191–202

[12] Kaestner A P, Hartmann S, Kühne G, Frei G, Grünzweig C, Josic L, Schmid F and Lehmann E H 2011 The ICON beamline—a facility for cold neutron imaging at SINQ *Nucl. Instrum. Methods* A **659** 387–93

[13] Calzada E, Gruenauer F, Mühlbauer M, Schillinger B and Schulz M 2009 New design for the ANTARES-II facility for neutron imaging at FRM II *Nucl. Instrum. Methods* A **605** 50–3

[14] Arif M, Hussey D S, Baltic E M and Jacobson D L 2015 Neutron imaging facility development and research trend at NIST *Phys. Proc.* **69** 210–7

[15] Garbe U, Randall T, Hughes C, Davidson G, Pangelis S and Kennedy S J 2015 A new neutron radiography/tomography/imaging station DINGO at OPAL *Phys. Proc.* **69** 27–32

[16] Morgano M, Peetermans S, Lehmann E H, Panzner T and Filges U 2014 Neutron imaging options at the BOA beamline at Paul Scherrer Institut *Nucl. Instrum. Methods* A **754** 46–56

[17] Tengattini A, Lenoir N, Andò E, Giroud B, Atkins D, Beaucour J and Viggiani G 2020 NeXT-Grenoble, the neutron and x-ray tomograph in Grenoble *Nucl. Instrum. Methods* A **968** 163939

[18] Hussey D S, Brocker C, Cook J C, Jacobson D L, Gentile T R, Chen W C, Baltic E, Baxter D V, Doskow J and Arif M 2015 A new cold neutron imaging instrument at NIST *Phys. Proc.* **69** 48–54

[19] Santodonato L, Bilheux H, Bailey B, Bilheux J, Nguyen P, Tremsin A, Selby D and Walker L 2015 The CG-1D neutron imaging beamline at the Oak Ridge National Laboratory high flux isotope reactor *Phys. Proc.* **69** 104–8

[20] Bücherl T *et al* 2011 NECTAR—a fission neutron radiography and tomography facility *Nucl. Instrum. Methods* A **651** 86–9

[21] Nelson R O, Vogel S C, Hunter J F, Watkins E B, Losko A S, Tremsin A S, Borges N P *et al* 2018 Neutron imaging at LANSCE—from cold to ultrafast *J. Imaging* **4** 45

[22] Mezei F 2007 New perspectives from new generations of neutron sources *C. R. Phys.* **8** 909

[23] Strobl M 2009 Future prospects of imaging at spallation neutron sources *Nucl. Instrum. Methods* A **604** 646–52

[24] Woracek R, Santisteban J, Fedrigo A and Strobl M 2018 Diffraction in neutron imaging—a review *Nucl. Instrum. Methods* A **878** 141–58

[25] Santisteban J R, Edwards L, Steuwer A and Withers P J 2001 Time-of-flight neutron transmission diffraction *J. Appl. Crystallogr.* **34** 289

[26] Shinohara T *et al* 2020 The energy-resolved neutron imaging system, RADEN *Rev. Sci. Instrum.* **91** 043302

[27] Kockelmann W, Zhang S Y, Kelleher J F, Nightingale J B, Burca G and James J A 2013 IMAT—a new imaging and diffraction instrument at ISIS *Phys. Proc.* **43** 100–10

[28] Strobl M 2015 The scope of the imaging instrument project ODIN at ESS *Phys. Proc.* **69** 18–26

[29] Hassina Bilheux K H and Scott Keener L D 2015 Overview of the conceptual design of the future VENUS neutron imaging beam line at the spallation neutron source *Phys. Proc.* **69** 55–9

[30] Wang S, Tan Z, Xiu Q, Zhen Z, Zhen H, Yu C, Yang L and Chen J Design of neutron and x-ray CT system in ERNI of CSNS *Proc. SPIE 12507, Advanced Optical Manufacturing Technologies and Applications 2022; and 2nd Int. Forum of Young Scientists on Advanced Optical Manufacturing (AOMTA and YSAOM 2022), 125072T (9 January 2023)*

[31] De Broglie M 1923 La relation $h\nu = \varepsilon$ dans les phénomènes photoélectriques *Atomes et Electrons* (Paris: Gauthier) pp 80–100

[32] Strobl M, Bulat M and Habicht K 2013 The wavelength frame multiplication chopper system for an ESS test-beamline and corresponding implications for ESS instruments *Nucl. Instrum. Methods* A **705** 74–84

[33] Schmakat P, Seifert M, Schulz M, Tartaglione A, Lerche M, Morgano M, Böni P and Strobl M 2020 Wavelength frame multiplication chopper system for the multi-purpose

neutron-imaging instrument ODIN at the European Spallation Source *Nucl. Instrum. Methods* A **979** 164467
[34] Cereser A *et al* 2017 Time-of-flight three dimensional neutron diffraction in transmission mode for mapping crystal grain structures *Sci. Rep.* **7** 9561
[35] Samothrakitis S, Raventós M, Čapek J, Buhl Larsen C, Grünzweig C, Tovar M, Garcia-Gonzalez M, Kopecek J, Schmidt S and Strobl M 2020 Grain morphology reconstruction of crystalline materials from Laue three-dimensional neutron diffraction tomography *Sci. Rep.* **10** 3724
[36] Busi M, Capek J, Polatidis E, Hovind J, Boilat P, Kockelmann W and Strobl M 2020 Frame overlap Bragg edge imaging *Sci. Rep.* **10** 14867
[37] Strobl M, Treimer W, Ritzoulis C, Wagh A G, Abbas S and Manke I 2007 The new V12 ultra-small-angle neutron scattering and tomography instrument at the Hahn-Meitner Institute *J. Appl. Crystallogr.* **40** s463–5
[38] Strobl M *et al* 2011 Time-of-flight neutron imaging at a continuous source. Part 1: using a scintillator CCD imaging detector *NIMA* **651** 149–55
[39] Strobl M, Bouwman W G, Wieder F, Duiff C, Hilger A, Kardjilov N and Manke I 2012 Using a grating analyser for SEMSANS investigations in the very small angle range *Physica* B **407** 4132–5
[40] Strobl M, Woracek R, Kardjilov N, Hilger A, Penumadu D, Wimpory R, Tremsin A, Wilpert T, Schulz C and Manke I 2012 Time-of-flight neutron imaging for spatially resolved strain investigations based on Bragg edge transmission at a reactor source *Nucl. Instrum. Methods* A **680** 27–34
[41] Strobl M, Tremsin A S, Hilger A, Wieder F, Kardjilov N, Manke I, Bouwman W G and Plomp J 2012 TOF-SEMSANS—time-of-flight spin-echo modulated small-angle neutron scattering *J. Appl. Phys.* **112** 014503
[42] Strobl M, Pappas C, Hilger A, Wellert S, Kardjilov N, Seidel S-O and Manke I 2011 Polarized neutron imaging—a spin-echo approach *Physica* B **406** 2415–8
[43] Kis Z, Szentmiklósi L, Belgya T, Balaskó M, Horváth L Z and Maróti B 2015 Neutron-based imaging and element-mapping at the Budapest Neutron Centre *Phys. Proc.* **69** 40–7
[44] Matsubayashi M, Kobayashi H, Hibiki T and Mishima K 2000 Design and characteristics of the JRR-3M thermal neutron radiography facility and its imaging system *Nucl. Tech.* **132** 309–24
[45] Zarebanadkouki M, Carminati A, Kaestner A, Mannes D, Morgano M, Peetermans S, Lehmann E and Trtik P 2015 On-the-fly neutron tomography of water transport into lupine roots *Phys. Proc.* **69** 292–8
[46] Dierick M, Vlassenbroeck J, Masschaele B, Cnudde V, Van Hoorebeke L and Hillenbach A 2005 High-speed neutron tomography of dynamic processes *Nucl. Instrum. Methods* A **542** 296–301
[47] Tötzke C, Kardjilov N, Lenoir N, Manke I, Oswald S E and Tengattini A 2019 What comes NeXT?—high-speed neutron tomography at ILL *Opt. Express* **27** 28640–8
[48] Boillat P, Carminati C, Schmid F, Grünzweig C, Hovind J, Kaestner A *et al* 2018 Chasing quantitative biases in neutron imaging with scintillator-camera detectors: a practical method with black body grids *Opt. Express* **26** 15769–84
[49] Carminati C, Boillat P, Schmid F, Vontobel P, Hovind J, Morgano M *et al* 2019 Implementation and assessment of the black body bias correction in quantitative neutron imaging *PLoS One* **14** e0210300

[50] Kis Z, Szentmiklósi L and Belgya T 2015 NIPS–NORMA station—a combined facility for neutron-based nondestructive element analysis and imaging at the Budapest Neutron Centre *Nucl. Instrum. Methods* A **779** 116–23

[51] Strobl M 2019 Neutron imaging *Advances in Neutron Optics* ed M L Calvo and R F Alvarez-Estrada (Boca Raton, FL: CRC Press) p 42

[52] Kozlenko D P, Kichanov S E, Lukin E V, Rutkauskas A V, Bokuchava G D, Savenko B N, Pakhnevich A V and Yu. Rozanov A 2015 Neutron radiography facility at IBR-2 high flux pulsed reactor: first results *Phys. Proc.* **69** 87–91

[53] Kallmann H 1948 Neutron radiography *Research* **1** 254–60

[54] Berger H 1966 Characteristics of a thermal neutron television imaging system *Mater. Eval.* **24** 475–81

[55] Vontobel P, Tamaki M, Mori N, Ashida T, Zanini L, Lehmann E H and Jaggi M 2006 Post-irradiation analysis of SINQ target rods by thermal neutron radiography *J. Nucl. Mater.* **356** 162–7

[56] Lehmann E and Vontobel P 2004 The use of amorphous silicon flat panels as detector in neutron imaging *Appl. Radiat. Isot.* **61** 567–71

[57] Lehmann E H and Boillat P 2023 Advances in scintillator screen technology for neutron imaging *Nucl. Instrum. Methods* A **1053** 168324

[58] Schillinger B, Chuirazzi W, Craft A *et al* 2022 Performance of borated scintillator screens for high-resolution neutron imaging *J. Radioanal. Nucl. Chem.* **331** 5287–95

[59] Kardjilov N, Dawson M, Hilger A, Manke I, Strobl M, Penumadu D, Kim F H, Garcia-Moreno F and Banhart J 2011 A highly adaptive detector system for high-resolution neutron imaging *Nucl. Instrum. Methods* A **651** 95–9

[60] Trtik P and Lehmann E H 2015 Isotopically-enriched gadolinium-157 oxysulfide scintillator screens for the high-resolution neutron imaging *Nucl. Instrum.* A **788** 67–70

[61] Chuirazzi W, Craft A, Schillinger B, Cool S and Tengattini A 2020 Boron-based neutron scintillator screens for neutron imaging *J. Imaging* **6** 124

[62] Morad V, McCall K M, Sakhatskyi K, Lehmann E, Walfort B, Losko A S, Trtik P, Strobl M, Yakunin S and Kovalenko M V 2021 Luminescent lead halide ionic liquids for high-spatial-resolution fast neutron imaging *ACS Photonics* **8** 3357–64

[63] Zboray R, Adams R, Morgano M and Kis Z 2019 Qualification and development of fast neutron imaging scintillator screens *Nucl. Instrum. Methods Phys. Res.* A **930** 142–50

[64] Boillat P *et al* Who made the noise? A systematic approach for the assessment of neutron imaging scintillators *Opt. Express* (in press)

[65] Poikela T *et al* 2014 Timepix3: a 65K channel hybrid pixel readout chip with simultaneous ToA/ToT and sparse readout *JINST* **9** C05013

[66] Losko A S, Han Y, Schillinger B, Tartaglione A, Morgano M, Strobl M, Long J, Tremsin A S and Schulz M 2021 New perspectives for neutron imaging through advanced event-mode data acquisition *Sci. Rep.* **11** 21360

[67] Trtik P and Lehmann E H 2016 Progress in high-resolution neutron imaging at the Paul Scherrer Institut—the neutron microscope project *J. Phys.: Conf. Ser.* **746** 012004

[68] Trtik P, Meyer M, Wehmann T, Tengattini A, Atkins D, Lehmann E H and Strobl M 2020 PSI 'Neutron Microscope' at ILL-D50 beamline—first results *Mater. Res. Proc.* **15** 23–8

[69] Williams S, Hilger A, Kardjilov N, Manke I, Strobl M, Douissard P, Martin T, Riesemeier H and Banhart J 2012 Detection system for microimaging with neutrons *J. Instrum.* **7** P02014

[70] Morgano M, Trtik P, Meyer M, Lehmann E H, Hovind J and Strobl M 2018 Unlocking high spatial resolution in neutron imaging through an add-on fibre optics taper *Opt. Express* **26** 1809
[71] Grünzweig C, Frei G, Lehmann E, Kühne G and David C 2007 Highly absorbing gadolinium test device to characterize the performance of neutron imaging detector systems *Rev. Sci. Instrum.* **78** 053708
[72] Tremsin A S and Vallerga J V 2020 Unique capabilities and applications of Microchannel Plate (MCP) detectors with Medipix/Timepix readout *Radiat. Meas.* **130** 106228
[73] Tremsin A S, Vallerga J V and Raffanti R 2018 Optimization of spatial resolution and detection efficiency for photon/electron/neutron/ion counting detectors with microchannel plates and Quad Timepix readout *J. Instrum. JINST* **13** C11005
[74] Woracek R, Krzyzagorski M, Markötter H, Kadletz P M, Kardjilov N, Manke I and Hilger A 2019 Spatially resolved time-of-flight neutron imaging using a scintillator CMOS-camera detector with kHz time resolution *Opt Express 2* **27** 26218–28
[75] Friedrich H, Wagner V and Wille P 1989 A high-performance neutron velocity selector *Physica* B **156 and 157** 547–9
[76] Peetermans S, Grazzic F, Salveminic F and Lehmann E 2013 Spectral characterization of a velocity selector type monochromator for energy-selective neutron imaging *Phys. Proc.* **43** 121–7
[77] Strobl M, Hilger A, Kardjilov N, Manke I, Treimer W and Riemke M 2008 Time-of-flight spectrum measurements at the neutron imaging facility CONRAD at the Hahn-Meitner-Institute *Neutron Radiography (8): Proceedings of the 8th world Conference (Gaithersburg, MD) (October 16–19, 2006)* ed M Arif (WCNR-8. Lancaster: Destech Publishing) pp 11–7
[78] Treimer W, Strobl M, Kardjilov N, Hilger A and Manke I 2006 A wavelength tunable device for neutron radiography and tomography *Appl. Phys. Lett.* **89** 203504
[79] Schulz M, Böni P, Calzada E, Mühlbauer M and Schilinger B 2009 Energy-dependent neutron imaging with a double crystal monochromator at the antares facility at FRM II *Nucl. Instrum. Methods Phys. Res.* A **605** 33–5
[80] Kaestner A P, Schluepbach D, Malamud F, Busi M and Strobl M 2023 Design and validation of a neutron double crystal monochromator for ICON *J. Phys. Conf. Ser.* **2605** 012011
[81] Treimer W, Ebrahimi O, Karakas N and Seidel S O 2011 PONTO—an instrument for imaging with polarized neutrons *Nucl. Instrum. Methods* A **651** 53–6
[82] Tremsin A S, Sokolova A V, Salvemini F, Luzin V, Paradowska A, Muransky O, Kirkwood H J, Abbey B, Wensrich C M and Kisi E H 2019 Energy-resolved neutron imaging options at a small angle neutron scattering instrument at the Australian Center for Neutron Scattering *Rev. Sci. Instrum.* **90** 035114
[83] Woracek R, Hoffmann T, Bullat M, Sales M, Habicht K, Andersen K and Strobl M 2016 The testbeamline of the European Spallation Source—instrumentation development and wavelength frame multiplication *Nucl. Instrum. Methods* A **839** 102–16
[84] Schober H *et al* 2008 Tailored instrumentation for long-pulse neutron spallation sources *Nucl. Instrum. Methods* A **589** 34–46
[85] van Well A A 1992 Double-disk chopper for neutron time-of-flight experiments *Physica* B **180–181** 959–61
[86] Arai M 2008 J-PARC and the prospective neutron sciences *Pramana J. Phys.* **71**
[87] Ramadhan R S, Kockelmann W, Minniti T, Chen B, Parfitt D, Fitzpatrick M E and Tremsin A S 2019 Characterization and application of Bragg-edge transmission imaging for

strain measurement and crystallographic analysis on the IMAT beamline *J. Appl. Crystallogr.* **52** 351–68

[88] Malamud F, Santisteban J R, Vicente Alvarez M A, Busi M, Polatidis E and Strobl M 2023 An optimized single-crystal to polycrystal model of the neutron transmission of textured polycrystalline materials *J Appl. Crystallogr.* **56** 143–54

[89] Sato H *et al* 2013 Upgrade of Bragg edge analysis techniques of the RITS code for crystalline structural information imaging *Phys. Proc.* **43** 186–95

[90] Makowska M, Theil-Kuhn L, Frandsen H L, Lauridsen E M, De Angelis S, Cleemann L N, Morgano M, Trtik P and Strobl M 2017 Coupling between creep and redox behavior in nickel—yttria stabilized zirconia observed *in situ* by monochromatic neutron imaging *J. Power Sources* **340** 167–75

[91] Siegwart M, Woracek R, Márquez Damián J I, Tremsin A, Manzi V, Strobl M, Schmidt T J and Boillat P 2019 Distinction between super-cooled water and ice with high duty cycle time-of-flight neutron imaging *Rev. Sci. Instrum.* **90** 103705

[92] Stuhr U, Spitzer H, Egger J, Hofer A, Rasmussen P, Graf D, Bollhalder A, Schild M, Bauer G and Wagner W 2005 Time-of-flight diffraction with multiple frame overlap Part II: the strain scanner POLDI at PSI *Nucl. Instrum. Methods* A **545** 330–8

[93] Zou Y B, Schillinger B, Wang S, Zhang X S, Guo Z Y and Lu Y R 2011 Coded source neutron imaging with a MURA mask *Nucl. Instrum. Methods* A **651** 192–6

[94] Tremsin A S, Bilheux H Z, Bilheux J C, Shinohara T, Oikawa K and Gao Y 2021 Calibration and optimization of Bragg edge analysis in energy-resolved neutron imaging experiments *Nucl. Instrum. Methods* A **1009** 165493

[95] Kropff F, Granada J R and Mayer R E 1982 *Nucl. Instrum. Methods Phys. Res.* **198** 515–21

[96] Dawson M, Kardjilov N, Manke I, Hilger A, Jullien D, Bordenave F, Strobl M, Jericha E, Badurek G and Banhart J 2011 Polarized neutron imaging using helium-3 cells and a polychromatic beam *Nucl. Instrum. Methods* A **651** 140–4

[97] Schulz M, Böni P, Franz C, Neubauer A, Calzada E, Mühlbauer M, Schillinger B, Pfleiderer C, Hilger A and Kardjilov N 2010 *J. Phys.: Conf. Ser.* **251** 012068

[98] Dhiman I *et al* 2017 Setup for polarized neutron imaging using *in situ* 3He cells at the Oak Ridge National Laboratory High Flux Isotope Reactor CG-1D beamline *Rev. Sci. Instrum.* **88** 095103

[99] Kardjilov N, Manke I, Strobl M, Hilger A, Treimer W, Meissner M, Krist T and Banhart J 2008 Three-dimensional imaging of magnetic fields with polarized neutrons *Nat. Phys.* **4** 399–403

[100] Busi M, Polatidis E, Sofras C, Boillat P, Ruffo A, Leinenbach C and Strobl M 2022 Polarization contrast neutron imaging of magnetic crystallographic phases *Mater. Today Adv.* **16** 100302

[101] Strobl M, Kardjilov N, Hilger A, Jericha E, Badurek G and Manke I 2009 Imaging with polarized neutrons *Physica* B **404** 2611–4

[102] Schulz M, Neubauer A, Mühlbauer M, Calzada E, Schillinger B, Pfleiderer C and Böni P 2010 Polarized neutron radiography with a periscope *J. Phys.: Conf. Ser.* **210** 112009

[103] Hiroi K, Shinohara T, Hayashida H, Parker J D, Oikawa K, Harada M, Su Y and Kai T 2017 Magnetic field imaging of a model electric motor using polarized pulsed neutrons at JPARC/MLF *J. Phys.: Conf. Ser.* **862** 012008

[104] Krist T, Kennedy S J, Hicks T J and Mezei F 1998 New compact neutron polarizer *Physica* B **241–3** 82–8

[105] Strobl M, Heimonen H, Schmidt S, Sales M, Kardjilov N, Hilger A, Manke I, Shinohara T and Valsecchi J 2019 Polarization measurements in neutron imaging *J. Phys. D: Appl. Phys.* **52** 123001

[106] Weinfurter H and Badurek G 1989 *Nucl. Instrum. Methods* A **275** 233

[107] Bazhenov A, Lobashev V M, Pirozhkov A and Slusar V N 1993 Am adiabatic resonance spin-flipper for thermal and cold neutrons *Nucl. Instrum. Methods* A **332** 534–6

[108] Sales M, Shinohara T, Korning Sørensen M, Knudsen E B, Tremsin A S, Strobl M and Schmidt S 2019 Three dimensional polarimetric neutron tomography—beyond the phase wrapping limit *J. Phys. D: Appl. Phys.* **52** 205001

[109] Shinohara T *et al* 2017 Polarization analysis for magnetic field imaging at RADEN in J-PARC/MLF *J. Phys.: Conf. Ser.* **862** 012025

[110] Pfeiffer F, Grunzweig C, Bunk O, Frei G, Lehmann E and David C 2006 Neutron phase imaging and tomography *Phys. Rev. Lett.* **96** 215505

[111] Strobl M, Grünzweig C, Hilger A, Manke I, Kardjilov N, David C and Pfeiffer F 2008 Neutron dark-field tomography *Phys. Rev. Lett.* **101** 123902

[112] Takano H, Wu Y, Samoto T, Taketani A, Takanashi T, Iwamoto C, Otake Y and Momose A 2022 Demonstration of neutron phase imaging based on talbot–lau interferometer at compact neutron source RANS *Quantum Beam Sci.* **6** 22

[113] Harti R P, Strobl M, Schäfer R, Kardjilov N, Tremsin A S and Grünzweig C 2018 Dynamic volume magnetic domain wall imaging in grain oriented electrical steel at power frequencies with accumulative high-frame rate neutron dark-field imaging *Sci. Rep.* **8** 15754

[114] Grünzweig C, Pfeiffer F, Bunk O, Donath T, Kühne G, Frei G, Dierolf M and David C 2008 Design, fabrication, and characterization of diffraction gratings for neutron phase contrast imaging *Rev. Sci. Instrum.* **79** 053703

[115] Kim Y, Kim D, Lee S, Kim J, Hussey D S and Wook Lee S 2020 Neutron grating interferometer with an analyzer grating based on a light blocker *Opt Express* **28** 23284–93

[116] Samoto T, Takano H and Momose A 2019 Gadolinium oblique evaporation approach to make large scale neutron absorption gratings for phase imaging *Jpn. J. Appl. Phys.* **58** SDDF12

[117] Neuwirth T, Backs A, Gustschin A, Vogt S, Pfeiffer F, Böni P and Schulz M 2020 A high visibility neutron grating interferometer to investigate stress-induced magnetic domain degradation in electrical steel *Sci. Rep.* **10** 1764

[118] Valsecchi J *et al* 2019 Visualization and quantification of inhomogeneous and anisotropic magnetic fields by polarized neutron grating interferometry *Nat. Commun.* **10** 3788

[119] Manke I *et al* 2010 Three-dimensional imaging of magnetic domains *Nat. Commun.* **1** 125

[120] Strobl M 2014 General solution for quantitative dark-field contrast imaging with grating interferometers *Sci. Rep.* **4** 7243

[121] Strobl M, Betz B, Harti R P, Hilger A, Kardjilov N, Manke I and Gruenzweig C 2016 Wavelength dispersive dark-field contrast: micrometer structure resolution in neutron imaging with gratings *J. Appl. Crystallogr* **49**

[122] Harti R P, Strobl M, Betz B, Jefimovs K, Kagias M and Gruenzweig C 2017 Sub-pixel correlation length neutron imaging: spatially resolved scattering information of micro-structures on a macroscopic scale *Sci. Rep.* **7** 44588

[123] Harti R P, Valsecchi J, Trtik P, Mannes D, Carminati C, Strobl M, Plomp J, Duif C P and Grünzweig C 2018 Visualizing the heterogeneous breakdown of a fractal microstructure during compaction by neutron dark-field imaging *Sci. Rep.* **8** 17845

[124] Kim Y, Valsecchi J, Kim J, Wook Lee S and Strobl M 2019 Symmetric Talbot-Lau neutron grating interferometry and incoherent scattering correction for quantitative dark-field imaging *Sci. Rep.* **9** 18973

[125] Strobl M, Harti R P, Grünzweig C, Woracek R and Plomp J 2018 Small angle scattering in neutron imaging—a review *J. Imaging* **3** 64

[126] Bacak M *et al* 2020 Neutron dark-field imaging applied to porosity and deformation-induced phase transitions in additively manufactured steels *Mater. Des.* **195** 109009

[127] Valsecchi J, Makowska M G, Gruenzweig C, Piegsa F M, Kim Y, Lee S W, Thijs M A, Plomp J and Strobl Decomposing M 2021 magnetic dark-field contrast in spin analyzed Talbot-Lau interferometry—a Stern–Gerlach experiment without spatial beam splitting *Phys. Rev. Lett.* **126** 070401

[128] Pushin D A *et al* 2017 Far-field interference of a neutron white beam and the applications to noninvasive phase-contrast imaging *Phys. Rev.* A **95** 043637

[129] Strobl M, Valsecchi J, Harti R P, Trtik P, Kaestner A, Gruenzweig C, Polatidis E and Capek J 2019 Achromatic non-interferometric single grating neutron phase imaging *Sci. Rep.* **9** 19649

[130] Busi M, Zdora M-C, Valsecchi J, Bacak M and Strobl M 2022 Cold and thermal neutron single grating dark-field imaging extended to an inverse pattern regime *Appl. Sci.* **12** 2798

[131] Busi M, Shen J, Bacak M, Zdora M-C, Jacopo V and Strobl M 2023 Multi-directional neutron dark-field imaging with single absorption grating *Sci. Rep.* **13** 15274

[132] Valsecchi J, Harti R P, Trtik P, Kaestner A, Strobl M, Gruenzweig C, Wang Z, Jemovs K and Kagias M 2020 Visualization and characterization of oriented microstructures through anisotropic small-angle scattering by 2D neutron dark-field imaging *Commun. Phys.* **3** 42

[133] Gähler R 2006 *A Certain Class of Beam Modulation Techniques and Its Potential Applications* (PNCMI Polarized Neutron School)

[134] Bouwman W G, Duif C P and Gähler R 2009 Spatial modulation of a neutron beam by Larmor precession *Physica* B **404** 2585

[135] Strobl M, Sales M, Plomp J, Bouwman W G, Tremsin A S, Kaestner A, Pappas C and Habicht K 2015 Quantitative neutron dark-field imaging through spin-echo interferometry *Sci. Rep.* **5** 16576

[136] Bouwman W G, Stam W, Krouglov T V, Plomp J, Grigoriev S V, Kraan W H and Rekveldt M T 2004 *Nucl. Instrum. Methods Phys. Res.* A **3529** 16

[137] Li F, Parnell S R, Hamilton W A, Maranville B B, Wang T, Semerad R, Baxter D V, Cremer J T and Pynn R 2014 Superconducting magnetic Wollaston prism for neutron spin encoding *Rev. Sci. Instrum.* **85** 053303

[138] Sales M, Plomp J, Habicht K, Tremsin A S, Bouwman W G and Strobl M 2016 Wavelength-independent constant period spin-echo modulated small angle neutron scattering *Rev. Sci. Instrum.* **87** 063907

[139] Pynn R, Fitzsimmons M R, Lee W T, Stonaha P, Shah V R, Washington A L, Kirby B J, Majkrzak C F and Maranville B B 2009 Birefringent neutron prisms for spin echo scattering angle measurement *Physica* B **404** 2582–4

[140] Parnell S R, Dalgliesh R M, Steinke N J, Plomp J and Well A A V 2018 RF neutron spin flippers in time of flight spin-echo resolved grazing incidence scattering (SERGIS *J. Phys.: Conf. Ser.* **1021** 012040

[141] Lehmann E H, Mannes D, Kaestner A P, Hovind J, Trtik P and Strobl M 2021 The XTRA option at the NEUTRA facility—more than 10 years of bi-modal neutron and x-ray imaging at PSI *Appl. Sci.* **11** 3825 16 pp

[142] https://ndt.net/events/DIR2015/Paper/58_Kaestner.pdf (2015)

[143] Kaestner A P, Hovind J, Boillat P, Muehlebach C, Carminati C, Zarebanadkouki M and Lehmann E H 2017 Bimodal imaging at ICON using neutrons and x-rays *Phys. Proc.* **88** 314–21

[144] Tengattini A, Atkins D, Giroud B, Ando E, Beaucour J and Viggiani G 2017 NeXT-Grenoble, a novel facility for neutron and x-ray tomography in Grenoble *3rd Int. Conf. on Tomography of Materials and Structures (Lund, Sweden) (26–30 June)* ICTMS2017–163

[145] LaManna J M, Hussey D S, Baltic E and Jacobson D L 2017 Neutron and x-ray tomography (NeXT) system for simultaneous, dual modality tomography *Rev. Sci. Instrum.* **88** 113702

[146] Doi K, Minakawa N, Motohashi H and Masaki N 1971 A trial of neutron diffraction topography *J. Appl. Crystallogr.* **4** 529

[147] Ando M and Hosoya S 1972 Q-switch and polarization domains in antiferromagnetic chromium observed with neutron diffraction tomography *Phys. Rev. Lett.* **29** 281

[148] Schlenker M and Baruchel J 1986 Neutron topography: a review *Physica* B **137** 309–19

[149] Ballhausen H, Abele H, Gähler R, Trapp M and Van Overberghe A Imaging with scattered neutrons arXiv:nucl-ex/0610045

[150] Tremsin A S, Kardjilov N, Dawson M, Strobl M, McPhate J B, Vallerga J V, Siegmund O H W and Feller W B 2011 Scatter rejection in quantitative thermal and cold neutron imaging *NIMA* **651** 145–8

[151] Oba Y, Shinohara T, Sato H, Onodera Y, Hiroi K, Su Y and Sugiyama M 2018 Imaging measurement of neutron attenuation by small-angle neutron scattering using soller collimator *J. Phys. Soc. Jpn.* **87** 094004

[152] Reimann T, Bauer A, Pfleiderer C, Böni P, Trtik P, Tremsin A, Schulz M and Mühlbauer S 2018 Neutron diffractive imaging of skyrmion lattice nucleation in MnSi *Phys. Rev.* B **97** 020406(R)

[153] Peetermans S, King A, Ludwig W, Reischig P and Lehmann E H 2014 Cold neutron diffraction contrast tomography of polycrystalline material *Analyst* **139** 5765

[154] Raventós M, Tovar M, Medarde M, Shang T, Strobl M, Samothrakitis S, Pomjakushina E, Grünzweig C and Schmidt S 2019 Laue three dimensional neutron diffraction *Sci. Rep.* **9** 4798

[155] Samothrakitis S, Buhl Larsen C, Capek J, Efthymios Polatidis M, Raventos M, Tovar S, Schmidt M and Strobl 2022 Microstructural characterization through grain orientation mapping with laue three-dimensional neutron diffraction tomography *Mater. Today Adv.* **15** 100258

[156] Lăcătuşu M-E *et al* 2021 A multimodal operando neutron study of the phase evolution in a graphite electrode arXiv:2104.03564 [cond-mat.mtrl-sci]

IOP Publishing

Neutron Imaging
From applied materials science to industry
Markus Strobl and Eberhard Lehmann

Chapter 20

Image processing and software

Anders Kaestner, Matteo Busi and Florencia Malamud

Modern neutron imaging uses digital acquisition systems that allow the measured radiographs to be stored as two-dimensional matrices containing pixels represented using either integers or floating-point numbers. The matrix shows the distribution of the neutrons captured by the primary detector, e.g., scintillator, imaging plate or pixel detector. The pixel values, however, generally do not represent the absolute number of neutrons arriving at the detector during the exposure due to detector efficiency. The registered number is, however, mostly proportional to the applied dose. Images in matrix format allow for processing and use in calculations that provide quantitative information about the measured object. The fact that the images are stored in digital form already at acquisition is a great advantage over analogue techniques. Analogue radiography and computational radiography would require a multi-step procedure, including film development or imaging plate scanning followed by scanning to obtain data that can be analyzed using image processing techniques.

Today, most neutron imaging experiments aim to quantitatively assess specific characteristics of the observed sample or process. The character of the assessment varies with the objectives of the experiments but can coarsely be divided into two categories: (a) measurements based on discrete objects. These measurements involve detecting and counting the number of items, followed by determining the size, shape, and absolute or relative location. The second category of quantification measures (b) gradual changes in the grey levels. These changes are often related to the physical properties of the sample, for example, based on changes in the attenuation coefficient that can tell the difference between different materials or the amount of a specific material. Some acquisition techniques like computed tomography and grating interferometry also require transformation from the radiographs in acquisition space to the images in observation space prior to the quantitative analysis. Quantification of shapes and items or grey level changes are not mutually exclusive. The two categories of quantification are often combined such that the identified

doi:10.1088/978-0-7503-3495-2ch20 © IOP Publishing Ltd 2024

discrete objects are used as masks for further analysis in terms of grey level variations within these regions.

20.1 Typical image processing workflows

The schematic outline in figure 20.1 shows the typical image processing and analysis workflow to process neutron images. The processing of data from all neutron imaging experiment types has the first tasks that focus on preparing the images for analysis in common. These tasks include normalization, artefact, and noise reduction. Some experiments also require transformations from acquired images to present the intended information. Quantitative analysis is the final task, which consists of several specific tasks. The analysis workflow is experiment specific, and sometimes it is even relevant to develop new analysis algorithms tailored to the needs of the analysis. Typical analysis operations and tasks are segmentation, classification, and regression, followed by statistical analysis and presentation of the results.

20.1.1 Normalization

Neutron imaging is based on the acquisition of radiographs that follow the Beer–Lambert law as already defined in chapter 2:

$$I(x, y, \lambda) = I_0(x, y, \lambda)e^{-\int_L \mu(l, \lambda)dl}. \tag{20.1}$$

Here, I_0, is the image of the incident neutron beam, and μ is the material's linear attenuation coefficient. I_0 is also called 'open beam' image, I_{OB}, or 'flat field' image, I_{FF}. The attenuation coefficient varies spatially with the position in the sample, depending on the local material composition. The integral is computed for the positions, l, along the straight line, L, between source and detector. The plain radiograph also includes information about the beam profile and modulations caused by the acquisition system. Therefore, it is more relevant to work with transmission images,

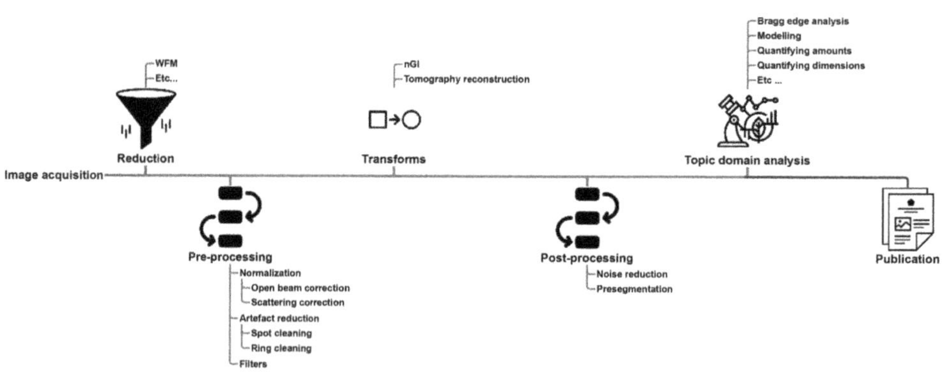

Figure 20.1. A typical image processing workflow for neutron imaging data.

$$T(x, y, \lambda) = \frac{I(x, y, \lambda)}{I_0(x, y, \lambda)}. \tag{20.2}$$

Both I_0 and μ depend on the wavelength of the incident neutrons, λ. Therefore, each image in a wavelength-resolved experiment must be normalized to its corresponding $I_0(x, y, \lambda)$ to obtain the transmission image, T, or optical thickness image, τ, defined as

$$\tau(x, y, \lambda) = -\ln(T(x, y, \lambda)) = \int_{\mathrm{L}} \mu(l, \lambda)dl. \tag{20.3}$$

There are also instrumentation contributions to the image intensity. The source may not always be stable, which results in intensity fluctuations in the images relative to I_0. The detector often has a bias introduced by thermal noise in the electronics. The normalization equation, therefore, takes the following form to correct for these deviations [1, 2]

$$T = \frac{D_0}{D}\frac{(I - I_{\mathrm{DC}})}{(I_0 - I_{\mathrm{DC}})}. \tag{20.4}$$

The pixel positions x and y are dropped here as this equation assumes pixel-wise operations. The bias introduced by the thermal noise in the detector is corrected by subtracting the so-called 'dark current' image, I_{DC}, or a scalar value corresponding to the average grey level of the dark current images. I_{DC} is acquired with neutron and light shutter closed and will thus only register the detector-specific noise and bias. The dose ratio $\frac{D_0}{D}$ is a scalar representing the ratio between the acquisition doses of the images I_0 and I. The dose ratio is essential at sources where the neutron flux varies, particularly for spallation neutron sources. The dose can be measured using a beam monitor like a neutron counting device or the time integral of the proton current for a spallation source. The acquisition dose can also be measured numerically directly in the image using the average grey level in a region never covered by any object. The latter method is sensitive to biases caused by effects like incoherent scattering background and detector lag, i.e., a time-delayed memory effect in the detector like the scintillator afterglow.

The noise contributions of the reference images propagate through equation (20.1). Therefore, a high signal-to-noise ratio (SNR) of both I_0 and I_{DC} is crucial for the normalization quality. The SNR is generally best increased by increasing the total dose of reference image acquisition. A single image with a long exposure time would achieve an improved SNR. In practice, the dose is increased by combining a series of reference images into one image per reference type. This approach is motivated by possible detector saturation and the ability to reject intensity outliers in the image caused by detector faults or gamma photons hitting the detector. The reference image stack is combined using either pixel-wise average, median, or weighted average [3]. The latter option has the advantage that it provides results comparable to the arithmetic mean and is also robust against outliers like the median.

20.1.2 Noise and artefact reduction

Noise is always present to some degree in any experiment, neutron imaging included as it relies on the neutron counting statistics which is Poisson distributed. Neutron images often contain biases and artefacts that have an impact on the analysis of the image contents. This unwanted information must be removed or suppressed to allow reliable image analysis of the experiment data. Images from a neutron experiment often have relatively low SNR compared to x-ray images. The low SNR results from the low neutron flux at most neutron sources and time constraints set by the observed process or even by available experiment time. Noise and artefact reduction are central topics of the processing workflow in neutron imaging. Section 20.4 describes different types of noise and artefacts and how image processing methods can reduce them.

20.1.3 Tomography reconstruction

Many neutron imaging experiments use computed tomography to obtain a volume representation of the sample. Tomography requires a transformation from acquisition space, projections from different observation angles to observation space, volume images describing the spatial distribution of attenuation coefficients in the scanned item. Methods for the tomographic reconstruction are described in more detail in section 20.3.

20.1.4 Experiment specific analysis

The previously described image processing steps provide a foundation for further investigation-specific analysis. The investigation or sample-specific analysis and the choice of analysis methods depend on the studied sample and the character of the relevant features. This part of the analysis provides answers to the objectives of the experiment. The choice of methods and workflow for the sample-specific analysis highly depends on the amount of data available. It may be sufficient and sometimes most efficient for single objects to work with interactive workflows. In contrast, larger datasets require well-planned scripting to solve the analysis task in a repeatable manner. A selection of techniques used to analyze neutron images is provided in section 20.5.

20.2 Computed tomography

Computed tomography has evolved into a routine method for neutron imaging. First tomographies were performed in the 1970s, but the method became more widespread with the development of digital detectors systems in the late 1990s [4]. It is a method to obtain three-dimensional information from a series of radiography projections, with coordinates u and v, acquired from different views of a sample that is rotated about a central axis [5]. The central axis is also referred to as the acquisition axis and is parallel to the v-axis of the projections and the z-axis of the reconstructed volume. Tomography is an indirect imaging method that requires an inversion of the Radon transform to transform the acquired projection data from

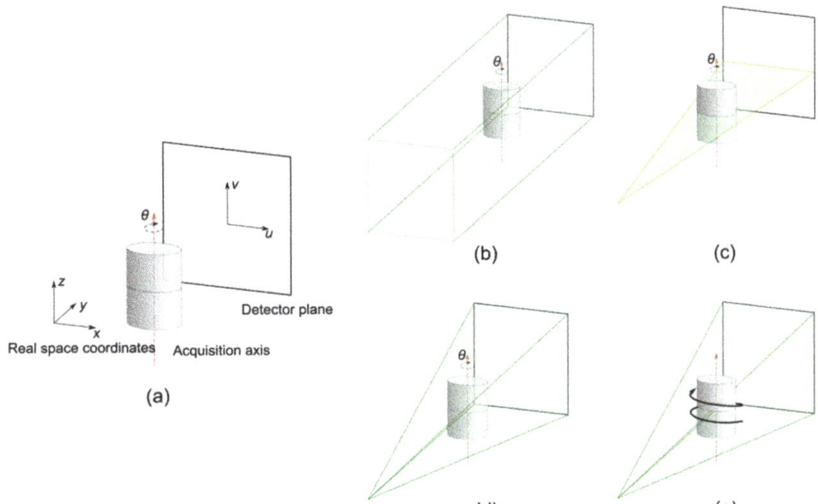

Figure 20.2. Coordinate system (a) and beam geometries for computed tomography; parallel beam (b), fan beam (c), cone beam (d), and cone beam with a helical scan (e).

acquisition space $p(u, v, \Theta)$ to observation space $f(x, y, z)$ in Cartesian coordinates, figure 20.2(a).

The v- and the z-axes are parallel, i.e., they represent the direction parallel to the rotation axis. The acquisition information needed to reconstruct a single slice of volume f is called a sinogram and is represented by the coordinates u and θ. The Radon transform mathematically describes the sinogram for parallel beam geometry

$$p(u, \theta) = \int_{-\infty}^{\infty}\int_{-\infty}^{\infty} f(x, y) \cdot \delta(x\cos\theta + y\sin\theta - u)dxdy \tag{20.5}$$

In this equation, we are interested in reconstructing the slice. The reconstruction is achieved by solving the equation to obtain f(x, y). Here $\delta(\,\cdot\,)$ is Dirac's delta function which is used to select the integration along the rays in the direction of θ at various detector positions u. The quantity of the reconstructed voxels are the local linear attenuation coefficients in the sub-volumes represented by the voxel and is measured in cm^{-1}. The pixel size of the projection is needed for the correct scaling of the attenuation coefficient. Therefore, it is of utmost importance to accurately measure the pixel size when the quantitative analysis is based on the attenuation coefficients.

The so-called filtered back-projection (FBP) algorithm is the most frequently used for tomography reconstruction of neutron projection data. This algorithm is based on the analytical inversion of the forward-projection equation above

$$f(x, y) = \int_{-\infty}^{\infty}\int_{-u_m}^{u_m} h(x\cos\theta + y\sin\theta - u) \cdot p(u, \theta)\, du\, d\theta \tag{20.6}$$

The reconstruction will consider detector positions between $\pm u_m$ and the sample must always stay inside these coordinates to avoid truncation artefacts. These appear as the elongation of structures along the axis perpendicular to the missing wedge direction, creating a directional blurring effect (pointy streaks radiating out in before and after small objects in the direction of the missing wedge). The convolution kernel $h(x)$ has high-pass characteristics and is an integral part of the inversion process. The ramp filter amplifies high frequency components of the spectrum, this is unfortunately also the part of the spectrum which is dominated by noise. Therefore, it is common to add an apodization filter window to the reconstruction filter to suppress noise. Typical filter windows are Ram-Lak, Hamming, Hann which all have low pass characteristics that reduce the impact of noise in the reconstructed image. The advantage of the filtered back-projection algorithm is that it provides the result in short time using only a single iteration to provide the result. The algorithm is however sensitive to severely under-sampled and to incomplete scans.

In neutron imaging, the beam geometry is generally considered to be parallel, figure 20.2(b). This assumption allows the use of trivial reconstruction algorithms for parallel geometry. This approximation is particularly valid for small samples placed close to the detector. The concept of placing the sample close to the detector is based on the collimation ratio L/D, see equation (2.1) in chapter 2, which allows to compute the degree of perspective distortion when the neutrons pass through the sample. However, with larger pieces, it can be worth considering using a reconstruction algorithm that includes the ray path of a divergent beam, figure 20.2(b) and (c), [6] to avoid the geometry artefacts. These artefacts appear as consequence of the large difference in distance between the two object surfaces, i.e., the ones closer to and more remote from the detector along the ray path.

The more general case of reconstruction geometry includes the beam divergence and is based on perspective projections. This case is often referred to a cone-beam geometry. The Feldkamp algorithm [7] is the most frequently used algorithm for cone-beam reconstruction. It is an approximate algorithm that works well for slight beam divergence and is defined for a scan trajectory in the plane perpendicular to the acquisition axis. The Feldkamp algorithm is only exact in the plane corresponding to the beam centre. It is worth considering using a helical acquisition trajectory to avoid the cone-beam artefacts. The helix trajectory provides information needed to avoid geometric distortion of the reconstructed objects when a beam with greater divergence. This choice is relevant for x-ray cone-beam acquisition which has much greater beam divergence than neutron beam. Helical scans require a reconstruction algorithm that includes the helical trajectory and the divergent beam, figure 20.2(e), such as the algorithm proposed by Katsevich [8]. This method provides an exact reconstruction of the observed object without the geometric distortions observed in volumes reconstructed with the Feldkamp algorithm.

A tomography scan of static samples is performed stepwise with small angular increments until at least 180° are covered by projection data. Scans covering an entire turn, i.e., 360° are ideally done with an odd number of projections. Using an odd number of projections fills the gaps between the projections from the first half of the turn with an additional projection. The advantage of full-turn scans is some reduction of the beam divergence artefacts in parallel beam reconstructions. This approach is valid for the slight beam divergence, often seen at neutron imaging instruments. The Nyquist theorem defines the number of projections needed for a well-sampled reconstruction. The theorem states that the number projections, N_p, relates to the number of pixels perpendicular to the acquisition axis, N_u, as

$$N_p = \frac{\pi}{2} N_u \approx 1.5 N_u \tag{20.7}$$

In practice, using as many projections as the theorem indicates is not necessary. The under-sampled data still provide useful reconstructions because the amplitude of the under-sampling artefacts is often much less than the noise amplitudes usually seen in neutron imaging, which are thus not visible. It is, however, worth noting that the number of projections should increase when small details are expected in the reconstructed image.

20.2.1 Iterative reconstruction techniques

The reconstruction inverse problem can also be described as a system of linear equations, e.g. [5], which are formed as discrete observations from equation (20.5).

$$\mathrm{A}x = p \tag{20.8}$$

The rows in the forward-projection matrix A describe the rays through the slice that produce the measured projection values in p. The reconstruction of the slice x is obtained by inverting A. The problem with this approach is that the forward-projection matrix of the equation system is enormous, sparse, and severely ill-posed. This combination makes it nearly impossible to solve the matrix inversion needed to reconstruct the volume by direct methods. The solution is to use regularized iterative methods that reduce the hardware requirements as they break down the forward-projection matrix to only work with parts of it simultaneously. The most known methods are the Algebraic Reconstruction Technique (ART), Simultaneous Algebraic Reconstruction Technique (SART), and Simultaneous Iterative Reconstruction Technique (SIRT). These methods are more robust and produce fewer reconstruction artefacts than the analytical reconstruction for cases with under-sampled data. Regularization methods based on total variation minimization are also proposed and provide good SNR performance. However, when too strong regularization is applied, the reconstructed volume often appears patchy with constant regions. An alternative approach to the matrix inversion methods is using model-based reconstruction methods based on statistical data models to iteratively refine the images by maximizing the *a posteriori* distribution of the data. Statistical methods are best suited for projection data with low SNR [9] models that can reject outliers like spots and lines in the sinogram to produce noise and artefact-free images [10].

Recent development leads toward using model-based and even machine learning methods that reduce reconstruction artefacts and, at the same time, drastically improve the quality and level of detail in the reconstructed data [11, 12].

20.2.2 Time series tomography

Dynamic processes that are studied using tomography require defining strategies for the acquisition to be able to follow the changes appearing in time to provide a time series of volumes. The problem with the tomography of processes is that the reconstruction requires that the sample does not change significantly for the duration of the entire scan; otherwise, motion artefacts will appear in the reconstructed image. Motion artefacts are not a problem for slow processes, particularly when the process can be temporarily stopped to provide a steady-state condition before changing to the next state. Increasing the frame rate of the tomography scans is, however, the only option when the process changes the material distribution or the structure shape considerably.

The increased frame rate naturally comes at the cost of reduced dose per volume and thus a reduction in SNR, i.e., noisier images. A counter action to increase the SNR is to work with larger pixels. Therefore, in the end, the highest frame rate for an experiment is decided by the detectability of relevant features, which includes feature contrast and dimensions versus the SNR needed to observe and quantify the process.

Time-series tomography can be acquired using different strategies depending on the sample dynamics. On-the-fly scans are mainly used for rapid processes. Here, the sample is continuously turned at constant speed in contrast to the stepwise increments normally used in neutron tomography. The detector frame rate must be adapted to the sample rotation to obtain sufficiently many projections.

An alternative to the on-the-fly approach which is more suited for relatively slow processes is to use series of nearly oblique acquisition angles determined by an irrational number like the golden number, ϕ [13] or its inverse using the following equation to determine the next acquisition angle in the sequence indexed by i

$$\theta_i = i \cdot \phi \cdot \pi \bmod \quad \pi. \tag{20.9}$$

The on-the-fly technique ideally includes synchronization of the image acquisition and the stage movement information on the instrument control system level. Still, it is also possible to identify the sub-scans of the rotations by analyzing the sinograms.

The projection data in time-series tomography scans are often severely under-sampled to allow capturing the process dynamics. Therefore, the temporal resolution is a trade-off against artefacts caused by under-sampling. Some parts of the sample are often static and do not change for the entire duration of the experiment. These structures can provide *a priori* information about the sample to allow reconstructions with much higher SNR and fewer artefacts [14]. A further approach is to include regularization between time frames [15]. It has been shown that it is possible to reconstruct time series tomography data with as few as 16 projections per volume, given the *a priori* information from the entire time series, and identify static regions versus altering regions [14].

20.2.3 Tomography with limited data

There are several types of limited data in computed tomography. The most frequent form is moderate under-sampling of the acquisition angles. This level of under-sampling is widespread in neutron imaging due to the long scan times required to obtain sufficient data to allow reconstruction with acceptable quality for the investigation. The number of projections in the scan is a trade-off between sufficient grey levels and SNR in each projection versus avoiding severe under-sampling artefacts in the reconstructed slices. The priority is often to obtain well-exposed projections instead of fulfilling the sampling theorem for projection data. The under-sampling typically lies in the order of two to five times relative to the number of projections indicated by the sampling theorem. The rule of thumb is that the smaller the features present in the object, the more projections are needed. The under-sampling still provides data to fill the Radon space with views from all directions.

In the previous example of limited data, it was always assumed that the scan covers at least a half-turn around the sample, a requirement for a complete tomography dataset. It is not always possible to fulfil this requirement, particularly when the sample shape gives geometric constraints. This usually happens when the sample has a high aspect ratio between the minor and major axes, such as in the case of slabs. The neutron beam cannot pass through the sample along the major axis, resulting in detector starvation, and the sample would need to be placed far away from the detector to allow a full rotation. To address this problem, two methods can be used: laminography [16, 17] and tomosynthesis [18] (figure 20.3).

Laminography involves tilting the acquisition axis in the beam direction and performing a full-turn scan. This method allows transmission through the slab, but there is a wedge of missing data in the Radon space. The tilting and rotation of the sample also require that the sample contents are fixed during the scan to avoid displacements causing motion artefacts. Sample stability is, for example, a problem for loosely packed soil samples, but the scan is feasible with proper sample preparation.

Tomosynthesis, on the other hand, involves mounting the sample as a normal tomography with the vertical acquisition axis parallel to the detector and the wide slab area parallel to the acquisition axis. The geometric constraints are addressed by reducing the scan range from 180° to a fraction of this arc. This technique is

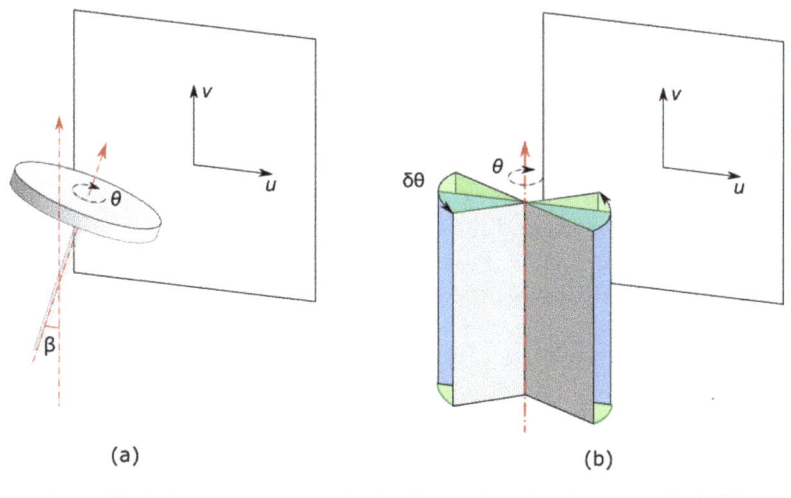

Figure 20.3. Scan arrangements for laminography (a) and tomosynthesis (b).

frequently used in microscopy [19] and mammography [20]. However, the tomosynthesis scan only covers a fraction of the views, resulting in a missing wedge in the sinogram.

The missing information from these scanning strategies appears as cross-talk between slices and in cone-shaped artefacts, resembling small hats, on top of small sample features [21]. Xu *et al* conclude that laminography has less data loss than tomosynthesis and that tomosynthesis has asymmetric artefacts due to the nature of the missing wedge. Despite these flaws in the reconstructed data, both methods are relevant to obtain depth information for samples that previously were only imaged using radiography.

Both laminography and tomosynthesis can be reconstructed using analytical reconstruction, but the use of iterative reconstruction techniques has proven to deliver images with less pronounced missing wedge artefacts [18].

Finally, not all samples with high aspect ratio between minor and major axis need the use of laminography or of tomosynthesis. Still, it is not entirely unproblematic. The noise in the reconstructed images from such a sample can be stronger in the direction of the thicker parts. These noise fluctuations can be reduced with use of an adaptive filter which is applied on the projections [22].

20.3 Image artefacts

Neutron images are often affected by different artefacts that sometimes severely impact the image quality and, thus, the following quantitative analysis of the data. In this section, we provide an overview of different image artefacts that are commonly occurring and suggest methods to correct them. Due to their different origins, the artefacts must be handled separately to address each type. The origin categories are noise (quantum noise and outliers), neutron matter interaction (scattering and beam hardening), sample dynamics, instrumentation (optics, scintillators, and instrument geometry), and algorithmic parameterization.

20.3.1 Noise in neutron images

Noise is present in any measurement; neutron imaging is no exception (figure 20.4(a)). Noise in neutron imaging is complicated to describe, as many noise sources also depend on each other. The Poisson distribution is dominant due to the counting noise of the neutron beam itself and the light converted by the scintillator. The most important consequence of this noise model is that the SNR is proportional to the number of captured neutrons and the number of photons emitted by the scintillator. Since the standard deviation of Poisson noise is equal to the square root of the number of captured neutrons N_n, the SNR is

$$\mathrm{SNR} = \frac{N_n}{\sqrt{N_n}} = \sqrt{N_n}. \tag{20.10}$$

Figure 20.5 shows the SNR from an experiment with increasing exposure times. The neutron flux was in the order of 2×10^7 neutrons cm^{-2} s^{-1}. The plot clearly shows the relation to equation (20.10).

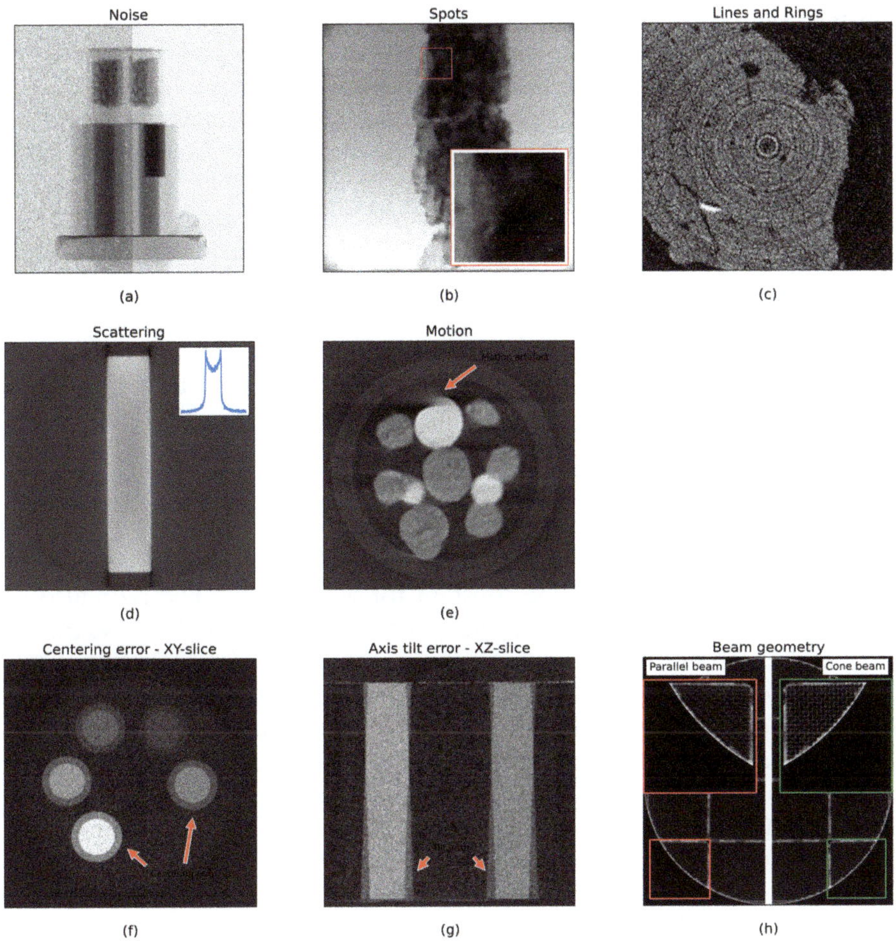

Figure 20.4. An overview of different degradations observed in neutron imaging data.

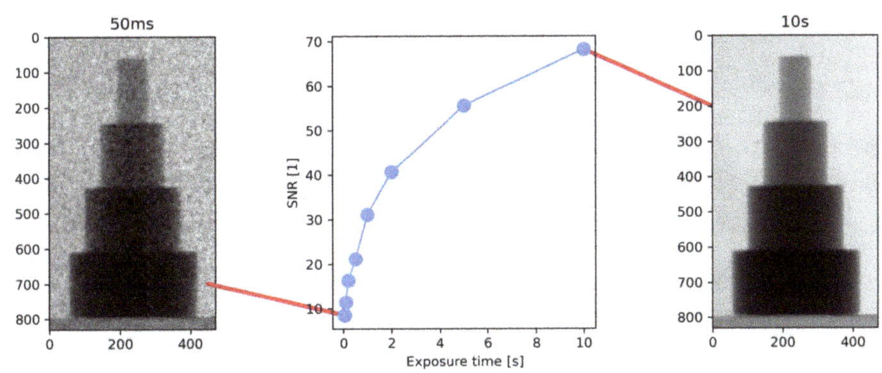

Figure 20.5. The SNR is proportional to the square root of the exposure time. The plot shows the SNR for neutron radiographs using exposure times in the interval 50 ms to 10 s.

This relation means that the SNR can be increased by increasing the number of captured neutrons. The increase can be achieved by changing neutron intensity, choice of scintillator, pixel size, and exposure time. The intensity is often hard to increase as it ultimately depends on the neutron source. So, unless there is an option to change to a larger neutron aperture, the remaining factors can easily be manipulated at experiment time.

The SNR is a scalar metric that helps to understand the ability to detect features but is mainly measured in a region with constant image intensity. It does not, however, tell anything about how the noise relates to different spatial frequencies. The detection quantum efficiency, DQE, is the ratio between the modulation transfer function, MTF, and the noise power spectrum, NPS [23, 24] as a function of the frequency, ξ, and provides more information about the ability to detect features of some specific size.

$$\mathrm{DQE}(\xi) \sim \frac{\mathrm{MTF}(\xi)^2}{\mathrm{NPS}(\xi)}. \tag{20.11}$$

The typical shapes of these functions are that the MTF signal decays for higher frequencies and the NPS would be constant for white noise. The NPS also decays slightly for higher frequencies. Given these base shapes, the DQE shows that higher frequencies, i.e., small details and rapid changes, are mostly more sensitive to noise than more widespread objects. This has the consequence that higher SNR is needed to confidently detect small and high frequency features in the image.

Neutron images are often acquired with a relatively low SNR. The noise often obscures the relevant features and prevents a reliable analysis based on observations in the raw data. The SNR can, fortunately, be increased using filters which are applied to the images after acquisition. The first approach is often to use convolution-based low-pass- or median-filters, which are fast and give good preliminary results. The disadvantage of these filters is that they also blur relevant features in the image simultaneously while reducing the noise component [25]. Consequently, tiny details with limited spatial extent like inclusions and thin layers may vanish in the filtering process. This is the case for low SNR images, as the needed filter window must be wide to reduce the noise.

There are, however, advanced filters designed to suppress noise and, at the same time, preserve image structures. They are based on differential equations inspired by, e.g., the diffusion equation applied on the grey levels in the image [26]. The filtering process is iterating over difference equations until the desired filter effect is achieved. A further denoising filter that has gained popularity is the non-local means filter [27] and more recently the BM3D-filter [28]. These filters provide good denoising performance but have the disadvantage of very long processing times, in particular, for three-dimensional data. More recently, a new class of denoising filters based on deep learning has been introduced [29], they may not improve the global SNR drastically beyond earlier methods, but have better fidelity to fine image details. The application of these filters in neutron imaging is so far limited.

20.3.2 Spots

Spots are the first and most obvious artefacts observed in neutron images. Figure 20.4(b) shows an example of spots in an image. The spots are outliers in the radiograph with limited spatial extents; most spots only extend over a few pixels. The spot shape varies from almost round to short lines. The grey levels of the spot pixels are generally significantly greater than the surrounding pixels.

There are two main approaches to reducing or removing spots from the radiographs: (a) minimizing the number of produced spots at experiment time, and (b) applying image processing methods to the acquired images. The spots are often caused by gamma photons from the sample or instrument environment. These photons originate from the neutron source, interactions in the sample environment, and the scintillator, in particular when the conversion material is Gd. Installation of shielding that minimizes the neutron beam to the detector region of interest reduces the number of spots at experiment time. The second type of shielding is to install gamma absorbers between the detector and sample/scintillator. These measures reduce the number of spots in the images, but some spots will remain in the images. These spots bias the analysis and must be removed using spot removal filters.

A trivial problem with the spots is that auto contrast and brightness settings for image viewers are scaling the levels incorrectly, resulting in dark images where the sample is too dark to observe. The more severe problem is that the spots strongly bias region measurements such as local averages, resulting in very unbalanced image histograms with most of the image information represented by very few histogram bins. Furthermore, spots are also the origin of line artefacts in tomography slices, figure 20.4(c). The lines come from the relation that a spot in the sinogram represents a line in the reconstructed slice. The slope of the line is defined by the acquisition angle of the projection where it appears. Thus, the impact of spots extends over far more pixels in reconstructed tomography images. In the extreme, with many line artefacts, they cannot be distinguished as lines anymore and appear as additional noise in the reconstructed slices.

Different filtering techniques can remove the spots. The easiest method is to apply a median filter to the images. The outlier rejecting property of the median filter does remove the spots, but at the same time, it also has a global low pass filtering effect, in particular when the filter kernel radius must be increased to remove also larger spots. Therefore, several correction methods use a two-step approach [1, 30] involving detection and replacement. The detection is based on thresholding a spot-enhanced image, typically a high-pass filter. The filter is tuned to enhance the outliers. The replacement step fills the detected spot regions with a smooth version of the original image. The smooth image is often a median filtered image with a wide filter window. The area of the filter window must be wider than at least twice the largest spot to provide a reliable replacement.

A different approach to reducing the spots is acquiring multiple images and combining them into a single image. The exposure time of each image is then reduced by the factor $1/N$ for N repeated acquisitions to correspond to the same acquisition dose. The cost of this approach is naturally more data and the readout time of the images is N times longer. The readout time can significantly impact the total experiment time, particularly when it is similar to the exposure time. The total experiment time for M images is then

$$t_{\text{experiment}} = M \cdot N \cdot (t_{\text{exposure}} + t_{\text{readout}}). \tag{20.12}$$

The repeated images can be combined using different methods. Averaging is the same as exposing a single image and has no outlier rejecting ability unless many images are used and should therefore not be used. The next option is to use average with sigma clipping, i.e., only considering data points inside a given confidence interval. Median is a frequently used method to combine images; it has good outlier rejecting properties but should be avoided for less than five projections. Finally, using a weighted average provides the best combined SNR and outlier rejection properties [3].

20.3.3 Rings

Ring artefacts are specific to tomographic slices. They appear as concentric rings or arcs around the rotation centre of the reconstruction, figure 20.4(c). The origin of the rings is a pixel or a cluster of pixels that is biased relative to its neighbourhood during the entire acquisition process, i.e., very similar to the spots. These pixel clusters draw a line parallel to the angle axis of the sinogram. Lines of this type in the sinogram appear as rings in the tomography slice. There are many reasons for these regions appearing in the projection data: (a) malfunctioning detector pixels, (b) particles on the detector chip, (c) damage to the scintillator, or (d) poorly prepared reference data in the normalization process such as insufficient spot removal quality or low SNR. The ring intensity and width can vary from barely visible rings to high-intensity wide rings. Weaker rings can often be neglected as they are removed in later denoising filtering steps. Strong rings, however, gravely interfere with the sample information in the reconstructed slices and must be corrected.

The removal of rings is an essential step in tomography reconstruction. They can be removed either before or after reconstruction. The sinogram is used when the rings are corrected before reconstruction. The removal is based on filtering in the direction parallel to the θ-axis of the sinogram. There are different approaches to performing this filtering: subtraction of outliers in the average sinogram profile [31], using a combination of a wavelet transform and high-pass filters [32] or applying a series of corrections targeting specific types of stripes in the sinogram [33]. Methods to remove the rings after reconstruction mainly perform a transform from Cartesian to polar coordinates. This transformation results in an image that resembles the sinogram because it is represented by the coordinates radius and angle. The rings are then detected similarly as in the sinogram, i.e., as lines parallel to the θ-axis. The detected rings or their replacement are then transformed back to Cartesian coordinates as a mask in the correction operation, e.g., [34].

20.3.4 Scattering and beam hardening

The measured signal in a neutron image experiment deviates from the value expected by the theory. A validation using the Beer–Lambert law and known material thicknesses produces greater attenuation than the measured signal. This deviation results in an underestimate of the attenuation coefficients and thus prevents reliable quantification based on the neutron attenuation. The deviation appears as a decreased intensity in the sample interior in tomography, figure 20.4(d), this effect is often referred to as *cupping artefact* due to the intensity profile shape, which resembles a cup. The origin of the cupping is explained by two different mechanisms of the neutron matter interaction, namely, beam hardening and scattering from sample interaction and instrument background. The scattering is the dominant effect in neutron imaging.

Beam hardening is a phenomenon originating from the fact that low energy radiation is more absorbed in the material, with the consequence that more high energy radiation is detected behind thicker part of the sample. Beam hardening is more prominent for x-rays than for neutrons. Several methods correct deviations from a straight-line relation between material thickness and the measured image intensity. Kang *et al* [35], describe a calibration procedure using samples with known material thickness to obtain a calibration curve that allows quantification of material thickness from neutron images. This method targets contributions from both beam hardening and scattering. It provides realistic values for the calibrated use case but does not generalize well. Another approach is to assume the deviation to be solely caused by beam hardening and use a polynomial correction of the image intensity values [34, 35].

The second contribution to biases in neutron images is caused by incoherent scattering from the sample itself and also from the instrument background. The contribution of this bias can be easily shown using opaque items to block the direct neutron beam [36, 37]. The image intensity behind the opaque object is expected to be zero for a non-scattering environment. In reality, a signal is measured behind the black bodies. The amplitude depends on the type and shape of the measured sample. Hassanein *et al* [36, 38] proposed using a large black body covering the entire specimen to measure the scattering contribution and used a correction procedure that includes material-related parameters for the sample and detector. This approach allowed correcting neutron radiographs and thus also tomography projections to obtain absolute water contents in soil samples, e.g., [39, 40]. However, a single black body cannot fully capture the shape of the scattered field of neutrons, and the correction method turned out to be sensitive to the chosen parameterization. Raventos *et al* [41] proposed a method based on the fact that the scattering contribution from a sample placed remotely from the detector is negligible and can be used to estimate the sample scattering for the sample positioned at the ideal sample position. The works of Hassanein *et al* and Raventos *et al* provided the fundamental knowledge to refine the scattering correction method further into the currently successful method.

The new approach with small black-body items distributed over the field of view was proposed by Boillat *et al* [37]. Carminati *et al* [42] provided a correction procedure and the validation demonstrating that the accuracy of the measured attenuation coefficients for water and different materials improved radically. This method requires two additional reference images which are acquired when a black-body grid is installed. These images are made to capture the scattering profile for the open beam image and for the sample images. The correction is integrated in the image normalization and includes beside the images in equation (20.4) also images of estimated scattering contribution from sample, $I^S_{n,\,\mathrm{BB}}$, and background $I^S_{\mathrm{BG},\,\mathrm{BB}}$. These need information from images of sample and open beam measured with that black body installed $I^*_{n,\,\mathrm{BB}}$ and $I^*_{\mathrm{OB},\,\mathrm{BB}}$ respectively. The explicit expression for the normalization then becomes.

$$\frac{\tilde{I}_n}{\tilde{I}_{\mathrm{OB}}} = \frac{I_n}{I_{\mathrm{OB}}}\cdot\frac{\mathrm{D}(I_{\mathrm{OB}})}{\mathrm{D}(I_n)}$$

$$= \frac{I^*_n - I_{\mathrm{DC}} - I^S_{n,\,\mathrm{BB}}\dfrac{\mathrm{D}\left(I^*_n - I_{\mathrm{DC}}\right)}{\mathrm{D}\left(I^*_{n,\,\mathrm{BB}} - I_{\mathrm{DC}} - \left(1-\frac{1}{\tau_{\mathrm{BB}}}\right)I^S_{n,\,\mathrm{BB}}\right)\tau_{\mathrm{BB}}}}{I^*_{\mathrm{OB}} - I_{\mathrm{DC}} - I^S_{\mathrm{BG},\,\mathrm{BB}}\dfrac{\mathrm{D}\left(I^*_{\mathrm{OB}} - I_{\mathrm{DC}}\right)}{\mathrm{D}\left(I^*_{\mathrm{OB},\,\mathrm{BB}} - I_{\mathrm{DC}} - \left(1-\frac{1}{\tau_{\mathrm{BB}}}\right)I^S_{\mathrm{BG},\,\mathrm{BB}}\right)\tau_{\mathrm{BB}}}}$$

$$\cdot\frac{\mathrm{D}\left(I^*_{\mathrm{OB}} - I_{\mathrm{DC}} - I^S_{\mathrm{BG},\,\mathrm{BB}}\dfrac{\mathrm{D}\left(I^*_{\mathrm{OB}} - I_{\mathrm{DC}}\right)}{\mathrm{D}\left(I^*_{\mathrm{OB},\,\mathrm{BB}} - I_{\mathrm{DC}} - \left(1-\frac{1}{\tau_{\mathrm{BB}}}\right)I^S_{\mathrm{BG},\,\mathrm{BB}}\right)\tau_{\mathrm{BB}}}\right)}{\mathrm{D}\left(I^*_n - I_{\mathrm{DC}} - I^S_{n,\,\mathrm{BB}}\dfrac{\mathrm{D}\left(I^*_n - I_{\mathrm{DC}}\right)}{\mathrm{D}\left(I^*_{n,\,\mathrm{BB}} - I_{\mathrm{DC}} - \left(1-\frac{1}{\tau_{\mathrm{BB}}}\right)I^S_{n,\,\mathrm{BB}}\right)\tau_{\mathrm{BB}}}\right)} \tag{20.13}$$

Carminati *et al* also showed that the beam hardening to some degree also contributes to the total bias. Thus, both corrections are required to obtain flat intensity profiles in tomography slices from a homogeneous sample.

20.3.5 Detector lag

The scintillators used to convert the captured neutrons to visible light have a fluorescent component that emits the light used to acquire the images. The emitted light is not instantaneous but has an afterglow that decays with time, t, following the sum of two exponential functions [43, 44].

$$\mathrm{L}(t) = C_0 + C_1 e^{-t/P_1} + C_2 e^{-t/P_2}. \tag{20.14}$$

The first exponential has a short duration, P_1, [45] while the second may linger for a longer time, P_2. The coefficients C_0, C_1, C_2 weight the contributions of each decay component. These parameters are not the same for different types of scintillators, but can differ much.

The rapid term is mostly only relevant for fast acquisitions, e.g., time-of-flight imaging or very rapid time series. There is also a more permanent effect in addition to these decaying effects, the so-called burn-in that leaves a permanent imprint in the scintillator. The burn-in is mainly observed for long-time exposure sessions using high flux neutron beams when ^{6}Li/ZnS scintillators are used. The afterglow will introduce a bias in the reconstructed attenuation coefficients, which can appear as ghost shapes in tomography reconstructions [46]. The ghosts are mainly visible for objects with a significant aspect ratio between the minor and major axes. A method to correct for detector lag would be to apply an IIR (infinite impulse response) filter

$$a_n = a_n - L(\tau_n) \cdot a_{n-1}. \tag{20.15}$$

Here, n is the index of the current image, and τ_n is the time between images a_n and a_{n-1}. This is a conceptual equation for the correction; it requires additional regularization to provide robustness against outliers. Detector lag is generally a second-order effect contributing to biases of less than a percent in most cases. In particular, experiment data from high flux sources can benefit from this correction since the burn-in and lag effects are more pronounced at these sources.

20.3.6 Motion artefacts

Motion artefacts appear when the sample changes shape or position during the acquisition. Small changes result in blurred images where image features are less sharp. In contrast, greater changes severely impact the image quality, making it very hard to analyze the images.

Tomography is even more strict; the general requirement is that the sample is not supposed to change for the entire scan duration. The reconstruction of projections that do not fulfil this requirement will result in snail shell-like artefacts, see figure 20.4(e), when the sample changes shape monotonically, e.g., in the cases when an object is growing or shrinking. This requirement can be fulfilled by reducing the exposure time until there are virtually no displacements in the projections and the experiment reaches a quasi-steady-state condition. An alternative method to reduce the impact of motion is to change the acquisition strategy from the most commonly used sequence scan with constant angular increments to determining the increments using the golden ratio [13] see also section 20.2.2. Using the golden ratio relaxes the conditions to allow dimensional changes within the sample in the order a tenth of a pixel between two projections.

The challenge of time series tomography acquisition is to balance the number of projections versus the total dose acquired for a timeframe. The duration of the time frame is defined as the change rate of the observed process and the SNR and number of grey levels needed to obtain images that reveal the expected process. Mostly, these boundary conditions lead to an under-sampled set of projections because a minimum of neutrons must be captured to provide projections that can be reconstructed. A further aspect is that there is an overhead for turning that sample and to read out the images. The under-sampling in projection space would result in bad image quality with low SNR and sampling artefacts. These problems can be handled using iterative reconstruction techniques [14, 15, 47] occasionally even in combination with machine learning techniques [12].

20.3.7 Errors introduced by incorrect geometry calibration

A geometric calibration is vital for the successful tomography reconstruction. In parallel beam geometry, there are three relevant parameters: centre of rotation, the tilt of the acquisition axis relative to the detector grid, and tilt in the plane parallel to the beam [2]. The effect of first two are demonstrated in figure 20.4(f) and (g). Misalignment of these parameters appears as unsharp edges for minor deviations from the true geometry; greater deviations will distort the reconstructed images severely but are also trivial to detect by visual inspection. Cone-beam geometry requires the additional parameters source-to-detector distance, source-to-sample distance, and the detector position where the beam is perpendicular to the detector plane. Errors in these parameters result in perspective distortions of the reconstructed objects.

20.3.8 Beam divergence

The beam divergence is not an artefact but a well-understood consequence of the beamline geometry, see section 2.1. The effect of the beam divergence is the so-called penumbra blurring. The spread, d, of this blurring can be calculated using the relation between the pinhole diameter, D, the distance from the pinhole to the detector, L, and the distance from the sample to detector, l,

$$\frac{l}{d} = \frac{L}{D} \tag{20.16}$$

The divergence angle, calculated as $\tan^{-1}(L/D)$, of most neutron imaging instruments is less than one degree, and the effect is often negligible when the sample is near the detector. Therefore, the penumbra effect in radiography can only be reduced by changing the experiment conditions, i.e., moving the sample closer to the detector or changing the beamline geometry. The only exception is when the sample is relatively thin. Then, it would be possible to use deconvolution to sharpen the image. In tomography experiments, the first assumption is that the beam is parallel, which allows the use of reconstruction algorithms for parallel beam geometry. The divergence can, however, play a role in cases when large samples with small features are studied, figure 20.4(h). The grid in the example is distorted to a degree where it is impossible to confidently recognize the structures. In such cases, it is needed to use cone-beam algorithms that will compensate for the beam divergence [6].

20.4 Analyzing neutron images and modelling

The analysis is the primary intention of the images. Each experiment will require its own analysis workflow, which depends on the questions to be addressed and how many images must be analyzed. With a single or few images, it can be adequate to use interactive methods, whereas larger data quantities would require automation to reduce the need for manual labour. Qualitative analysis of a single or few samples is less strict and more manual interaction in the analysis can be tolerated. In quantitative analysis it is essential to promote repeatability in the analysis workflow and rely as little as possible on manual choices during the analysis.

20.4.1 Visual analysis

Visual analysis is mainly a qualitative method used to explore and visualize the contents of an object. The aim is, in many cases, to provide eye-catching renderings for presentations and publications. In the initial stages of the analysis, it can also be helpful to visualize volumes to explore and obtain a first understanding of the object composition and complexity. This analysis provides input to decide how to proceed with the quantitative analysis. Visual analysis is best suited for a few images as the amount of manual work can be overwhelming when many images are to be analyzed.

Advanced visualization tools can also perform quantitative analysis. These tools provide means to interactively or automatically segment images before the analysis. The segmentation can be done with different approaches: manual painting or region mark-up, semi-automated where the user provides segmentation seed points and fully automated segmentation. Typical analysis tasks are, for example, fault detection, dimensional analysis on a few objects, or a comparison between different items.

20.4.2 Interpreting the grey levels

In many cases, intensity values or grey levels in the image have a physical meaning unless the image is a product of a segmentation or classification operation. The physical meaning in the context of neutron imaging could be that the pixel value is proportional to, for example, the number of captured neutrons, the transmission, the optical thickness, or the attenuation coefficient. These values are often related to other quantities like density or mixing ratios on the sample. The grey level analysis aims to quantify the variations in pixel values to obtain information about the sample or a process. Therefore, the images must be corrected for biases and artefacts to avoid systematic and random errors in the analysis, see section 20.4.

Quantifying the amount of water or hydrogen is one of the most frequent applications of grey level analysis in neutron imaging. The analysis is done in both 2D and 3D and aims to quantify the local water content in different sample regions. The optical thickness images, equation (20.3), are used in radiographs, where the water content is determined by dividing the normalized image or the target region by the attenuation coefficient of water. In some cases, a calibration procedure is needed to map water contents to the observed pixel values [35]. This calibration is in particular needed when the normalization lacks correction for biases introduced by scattering or for different neutron spectra. The quantified water distribution is used in further application-specific modelling to describe, for example, transport or diffusion processes in the samples.

Applications of quantitative grey level analysis are found in a variety of research domains like soil hydrology/root analysis and geology, Part III, and electrochemistry Part IV.

20.4.3 Shape and content analysis

An alternative method to using the grey levels is identifying regions with similar properties based on, for example, the pixel values. The first step of this analysis is to segment the image based on its grey levels. Image segmentation is a field of research of its own with methods spanning from histogram-based methods to complex machine learning techniques that besides grey levels also take features in the pixel neighbourhood into account. The choice of method depends on the information content in the images and also on the number of images to analyze.

The segmented images have either two or a few levels representing the different categories present in the data. The categories can be air, container, sample, materials in the sample, and possibly a liquid. The segmentation can be either manual/guided or fully automated. The segmentation method is often application-specific and the strategy varies depending on which features to detect, their contrast to the neighbourhood, how well resolved they are, the SNR, etc. Bi-level images are used for single material objects or to provide a mask for the area of interest for further analysis. The character of the analysis in the segmented regions is application-specific. It could be grey level analysis, e.g., water content within the segmented regions, or structural analysis of items found in the images, e.g., counting the number of faults or their volume distribution in a metal cast.

Interactive segmentation of neutron images uses tools allowing the user to navigate the data and mark-up relevant regions. The mark-up may be guided by region-growing algorithms that reduce the amount of manual labour. Interactive segmentation is mainly used with a single or few images as the manual work is very time-consuming and less reproducible than automated methods. Another motivation to use manual segmentation is when the structures in the image are too distorted by starvation, beam hardening, and motion artefacts. In such cases, choosing the manual approach over the development needed to perform an automated segmentation may be less demanding. The automated segmentation produces segmented images with little to no interaction. Some are unsupervised and only need the number of expected classes to detect, e.g., many histogram-based methods [48] or *k*-means [49], while other methods are supervised and require prelabeled training data to build a model for the segmentation.

The aim of the segmentation algorithms is to be as accurate as possible, yet there are often some misclassified pixels caused by outliers or regions where the classification confidence is low. Even though these pixels can be removed using morphological filters like opening and closing [50], it is often beneficial to use high SNR images as input to the segmentation algorithm.

The segmented items can be counted, and their positions, shapes, and dimensions can be measured. The items and regions can also be used as masks to quantify the information carried by the grey levels.

20.4.4 Multi-modal analysis

In recent years it has become more common to add a secondary radiation source to neutron imaging beamlines [51–53]. The purpose of using images from radiation provided by a second source, also called modality, is to provide new information about the sample that enhances the accuracy and confidence of the analysis.

The multi-modal analysis combines information from multiple modalities to enhance the accuracy of the results. The purposes of adding a modality are:

Filling in the gaps—the element composition of some objects can result in detector starvation depending on the amount. A second modality may then be used to reduce artefacts and also to fill in the structural gaps. A good example for this is the investigation of the sword from the lake of Zug [54]. The sword was made of a material combination of metals and wood, for each of which neutrons and x-rays have their disadvantages and advantages, respectively.

Improve confidence of change detection—in systems where two parameters change simultaneously, e.g., swelling and shrinkage of a porous medium as a consequence of altering water content. The purpose here is to use one modality to detect structural changes, while the other detects changes in water content; many applications are found in chapter 9.

The bi-variate histogram is the most frequently used reduced representation of bimodal data in neutron imaging. It is a two-dimensional histogram that counts the frequency of colocation pairs of attenuation coefficients in the images. This information can be used for segmentation purposes by manually marking up and assigning class values to regions of the histogram [52]. Automated techniques based on clustering, Gaussian mixture models, and deep learning can also be used. Supervised methods require training data which is hard to come by without intense and time demanding manual labour. These methods are, therefore, mainly relevant for series of similar data and for data with similar characteristic features. The training may then be recycled and, thus, save work and experiment time.

20.4.5 Time series analysis

The study of dynamic processes results in time-structured data that needs to be analyzed to demonstrate and prove causality of a system. A time series represents a sequence of data where one point shows the change relative to the previous points. In general terms, this does not mean that only time is used to index the series but it could also be for example changes in load conditions provided the process is observed at steady state for each point. This analysis is very closely related to the observed process. Theory categorizes the components of a time series into the following four categories [55]:

Trend—The long-run development of the process is described by the trend. The trend follows a continuous raise or decrease of the measured quantity.

Seasonality—The measured quantity alternates slowly between low and high values with a recognizable period.

Cyclic—The cyclic processes also alternate between low and high values but do not repeat with a specific period.

Residuals—Unexpected outliers or anomalies that cannot be described by the previous three categories. It could be the aim of the experiment to detect these irregularities, in other cases they are just outliers that need to be suppressed to avoid prediction errors. The residuals also include the measurement noise.

An experiment often involves a combination of these components. Time series experiments are often controlled by external factors like triggering pulses that start the process and also, for example, applied temperature gradients or mechanical load.

Time series in neutron imaging is obtained using both radiography and tomography. The series obtain an additional dimension to represent time. Therefore, time series of images either produce 3D or 4D data. Operations like denoising and artefact reduction can often with benefit be performed including the temporal axis to gain from information provided by the additional dimension.

Images in time series often have a static fraction and one representing the time dependent changes. An initial step of the analysis is therefore often to identify the region to analyze by some means of segmentation to reduce the amount of data and analysis complexity. The measured quantities can be either changes in grey levels or changes in shape and position.

Examples of applications of time series analysis in neutron imaging are modelling of water flow in soil and roots [56], optical flow in plants [56], strain fields caused by mechanical loading [57, 58], and bubble tracking in liquid metals [59].

20.4.6 Analysis of Bragg edges

Several tools have been developed over the years to analyze neutron Bragg edge imaging data and to extract quantitative microstructural information from wavelength-dependent neutron transmission experiments, including density maps, phases, lattice parameters, crystallite size, strain maps and crystallographic texture information [60–63]. In particular, different toolkits have been developed for the analysis of Bragg edge imaging data obtained at time-of-flight (ToF) imaging instruments, including data reduction and processing, as *BEAn* (Bragg Edge Analysis, [64]) software, the Bragg Edge User Interface, *Ibeatles* [65] or the Time-of-Flight neutron Imaging toolkit (*ToFImaging*, [66, 67]). On the other hand, diverse fitting routines including single edge fitting (TPX EdgeFit, Tremsin A 2017 personal communication) and whole pattern analysis approaches have been implemented in several fitting codes. Among others, the software package *BETMAn*, (Bragg-Edge Transmission Measurements Analysis [68]), the *RITS* program (Rietveld Imaging of Transmission Spectra [69]) and the *BEATRIX* code (Bragg Edge Analysis for Transmission Imaging Experiments [70]) implemented a Rietveld-type fitting approach to extract bulk crystallographic information in crystalline materials from for TOF-resolved neutron imaging experiments. The codes calculate the neutron transmission spectra based on crystalline theoretical models and fit them into experimental data using a nonlinear least-squares fitting algorithm, allowing for the quantification of crystalline phases, preferred orientation, crystallite size and crystal lattice spacing. A similar approach has also been implemented to study single and oligocrystalline materials [71, 72], employing a full pattern analysis routine to study lattice parameters and lattice misfits, mosaicity and crystalline orientation with hundreds of micrometers spatial resolutions. An individual [73, 74] and multiple Bragg-dip profile fitting analysis [75] was also implemented to study crystal orientation, mosaic spread and crystalline defects.

20.4.7 Dark-field analysis

Regardless of whether neutron grating interferometry or non-interferometric single-grating datasets are used, raw data alone is of limited usefulness. Dedicated reduction procedures are required to reconstruct attenuation, differential phase, and dark-field images. The most commonly used technique involves a phase-stepping procedure of one of the gratings, resulting in a dataset consisting of a stack of frames that can be used to analyze the interference pattern at each pixel of the imaging detector. TaPy [76] is a Python module used for data reduction of grating interferometer data with x-rays and neutrons. It is designed for experiments with Talbot–Lau interferometers and can directly retrieve attenuation, differential phase, and dark-field images. As of today, since the technique started to raise its popularity and use at different neutron facilities, more advanced software packages for data reduction, including graphic user interface are currently under development.

In the case where a single-grating approach is employed, a single image is typically only available for the data reduction. In this case, an analyzing window with a size corresponding to the grating period is typically raster-scanned through the image to calculate mean intensity and visibility at each pixel, allowing for the calculation of attenuation and isotropic dark field [77, 78]. In the case where the grating pattern allows for the detection of anisotropic scattering, a 2D Fourier-based approach can be employed to retrieve the multi-directional differential phase and dark-field as well [79].

In all cases, the data still suffers from the other aspects presented in this chapter, thus procedures like normalization, noise correction etc must be employed either before or following the reduction into attenuation, differential phase and dark-field contrast images. Their analysis also includes other aspects presented in this chapter such as segmentation, tomographic reconstruction, interpretation etc. In general, what is most valuable when using these techniques is the dark-field contrast, which can provide qualitative information about the microstructural properties of the samples, at a length scale below the detector's resolution. When the distance between the sample and the analyzing grating or detector is scanned in the experiment (autocorrelation length scan), the dark-field plot versus autocorrelation length can be treated similar to a spin-echo ultra-small-angle neutron scattering dataset. Thus, typical SANS software analysis toolkits can be used for modelling and characterization of the sample's microstructure. As of today, the most notable software packages include SasView [80], SASfit [81, 82] and the NCNR Igor Pro macro package [83].

20.5 Validation and error analysis

20.5.1 Validation of the analysis

The nature of image processing is to modify the images in a way that they reveal relevant information. Here, the word 'modify' must be put in particular focus as you have to be confident that the modifications only enhance the relevant information like objects or the progress of the process. The other purpose is to suppress unwanted information like noise and artefacts. The image processing task mostly consists of a series of operations performed in a sequence. Each step is essential for the result but may also introduce new uncertainties and errors. Trivial qualitative tools for this analysis are based on visual inspection of the processed images and the difference between images before and after processing. The human eye is good at seeing whether the modifications make sense. This evaluation can be complemented by comparing the image histograms before and after processing to detect biases introduced by the processing. The qualitative assessments are good in the initial exploration and development phase. Later, it is crucial to quantify the algorithm's performance to determine how reliable the analysis algorithm is as the quantification itself.

The best method to validate your analysis algorithm is to perform a sensitivity analysis involving known data, the ground truth, degraded by blurring, noise, and artefacts. The algorithm response is then compared with the ground truth using different metrics. The metric type depends on the information you want to measure; deviations in grey level in continuous images or classes in segmented images. The most frequently used metric between two grey level images f and g is the mean squared error

$$\mathrm{MSE} = \frac{1}{N}\sum_{i\in\Omega}(f_\mathrm{i} - g_\mathrm{i})^2, \tag{20.17}$$

for the set of all pixels, Ω. The mean absolute error is a similar metric which is also used

$$\mathrm{MAE} = \frac{1}{N}\sum_{i\in\Omega}|f_\mathrm{i} - g_\mathrm{i}|. \tag{20.18}$$

These metrics show average deviations between two images. This can, however, be a blunt tool and should be combined with the mean structure similarity index metric (MSSIM) to be able to tell how the images have been degraded. The MSSIM was developed to measure image degradation by compression or image transmission, but it also useful to measure the performance of filtering techniques [84].

$$\mathrm{MSSIM}(f, g) = \frac{1}{N}\sum_{i\in\Omega}^{N}\mathrm{SSIM}(f_i, g_i), \tag{20.19}$$

which is based on the pixel-wise similarity metric

$$\mathrm{SSIM}(f, g) = \frac{\left(2\mu_f\mu_g + C_1\right)\left(2\sigma_{fg} + C_2\right)}{\left(\mu_f^2 + \mu_g^2 + C_1\right)\left(\sigma_f^2 + \sigma_g^2 + C_2\right)}. \tag{20.20}$$

Here, μ_f and μ_g are local averages and σ_f, σ_g and σ_{fg} are variances and covariances. The constants C_1 and C_2 are scalar values defined by the grey level intensity dynamic and should be small relative to the average intensity.

The performance of segmentations is often measured by categorizing the segmented pixels as *true positive* (TP) and *true negative* (TN) for the correctly classified pixels and as *false positive* (FP) and *false negative* (FN) for misclassified pixels. Counting the pixels in each category gives an overview on how well the segmentation performed. One way to visualize the results is to use a confusion matrix. A different option is to compute the sensitivity or recall of the segmentation

$$\text{recall} = \frac{\text{TP}}{\text{TP} + \text{FN}} \tag{20.21}$$

and precision

$$\text{precision} = \frac{\text{TP}}{\text{TP} + \text{FP}}. \tag{20.22}$$

These are all scalar metrics that provide global feedback about the segmentation performance. Often, it is, however, more interesting to measure how well the method performs near the items. For this purpose it makes more sense to compute a histogram of false positives and false negatives as function distance from the ground truth.

The performance evaluation workflow involves a test matrix of algorithm control parameters and variations in image properties like item shapes, sizes, and orientations. The test images should also have variations in acquisition parameters, like SNR, contrast variations, and blurring. It is mostly not possible to collect experiment data with all variation needed for a performance evaluation. The main interest of the beam time is to obtain the data relevant for the investigation and not for testing algorithms. Therefore, many of these variations must be simulated to make the performance. With simulations, it is easier to provide the needed data with sufficiently many repeated images for each parameter set to allow a statistical analysis. A comprehensive performance evaluation will allow identifying the working range of the proposed image processing and analysis algorithm. It is even possible to provide feedback to the experiment which image quality is required to allow a reliable analysis when the evaluation if done before the experiment.

20.5.2 Uncertainty analysis

There is always an uncertainty associated with every measurement. Images seem at first glance to be quite absolute with pixels arranged in a discrete grid. This, however, is misleading; the detector grid and its pixels have a metric dimension that has been measured in some way. Many quantitative investigations, e.g., area and volume measurements based on counting the pixels, rely on the pixel size to determine measured value. Also, the determination of water content from the grey value depends on the pixel size both in 2D and in 3D. In radiography, the water volume is computed using the pixel size squared times the thickness derived from the grey level. The attenuation coefficient of a voxel in a tomography volume is scaled by the pixel size. These are only simplified examples of how the uncertainty of a single measurement propagates in the quantitative analysis. The theoretical formula for uncertainty propagation is defined as

$$\sigma_f(x, y)^2 = \left(\frac{\partial f}{\partial x} \cdot \sigma_x\right)^2 + \left(\frac{\partial f}{\partial y} \cdot \sigma_y\right)^2 + \underbrace{\frac{\partial f}{\partial x} \cdot \frac{\partial f}{\partial y} \cdot \overline{\partial_{xy}}}_{x \text{ and } y \text{ uncorrelated?}} \tag{20.23}$$

Where x and y are two variables with their uncertainties σ_x and σ_y. The last term for cross-correlation between the two measurements can be omitted when the measurements are uncorrelated. This equation works well for direct relations, but can soon evolve into complicated expression. Computed tomography is one example where the uncertainty propagation chain is very hard to express explicitly. Instead, some authors use simulations to analyze the uncertainties in reconstructed images [85].

20.6 Choosing software and data formats

Software tools are essential for the work with digital images. It can be anything from a plain image viewer for qualitative assessments of the images at the time of the experiment or advanced algorithms that provide tables of information quantified from the images. Imaging is a visual method, therefore, it is often relevant to have tools that are able to present the images visually. This is often straightforward for 2D and 3D, but more challenging for higher dimensional data.

These matrices are stored using different file formats and number representations. The images are mostly processed in floating-point format with 32- or even 64-bits to avoid rounding errors. These representations require a lot of memory and storage space. Therefore, it is common to transfer the data to 8- or 16-bits integers within the used dynamics. Using 8-bits has the advantage of being the least space demanding, however, it may not be able to preserve small gradual intensity changes in high dynamic range images. The human eye can work with up to 720 levels using good equipment [86] and even more if pseudo colour maps are used. The pragmatic choice is mostly to use 16-bit integer to store the images. With this dynamic range, it is still possible to rescale the images to floating point with reasonable quality if needed. Currently, the formats FITS and TIFF are most frequent. These formats mainly store the images in separate files. When the number of images in a series increases, it is more efficient to collect the images in a single file. Storing in a single file has many advantages for the data processing as it reduces the overhead for reading and copying files. The drawback of using large files is that they require software to conveniently inspect, extract, and work with individual files. Both FITS and TIFF support the use of multi-frame data but the better choice is to use a file format that is designed for large data volumes. Currently, HDF5 is the most suited foundation for such format. HDF5 provides the primitives for structured storage of large data files and its associated meta data, e.g. instrument description, scan parameters, previous processing etc. A derivative of this format is the NeXus-format [87] which is designed for the use with data from large-scale neutron and x-ray facilities.

There is a multitude of options available for processing images. The choice of which kind of application or toolbox to use depends on different factors and you have to ask yourself many questions:

Type of image data to analyze. Are there only some few radiographs or is it maybe a complex combination of tomography data from different modalities in a time series? How good is the contrast or how clear are the changes in contrast between different features in the images?

Which information to retrieve from the images. Are discrete objects or gradual changes in grey levels relevant? Perhaps even a combination of the two is needed to obtain the results. How well-resolved are the features you want to quantify?

Experience and personal preferences of the person who will perform the analysis. Some people prefer and are maybe only able to work with interactive tools while other people with a different background prefer tools based on scripting. This may seem like a trivial requirement, but it is of utmost importance. If your data only can be analyzed using a method you are not able to use due to lack of expertise then there will be long delays until the data can be analyzed, if at all. There may on the other hand be project partners who are able to help you getting started or maybe even provide tools that match your preferences.

Availability of tools suited for the task. It is often a better choice to use available tools that people near you already have experience of working with. These tools can be commercial, open source, or maybe even tools developed in your group that are tailored to the specific needs of your analysis. If such tools are available, it is likely to be the best choice to use these tools before looking for new alternatives.

Recycling of workflows is maybe not on the top of the list, but is relevant to consider if you or other people in your group are planning to perform more experiments of the same type. Then, some choices are better at promoting repeatability than other tools.

The most frequently used tool for early analysis of neutron images is the open-source tool ImageJ [88–90]. It provides a simple user interface for inspecting and modifying images interactively. More complicated and repeated tasks can be implemented using scripts or additional processing plugins. An alternative to GUI based tools is to use scripting languages. Here, the trend goes towards using Python-based analysis in the form of scripts or using Jupyter notebooks to achieve a basic level of interaction. Dedicated tools have been developed for tomography reconstruction, e.g., [1], and recently many neutron imaging facilities develop and provide their own user interfaces, often based on the same reconstruction toolbox, ASTRA [91]. Visual analysis is a next step and many commercial vendors provide full-featured interactive tools for many advancesd analysis tasks.

Image analysis is a wide topic and a field of research of its own with several books and journals covering specific aspects of the analysis that go far beyond what can be covered in this book. Here we mainly focus on details specific to neutron imaging, i.e., the initial steps of processing and analysis that are closely related to the neutron sample interaction of the artefacts that needs to be identified and corrected for.

References

[1] Kaestner A P 2011 MuhRec—a new tomography reconstructor *Nucl. Instrum. Methods Phys. Res.* A **651** 156–60

[2] Bradley J, Pooley D E and Kockelmann W 2021 Artifacts and quantitative biases in neutron tomography introduced by systematic and random errors *J. Instrum.* **16** P01023

[3] Kaestner A 2020 Methods to combine multiple images to improve quality *Neutron Radiography—WCNR-11, Materials Research Forum LLC* 193–7

[4] Brenizer J S 2013 A review of significant advances in neutron imaging from conception to the present *Phys. Procedia* **43** 10–20

[5] Buzug T 2008 *Introduction to Computed Tomography: From Photon Statistics to Modern cone-Beam CT* (Berlin: Springer)

[6] Kaestner A P, Grünzweig C and Lehmann E H 2012 Improved analysis techniques for new applications in neutron imaging *Proc. 18th World Conf. on Nondestructive Testing* pp 1–10

[7] Feldkamp L A, Davis L C and Kress J W 1984 Practical cone-beam algorithm *J. Opt. Soc. Am.* A **1** 612

[8] Katsevich A 2004 An improved exact filtered backprojection algorithm for spiral computed tomography *Adv. Appl. Math.* **32** 681–97

[9] Brown J M C, Garbe U and Pelliccia D 2019 Statistical image reconstruction for high-throughput thermal neutron computed tomography *' Nucl. Instrum. Methods Phys. Res.* A **942** 162396

[10] Kazantsev D *et al* 2017 A novel tomographic reconstruction method based on the robust student's t function for suppressing data outliers *IEEE Trans. Comput. Imaging* **3** 682–93

[11] Micieli D, Minniti T, Evans L M and Gorini G 2019 Accelerating neutron tomography experiments through artificial neural network based reconstruction *Sci. Rep.* **9** 2450

[12] Venkatakrishnan S V, Mohan K A, Ziabari A K and Bouman C A 2022 Algorithm-driven advances for scientific CT instruments: from model-based to deep learning-based approaches *IEEE Signal Process. Mag.* **39** 32–43

[13] Kaestner A P, Münch B, Trtik P and Butler L G 2011 Spatio-temporal computed tomography of dynamic processes *Opt. Eng.* **50** 123201

[14] van Eyndhoven G *et al* 2015 An iterative CT reconstruction algorithm for fast fluid flow imaging *IEEE Trans. Image Process.* **24** 4446–58

[15] Kazantsev D *et al* 2015 4D-CT reconstruction with unified spatial-temporal patch-based regularization *Inverse Probl. Imaging* **9** 447–67

[16] Helfen L *et al* 2011 Neutron laminography—a novel approach to three-dimensional imaging of flat objects with neutrons *Nucl. Instrum. Methods Phys. Res.* A **651** 135–9

[17] Rudolph-Mohr N, Bereswill S, Tötzke C, Kardjilov N and Oswald S E 2021 Neutron computed laminography yields 3D root system architecture and complements investigations of spatiotemporal rhizosphere patterns *Plant Soil* **469** 489–501

[18] Levakhina Y 2014 *Three-Dimensional Digital Tomosynthesis* (Wiesbaden: Springer Fachmedien Wiesbaden)

[19] Fernandez J-J 2012 Computational methods for electron tomography *Micron* **43** 1010–30

[20] Tirada N *et al* 2019 Digital breast tomosynthesis: physics, artifacts, and quality control considerations *RadioGraphics* **39** 413–26

[21] Xu F, Helfen L, Baumbach T and Suhonen H 2012 Comparison of image quality in computed laminography and tomography *Opt. Express* **20** 794

[22] Kachelrieß M, Watzke O and Kalender W A 2001 Generalized multi-dimensional adaptive filtering for conventional and spiral single-slice, multi-slice, and cone-beam CT *Med. Phys.* **28** 475–90

[23] Boillat P *et al* 2022 Who made the noise? A systematic approach for the assessment of neutron imaging scintillators *Opt. Express* (in press)
[24] Cunningham I A, Westmore M S and Fenster A 1994 A spatial-frequency dependent quantum accounting diagram and detective quantum efficiency model of signal and noise propagation in cascaded imaging systems *Med. Phys.* **21** 417–27
[25] Hungler P C, Bennett L G I, Lewis W J, Bevan G and Metzler J 2013 Comparison of image filters for low dose neutron imaging *Phys. Procedia* **43** 169–78
[26] Kaestner A, Lehmann E and Stampanoni M 2008 Imaging and image processing in porous media research *Adv. Water Resour.* **31** 1174–87
[27] Buades A, Coll B and Morel J-M 2005 A non-local algorithm for image denoising *IEEE Computer Society Conference on Computer Vision and Pattern Recognition (CVPR'05)* (Piscataway, NJ: IEEE) pp 60–5
[28] Lebrun M 2012 An analysis and implementation of the BM3D image denoising method *Image Process. On Line* **2** 175–213
[29] Elad M, Kawar B and Vaksman G 2023 Image denoising: the deep learning revolution and beyond—a survey paper *SIAM J. Imag. Sci.* **16** 1594–654
[30] Li H, Schillinger B, Calzada E, Yinong L and Muehlbauer M 2006 An adaptive algorithm for gamma spots removal in CCD-based neutron radiography and tomography *Nucl. Instrum. Methods Phys. Res. Sect.* A **564** 405–13
[31] Sijbers J and Postnov A 2004 Reduction of ring artefacts in high resolution micro-CT reconstructions *Phys. Med. Biol.* **49** N247–53
[32] Münch B, Trtik P, Marone F and Stampanoni M 2009 Stripe and ring artifact removal with combined wavelet-Fourier filtering *Opt. Express* **17** 8567–91
[33] Vo N T, Atwood R C and Drakopoulos M 2018 Superior techniques for eliminating ring artifacts in x-ray micro-tomography *Opt. Express* **26** 28396
[34] Yang Y, Zhang D, Yang F, Teng M, Du Y and Huang K Post-processing method for the removal of mixed ring artifacts in CT images**28** 30362
[35] Kang M *et al* 2013 Water calibration measurements for neutron radiography: application to water content quantification in porous media *Nucl. Instrum. Methods Phys. Res.* A **708** 24–31
[36] Hassanein R, Lehmann E and Vontobel P 2005 Methods of scattering corrections for quantitative neutron radiography *Nucl. Instrum. Methods Phys. Res.* A **542** 353–60
[37] Boillat P *et al* 2018 Chasing quantitative biases in neutron imaging with scintillator-camera detectors: a practical method with black body grids *Opt. Express* **26** 15769–84
[38] Hassanein R, de Beer F, Kardjilov N and Lehmann E 2006 Scattering correction algorithm for neutron radiography and tomography tested at facilities with different beam characteristics *Phys. B Condens. Matter* **385–6** 1194–6
[39] Carminati A, Kaestner A, Hassanein R, Ippisch O, Vontobel P and Flühler H 2007 Infiltration through series of soil aggregates: neutron radiography and modeling *Adv. Water Resour.* **30** 1168–78
[40] Vasin M *et al* 2008 Drainage in heterogeneous sand columns with different geometric structures *Adv. Water Resour.* **31** 1205–20
[41] Raventos M, Harti R P, Lehmann E and Grünzweig C 2017 A method for neutron scattering quantification and correction applied to neutron imaging *Phys. Procedia* **88** 275–81
[42] Carminati C *et al* 2019 Implementation and assessment of the black body bias correction in quantitative neutron imaging *PLoS One* **14** e0210300

[43] Mail N, O'Brien P and Pang G 2007 Lag correction model and ghosting analysis for an indirect-conversion flat-panel imager *J. Appl. Clin. Med. Phys.* **8** 137–46
[44] Siewerdsen J H and Jaffray D A 1999 A ghost story: spatio-temporal response characteristics of an indirect-detection flat-panel imager *Med. Phys.* **26** 1624–41
[45] Hildebrandt M, Mosset J-B and Stoykov A 2018 Evaluation of ZnS: ^{6}LiF and ZnO: ^{6}LiF scintillation neutron detectors readout with SiPMs *IEEE Trans. Nucl. Sci.* **65** 2061–7
[46] Schillinger B and Grazzi F 2015 Artefacts in neutron CT—their effects and how to reduce some of them *Phys. Procedia* **69** 244–51
[47] Kazantsev D *et al* 2016 Temporal sparsity exploiting nonlocal regularization for 4D computed tomography reconstruction *J. Xray Sci. Technol.* **24** 207–19
[48] Glasbey C A 1993 An analysis of histogram-based thresholding algorithms *CVGIP, Graph. Models Image Process.* **55** 532–7
[49] Russ J C and Neal F B 2018 *The Image Processing Handbook* (Boca Raton, FL: CRC Press)
[50] Soille P 2004 *Morphological Image Analysis* (Berlin: Springer)
[51] Kaestner A P *et al* 2017 Bimodal imaging at ICON using neutrons and x-rays *Phys. Proc.* **88** 314–21
[52] LaManna J M, Hussey D S, DiStefano V H, Baltic E and Jacobson D L 2020 NIST NeXT: a system for truly simultaneous neutron and x-ray tomography *Hard X-Ray, Gamma-Ray, and Neutron Detector Physics XXII* ed M Fiederle, A Burger and S A Payne (Bellingham, WA: SPIE) p 24
[53] Tengattini A, Lenoir N, Andò E and Viggiani G 2021 Neutron imaging for geomechanics: a review *Geomechan. Energy Environ.* **27** 100206
[54] Mannes D, Schmid F, Frey J, Schmidt-Ott K and Lehmann E 2015 Combined neutron and x-ray imaging for non-invasive investigations of cultural heritage objects *Phys. Procedia* **69** 653–60
[55] Review of Economics and Statistics 1919 General considerations and assumptions *Rev. Econ. Stat.* **1** 5
[56] Matsushima U, Herppich W B, Kardjilov N, Graf W, Hilger A and Manke I 2009 Estimation of water flow velocity in small plants using cold neutron imaging with D_2O tracer *Nucl. Instrum. Methods Phys. Res.* A **605** 146–9
[57] Le Cann S *et al* 2017 Characterization of the bone-metal implant interface by digital volume correlation of *in situ* loading using neutron tomography *J. Mech. Behav. Biomed. Mater.* **75** 271–8
[58] Hall S A 2013 Characterization of fluid flow in a shear band in porous rock using neutron radiography *Geophys. Res. Lett.* **40** 2613–8
[59] Birjukovs M *et al* 2021 Resolving gas bubbles ascending in liquid metal from low-SNR neutron radiography images *Appl. Sci.* **11** 9710
[60] Sato H 2017 Deriving quantitative crystallographic information from the wavelength-resolved neutron transmission analysis performed in imaging mode *J. Imaging* **4** 7
[61] Ramadhan R S *et al* 2019 Characterization and application of Bragg-edge transmission imaging for strain measurement and crystallographic analysis on the IMAT beamline *J. Appl. Crystallogr.* **52** 351–68
[62] Santisteban J R, Edwards L, Fizpatrick M E, Steuwer A and Withers P J 2002 Engineering applications of Bragg-edge neutron transmission *Appl Phys A Mater Sci Process* **74** s1433–6
[63] Santisteban J R *et al* 2012 Texture imaging of zirconium based components by total neutron cross-section experiments *J. Nucl. Mater.* **425** 218–27

[64] Liptak A, Burca G, Kelleher J, Ovtchinnikov E, Maresca J and Horner A 2019 Developments towards Bragg edge imaging on the IMAT beamline at the ISIS pulsed neutron and muon source: BEAn software *J. Phys. Commun.* **3** 113002

[65] Bilheux J 2023 https://jeanbilheux.pages.ornl.gov/project/ibeatles/

[66] Busi M and Carminati C ToF imaging—tool package for neutron time of flight analysis (https://neutronimaging.github.io/ToFImaging/)

[67] Busi M *et al* 2021 Nondestructive characterization of laser powder bed fusion parts with neutron Bragg edge imaging *Addit. Manuf.* **39** 101848

[68] Vogel S 2000 A Rietveld-approach for the analysis of neutron time-of-flight transmission data *PhD Thesis* Christian-Albrechts University, Kiel

[69] Sato H, Kamiyama T and Kiyanagi Y 2011 A Rietveld-type analysis code for pulsed neutron Bragg-edge transmission imaging and quantitative evaluation of texture and microstructure of a welded α-iron plate *Mater. Trans.* **52** 1294–302

[70] Minniti T 2019 Bragg edge analysis for transmission imaging experiments software tool: BEATRIX *J. Appl. Crystallogr.* **52** 903–9

[71] Malamud F and Santisteban J R 2016 Full-pattern analysis of time-of-flight neutron transmission of mosaic crystals *J. Appl. Crystallogr.* **49** 348–65

[72] Malamud F, Santisteban J R, Gao Y, Shinohara T, Oikawa K and Tremsin A 2022 Non-destructive characterization of the spatial variation of γ/γ' lattice misfit in a single-crystal Ni-based superalloy by energy-resolved neutron imaging *J. Appl. Crystallogr.* **55** 228–39

[73] Sakurai Y, Sato H, Adachi N, Morooka S, Todaka Y and Kamiyama T 2021 Analysis and mapping of detailed inner information of crystalline grain by wavelength-resolved neutron transmission imaging with individual Bragg-dip profile-fitting analysis *Appl. Sci.* **11** 5219

[74] Strickland J *et al* 2020 2D single crystal Bragg-dip mapping by time-of-flight energy-resolved neutron imaging on IMAT@ISIS *Sci Rep.* **10** 20751

[75] Sato H *et al* 2017 Inverse pole figure mapping of bulk crystalline grains in a polycrystalline steel plate by pulsed neutron Bragg-dip transmission imaging *J. Appl. Crystallogr.* **50** 1601–10

[76] Harti R P and Valsecchi J 2017 nGIMagic—TaPy http://doi.org/10.5281/zenodo.803046

[77] Strobl M *et al* 2019 Achromatic non-interferometric single grating neutron dark-field imaging *Sci. Rep.* **9** 19649

[78] Busi M, Zdora M-C, Valsecchi J, Bacak M and Strobl M 2022 Cold and thermal neutron single grating dark-field imaging extended to an inverse pattern regime *Appl. Sci.* **12** 2798

[79] Valsecchi J *et al* 2020 Characterization of oriented microstructures through anisotropic small-angle scattering by 2D neutron dark-field imaging *Commun. Phys.* **3** 42

[80] Doucet M *et al* 2020 SasView https://www.sasview.org latest access 2024-03-08

[81] Breßler I, Kohlbrecher J and Thünemann A F 2015 SASfit : a tool for small-angle scattering data analysis using a library of analytical expressions *J. Appl. Crystallogr.* **48** 1587–98

[82] Kohlbrecher J and Studer A 2017 Transformation cycle between the spherically symmetric correlation function, projected correlation function and differential cross section as implemented in *SASfit J. Appl. Crystallogr.* **50** 1395–403

[83] Kline S R 2006 Reduction and analysis of SANS and USANS data using IGOR Pro *J. Appl. Crystallogr.* **39** 895–900

[84] Zhou Wang and Bovik A C 2009 Mean squared error: love it or leave it? A new look at Signal Fidelity Measures *IEEE Signal Process Mag.* **26** 98–117

[85] Ferrucci M, Ametova E and Dewulf W 2021 Monte Carlo reconstruction: a concept for propagating uncertainty in computed tomography *Meas. Sci. Technol.* **32** 115006

[86] Kimpe T and Tuytschaever T 2007 Increasing the number of gray shades in medical display systems—how much is enough? *J. Digit. Imaging* **20** 422–32

[87] Maddison D R, Swofford D L and Maddison W P 1997 Nexus: an extensible file format for systematic information *Syst. Biol.* **46** 590–621

[88] Schneider C A, Rasband W S and Eliceiri K W 2012 NIH image to ImageJ: 25 years of image analysis *Nat. Methods* **9** 671–5

[89] Rueden C T *et al* 2017 ImageJ2: ImageJ for the next generation of scientific image data *BMC Bioinf.* **18** 529

[90] Schindelin J *et al* 2012 Fiji: an open-source platform for biological-image analysis *Nat. Methods* **9** 676–82

[91] van Aarle W *et al* 2015 The ASTRA toolbox: a platform for advanced algorithm development in electron tomography *Ultramicroscopy* **157** 35–47

www.ingramcontent.com/pod-product-compliance
Ingram Content Group UK Ltd.
Pitfield, Milton Keynes, MK11 3LW, UK
UKHW051909100926
8556IPUK00005B/816

9 780750 334938